John Harvey
David Nichols

A. W. Brown.   Frank Billett

A. P. Mathison

Nancy Jane Lane   Owen L Thomas.

W. Holmes

Doreen Ashhurst.

Keith Ross

B. Casselman   Gudersham

A. E. JANS.

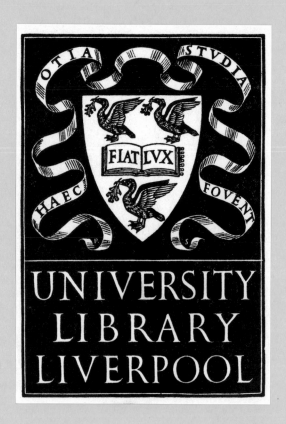

Cell Structure and its Interpretation

Photograph by G. B. David

JOHN RANDAL BAKER
Fellow of the Royal Society, Honorary Fellow and Past President of the Royal Microscopical Society,
and Reader Emeritus in Cytology in the University of Oxford

Cell Structure and its Interpretation

Essays presented to John Randal Baker F.R.S.

edited by

S. M. McGee-Russell

Member of Scientific Staff, Medical Research Council, Virus Research Unit, Medical Research Council Laboratories, Carshalton

and

K. F. A. Ross

Principal Research Associate in Neurology, Department of Medicine, University of Newcastle upon Tyne

Edward Arnold (Publishers) Ltd.   London

© EDWARD ARNOLD (PUBLISHERS) LTD. 1968

First published 1968

SBN: 7131 2148 3

Printed in Great Britain by
William Clowes and Sons, Limited, London and Beccles

Editorial Preface

The idea of compiling a collection of essays to honour John Baker originated eight years ago with Dr. G. B. David, who, in 1959, suggested preparing a book in time for John Baker's sixtieth birthday in 1960. This was impractical, because of the short time available in which to get in touch with possible contributors and for producing such a work, and the idea was shelved. Three years later, however, one of the present editors (K.F.A.R.), wrote to nearly fifty of John Baker's former research students and associates to ask them whether they would be prepared to contribute to such a volume. This met with an extremely enthusiastic response, which reflected the high regard in which John Baker was held by everyone who had worked in close contact with him. Not only was the production of the book unanimously approved, but no less than twenty-nine of the scientists who had been approached agreed to write articles for it. The firm of Edward Arnold Ltd. undertook to publish their contributions and we agreed to compile and edit them, so as to form a book with a coherent theme.

The essays in this book provide illustrations of the kinds of experimental approach to cytological studies that John Baker has always advocated, written by many of those who have been most influenced by him as a scientist. When he entered this field, more than thirty years ago, cytological studies were, for the most part, being pursued in a very unscientific manner. Great emphasis was, for example, placed on producing aesthetically pleasing microscopical preparations coloured with brilliant aniline dyes, in which cellular structures were often displayed in an unselective and almost uninterpretable form. Many of the methods used were chemically irrational and conducive to the production of gross artifacts. The central theme of all John Baker's cytological work has been his preference for methods capable of some measure of scientific interpretation and his insistence on attempting this. The title of this book therefore suggested itself. Each contributor was asked to write his article with the title in mind.

The resulting collection of essays covers a wide variety of cytological topics. Physiological cytology is a broad field of widening application, but no broader than the interests of John Baker. The pattern of the articles in this book bears witness to this. At the beginning we have placed the articles which discuss problems in fixation, colouring and histochemistry. These are usually associated with qualitative approaches to the interpretation of cell structure. Then there is a small, but not unimportant section, in which some examples are given of information to be derived from quantitative techniques. This sort of information often illuminates the more dynamic aspects of cytology. The final series of articles illustrates how the study of cell structure can lead to the interpretation of specialized cellular activities.

Two-thirds of the articles deal with structures which can be seen by light microscopy, and the remaining one-third with the far finer structures revealed at the electron microscope levels of resolution. No attempt has been made to separate these two aspects. It is the editors' intention, rather, to emphasize the problems of the comparative experimental approach, to indicate the interpretations that are common to both these fields, and to show how the seeming gap between light and electron microscope studies can be, and is being, bridged.

Although the subject matter of this book is confined to cytological topics, it should be pointed out that cytology has, in fact, been only one of John Baker's many interests. The full bibliography published at the end of this book gives some indication of these. His notable contributions to microscopy include two important recent papers on the function of the human eye in microscopy, and on the design of the biological student's microscope. Almost all his earliest work was devoted to topics connected with reproduction and breeding cycles. It is now regarded as having had a considerable influence on the subsequent development of this field by others. The benefit of his contributions to the development of effective chemical contraceptives is today beginning to be appreciated, when, at last, the urgency of the need to limit the rise in world population has been generally realized. His activities in the cause of freedom in science helped to inform well-intentioned, but ignorant people of the true nature of scientific work.

Mention should also be made of his remarkable series of papers on 'The cell theory, a restatement, history and critique' published in the *Quarterly Journal of Microscopical Science* between 1948 and 1953. It is no exaggeration to say that, as works of historical scholarship, they are probably unsurpassed in all modern cytological literature. We believe that they would now be widely appreciated by cell biologists and others if they were to be republished as a short book. Another work of John Baker's, which unquestionably should now be republished, is his 'Principles of Biological Microtechnique' (which is *not*, as some who may have not read it have supposed, an expanded version of his more famous 'Cytological Technique'). It was singularly unfortunate that this extremely valuable criticism of the long-established histological and cytological procedures involving fixation and staining, should have been published just ten years ago; for, at that time, cell biologists might have been forgiven for thinking (as many undoubtedly did) that the then new techniques of interference microscopy, autoradiography, fluorescent antibody labelling, mass measurement by Cartesian diver, and indeed electron microscopy, would soon render the examination of fixed and stained preparations of cells with the ordinary light microscope almost an obsolete and unnecessary procedure. John Baker was percipient enough to know that this was nonsense, and the well-thumbed copies of this book, now regrettably almost unobtainable, attest, once again, to the value of his judgements. He has also written valuable and highly entertaining articles on the correct use of English and the importance of clarity of expression in scientific writing. In Appendix III in this book we have placed the document which we think illustrates, better than any other, his outstanding qualities and character as a supervisor of research students.

It is with great sorrow that we have just heard of the untimely death of one of this book's most distinguished contributors: the late Professor C. F. A. Pantin, F.R.S., who has been a close friend of John Baker's for many years.

Carl Pantin was associated with John Baker in editing the *Quarterly Journal of Microscopical Science* from 1946 to 1961. His, rightly, is the first contribution in this book. His outstanding intellect and pleasant personality will be greatly missed.

It is of considerable interest to recall the words that John Baker and Professor Pantin used in their joint editorial in 1946 on assuming the editorship of the *Quarterly Journal of Microscopical Science*:

'There is another class of work in which the Editors are anxious to see a very considerable expansion. The Journal has been since its early days a medium for the publication of what may broadly be called cellular biology. Lankester himself was a pioneer in some parts of this field. A period now lies before us when great advances are to be expected in our interpretation of cellular organization and in our knowledge of the fine structure of biological systems, especially in the fields of strict histochemistry and cytochemistry—the biochemistry and biophysics of cell structure *in situ*. The Editors wish to give every encouragement to contributions in these fields of enquiry.'

They were right, and their self-appointed task was most admirably performed.

S. M. McGee-Russell

K. F. A. Ross

January 1967

Contributors

Doreen E. Ashhurst, D.Phil.,
Lecturer in Anatomy,
Department of Anatomy,
The Medical School,
University of Birmingham,
Birmingham, 15., England.

Frank S. Billett, Ph.D.,
Senior Lecturer,
Department of Zoology,
University of Southampton,
Southampton, SO9 5NH, England.

Savile Bradbury, D.Phil.,
Fellow of Pembroke College, Oxford and
University Lecturer in Human Anatomy,
Department of Human Anatomy,
South Parks Road,
Oxford, England.

A. W. Brown, B.Sc.,
Member of Scientific Staff, Medical Research
    Council,
Neuropsychiatric Research Unit,
M.R.C. Laboratories,
Woodmansterne Road,
Carshalton,
Surrey, England.

Wim. C. de Bruijn, Drs.,
Head of Electron Microscopy Section,
Het Centraal Pathologische Laboratorium,
Zienkenhuis Dijkzigt,
Rotterdam,
Holland.

H. G. Callan, D.Sc., F.R.S.,
Professor of Zoology,
Department of Zoology,
The University,
St. Andrews,
Fife, Scotland.

W. G. Bruce Casselman, M.D., Ph.D.,
Medical Director,
Geigy (Canada) Limited,
4984 Place de la Savane,
Montreal 9, P.Q.,
Canada.

Joe Chayen, D.Sc.,
Head of the Division of Cellular Biology,
The Mathilda and Terence Kennedy Institute
    of Rheumatology,
Bute Gardens,
Hammersmith,
London, W.6. England.

John Tung Yang Chou, D.Phil.,
Research Professor,
Cytologisches Labor,
Universitäts Hals,-Nasen,-Ohren,-Klinik
    Johann Wolfgang Goethe Universität.
Ludwig-Rehn Strasse 14,
6000 Frankfurt-am-Main,
West Germany.

Shou-Hwa Chuang, D.Phil.,
Senior Lecturer,
Department of Zoology,
University of Singapore,
Singapore.

John D. Currey, D.Phil.,
Senior Lecturer,
Department of Biology,
University of York,
Heslington,
York, England.

George B. David, D.Phil.,
Senior Research Associate and Lecturer
Department of Biological Sciences,
State University of New York at Albany,
Albany, N.Y. 12203,
U.S.A.

William Galbraith, M.A.,
Research Scientist,
Chester Beatty Research Institute,
Royal Cancer Hospital,
Fulham Road,
London, S.W.3. England.

Dan J. G. Goldstein, M.B.,
Senior Lecturer,
Department of Human Biology and Anatomy,
University of Sheffield,
Research Laboratories,
3 Clarkehouse Road,
Sheffield 10, England.

M. A. Gorycki, M.S.,
Research Associate,
Department of Cell Biology,
Institute for Muscle Diseases Inc.,
515 East 71st Street,
New York, N.Y. 10021,
U.S.A.

E. G. Gray, Ph.D.,
Professor of Cytology,
Department of Anatomy,
University College,
University of London,
Gower Street,
London, W.C.1. England.

Jennifer M. Gregory (née Byrne), D.Phil.,
at present a housewife with small children,
79 Kingsgate Street,
Winchester,
Hants., England.

Gottwalt Christian Hirsch, Dr.,
Professor ordin. emer. der Cytologie,
I. Zoologisches Institüt der Universität
  Göttingen,
34 Göttingen, Berliner Str. 28,
West Germany.

William Holmes, D.Phil.,
Fellow of St. John's College, Oxford and
University Lecturer in Zoology,
Department of Zoology,
Parks Road,
Oxford, England.

D. E. Jans, Drs.,
Research Associate, University of Newcastle
  upon Tyne,
Muscular Dystrophy Research Laboratories,
Newcastle General Hospital,
Newcastle on Tyne, 4, England.

Nancy J. Lane, D.Phil.,
Head of Electron Microscopy Section,
A.R.C. Research Unit,
Department of Zoology,
Downing Street,
Cambridge, England.

Sudarshan K. Malhotra, D.Phil.,
Professor,
Biological Sciences Electron Microscopy Lab-
  oratory,
University of Alberta,
Edmonton,
Alberta,
Canada.

Sam M. McGee-Russell, D.Phil.,
Senior Research Associate and Lecturer,
Department of Biological Science,
State University of New York at Albany
Albany, N.Y. 12203,
U.S.A.

R. B. McKay, Ph.D.,
Technical Officer,
Geigy (U.K.) Ltd.,
Pigments Division,
Hawkhead Road,
Paisley,
Renfrewshire,
Scotland.

J. F. A. McManus, M.D.,
Executive Director,
Federation of American Societies for Experi-
  mental Biology,
9650 Rockville Pike,
Bethesda,
Maryland 20014,
U.S.A.
*formerly Professor of Pathology, Combined Degree
  Program in Medical Education, University of
  Indiana.*

Geoffrey A. Meek, Ph.D.,
Senior Lecturer,
Department of Human Biology and Anatomy,
University of Sheffield,
Research Laboratories,
3, Clarkehouse Road,
Sheffield 10, England.

Winfield S. Morgan, M.D.,
Director of Laboratories,
Department of Pathology,
Aultman Hospital,
625, Clarendon Avenue S.W.,
Canton, Ohio 44710,
U.S.A.
formerly Professor of Pathology, Metropolitan General Hospital, Western Reserve Medical School, Scranton, Ohio.

David Nichols, D.Phil.,
Fellow of St. Peter's College, Oxford and
University Lecturer in Zoology,
Department of Zoology,
Parks Road,
Oxford, England.

The late Carl F. A. Pantin, D.Sc., F.R.S.
formerly Professor of Zoology, University of Cambridge.

Roy MacLean Park, M.A., A.L.A.,
Assistant Librarian,
Ministry of Transport,
St. Christopher's House,
Southwark Street,
London, S.E.1., England.

Aleksandra Przełęcka, Dr.,
Docent (Associate Professor),
Head of the Cytochemical Laboratory,
Department of Biochemistry,
Nencki Institute of Experimental Biology,
3, Pasteur Street,
Warsaw 22,
Poland.

Keith F. A. Ross, D.Phil.,
Principal Research Associate, University of Newcastle upon Tyne, and
Head of the Cell Biology Section,
Muscular Dystrophy Research Laboratories,
Newcastle General Hospital,
Newcastle upon Tyne, 4, England.

S. Ahmad Shafiq, D.Phil.,
Assistant Member,
Department of Cell Biology,
Institute for Muscle Diseases Inc.,
515, East 71st Street,
New York, N.Y. 10021,
U.S.A.

Bhupinder N. Sud, Ph.D., D.Phil.,
Assistant Member,
Research Laboratories,
Albert Einstein Medical Center,
York and Tabor Roads,
Philadelphia, PA 19141,
U.S.A.
also Reader in Cytology, University of the Panjab, India.

Owen L. Thomas, M.D., D.Phil.,
Professor of Zoology,
Department of Zoology,
University of Natal,
P.O. Box 375,
Pietermaritzberg,
Natal,
South Africa.
formerly Professor of Zoology at the Universities of Alabama and Malta.

John Z. Young, D.Sc., F.R.S.,
Professor of Anatomy,
Department of Anatomy,
University College,
University of London,
Gower Street,
London, W.C.1., England.

Contents

Frontispiece ii

Preface v

Contributors vii

PART 1

THE QUALITATIVE INTERPRETATION OF CELL STRUCTURE

Aspects of Fixation and Microtomy

1 Fixation and the Display of Cellular Structure 3

    C. F. A. PANTIN

2 Freeze-Substitution and Freeze-Drying in Electron Microscopy 11

    S. K. MALHOTRA

       Introduction
       Materials and Methods
       Results
       Differences in Preservation by Freeze-Substitution and Freeze-Drying
       Discussion
       Summary

3 On Using a Cryostat 23

    G. B. DAVID and A. W. BROWN

       Introduction
       On the Design of Cryostats
       Methods of Handling Unfixed Tissues
       Methods of Handling Fixed Tissues
       Envoi
       Summary
       Acknowledgements

4 Sea Water and Osmium Tetroxide Fixation of Marine Animals 51

    S. H. CHUANG

       Introduction
       Materials
       Method
       Results
       Discussion
       Summary
       Acknowledgements

The Display of Structure with the Light Microscope by inducing Differences of Colour and Contrast

A. Dyestuffs and Cells

5   Some Cytologically Relevant Aspects of the Physical Chemistry of Dyes   .   .   .   59
        R. B. McKay
            Properties of Dissolved and Adsorbed Dyes
            Metachromasy
            Summary

6   On the Affinity of Dyes for Histological Substrates   67
        D. Goldstein
            The Definition of Affinity
            Qualitative Recognition of Differences in Affinity
            The Quantitative Measurement of Affinity
            Miscellaneous Data Relevant to the Affinity of Dyes for Histological Substrates
            Conclusion
            Acknowledgements

7   The Biological Activity of Neutral Red   79
        W. S. Morgan
            'Golgi Techniques' and Neutral Red Granule Formation
            Cellular Basiphilia and Capacity for Neutral Red Granule Formation
            Concerning the Fate of Neutral Red in the Body
            Histological Studies of the Liver
            Electron Microscopic Studies of Neutral Red Granules
            Discussion

8   Cellular Response to Vital Staining   87
        Jennifer M. Gregory (née Byrne)

B. Impregnation and Empiricism

9   Empiricism—Silver Methods and the Nerve Axon   95
        W. Holmes

10 Intracellular Canals   103
        O. L. Thomas

The Display of Structure with the Electron Microscope by inducing Differences of Contrast

11 Image and Artifact—Comments and Experiments on the Meaning of the Image in the
    Electron Microscope   115
        S. M. McGee-Russell and W. C. de Bruijn
            Introduction
            Materials and Methods
            Dehydration and Extraction
            Fixation and Contrasting
            Conclusion
            Summary

12 The Development of Specific Histochemical Methods for the Localization of Substances
with the Electron Microscope 135
    S. BRADBURY
        Introduction
        Enzymes
        Nucleic Acids
        Protein Groupings
        Carbohydrates
        Noradrenalin and 5 Hydroxytryptamine
        Limitations of the Present Methods
        Future Prospects
        Summary

The Histochemistry and Organization of Intracellular Structures and Cell Products

13 The Histochemistry of Phospholipids and its Significance in the Interpretation of the
Structure of Cells 149
    J. CHAYEN
        The Demonstration of Phospholipids
        Baker's Acid Haematein Method
        Bound Lipids
        Lipid–Protein Structures
        The Structural Significance of Lipid–Protein Complexes
        Envoi

14 Phospholipids and Nuclear Structure 157
    ALEKSANDRA PRZEŁĘCKA
        Introduction
        Survey of Results Reported in Biochemical Literature
        Survey of Results Reported in Cytochemical Literature
        Phospholipids in the Nuclei in the Follicular Vesicle of Galleria Mellonella
        Discussion

15 Lipochondria, Neutral Red Granules and Lysosomes: Synonymous Terms? . . . 169
    NANCY J. LANE
        Introduction
        Lipochondria and Lysosomes in Invertebrate Nerve Cells
        Lipochondria and Lysosomes in Vertebrate Neurons
        Lipochondria and Lysosomes in Cultured Animal Cells
        Intracellular Sites of Vital Dye Uptake
        Concluding Remarks
        Summary
        Acknowledgements

16 The Method of Combined Observations with Light and Electron Microscopes Applied to
the Study of Histochemical Colourations in Nerve Cells and Oocytes 183
    S. M. McGEE-RUSSELL
        Introduction
        Part I: The Periodic Acid Schiff Technique Applied to Tissues in Plastic for Combined
            Observations
        Materials and Methods
        Results
        Part II: Histochemical Methods for Calcium Applied to Plastic Sections
        Materials and Methods
        Results
        Summary
        Acknowledgements

17 Changing Concepts of the Connective Tissues 209
    J. F. A. McMANUS

18 Apparent Intracellular Collagen Synthesis 225
    G. A. MEEK
        Introduction
        Results

19 Fibroblasts—Vertebrate and Invertebrate 237
    DOREEN E. ASHHURST
        Vertebrate Fibroblasts
        Invertebrate Fibroblasts
        Conclusions
        Acknowledgements

20 The Secretion, Structure and Strength of Echinoderm Calcite 251
    D. NICHOLS and J. D. CURREY
        Introduction
        General Nature of the Echinoderm Skeleton
        Chemical Composition
        Crystallography
        Secretion and Absorption of the Skeleton
        The Problem and Methods of Study
        Fine Structure
        Echinoderm Calcite as a Structural System
        Conclusions
        Summary
        Acknowledgements

PART 2

THE QUANTITATIVE INTERPRETATION OF CELL STRUCTURE

21 Quantitative Histochemistry: Cell Structure revealed through Cell Function . . . 265
    J. CHAYEN
        Introduction
        Densitometry
        Enzyme Histochemistry: 'Manifest' Activity as compared with 'Latent' Activity
        Quantitative Enzyme Histochemistry
        Stoichiometry
        The Use of Unfixed Sections
        The Effect of Damage on the Permeability of Mitochondria
        Conclusion
        Summary
        Acknowledgements

22 The Study of Cell Differentiation by Quantitative Microscopic Methods . . . 275
    K. F. A. ROSS and D. E. JANS
        Problems Connected with Quantitative Cytological Studies
        The Processes of Cell Differentiation, illustrated particularly by Developing Muscle Tissue
        Changes in the Dry Mass of Living Myoblast Nucleoli during Differentiation as Measured
          by Interference Microscopy
        A Comparison of the Dry Mass of Nucleoli in Myoblasts from Normal and Dystrophic
          Mice

A Densitometric Technique for Measuring Changes in Cytoplasmic RNA in Differentiating
    Myoblasts
The Changes in Cytoplasmic RNA Found in Differentiating Myoblasts, and their Possible
    Interpretation
Summary
Acknowledgements

23 A Method for the Assessment of Surface Roughness in Landschütz Ascites Tumour Cells .   305
       W. GALBRAITH
          Introduction
          Apparatus
          Experimental Procedure
          Calculations
          Conclusions
          Summary
          Acknowledgements

PART 3

THE QUALITATIVE INTERPRETATION OF SPECIALIZED CELL ACTIVITIES

Unusual Cells

24 Some Observations on the Structure and Function of the Reissner's Membrane of the Inner
     Ear   315
       J. T. Y. CHOU
          Introduction
          Materials
          Results
          Discussion

25 The Mid-Gut Epithelium of Insects   321
       B. N. SUD
          Principal Cells
          Goblet Cells
          Regenerative Cells
          Peritrophic Membrane
          Summary

Developmental Activities, Unusual Nuclei, and Mitochondria

26 Cellular Differentiation in Ectodermal Explants from Amphibian Gastruiae .   .   341
       F. S. BILLETT
          Introduction
          Methods
          Observations on the External Appearance and Survival of the Explants
          Cytological Observations
          Discussion
          Summary

27 A Problem Posed by the Structure of Lampbrush Chromosomes 357
       H. G. CALLAN

28 Mitochondria in Differentiation and Disease 361
       S. A. SHAFIQ and M. GORYCKI
          Mitochondria in Differentiation
          Mitochondria in Disease

Degeneration of Cell Structures

29 The Electron Microscopy of Experimental Degeneration in the Octopus Brain . . 371
       E. G. GRAY and J. Z. YOUNG
          Introduction
          Methods
          Observations
          Discussion
          Summary
          Acknowledgements

30 Initial *Post-mortem* Changes in Mitochondria and Endoplasmic Reticulum . . . 381
       S. K. MALHOTRA
          Introduction
          Results
          Discussion
          Summary

PART 4
THE FUTURE FOR THE INTERPRETATION OF CELL STRUCTURE

31 The Golgi Field and Secretion: An Example of the Dynamic Interpretation of Cell Structure 395
       G. C. HIRSCH

APPENDICES

   I John Baker, Cytologist: an Appreciation 407
          W. G. B. CASSELMAN
  II Bibliography of the published writings (1924–68) of John Randal Baker 409
          R. M. PARK
 III To advanced students about to undertake research under my supervision 419
          J. R. BAKER
Index 421

Part 1

The Qualitative Interpretation of Cell Structure

Chapter 1

Fixation and the Display of Cellular Structure

C. F. A. PANTIN

> Whenever it is our intention to study objects that are visible with the light-microscope, it is reasonable to start with the living cell, to study the effects of fixation on its visible constituents, to analyse their composition by cytochemical methods, and then to explore their fine structure by the use of the electron microscope. A many-sided attack of this sort on a single kind of cell is much more likely to give valuable results than the independent study of different kinds of cells by different kinds of cytologists.
>
> J. R. BAKER[1]

The advent of new special techniques, particularly that of electron microscopy, is increasing our information about cells and tissues at an astounding speed. It has been easy to forget that the most certain knowledge of the nature of living systems is to be obtained, not through the development of such methods alone, but by establishing congruence between these and many different methods of study; and that our job is to understand the living system and not to confuse this with dead artifacts we produce from it, however valuable may be the information which such artifacts can give. That is made very clear by this quotation from John Baker. It is particularly because he has kept this principle so clearly before us, when in our haste we may overlook it, that we all stand so much in his debt.

But the physical nature of the living system is by no means a simple problem. To consider this at the cellular level alone is somewhat arbitrary, and in this brief essay I want to consider a range of problems, from those of the gross organism to those of the molecules of which it is composed.

In the first place, when we fix a cell or tissue, and pronounce the fixation to be 'good', we imply that what we see corresponds to the living system with the minimum degree of visible alteration. Yet alteration there has certainly been, and an obvious result would seem to be that well-fixed tissue, suitably impregnated, will last in that state indefinitely, whilst the living cells and tissues are normally fugitive structures which do not last in that way. Unlike the fixed preparation, the active living cell is dynamic: it is in fact an open steady state maintained by the passage through it of matter and of energy. Even the structures we look upon as permanent are liable to change. Thus Libby and his co-workers[12] point out that as a result of thermo-nuclear weapon tests the C^{14} concentration in the atmosphere has very recently risen to double the value for the year 1890. This increase enables us to gauge the rate of turnover of carbon in tissues. In some, such as the human brain, there is a rapid turnover in the proteins and to a slightly less extent in the lipids, so that the C^{14} concentration reflects that of the atmosphere only a matter of months previously. Collagen and cartilage from 70-year adults on the other hand shows no significant change.

At the other extreme from man, we may take the remarkable structures described by Grimstone[6] in the complex flagellates of the genus *Trichonympha*. In these, the parabasal bodies appear to consist of well-defined stacks of flattened sacs. But the system is continually gaining fresh sacs from the nuclear membrane, and losing them as cytoplasmic vesicles in a steady-state system. For all its well-defined appearance, the cell organelle is not a permanent structure, even at the moment of observation and certainly not during ontogeny.

But potentially living systems are not purely open steady states in which the perceptible material is in a continual flux, as it is, for instance, in a thunderstorm. Nor are they merely made up from well-organized, substantially enduring parts, such as the collagen of our tissues already mentioned. In living systems the essential material may exist either as an inert object, or as part of biochemically dynamic systems through which matter and energy are continuously passing. This is very clear among those borderline systems, the viruses, which can persist unchanged in an inert extra-cellular form and may even become crystalline; but which can also infect a host-cell and parasitize some part of its stream of metabolic activity so as to reproduce more virus material (Caspar and Klug).[5] In viruses, this is not perhaps surprising, for the material of the inert phase is essentially DNA or RNA, usually protected by protein. The ordered organization of the nucleic acid necessarily remains intact, for upon this the continued identity of the specific virus depends. Thus here, even at the first appearance of properties which we associate with living systems, there is a stable molecular structure, which is preserved from one generation to another. Even in the replicative stage, the nucleic acid, though actively concerned with synthesis and breakdown of material available in the host cell, presumably remains intact as a nucleic acid molecule.

But what is very surprising is the power of survival of cells themselves, and indeed of entire organisms, in the metabolically inert state of 'anabiosis'—or better still, to use Keilin's term,[10] 'cryptobiosis'. Extreme dehydration, or cold suitably applied, can throw even complex multi-cellular organisms into an inert state which can be endured for very long periods, and which is reversible. Hinton[9] showed how certain chironomid larvae can be brought to a metabolic standstill by dehydration, and that in this state they can survive temperatures of $-270°C$ and even brief exposure to $+200°C$. Sometimes dehydration may even be far from complete, as when an inert cryptobiotic state is produced by cold in organisms with a higher content of water, accompanied by vitrification, or even by the crystallization, of ice, whilst the inert state is nevertheless still reversible.[19]

Such reversibility of the inert state is much more remarkable in complex organisms such as chironomid larvae, and Rotifera, than in viruses. For in the developed higher organisms the potentialities of the DNA have been fully deployed (in ways we do not yet understand) to give rise to cell organelles, and to functional tissues such as nerve or muscle. As Hinton points out, the recovery of the tissue at normal temperatures and water content may even permit reflex responses, so that the sensori-neuro-motor machinery has survived intact. Even broken fragments of the inert organism will recover for a time, and death can, at least sometimes, be ascribed to gross mechanical damage, rather than to disruption of the biochemical machinery. Yet it is the apparently fugitive structure of cells and their membranes which one might have expected to fail under cryptobiotic conditions. Unlike the stable molecular organization of DNA, cell surfaces, including those of cell organelles such as mitochondria, are not in general fixed, 'solid' structures, but rather interfaces between fluids, with associated two-dimensional orientated molecular layers, which can also exhibit the character of fluids. Models such as those described by Beament,[4] in considering the water relations of the superficial lipids of insects, necessarily involve both two-dimensional solid, and liquid states.

This discussion of the inert state in viruses and in organisms leads to an important conclusion concerning fixation. A living cell can be classed as an open steady state; that is: a material system of constant form which, however, is maintained by the continuous passage through it of matter and of energy (cf. Bayliss).[3] In general any attempt to 'fix' such states destroys them. The fluid components of a thunderstorm could not be preserved in a fixed inert condition without destruction of the system. The same is true of a whirlpool in the bed of a river, or a bunsen burner with its flame, notwithstanding their solid elements. But as we have seen, in some organisms, at least the essential molecular machinery of nucleic acid, of the cell itself, and even of its functional organization as a tissue, can pass reversibly from a steady state to an enduring inert state. This is as near 'perfect fixation' as living material can be expected to undergo. Of course we cannot suppose that the molecular configuration is precisely the same in the 'living' and in the cryptobiotic states; but their reversibility demands that their molecular configurations must be very closely related. Hence we may conclude: notwithstanding the fact that a normal organism is an open steady state in which seemingly permanent structure may be fugitive, there are also elements of structure sufficiently permanent at all levels to warrant the use of fixation as a method of gaining knowledge about cell structure.

The existence in some cases of the cryptobiotic state at the molecular, the cellular, and even at the level of the functional organism itself suggests that the goal of 'perfect fixation' may not be quite so chimerical as it would seem. Our problem is in fact further simplified, because, in any investigation, fixation is generally aimed at the preservation, not of every feature of living tissue, but of some particular level of organization. In Baker's work, fixation has been particularly used to preserve the cell and its organelles. But it is allowable to extend the term 'fixation' to inert preservation at other levels of organization.

It will be seen that we are here most concerned with preservation and fixation of an image at a particular level of organization; in this case, that of the cell. For convenience we can distinguish three such levels, at each of which different kinds of biological problems are presented: that of gross organization, that of cellular structure, that of molecular structure. The problems at each level may be more or less complete in themselves, not directly raising considerations at other levels; and the rules which govern structure at each of these levels and the problems raised are not necessarily the same.

We may preserve and fix the gross structure of the vascular system of a dogfish. The anatomical problems raised here are of two sorts, with both of which the comparative anatomists of the last century were particularly concerned. Our well-preserved specimen may, by comparison with those of other species, bring us to conclusions about homology of plan. The second class of problem concerns the function of the parts. I have discussed these matters recently.[15] It will be noted that a single part of the gross structure such as a bone or an appendage can allow us to infer a great deal about its origin and function; whence it came and what it is for. It will also be noticed that these inferences are strongest with the gross structure and become weaker as we pass downwards in the scale of dimensions. The heart and main vascular branches of a mammal are remarkably stereotyped, but variation becomes greater and greater as we pass to peripheral vessels and finally to the capillaries. Homologous, analogous, and functional features are, as it were, imposed 'from the top' (the higher levels of integration), and are not implicit in the characters of sub-units. Such problems, and the morphogenetic rules they disclose, were apparent, and could be subjected to scientific analysis, long before the existence of 'molecular biology', or even of useful knowledge of cellular structure; though ultimately all must be related together.

We have good reason to ascribe both homology and functional adaptation to the operation of natural selection upon species, forcing them to be modified to meet certain classes of engineering solution which the organism must attain if it is to work as a machine.[13] Whether the selected engineering solution requires the development of an eye, or any other functional structure, just how this structure is attained is of secondary importance providing the need is met. We may note parallel engineering principles in the following case, where the functional requirement is met by entirely different structural organizations. In *Sacculina*, the Cypris larva, about $\frac{1}{16}$ mm long, has a sac-like body containing undifferentiated reproductive tissue. The larva attaches to the host-crab by its tubular antenna through which this tissue is forced, to form a growing mass of cells which feed on the tissues of the host. A remarkably parallel functional organization is to be found at the molecular level of organization. In the Phages we have complex viruses which parasitize bacteria. The main body is a sac about 10^{-4} mm long containing DNA which is forced through a tube, which apparently undergoes a contractile molecular deformation when attached to the surface of the bacterium, thereby forcing the nucleoprotein into its body. As machines, the two parasites are remarkably similar in principle, though there is an enormous difference in scale, and the way the engineering requirement has been achieved is quite different in the two cases.

The same principle is to be seen in the evolution of contractile tissue. There is a remarkably animal-like machinery developed in many insectivorous plants, as Charles Darwin himself noted. In *Dionaea muscipula*, flies alighting on the leaves excite sensory hairs which in turn excite all-or-nothing impulses.[14] These impulses excite rapid closure of the leaves, entrapping the fly. But the contractile machinery producing this closure is quite unlike that which seems universal for animal muscle. The wholly analogous function in *Dionaea* is produced by adaptation of a characteristic piece of plant machinery, turgor loss.[17]

Within the range of animal muscles, the repeated independent evolution of striated muscle, in which the striae seem to represent the same functional mechanico-chemical differentiation, shows that the evolution of the animal response to the need for contraction often makes use of essentially the same 'pre-adaptive' cellular machinery. Whether all the different kinds of striated muscle are built up *in detail* from the same ultra-structural parts seems unlikely, but this, surprising as it may seem, is, I believe, not yet known.

Clearly, structures with analogous function may present two distinct reasons for similarity. There is the enforced adaptive requirement to achieve a particular function, depending on the operation of natural selection on the properties of the evolutionary ancestral species; but there is also the limitation that the evolutionary ability to provide an appropriate adaptive structure is rigidly limited by the properties of matter and energy. Form is in fact determined by the interaction of two quite distinct classificatory principles, one determined by the ancestral history of the species, the other by what one may call an engineering classificatory system based directly upon what is possible in this Universe; a multi-dimensional classification which is much harder to grasp than the evolutionary one.

At the present time, the interaction of these two classificatory principles is best to be seen in work at the cellular level, upon nematocysts and related structures. It has long been known that remarkably similar structures of this sort appear sporadically throughout the animal kingdom.[13] Although some have tried to infer general evolutionary relationships from such cases, the inferences do not stand up to the great weight of evidence from other sources. Nevertheless, from the time of the work of Weill[20] there has been evident within the Cnidaria a clear correlation between certain kinds of nematocysts and certain supposedly evolutionary taxonomic groups. Nor, when one considers the appearance of directed complexity in the development of nematocysts of *Hydra* as described by Slautterback,[18] is it easy to avoid supposing that we are observing machinery selected from alternative possibilities, rather than a single engineering device common to all nematocyst-bearing organisms. This is indeed borne out by the invaluable discussion of Picken and Skaer,[16] who point out that more detailed knowledge of the ultra-structure of nematocyst-like structures tends to separate by significant detail the widespread resemblances of those derived from different classes of organism.

Confusion due to the interaction of two systems of classification, the evolutionary, and what I have termed the 'engineering', can occur at all structural levels. But it is most particularly at the cellular level that this interaction—and confusion—is greatest. A well-defined morphological feature, to which, at a higher level, we should unhesitatingly, and correctly, attribute evolutionary taxonomic significance, may, in fact, be attributable to physical principles, and the peculiar properties of matter. Functional analogy, like homology, is thus imposed 'from above'. Both represent different consequences of the operation of natural selection. They give the clearest problems and answers at the higher levels of organization; and these problems and answers are often complete in themselves, without resort to ultra-structural and ultra-physiological explanations. At a gross homological level all that we need to know is that the muscle is there and that it contracts.

In contrast to these instances, we may consider some of the more obvious problems of living organization at the molecular level. It has been a common biochemical practice to comminute living tissue, whereby all tissue and cellular organization is lost, so that we can isolate and detect the enzymes directing and controlling metabolism. Here, we have deliberately destroyed information obtainable from higher levels of organization. The results may be valuable indications of the metabolic pathways, and again, give answers to questions complete in themselves at this chemical level. In a similar way, useful information can be gained from the study of the individual machines of a factory which has been destroyed, when direct knowledge of their organization within it has been lost. Of course, the tendency for each level to present some questions which can be answered in this way, by no means covers *all* the questions, many of which necessarily involve other levels of organization. The gross features of a species, upon which natural selection acts, arise through morphogenesis by cellular growth and differentiation; valuable as the information gained from enzymatic studies of ground-up cells may be, still more is to be gained by, for instance, differential centrifugation, by which enzyme activities can be associated with particular cell organelles, and by which we begin to see the metabolic activities follow significant directed sequences.[11]

It is, of course, by no means only living organisms which present different, and more or less independent problems at different levels of complexity of organization, though it is in living things that these differences become especially important. Let us consider the kinds of objects of which, one way or another, we become aware. Starting arbitrarily with molecules, we can go downwards to atoms; from these to the protons, electrons, mesons and the ever-increasing family of 'ultimate' particles into which modern physics analyses what for a time we once considered to be indivisible units. We can also go upwards to higher orders of structure; crystals, thunderstorms, rivers, organisms, and so on.

Within these orders of structure the molecule stands at an important boundary. Two simple, small molecules of the same substance, two hydrogen atoms, or two electrons, are each indistinguishable from the other by individual features. There are no individual differences, and the species in absolutely defined. Objects of higher order than this, such as chairs, hills, or living organisms, each have individual characteristics. They fall into species, but the boundaries of the species are not sharply defined. In living organisms we may find another distinction. Since we can neither distinguish not set a mark upon an individual atom or molecule, we can never say unequivocally that it came from such and such a source; and we can never ask, 'What is a C^{12} atom for?' in the same way as we can ask this question about a molar tooth. An atom by itself gives us no such evidence of whence it came or of any contribution to function.

Another important difference between complex structures and those of molecules and lower levels is that the structures of higher order undergo denudation and wear out, whether they are hills or men. In contrast, the individual small molecule or atom remains unchanged, unless and until it undergoes destruction into fragments with wholly different properties. This feature is particularly important for living organisms as we know them, for though, in the adult stage, they are higher-order systems which undergo attrition, during reproduction they pass through a stage in which future individual development is coded in the molecular organization of specific nucleic acid molecules.

Because of the indistinguishable character of the atomic components in molecules, essential individual characteristics can, at this level, be reproduced unchanged in a manner impossible in gross material objects. All the same, because of their complexity, two nucleic acid molecules can resemble each other very closely in form and in potentialities. and yet differ in some minor degree in their molecular constituents. It is precisely because species of organisms include closely related but not absolutely identical sets of reproducing individuals that they are endowed with the features upon which natural selection can act. Natural selection can be effective because we are not concerned with identical individuals, but with a variable class: once we pass below the molecular boundary we concern ourselves with identical and unmarkable individuals, and there is nothing for selection to act upon.

The deployment of the potential morphogenetic processes of a species of DNA molecules (however that deployment will be found to take place) thus bridges the gap between ontogenetic morphogenesis and phylogenetic organization at grosser levels. As I have said, it is at the level of organization of the cell and its proteins that we pass from one set of morphological rules to another. It is in the study of cells that notions such as adaptation, homology, and so on may, therefore, be misleading.

I have tried to show in this discussion how the problems raised by the structure of organisms differ for those at higher levels than the molecular boundary, and for those below it. The peculiar position of the cytologist is that the structures with which he is concerned—cells—are astride this boundary. The significance of fixation on the two sides of this boundary is not the same. For the higher levels of structure, what are the requirements of good fixation? They are, in fact, not very exacting. If we wish to establish the homologies or the functions of an anatomical structure, simple formalin fixation may well be adequate for the necessary dissections. Even here requirements are more exacting as we proceed to lower levels of structure. To show the pattern of the vertebrate vascular system is easier than to demonstrate that the vessels are patent in life. Special fixatives are required if we are to demonstrate this for a vertebrate artery and, say, show that it is untrue for the haemal system of an echinoderm.

At a cellular level the demonstration of function is more exacting still. This is particularly true of histological work on the nervous system. The individual cells, the nerve-cells, here determine the pattern of organization in a way that the cells of more or less homogeneous tissues, such as muscle and liver, do not. To show that a cell is without doubt a nerve-cell thus becomes an important matter, even though, at this histological level, precise preservation of the image of cell-contents, close to that of the living cell, is still not necessarily so important. In discussing artifacts and the nerve-net in Anthozoa, Batham and others[2] proposed the following structural criteria:

(1)   The cells should stain by a variety of techniques in the same manner as nerve in other animals; and they should not stain in ways characteristic of other fibrous structures, such as collagen.
(2)   Their cytological structure should be consistent with that of nerve-cells.
(3)   Their anatomical relations should be consistent with the physiological evidence.

To this we may add:
(4)   The structure should if possible be identified in the living state.

When one passes to still lower levels, in fact to those near the molecular boundary, two new classes of error arise. There are those due to our transition to the use of very different kinds of instruments. This was discussed by Grimstone and his colleagues[8] as long ago as 1958, and may be quoted:

Before accepting electron micrographs as evidence of the structure of living cells, two classes of error must be considered. We must show that the appearance we examine is not of instrumental origin, as in optical diffraction or interference; and we must show that it is not an artifact of fixation or subsequent treatment. In the light microscope the appearance of double membranes at cell surfaces or the tubular appearance of mesogloeal fibrils might raise such suspicion of a diffraction effect. But considering the dimensions of the instrument we use and the exceedingly short equivalent wavelength of the electron beam (about 0·041 Å) it becomes clear that we cannot attribute these appearances in the structures we are considering to an effect analogous to optical diffraction. Effects analogous to optical interference are less easy to dismiss, and, as is well known, are sometimes apparent in electron micrographs. There is, however, a test which can be applied. False images of instrumental origin will vary with instrumental conditions, that is with the applied magnification and with the focus; *and such false images will be apparent in all objects of the same dimensions in the object examined.*★ Now the dimensions of the features of the cell membranes and of the mesogloeal fibrils which we are discussing are independent of magnification by the instrument and of the focus. Moreover, the appearance of double structure is quite evidently absent in many minute features in micrographs which clearly illustrate it in the cell membrane; whilst a tubular appearance like that of the mesogloeal filament is absent in the cross-section of other filaments and of granules of the same size, and in the same micrograph. We may therefore conclude that these features are not instrumental artifacts.

The possibility of histological artifact is still more difficult to eliminate. Grosser features can be directly identified with structures seen optically in the living cell, but finer detail cannot receive this direct confirmation. However, there is strong circumstantial evidence for the reality of some electron-micrographic structures in the living cell. The fact that the mesogloeal fibres are seen after both osmic and formol fixation lends probability to the supposition that the appearance is not an artifact. Whether they are hollow tubes, as they appear to be after phosphotungstic treatment, or whether for some reason the cortex of the fibres takes stains differentially, cannot at present be determined; though we may note that the very different method of metallic shadowing suggests that the rather similar fibres of vertebrate collagen may be tubes.[21]

Further evidence for the reality in life of electron-micrographic structures may be gained by what may be termed 'the principle of functional connections'. If some feature of a structure seen in an electron micrograph fulfils, and alone fulfils, an evident functional requirement inferred on other grounds there is presumptive evidence for its existence in life. Thus the known contractility of the muscle-fibres implies the existence of contractile structures in them which are in turn attached, directly or indirectly, to the underlying mesogloea. The muscle filaments seen under the electron microscope fulfil this requirement and there is no other evident structure which could do so. On the other hand, any feature that seems attributable to the histological treatment which the specimen has received is a presumptive artifact. That is seen in shrinkage, disruption of membranes, or regional differences in fine structure correlated with the direction of penetration of histological reagents.

But in addition to confusion due to change of instrumental method, errors are also apt to arise from confusion about patterns of structure, to which we are accustomed, at the normal higher levels, to attribute homologies or functional significance; but which may have, at the near molecular level, a spurious validity. As Grimstone[7] noted, similarities at the lower level of organization do not necessarily

★ This part of the quotation has been italicized by the Editors who consider that it deserves emphasis.

imply common ancestry. Independent origin due to the similarly organized patterns arising from common physical principles, such as those of crystallization, may become highly significant at this level. Finally we may note that, especially with biological material, the appearances at high magnifications are apt to deceive us: we do not always allow for relative changes in the dimensions M, L, and T (mass, length, and time) which experience has taught us to expect from our study of the everyday world. But further, perception itself may become misleading. In the gross world we have learned to attribute to repetitive patterns and other 'organized images' an improbability which is interpreted by us as indicative of an evolutionary, or other, relationship, or of a functional significance which we are always trying to insert into the patterns we perceive. But below the cellular level all such appearances may be normal features of the kind of environment we live in—as are the intricate ordered patterns of 'frost flowers' in the ice upon a window-pane.[13]★

The object of fixation differs, therefore, in important respects with the level of analysis of the living organization with which we are concerned. Its interpretation likewise differs greatly at different levels, and raises fundamental questions about the properties of material systems. Fortunately we may—with John Baker—ask a simple and direct question: how far does the result of fixation give us an image of the living system? As I have shown, the fact that we are contriving to substitute a fixed image for a dynamic system is not the absolute stumbling-block which it would first appear. This is fortunate, because fixation necessarily alters the chemical constituents of the cell. This necessarily engenders potential artifacts; and, without special precautions, may generate artifacts of a higher order in the higher levels of organization. For example, coagulation of the blood in vessels by fixation destroys the gross evidence that we are dealing with a dynamic system of fluid conduction. All the same, a well-chosen method of fixation can give valuable evidence of the structural organization of a cell-organelle, or of any higher level of structure.

But this should not lead us to overlook the valuable information that artifacts themselves can give us; so long as we do not jump to the conclusion that the image we see is identical with something visible in life. This again is admirably illustrated in the work of John Baker. If we may take his essay of 1958 with which this discussion began, he gives there an example of analysing the nature of the spherical objects to be seen in the cytoplasm of the ganglion cells of *Helix aspersa*. He notes that various evidence suggests that these spheres contain phospholipid and water. Baker points out that consideration of the consequences of fixation upon orientated layers of phospholipid could well lead to sundering of continuous spherical phospholipid layers at the surface of globules, and thereby lead to a false picture of the living structure. The artifacts produced by various treatments of the living substance nevertheless can, after all, give one direct information about the living system, provided that the nature of the artifacts can be interpreted consistently with all other evidence. This evidence should include, particularly, biochemical and biophysical data, and the morphological evidence at *all* levels of microscopical investigation.

SUMMARY

1. The living cell is normally an open steady state, the form of which depends upon the passage of matter and energy through it. A fixed preparation is, necessarily, not dynamic, and thus differs essentially from the living cell or tissue.

2. Nevertheless, living things are not simple steady states. Certain structures endure for very long periods. Moreover, not only do we find that viruses can survive extra-cellularly for indefinite periods, but, developed higher organisms can become functionally and metabolically inert in the 'anabiotic' state. It is argued that in the light of such instances fixation can, after all, give us direct knowledge of cell-structure.

★ [*Editorial note:* The reader is nevertheless referred to a remarkable article by Makio Mayawura (*Science* 1966) on haemoglobin, in which it is suggested that the change of a valyl–valyl bonding alone produces the most widespread and striking changes at higher levels, including the shape and response to environmental change of whole blood cells, and the phenotypic, and genotypic correlates of the sickle-cell phenomenon.]

3. Fixation presents distinct problems at different levels: the gross level, the cellular, and the molecular. Homology and functional analogy concern the gross level. Both are dominated by natural selection and the limited potentialities of material systems.

4. Interpretation of structure at the cellular level is particularly difficult because of the interaction upon structure of two distinct classificatory systems; one of evolutionary origin, the other stemming from the 'engineering possibilities' in the properties of matter.

5. The molecule stands at a very important boundary. Above this, grosser objects have individual characteristics which betray their origin, and which in the parts of organisms may betray a seeming 'purpose'. Below this, there are no individual marks of distinction. A carbon atom or a simple molecule can tell us nothing of its origin, let alone its 'purpose'. But complex molecules, particularly the nucleic acids, may be members of a related class, with the power of replication. This at once confers the necessary conditions required for the operation of natural selection.

6. Understanding at the cellular level can, nowadays, be widened valuably, by a multi-disciplinary approach, which encompasses problems at many different levels of analysis, and which attempts to find answers satisfactory to each level, and related to all levels.

REFERENCES

1. BAKER, J. R. 1958. Fixation in cytochemistry and electron-microscopy. *J. Histochem. Cytochem.* **6**, 303–8
2. BATHAM, E. J., PANTIN, C. F. A., and ROBSON, E. A. 1961. The nerve-net of *Metridium senile*: Artifacts and the nerve-net. *Quart. J. Micr. Sci.* **102**, 143–56
3. BAYLISS, L. E. 1959, *Principles of General Physiology*. Volume 1. London: Longmans, Green & Co.
4. BEAMENT, J. W. L. 1964. The active transport of water: Evidence models and machinery. *14th Symposium of Soc. Exp. Biol.*, pp. 273–98. Cambridge: University Press
5. CASPAR, D. L. D. and KLUG, A. 1962. Physical principles in the construction of regular viruses. *Cold Spring Harbor Symposia on Quantitative Biology* **27**, 1–24
6. GRIMSTONE, A. V. 1959a. Cytoplasmic membranes and the nuclear membrane in the flagellate *Trichonympha*. *J. Biophys. Biochem. Cytology* **6**, 369–78
7. —— 1959b. Cytology, homology and phylogeny—A note on 'Organic Design'. *Amer. Naturalist* **93**, 273–82
8. ——, HORNE, R. W., PANTIN, C. F. A., and ROBSON, E. A. 1958. The fine structure of the mesenteries of the sea-anemone *Metridium senile*. *Quart. J. Micr. Sci.* **99**, 523–50
9. HINTON, H. E. 1960. Cryptobiosis in the larva of *Polypedilum vanderplanki* Hint. (Chironomidae). *J. Ins. Physiol.* **5**, 286–300
10. KEILIN, D. K. 1958. The problem of anabiosis or latent life: History and current. *Proc. Roy. Soc. B* **150**, 150–91
11. KIRKLAND, R. J. S. 1958. Particles in plant and animal cells. *Adv. Sci.* **14**, 353–64
12. LIBBY, W. F., BERGER, R., MEAD, J. F., ALEXANDER, G. V., and ROSS, J. F. 1964. The replacement rates of human tissue from atmospheric radio-carbon. *Science* **146**, 1170–2
13. PANTIN, C. F. A. 1951. Organic design. *Adv. Sci.* **8**, 138–49.
14. —— 1965. Capabilities of the coelenterate behaviour machine. *Amer. Zoologist* **5**, 581–9
15. —— 1966. Homology, analogy and chemical identity in the Cnidaria. *Symp. Zool. Soc. London* **16**, 1–17. London: Academic Press
16. PICKEN, L. E. R. and SKAER, R. J. 1966. A review of researches on nematocysts. *Symp. Zool. Soc. London* **16**, 19–50. London: Academic Press
17. RUHLAND, W. (ed.) 1959. *Handbuch der Pflanzenphysiologie*, Vol. 17. Berlin: Springer-Verlag
18. SLAUTTERBACK, D. B. 1961. *The Biology of Hydra and of some other Coelenterates*, p. 77. Coral Gables, Florida: University of Miami Press
19. SMITH, A. U. 1958. The resistance of animals to cooling and freezing. *Biol. Revs.* **33**, 197–253
20. WEILL, R. 1934. *Contribution à l'étude des cnidaires et de leurs nématocystes*. Vol. II, 349–701. Paris: Les Presses Universitaires de France
21. WYCKOFF, R. W. G. 1952. The fine structure of connective tissues. *Proc. 3rd Josiah Macy Conference on Connective Tissue*, p. 38. Caldwell, N.J. (Macy Foundation)

Chapter 2

Freeze-Substitution and Freeze-Drying in Electron Microscopy[*]

S. K. MALHOTRA

INTRODUCTION

Our present knowledge of the structure of the cell has come mainly from the use of the excellent preparative procedures that are now available for investigations of biological materials by electron microscopy. Methods that are used routinely for this purpose often involve fixation of living material by direct treatment with chemical agents, like osmium tetroxide. This is well known to retain life-like preservation at anatomical and molecular levels. This information has accumulated from study of the structure of such organized membranous systems as the myelin sheath and myelin figures, that have been investigated profitably by independent methods, like x-ray diffraction and polarizing optics.[4,5,6] In cases where direct evidence on the fine structure of living systems is not available, use of indirect criteria for the evaluation of electron micrographs becomes a matter of necessity. It is now generally believed that electron microscopy reveals as life-like a representation of cell structure as is possible with the current technical procedures. It is, however, not yet certain that every feature of the living tissue is reliably represented in micrographs obtained in this way.

Freezing, when instantaneous or quick enough to avoid formation of ice-crystals, has obvious advantages over chemical fixation for the preservation of biological materials. Frozen tissues may be investigated by electron microscopy after they have been subjected to suitable treatments to achieve dehydration (such as drying by sublimation, or substitution with organic solvents) and embedding, without the need to subject living tissue to the direct action of chemical fixatives. The possibility has been considered that stimulation and alterations caused by the application of chemical fixatives might induce movement of extracellular electrolytes, sodium and chloride, into cells in central nervous tissue (accompanied by water to maintain osmotic equilibrium), thus giving an artificial picture of the organization.[28] The results of experiments described by Van Harreveld and his colleagues[28] suggest that such a movement of materials is prevented, or at least very much reduced, if the tissue is rapidly frozen immediately after the animal is killed. This results in a more reliable preservation of cellular elements and their spatial relationship.

This paper deals with differences from the commonly recognized patterns of some of the membranous systems encountered in electron micrographs of tissues prepared by freeze-substitution and freeze-drying. Though feasible evidence has been presented, which indicates that the preservation obtained by freeze-substitution or freeze-drying may be more representative of *in vivo* organization, there are difficulties in accepting certain appearances, especially those pertaining to the spatial relationships of membranes. Even when tissues are initially satisfactorily frozen, vitreously frozen ice may recrystallize during substitution, and induce alterations in the tissue.

* Supported by grants from the National Science Foundation (GB2055)

MATERIALS AND METHODS

The techniques of freeze-substitution and freeze-drying have been applied to preparation of various tissues, mostly from the nervous systems of mouse or rabbit. However, the results obtained are of more general interest. The pancreatic exocrine cells constitute a homogeneous tissue convenient for investigation by electron microscopy, and this tissue has therefore been used as a test-object, before extending the use of the above techniques to other tissues.

The method used for quick freezing involves bringing the tissue in contact with a polished silver surface cooled to about $-200°C$ by liquid nitrogen under reduced pressure (see reference 28 for details). A stream of cold, dried, helium gas is passed over the silver surface to prevent condensation of air and water upon it. The tissue to be frozen is lowered onto the silver surface at a controlled velocity of 30 to 40 cm/sec. The central nervous system is very sensitive to mechanical disturbances, and in order to minimize such interference, this tissue is frozen while it is still intact in the skull. A detailed procedure adopted for this purpose is given elsewhere.[28] Briefly, a suitable brass strip is inserted into the mouth of the mouse, this strip being first fixed to a wooden stick at such an angle that the required part of the brain makes first contact with the silver surface. The bones that cover the required part of the brain are gently removed and the wooden stick carrying the head is lowered onto the precooled silver surface. The roots of the dorsal ganglia and retina of the rabbit, and pancreas of the mouse, have been frozen by excising small pieces and placing them on a block of agar kept in a small trough of aluminium foil stuck (by epoxy resin) to one end of the wooden stick that carries the tissues to the silver surface for freezing.

After freezing, the tissue is either subjected to substitution fixation[14, 28] or to drying in vacuum. The substitution is done for two days at $-85°C$ in absolute acetone containing 2 per cent osmium tetroxide. It has been estimated that an approximately 50 to 100 μ thick layer can be substituted in 2 days. The tissue is then gradually warmed to room temperature, by first keeping it at $-25°C$ for about 2 hr, and then at $4°C$ for another 2 to 3 hr, while still in the substitution medium. It is then cut into the small pieces necessary for electron microscopy, as soon as it is brought to room temperature. These pieces are washed in absolute acetone for 3 to 4 hr before embedding in Maraglas[7] or Araldite.[12] It has been pointed out[28] that the colour of the tissue does not change during substitution at $-85°C$ and therefore chemical fixation does not seem to occur at this temperature. The tissue undergoes blackening when it is kept at $-25°C$, which indicates that it has been fixed by OsO_4 before it is brought to room temperature.

For freeze-drying, the frozen tissue is first kept at the temperature of solid CO_2 ($-79°C$) for 7 days while it is being dried in an apparatus evacuated to a pressure of about 10^{-3} mm, using a mercury diffusion pump, backed by a mechanical pump. This period of drying seems to be adequate to dry an approximately 200 μ thick layer of tissue. The tissue is subsequently dried in vacuum for 2 days at $-25°C$, followed by 2 days at -5 to $10°C$, and one day at room temperature. The tissue is then brought to atmospheric pressure and treated for 24 hr with OsO_4 vapour. It is then impregnated with propylene oxide in vacuum, and embedded in Maraglas. The embedding is done after the vacuum has been broken. For drying central nervous tissue, the required part is separated from the frozen brain by using an electric circular saw. The brain is prevented from warming up by keeping it under liquid nitrogen.

Details of fixatives used for control studies are given in the legends of the illustrations.

Sections were cut on an LKB Ultrotome, and stained with lead citrate[20] or uranyl acetate (aqueous) followed by staining with lead,[17] before examination with a Philips-200 electron microscope.

RESULTS

The useful tissue prepared by the technique of freeze-substitution, or freeze-drying, is limited to a superficial layer, often only about 10 μ thick, as evidenced by lack of recognizable ice-crystals in electron micrographs of such layers. In deeper layers ice-crystal formation often causes disorganization of tissue structures, and such material therefore seems to be unsuitable for critical study.★ The presence of relatively thick connective tissue coverings, pia, and large blood vessels on the surface, further reduces

★ Although the preservation is unsatisfactory in deeper layers, the cell inclusions show essentially the same structure as in the ice-crystal-free region.

the ice-crystal-free region of the tissue. Tissues with such overlying structures are not best suited for freezing.

Electron microscopy of freeze-substituted tissues

In electron micrographs, ghosts of ice-crystals are recognized as electron-transparent areas distinctly demarcated from each other by thin electron-dense regions, which presumably represent protoplasm concentrated through separation of pure water to form crystals. Such crystals are generally not apparent in superficial layers of tissues, which could therefore be considered well frozen. It is not easy to explain why ghosts of ice-crystals can be sometimes observed in the nucleoplasm, in superficial layers which otherwise look satisfactorily frozen. Electron micrographs of superficial layers of tissue without recognizable ice-crystal formation, easily meet the usual criteria adopted for evaluation of satisfactory preservation. These criteria are the demonstration of sharply demarcated, continuous plasma membranes, clearly recognizable membranous inclusions, and uniformly well-preserved ground cytoplasm. The triple-layered unit membrane structure of the plasma membranes (which may be geometrically symmetrical, as at the lateral and basal surfaces of the pancreatic exocrine cells, or asymmetrical, as at the luminal end of the pancreatic exocrine cells, when the adjacent plasma membranes do not form tight junctions), and of the membranes of the Golgi apparatus, and endoplasmic reticulum, is clearly demonstrated in suitable micrographs.[14, 28] The unit membrane (~ 95 Å) that surrounds the zymogen granules in the pancreatic tissue, also appears asymmetrical, the dense line on the cytoplasmic side being slightly thicker than the inner dense line.[14] Specific differences in the chemical composition of the plasma membranes are suggested by variations in the appearance of the unit membrane structure in micrographs. For example, the unit membrane structure of the plasma membrane of the non-myelinated axons in the dorsal root ganglion of the rabbit,[15] and in the molecular layer of the cerebellar cortex of the mouse,[28] is not clearly demonstrated, even after staining sections in uranyl acetate followed by lead, or in lead citrate. Yet the same technique of freeze-substitution and section staining shows symmetrical unit membrane structure in the plasma membranes of non-myelinated axons in the central white matter.[16]

The fine structural details of plasma membranes and cellular inclusions seen by freeze-substitution are essentially comparable to those seen by routine fixation with OsO_4. There is, however, a noteworthy difference in the structural pattern of mitochondria (Fig. 2.1, A and B). The membranes of the cristae are represented by three dense lines separated by two electron transparent lines, making a five-layered pattern (120 to 130 Å in thickness). The middle dense line and the two pale lines measure about 25 to 30 Å each in thickness. The outer dense lines are not always conspicuous in the micrographs and are slightly thinner than the middle dense lines. The continuity between the membranes of the cristae, and the membranes that delimit a mitochondrion from the ground cytoplasm, is shown in many micrographs. In favourably sectioned mitochondria, the delimiting membranes also show the same five-layered pattern that is seen in the membranes of the cristae (Fig. 2.1, B). This structure of the mitochondrial membrane system is consistently observed in a variety of tissues prepared by freeze-substitution, e.g. nervous and non-nervous cells in the cerebellar and cerebral cortices, and the medulla oblongata of the mouse, the retina of the rabbit, and pancreatic exocrine cells and liver cells of the mouse. The five-layered pattern seen with freeze-substitution differs from the commonly known seven-layered pattern (two triple-layered unit membranes separated by an electron-transparent gap, ~ 100 Å wide) of the mitochondrial membranes in as much as the usual gap (henceforth referred to as the gap) between the two unit membranes is not shown. A complete or almost complete obliteration of the gap between the two membranes results in a close approximation ('fusion') of the apposing surfaces, thus producing a five-layered pattern in electron micrographs. The structure and formation of the tight junction (*zonula occludens*) observed at the apposing surfaces of adjacent cells (see e.g., references 2, 22), and of a lamella of myelin sheath (see e.g., reference 21) is very similar to the five-layered mitochondrial membrane pattern encountered in micrographs of freeze-substituted material.

It has been remarked previously[13, 14] that a few of the cristae in mitochondria in freeze-substituted material may show the seven-layered pattern commonly seen after routine fixation in OsO_4. The seven-layered pattern is mostly shown along a small part of the length of the cristae while the rest of the length

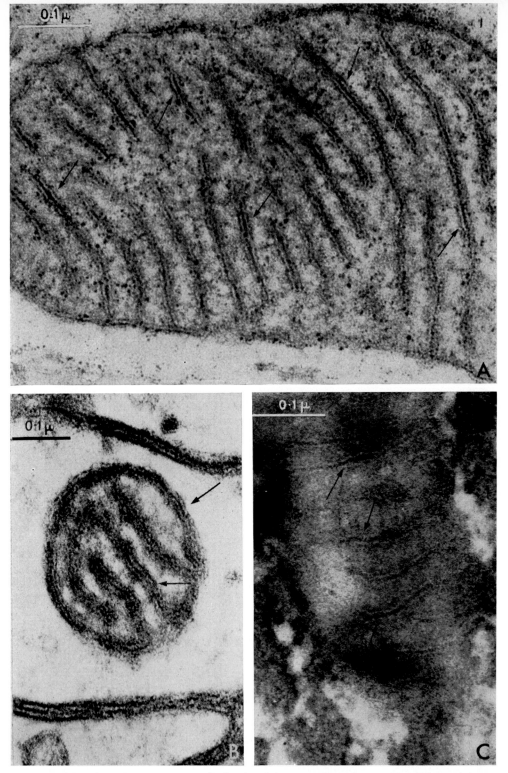

Fig. 2.1 A and B. Show the appearance of mitochondria as consistently seen in sections of tissues prepared by a freeze-substitution technique. The five-layered pattern of the membranes of the cristae and of the limiting membranes is clearly shown (arrows). In B, five-layered tight junctions between apposed plasma membranes are also seen, above and below the mitochondrion. A and B from cerebellar cortex of mouse.

C shows the appearance of a mitochondrion seen in sections of tissues prepared by freeze-drying technique. The mitochondrial membrane system shows essentially the same structure as seen by freeze-substitution. The middle dense line of the five-layered pattern is clearly shown (arrows) but the two outer dense lines on the matrix side are not easily discerned. From pancreatic exocrine cells.

of the crista shows the five-layered pattern (see Figs. 30.3 and 30.4 on pp. 385 and 387). The number of such mitochondria showing double-featured cristae is very small (a rough estimate shows less than 4 per cent in over 100 micrographs of pancreas), but this appearance is more common in cerebral cortex than in other tissues studied by freeze-substitution.

Another noteworthy difference between preservation by freeze-substitution, and by direct fixation, is shown in the structure of the parallel arrays of disks, which are formed from infoldings of the plasma membranes in the outer segments of the retinal photoreceptors.[4,18,19] In freeze-substituted material, the membranes of the disks mostly show a five-layered pattern (~ 140 Å) formed by fusion of two unit membranes (Fig. 2.3, D). At places where there is a separation of the two membranes along the length of the disks, the latter show a symmetrical triple-layered structure. At the periphery of the disks (where the membranes undergo a 180° bend) there is always a narrow lumen present as the membranes form a hairpin loop, perhaps to avoid sharp bends. (The molecular structure of membranes probably limits the degree of curvature they can undergo.[18]) The adjacent disks are separated from each other by approximately 100 Å. In retina routinely fixed by OsO_4 (Fig. 2.3, C), each unit membrane of the disk is represented by a single dense line (~ 50 Å), and there is no evidence in these micrographs of the presence of a middle dense line formed by the fusion of apposing surfaces of the two plasma membranes which is clearly demonstrated in freeze-substituted material. In contrast with the poorly preserved and featureless membranes of the disks after routine OsO_4 fixation, the membranes show evidence of the presence of minute, electron-dense particles in freeze-substituted material. These particles appear mostly to be associated with the dense lines of the membranes, in the micrographs. Fernández-Morán[4] has shown similar particles in the frog retina, prepared by specially developed techniques for low-temperature electron microscopy, and considers them to be related to the photopigment.

Other minor differences from routinely fixed tissue have been noticed in the preservation of the dorsal root ganglia of the rabbit[15] and should be mentioned. The structure of the myelin sheath seen in electron micrographs may be compared with the data available from x-ray diffraction studies of fresh nerve, to evaluate the quality of preservation. The major dense lines of the myelin sheath are wider (~ 50 Å compared with ~ 30 Å), and their radial repeat period is also larger (~ 150 Å as against ~ 120 Å) in freeze-substituted nerves, as compared with those seen after fixation in isotonic solutions of OsO_4 buffered at pH 7·2 to 7·4 (Fig. 2.3, A and B). The extracellular space between the non-myelinated axons, and their associated Schwann cells, and between the apposed Schwann cell membranes, is wider in freeze-substituted material (200 to 250 Å against 100 to 150 Å).

Differences in preservation by freeze–substitution and freeze-drying

The contrast is generally lower in electron micrographs of freeze-dried material, and this is presumably due to the complete, or almost complete, absence of loss of any material, which might (in freeze-substitution) conceivably occur during substitution in acetone. However, the structural details of plasma membranes and membranous inclusions do not greatly differ in the two types of preparation. The matrix of the mitochondria is extremely dense in freeze-dried tissue, and this makes it difficult to study their membranous system. In suitable micrographs the membranes of each of the cistae are represented by a dense line (~ 35 Å wide) with a less dense line of approximately the same width on either side (Fig. 2.1, C). The middle dense line represents the fused surfaces of the two apposing membranes, and in this respect the preservation is similar to that in mitochondria in freeze-substituted tissues. This appearance of the mitochondrial membranous system in freeze-dried material is very similar to that described by Sjöstrand and Elfvin[25] in the pancreas, prepared by a freeze-drying technique. The triple-layered pattern seen in freeze-dried material differs from the five-layered pattern shown after freeze-substitution, in the apparent absence of the two outer dense lines. There are some indications that the cristae are sharply delimited from the matrix by the presence of thin dense lines, like those seen in freeze-substituted material, but they are not easy to demonstrate in electron micrographs because of the lack of contrast in freeze-dried tissue.[25]

The electron micrographs of pancreatic exocrine cells prepared by freeze-drying show an aspect of the structure of the endoplasmic reticulum which has not been observed by freeze-substitution, or other

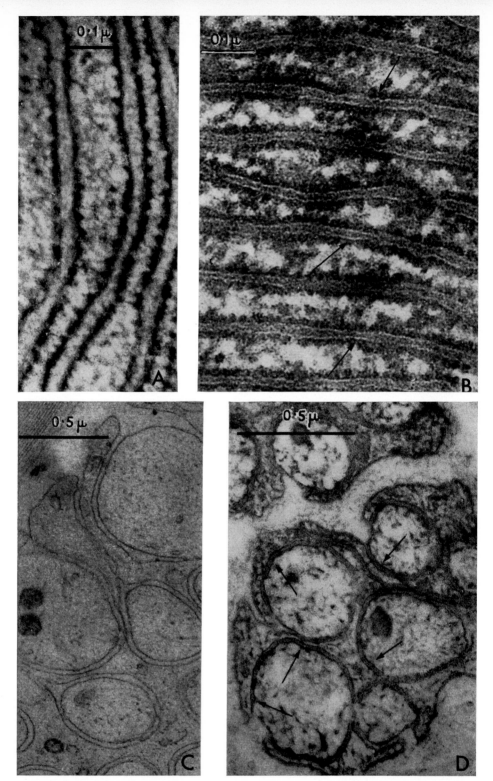

Fig. 2.2 Comparison of the endoplasmic reticulum in sections of tissue prepared by freeze-substitution (A) and freeze-drying (B). The lumen of the endoplasmic reticulum, which is regularly seen after freeze-substitution, is absent in many places in micrographs of freeze-dried material, resulting in the formation of a five-layered pattern (arrows in B). Note the presence of discrete ribosomes associated with the endoplasmic reticulum, in *both* types of preparations. From pancreatic exocrine cells.

C and D show a comparison of the non-myelinated nerve fibres and their associated Schwann cells in the dorsal roots of rabbit prepared by freeze-substitution (C) and freeze-drying (D). Note that the extracellular space between the axons and their Schwann cells, and between the apposed Schwann cell membranes which is present in freeze-substituted preparations is absent from freeze-dried preparations, resulting in the formation of tight junctions.

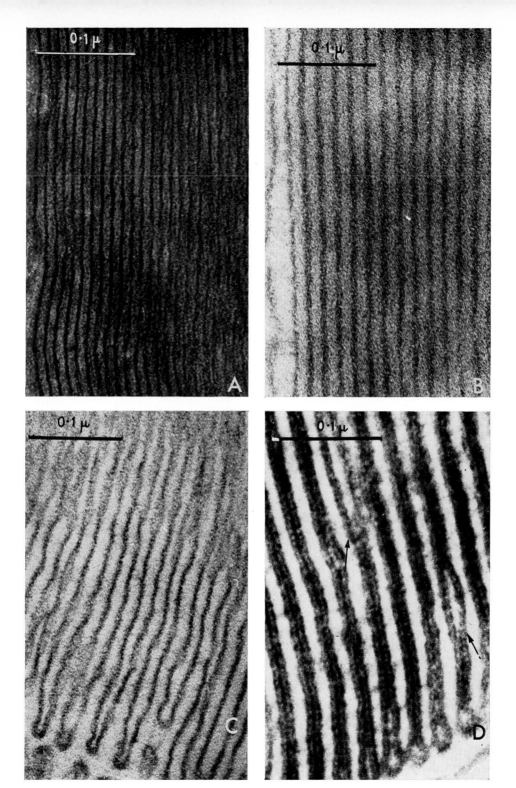

Fig. 2.3 Comparison of the myelin sheath in nerve fibres from the dorsal roots of rabbit fixed by immersion in isotonic solution of OsO_4 (pH 7·2 to 7·4) (A), and prepared by freeze-substitution (B). The radial repeat period of the lamellae is wider in the freeze-substituted roots than in those fixed in OsO_4 (145 to 155 Å against 100 to 120 Å). The major dense lines are also wider in freeze-substituted preparations ($\sim$50 Å against $\sim$30 Å).

C and D show appearance of the disks in the outer segments of the rods of the retina of rabbit fixed directly by immersing in buffered (pH 7·4) solution of OsO_4 (C), and prepared by freeze-substitution (D). Arrows in D indicate where the apposing membranes of the disks have separated from one another. Minute electron-dense particles are seen at many places on the membranes in D.

methods of tissue preparation routinely used in electron microscopy. The cisternae of the endoplasmic reticulum show a prominent dense line in the middle of their cavity at many places (Fig. 2.2, B). This middle dense line represents closely apposed inner surfaces of the cisternae, so that if a very narrow lumen exists in the cisternae, it is not easily demonstrated by electron microscopy. At other places, where a distinct lumen is shown in micrographs of freeze-dried material, its contents are denser than in comparable micrographs of freeze-substituted tissue. The extent of the lumen of the endoplasmic reticulum is known to vary with the physiological activity of the cell. Whether it may at time be obliterated completely, as suggested by the electron microscopy of freeze-dried tissue, is not known. The possibility that the close apposition of the endoplasmic reticulum membranes represents a somewhat shrunken appearance, due to the preparative procedure, should be borne in mind.*

A comparison of the roots of the dorsal ganglia of the rabbit shows that the 200 to 250 Å wide extracellular space consistently seen between the non-myelinated axons and the associated Schwann cell membrane, in micrographs of freeze-substituted roots (Fig. 2.2, C), is completely absent from micrographs of freeze-dried nerves (Fig. 2.2, D). The latter show typical five-layered tight junctions, formed by fusion of plasma membranes of axons and their associated Schwann cells, or apposed membranes of Schwann cells themselves.

DISCUSSION

Myelin sheath is one of the few organized components, amongst biological membranous systems, with a structure which has been fairly well investigated. The existence of a radial structure, regularly repeating at a distance of 180 to 185 Å, has been shown by low angle x-ray diffraction of fresh peripheral mammalian nerve.[4,5,6] It has also been demonstrated that routine fixation by OsO_4, and the subsequent preparative procedures required in electron microscopy, reduce the radial repeat period to 120 to 130 Å.[6] The demonstration of a larger radial repeat period (~ 150 Å) in electron micrographs of freeze-substituted dorsal roots, than in material routinely fixed by OsO_4, is an indication that the in vivo distribution of water may be more faithfully preserved by freeze-substitution, than by direct chemical fixation, at least in the myelin sheath. This finding is further supported by the preservation of a more or less complete triple-layered unit membrane structure in the disks of the outer segments of the retinal photoreceptors, in contrast to the incomplete preservation of these membranes in OsO_4-fixed tissue.[4,18]

However, in the absence of any direct evidence on the structure of the mitochondrial membranous system in living cells, it is difficult to decide from the evidence presented in this paper whether the complete or almost complete absence of the gap in mitochondria prepared by freeze-substitution or freeze-drying represents a life-like situation. The five-layered pattern (formed by fusion of two apposing membranes) in mitochondria has not commonly been described. On the contrary, the presence of a gap of ~ 50 to 100 Å width between two membranes is considered to be a regular feature of mitochondria, and the presence of fluid in this gap is assumed to be essential for ions to be able to move freely along the length of the cristae.[8,22] Such a role will require the existence of at least minute water channels. It is therefore quite likely that a narrow channel exists in vivo between the two apposed membranes, but its dimensions are too small to be easily visualized in electron micrographs of frozen tissues. It is also possible that such channels are temporarily produced between two closely apposed membranes to allow free movement of ions. According to the generally accepted evidence on the molecular structure of biological membranes, the middle dense line in the five-layered pattern represents sites of the polar ends of two

* In freeze-drying, as the water is removed from the frozen tissue directly into a gaseous state, without any intervening liquid phase, there is little scope for redistribution of materials in the tissue, or for shrinkage during drying. The possibility that alterations might occur during impregnation with propylene oxide and/or embedding medium is difficult to rule out.

[Editorial note: The reader is referred to the article by De Bruijn and McGee-Russell (pp. 115–132) for direct consideration of this possibility, and for discussion of experimental approaches, which may provide some evidence on the effects of the final reagents used in embedding. In the opinion of the editors, the possibility raised here by the author is not merely difficult to rule out—it must definitely not be ruled out. It must be studied experimentally.]

apposing bimolecular leaflets of lipid molecules, and thin layers of non-lipid material spread on their surfaces (see, e.g., reference 26). A channel of watery nature could be produced physiologically by taking up fluid at the apposed polar surfaces.

It is not inconceivable that chemical fixation of living tissues by OsO_4 may induce changes which would artificially separate the two closely apposed membranes in mitochondria, and produce a gap of watery nature, or enlarge it, if one already exists. The intracellular fluid contains a number of swelling agents, Ca^{2+}, phosphate, and free fatty acids. Their swelling action could be negated by such components as ATP and Mg^{2+}. The results of a dynamic balance of these opposing forces determine the volume and configuration of mitochondria within a living cell.[11] Any interference in this balance is likely to induce changes in the volume and shape of the mitochondria. For example, it has been demonstrated that the gap between the two membranes in mitochondria becomes discernible in electron micrographs of tissues prepared by the same freeze-substitution technique as used in this investigation, if the freezing of tissues is delayed, after decapitation of the animal.[13] This subject is further discussed in Chapter 30 in this book. The presence of the usual gap, in a few of the cristae, in mitochondria in electron micrographs of freeze-substituted tissues, is perhaps an indication of the extreme sensitivity of the structure of these membranous systems (compare with references 3, 9). The occurrence of such cristae with the five-layered pattern more often in cerebral cortex, than in other tissues investigated, would be in accordance with the well-known belief that the central nervous tissue is highly susceptible to alterations due to mechanical disturbances.

Apart from the observations by the present author,[13, 14] Sjöstrand seems to be the only one who has seriously considered the possibility that the mitochondrial membranes may have a five-layered pattern in life, although Afzelius[1] and Robertson[22] have also pointed out the absence of the gap in mitochondria in tissues fixed by permanganate. Sjöstrand[24] frequently observed the two mitochondrial membranes in mutual contact in $KMnO_4$-fixed tubular cells of the kidney, and attributed the appearance of the gap to an artifact of fixation, commonly seen in OsO_4-fixed tissue. Sjöstrand and Elfvin[25] have also recorded that the gap in mitochondria is regularly absent from pancreatic exocrine cells prepared by freeze-drying. Also, in formaldehyde-fixed tissues (with no post-fixation in OsO_4) the absence of a prominent gap between the two mitochondrial membranes is conspicuous, and the appearance of these membranes is similar to that seen after freeze-drying (see Fig. 16 of reference 13). Post-fixation by OsO_4 seems to result in the appearance of the gap in mitochondria.

Whereas the fusion of apposed membranes, seen in mitochondria by both freeze-substitution and freeze-drying, may represent a life-like situation, it is not easy to decide whether the tight junctions between plasma membranes of non-myelinated axons and their associated Schwann cells, and between apposing Schwann cell membranes, seen in the dorsal roots of the rabbit, prepared by freeze-drying, depict an *in vivo* situation. In freeze-drying the tissue is dried by molecular distillation, thus eliminating chances of redistribution of chemical constituents, which may occur during substitution in acetone in the freeze-substitution technique. One would therefore expect a more reliable preservation by freeze-drying than freeze-substitution. However, impregnation in propylene oxide, and subsequent embedding, might have caused shrinkage of the tissue, to the extent that the 200 to 300 Å extracellular space, encountered in between non-myelinated axons and their associated Schwann cells prepared by freeze-substitution, completely disappears from electron micrographs of freeze-dried nerves. The existence of extracellular channels, at least 150 Å wide, has been demonstrated in the peripheral nervous system by Rosenbluth and Wissig,[23] who have observed incorporation of ferritin by neurons in the spinal ganglion of the toad. These observations would indicate that the tight junctions in the freeze-dried preparations of the dorsal roots are, in fact, artifacts.* It should, however, be remarked that the uptake of ferritin by tissues may not necessarily be associated with the *in vivo* existence of watery channels leading to the

---

* If the tight junctions seen in the freeze-dried dorsal roots are artifacts, they are without functional significance compared to, for example, electrical synapses.[15] A process of shrinkage that accounts for the collapse in freeze-drying of extracellular channels at least 150 Å wide, could also explain the almost complete obliteration of the narrow lumen of the endoplasmic reticulum in many places, in pancreatic exocrine cells (Fig. 2.2, B) prepared by the same technique as was used for dorsal roots.

cells. Such observations should be cautiously interpreted, in view of the results of experiments conducted on the ciliary epithelium, which suggest that the permeability of membranes is drastically increased by OsO_4 fixation, so that ferritin particles freely diffuse across the membranes after fixation.[27] The problem of the extent of the extracellular space, if any (see references 10, 28), in the nervous system, may best be resolved by the application of suitable physiological methods, e.g., measurement of electrical impedance, or determination of distribution of extracellular ions, sodium or chloride, in conjunction with electron microscopy, as has been attempted, for example, by Van Harreveld.[28] But the limitations of these methods do not yet allow a definitive conclusion.

There is another aspect to the interpretation of the preservation of cellular material, that should be mentioned in this paper. Sjöstrand and Elfvin[25] have described how the ribosomal material is uniformly distributed over the surface of the endoplasmic reticulum in the pancreatic exocrine cells prepared by freeze-drying and suggest that the ribosomal particles seen in OsO_4-fixed tissue may be artifacts. They have further considered that subdivision into discrete particles occurs in frozen tissue as a result of dehydration of the cytoplasm, due to the formation of ice crystals. The presence of discrete ribosomes seen in electron micrographs of freeze-substituted tissue (Fig. 2.2, A) could be caused by a possible reconstitution of cellular materials, during substitution and/or fixation in OsO_4. However, their existence in electron micrographs of frozen-dried material, where there is no obvious evidence of ice crystal formation (Fig. 2.2, B) is difficult to explain, if the interpretation given by Sjöstrand and Elfvin is correct.

SUMMARY

Freezing, when rapid enough to avoid formation of ice-crystals, has obvious advantages for the preservation of living biological materials over direct fixation by chemical agents. Frozen materials, when suitably dehydrated, either by substitution with an organic solvent as in freeze-substitution or by sublimation as in freeze-drying, can be sectioned routinely and examined in electron microscope. Such techniques have been developed with a view to attempting preservation of *in vivo* distribution of water in animal tissues. Initially these techniques were used for investigation of the extracellular space in central nervous tissue by electron microscopy. The results obtained in this way were consistent with the data available from impedance measurements; and these results suggested that the preservation seen in electron micrographs prepared by freeze-substitution may be more lifelike than that achieved after direct fixation by solutions of osmium tetroxide. The appearance of mitochondrial membranes, in particular, was strikingly different from that commonly seen after fixation by osmium tetroxide; it appeared as if the two apposed membranes, both in the cristae and at the surface of the mitochondria, were 'fused' with one another as in tight-junctions. Further understanding of the organization of these membranes is, however, necessary before extensive conclusions may be drawn.

REFERENCES

1. AFZELIUS, B. J. 1962. Chemical fixatives for electron microscopy. *Symp. Intern. Soc. Cell Biol.* **1**, 1
2. FARQUHAR, M. G. and PALADE, G. E. 1963. Junctional complexes in various epithelia. *J. Cell Biol.* **17**, 375
3. FAWCETT, D. W. 1964. *Modern Developments in Electron Microscopy* (edited by B. M. Siegel), p. 257. New York: Academic Press
4. FERNÁNDEZ-MORÁN, H. 1962. Cell-membrane ultrastructure. *Circulation* **26**, 1039
5. FINEAN, J. B. 1961. *Chemical Ultrastructure in Living Tissues.* Springfield: Charles C. Thomas
6. —— and ROBERTSON, J. D. 1958. Lipids and the structure of myelin. *British Med. Bull.* **14**, 267
7. FREEMAN, J. A. and SPURLOCK, B. O. 1962. A new epoxy embedment for electron microscopy. *J. Cell Biol.* **13**, 437
8. GREEN, D. E. 1964. The mitochondrion. *Sci. Amer.* **210**, 63
9. ITO, S. 1962. Light and electron microscopic study of membranous cytoplasmic organelles. *Symp. Intern. Soc. Cell Biol.* **1**, 129
10. KARLSSON, U. and SCHULTZ, R. L. 1965. Fixation of the central nervous system for electron microscopy by aldehyde perfusion. *J. Ultrastruct. Res.* **12**, 160

11. LEHNINGER, A. L. 1964. *The Mitochondrion*. New York: W. A. Benjamin, Inc.
12. LUFT, J. H. 1961. Improvements in epoxy resin embedding methods. *J. Cell Biol.* **9**, 409
13. MALHOTRA, S. K. 1966. A study of structure of the mitochondrial membrane system. *J. Ultrastruct. Res.* **15**, 14
14. —— and VAN HARREVELD, A. 1965. Some structural features of mitochondria in tissues prepared by freeze-substitution. *J. Ultrastruct. Res.* **12**, 473
15. —— and ——. 1965. Dorsal roots of the rabbit investigated by freeze-distribution. *Anat. Rec.* **152**, 283
16. —— and ——. 1966. Distribution of extracellular material in central white matter. *J. Anat.* London, **100**, 99
17. MILLONIG, G. 1961. A modified procedure for lead staining of thin sections. *J. Biophys. Biochem. Cytol.* **11**, 736
18. MOODY, M. F. 1964. Photoreceptor organelles in animals. *Biol. Rev.* **39**, 43
19. NILSSON, S. E. G. 1964. Receptor cell outer segment development and ultrastructure of the disk membranes in the retina of the tadpole. (*Rana pipiens*). *J. Ultrastruct. Res.* **11**, 581
20. REYNOLDS, E. S. 1963. The use of lead citrate at high pH as an electron-opaque stain in electron microscopy. *J. Cell Biol.* **17**, 208
21. ROBERTSON, J. D. 1961. Cell membranes and the origin of mitochondria. *Proc. 4th Neurochem. Symp.*, p. 497
22. ——. 1964. *Cellular Membranes in Development*, edited by M. Locke, p. 1. New York: Academic Press
23. ROSENBLUTH, J. and WISSIG, S. L. 1964. The distribution of exogenous ferritin in toad spinal ganglia and the mechanism of its uptake by neurons. *J. Cell Biol.* **23**, 307
24. SJÖSTRAND, F. S. 1963. A new ultrastructural element of the membranes in mitochondria and of some cytoplasmic membranes. *J. Ultrastruct. Res.* **9**, 340
25. —— and ELFVIN, L. G. 1964. The granular structure of mitochondrial membranes and of cytomembranes as demonstrated in frozen-dried tissue. *J. Ultrastruct. Res.* **10**, 263
26. STOECKENIUS, W. 1962. Structure of the plasma membrane. *Circulation* **26**, 1066.
27. TORMEY, J. McD. 1965. Artifactual localization of ferritin in the ciliary epithelium *in vitro*. *J. Cell Biol.* **25**, 1
28. VAN HARREVELD, A., CROWELL, J., and MALHOTRA, S. K. 1965. A study of extracellular space in central nervous tissue by freeze-substitution. *J. Cell Biol.* **25**, 117

Chapter 3

On Using a Cryostat

G. B. DAVID and A. W. BROWN

INTRODUCTION

The 'cryostat' used in histochemistry is a modified microtome, which is operated within an insulated cold chamber, to cut sections of frozen tissues. External controls are sometimes fitted to permit the operation of the microtome from outside the chamber; the temperature of the chamber is usually controlled by a thermostat, and accessories of varying complexity are used to manipulate the frozen tissues and sections. The term 'cryostat' in this context is an unfortunate one: the refrigerated chamber might be described as a cryostat, in the sense that an accurately controlled oven might be described as a thermostat, but what about the microtome? Pearse[67] suggested the term 'cold microtome' instead, though this seems hardly an appropriate name for so complex an instrument. At least one American manufacturer describes his products as 'cryotomes', but it is unlikely that this or any other new name for these instruments will ever now gain universal acceptance.

The cryostat was introduced into microtechnique in 1938 by Linderstrøm-Lang and Mogensen,[55] as a simple method of cutting, at a uniform thickness, series of undistorted and flat sections of *unfixed* tissues. Before cryostats became available, sections of unfixed tissues were usually cut with the Schultz-Brauns microtome.[72,73] This was a modified routine freezing microtome, in which both the tissue and the knife were cooled by the expansion of liquid CO_2; it was not possible to maintain a constant temperature in any part of the system. The sections had to be supported and guided by a hand-held brush during the actual cutting. Somewhat improved versions of the Schultz-Brauns instrument have been described quite recently.[1,74] These devices were extremely awkward to use, and the quality of the resulting sections was not high. Linderstrøm-Lang and Mogensen[55] devised the first rational method of cutting frozen sections. A Minot microtome was put in an insulated glove-box provided with double-glazed windows, and portholes for the removal of the sections. The glove-box was cooled by CO_2, vaporized from a container of solid CO_2 at the back of the chamber, by a fan and heater under thermostatic control. The microtome was modified by providing a glass plate in front of the microtome-knife, parallel to its upper cutting facet (the *anti-roll plate*). Most of the problems of freezing microtomy were thus solved: accurate microtomes such as the Minot could be used; the sections remained frozen and flat, sliding between the knife and the anti-roll plate; thermal effects on section thickness, etc., could be largely discounted; and ribbons of sections could be obtained with the ease traditionally associated with the cutting of paraffin-embedded material. The best cryostats of today are merely improvements of the original Danish instrument: Coons and his associates[24] replaced the CO_2-cooling system by a mechanical refrigerator; Hydén[46] added built-in means for drying unfixed frozen sections under vacuum; David and Steel[30] developed a programmable servo-drive for the microtome; Pearse[3,67] arranged the controls of his instrument in such a way that they could all be operated from outside, thus enabling optimum cutting conditions to be maintained for long periods. Frozen sections of the order of 0.3 μ-thick were obtained with a slightly modified Pearse cryostat as early as 1960.[29] More complex instruments, suitable for producing serial sections thin enough for electron microscopy, are now at the prototype stage.[14,35,77] An improved cryostat of the Pearse type is illustrated in use in Fig. 3.1.

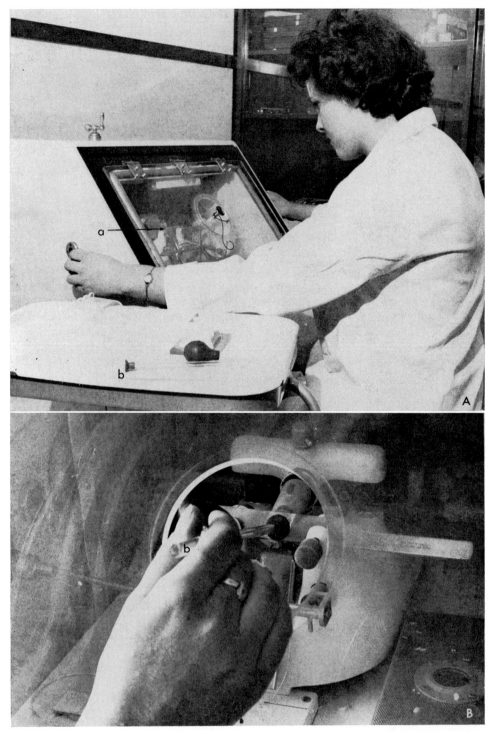

Fig. 3.1 Modified Pearse cryostat at the Carshalton Laboratories. In A, the main features of the instrument can be seen. The seated operator is controlling the pressure of the anti-roll plate with the left hand and the cutting stroke with the right. The flat section can be seen at *a*. When a dry coverglass is brought near the section with the suction device *b*, after opening the porthole shown in B, the section flattens itself on the coverglass by electrostatic attraction. It is then transferred to the freeze-drier (not visible in this figure), or placed in a holder.

Linderstrøm-Lang and Mogensen[55] required frozen sections primarily for microchemical work. Perhaps for this reason, twenty years ago cryostats were virtually unknown to histochemists outside the Carlsberg laboratories; ten years later they were surpassed only by the larger electron microscopes as status symbols of the modern cytological laboratory; today, cryostats have become so commonplace that one of us has recently counted not less than six different models of cryostats being used at the same time in a single pathological laboratory (in Texas). The rise in the use of the cryostat was brought about in the 1950's by the development of two kinds of histochemical methods requiring the use of frozen unfixed sections: the fluorescent antibody technique of Coons, Leduc, and Kaplan;[24] and enzymological methods, such as Lowry's microchemical techniques,[56-58] and the tetrazolium techniques for histochemical localization of the sites of action of oxidoreductases.[28,31,67] The use of cryostats for cutting better frozen sections of fixed tissues has also been advocated.[29] Using a cryostat is becoming almost as uncritical a routine in histochemistry, as the cutting of paraffin sections with a Minot microtome became in microscopical anatomy. It is therefore desirable to examine carefully some aspects of the methodology of handling tissues in cryostats.

The purposes of this paper are: (i) To discuss the design of cryostats in the light of the precision and ease with which they can be used to produce thin sections of unfixed and fixed tissues. (ii) To examine critically the theory and practice of methods of preparing unfixed tissues for cryostat microtomy. (iii) Briefly to discuss the uses of cryostats for sectioning fixed tissues. (iv) To describe for the first time some technical improvements related to (ii) and (iii). It is not intended to review the literature comprehensively. Nothing will be said of methods, such as the fluorescent antibody technique, of which we have had no personal experience.

ON THE DESIGN OF CRYOSTATS

Three kinds of cryostats are generally available today: the Coons type,[29] manufactured by Harris International, Cambridge, U.S.A., Prestcold (Oxford) and Dittes (Heidelberg); the Pearse type,[3,67] produced by South London Electrical Engineering and Bright Refrigeration (London); and the 'open-top' type, originally described by Chang and others,[19-21] and now mass-produced by American Optical, Lipshaw, Harris-International, South London Electrical Engineering, and other companies. The three types are refrigerated in the same manner; a single-stage compressor feeds expansion coils on three or more walls of the insulated chamber. When the chamber is closed and dry, temperatures of the order of -10 to $-30°C$ can be maintained thermostatically with fluctuations from place-to-place in the chamber, and over time, not exceeding about $\pm 1°C$. The Coons type is usually provided with glove-holes, through which the microtomist inserts his arms to actuate the microtome and handle the sections. In the model made by Dittes, the microtome is actuated by a wheel outside the case, requiring either two operators working in close synchrony or a three-armed microtomist. Jung or International microtomes are used in these. The Pearse type (Fig. 3.1) is operated entirely from outside the chamber; it is provided either with a Cambridge rocker microtome, or an International-type Minot. The open-top cryostats, as the name implies, rely on cold air being heavier than warm air to maintain a 'constant' temperature within the chamber.

Environmental control and the microtome chamber

There are two conflicting conditions to be considered in the environmental control of the cryostat. In view of the different thermal expansion coefficients of the tissue, the knife, and different parts of the microtome, and the known variation in cutting properties of the tissue with changes in temperature, it would seem important to keep all objects in the cryostat chamber at precisely the same temperature. This situation will be referred to as 'homoeothermic conditions'. Certain theoretical considerations[80] suggest that, for any given tissue, the block, the knife, and the air in the chamber should be at different temperatures. This situation will be referred to as 'heterothermic conditions'.

Theoretical considerations notwithstanding, perfectly satisfactory routine sections of most materials can be cut under homoeothermic conditions, with the cryostat thermostat set at about $-20°C$. It is only when sectioning 'difficult' tissues, or attempting to cut sections thinner than about 2 μ, that hetero-thermic conditions need be considered. It should perhaps be added that it is almost impossible to secure homoeothermic conditions with the open-top cryostats. In our experience, these cryostats are best avoided when *cytologically adequate*, as opposed to *histologically adequate*, sections are needed.

The theoretical basis for the use of heterothermic conditions was established by Thornburg and Mengers,[80] and largely confirmed experimentally in our laboratories.[32] Microtomy of a frozen tissue is a different process from the shearing of, say, a paraffin or a methacrylate section. The frozen block is cut by melting, much as a block if ice can be sliced by pressure with a taut steel wire. Thornburg and Mengers, indeed, proposed a physical model of cryostat sectioning in which they assumed that unfixed tissues were, for all practical purposes, embedded in frozen water. On this basis, they calculated that a block of tissue, having a cutting resistance of 20 g per cm width at $-20°$C, would dissipate 1.96×10^4 ergs. Taking into account the heat of fusion of ice, a layer of tissue some 650 Å thick would be melted, to be refrozen in the wake of the cutting. This physical model is illustrated diagrammatically in Fig. 3.2 (modified form[80]).

Several important consequences follow from this. In the case of unfixed tissue, the molten layer can be assumed to have been morphologically destroyed, and at best to have been transformed into a struc-

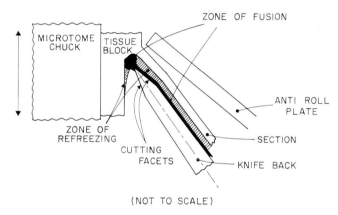

(NOT TO SCALE)

Fig. 3.2 Diagrammatic representation (not to scale) of the thawing and freezing processes by which tissues are sliced in the cryostat.

tureless surface film, which we shall call the *zone of fusion*. This surface artifact should be kept as thin as possible, for histochemical as well as purely mechanical reasons.

What considerations affect the depth of the zone of fusion ? (i) The warmer the block and the knife, obviously, the deeper the zone of fusion, but; (ii) the colder the block, the greater the cutting resistance encountered and hence the energy dissipated, and the deeper the zone of fusion becomes. (iii) The colder the knife, the greater becomes the proportion of the energy dissipated in the cutting that can be absorbed by the knife and thus conducted away from the tissue, therefore indirectly reducing the depth of the zone of fusion. (iv) The colder the knife and the anti-roll plate, the more quickly the molten layer is refrozen, and the more rigid the section becomes. The more rigid the section is, the better it can with-stand the frictional resistance to its progress between the upper surface of the knife and the anti-roll plate, and the thinner the section may be without crumpling. (v) Much is also to be gained by keeping the frictional resistance of the facets of the knife and of the lower surface of the anti-roll plate to a minimum.

An attempt was made to examine the validity of these conclusions experimentally in our laboratories.[32] A Pearse-type cryostat was modified by cooling the knife, the tissue, and the anti-roll plate independently by means of a battery of 'frigistor' semiconductor junctions (Peltier-effect thermo-electric heat-sinks). It was found, in accordance with expectations,[80] that optimum cutting conditions could be obtained when the knife and anti-roll plate were some 30°C colder than the air. Varying the temperature of, say, the knife and the anti-roll plate, also affected the temperature of the block and of the air in the vicinity of the block, and the experiments were not decisive. Different apparent temperatures seemed to be optimum for different tissues (testis required much lower temperatures than brain or liver). Means of measuring the actual depth of the zone of fusion were not available to us, and quantitative data cannot be given. One point that emerged was the extreme importance of avoiding condensation inside the chamber. A thin film of ice on the knife or the anti-roll plate effectively prevented the preparation of even comparatively thick sections (3–5 μ). Condensation becomes a most troublesome problem under heterothermic conditions. It is avoided in the Pearse-type cryostats by keeping the port-hole closed except for the removal of sections, by placing within the chamber several containers of silica gel (one of them is visible at the lower right of Fig. 3.1B), and wiping the knife and anti-roll plate frequently with acetone or methanol/ether at cryostat-chamber temperature. However uncertain our measurements might be, it remains true that, under controlled heterothermic conditions, 1 cm² blocks of unfixed tissues can be sectioned easily and adequately at 0·3–1·0 μ. This capability should encourage others to carry out more work along these lines.

A further reason why more work is needed is this: under homoeothermic conditions, the ideal cutting temperature is seldom lower than $-20°C$. This is considerably higher than the temperature at which protein solutions, and presumably tissues, are homogeneously frozen. To prevent the concentration of solutes by partial freezing, one ought to think in terms of keeping the block at, perhaps, $-50°C$ or even $-70°C$. For favourable tissues, this requires a knife temperature of about $-100°C$.

The microtome

Two types of microtomes are commonly used in cryostats; modified Cambridge rockers in the majority of Pearse-type instruments, and modified Minots in the remaining types.

In Minot microtomes, rotation of the handle, usually attached to a dynamically balanced flywheel, imparts a reciprocating vertical movement of the specimen block, by means of a cam; the vertical velocity of the block is related to the angular velocity of the handle by the sine law. Rotation of the handle also operates a ratchet-wheel by means of a pawl, and this actuates a fine micrometer screw that advances the object-holder between cuts. The knife is stationary and it is mounted nearly vertically, facing the tissue block. Three precision movements are involved: the ratchet-wheel and micrometer screw, and two dovetail slides. One of these carries the specimen-holder on its reciprocating vertical movement about the knife, and the other advances it by the thickness of one section between cuts. Minot microtomes are admirable instruments for the cutting of paraffin sections, but certain difficulties must be faced when they are used in cryostats.

The first of these is that many precision parts of ordinary Minots are made of carbon steels; these are rapidly corroded during the defrosting processes that must be carried out periodically. Corrosion is avoided in current instruments by building the entire microtome of stainless steel and brass. Even forgetting cost, this is not always satisfactory, because of machining and similar difficulties. But there are more severe complications: the dovetail slides of Minots must be made to very fine tolerances indeed. If the slides are adjusted for the instrument to work well at $-20°C$, it will stiffen enormously at $-40°C$. In view of the practical difficulty of setting clearances accurately at any temperature, this means that one must decide once and for all in what range of temperatures ($\pm 3°C$) one is going to work. There is also the problem of keeping the slides well-lubricated and free of ice at whatever temperature is chosen.

In our experience, the ice and lubrication problem is best solved by keeping the cryostat dry and avoiding lubricants. Metal parts can be impregnated successfully with polytetrafluoroethylene (PTFE), either by vaporizing a piece of PTFE in a carbon-coating evaporator, or by spraying all bearings and slides with PTFE suspended in Freon 12 (this is sold by DuPont de Nemours and Co., Wilmington, U.S.A., under the trade name of 'fluoroglide'). When these precautions are taken, Minots can be used to obtain excellent sections; this is particularly true of the so-called 'ultra-thin sectioning' models of the American Optical Co., Buffalo, N.Y., No. 820, and the International 'Custom' instruments.

The Cambridge rocker is a much simpler instrument, and it works on a kinematic principle that obviates the need for dovetail slides and other precision parts (other than the micrometer screw and ratchet-wheel). Its chief disadvantage is that the reciprocating movement of the specimen-holder about the knife is in an arc, rather than vertically. A modified rocker is shown in Fig. 3.1, A and B. The arc movement means that really flat sections of large blocks cannot be cut, but this disadvantage can be neglected if the newer 'rotary-rocker' microtomes (Cambridge Instruments, Ltd, London) are used: the arcs of these have a radius of 30 cm. One disadvantage of the rockers is the ease with which the ratchet-wheel can be damaged. In the standard model, rotating the wheel by one tooth corresponds to advancing the specimen by $0.35\,\mu$, and sections thinner than this cannot be cut. If a 1000-tooth wheel is fitted instead, the specimen can be advanced by increments of about $0.07\,\mu$. In summary, either type of microtome is capable of producing excellent results, if properly used; the choice is largely a matter of personal preference.

Both types of microtomes are equally affected by the accuracy with which the true cutting velocity is controlled. As can be predicted from the physical considerations discussed in connection with the zone of fusion (pp. 26–27), it is particularly important to maintain a constant velocity when cutting unfixed tissues. The ideal velocity varies from tissue to tissue and cryostat to cryostat. It is, at least, subject to the following factors, adduced from considerations of the physical process[80] and confirmed empirically:[30, 32] (i) the greater the velocity, the greater the energy dissipated in a given time, and the greater the friction between knife and section, resulting in an increase in the depth of the zone of fusion (see Fig. 3.2, p. 26). However: (ii) the lower the velocity, the greater the expansion of the block of tissue, and the more troublesome become the random expansions and contractions of various parts of the system due to vibration and to thermal fluctuations. In practice, velocities of 0.1 to 5 mm/sec will be found suitable for all types of specimens. It is equally important to keep constant the timing of the remainder of the cutting cycle, whilst the specimen-carrier is being returned to its starting position and advanced for cutting the next section, and whilst the section already cut is being removed. A certain rhythm of work has to be developed if one is to use any cryostat effectively: Pearse-type models are particularly convenient in this respect, but even then much skill and a certain flair for this kind of work are needed before perfect results are obtained consistently with manually driven microtomes. The whole process is greatly facilitated when the microtome is driven by a programmable servo-system.[30] This system can be controlled by a function-generator (excellent ones are commercially available). In this way, for example, the David-Steel microtome can be set, to cut a section at 0.5 mm/sec, return the specimen to the starting position at 20 mm/sec (thus preventing the specimen from striking the back of the knife during the upstroke), pause for 5 sec whilst the section is collected, and start again.

The knife

The problems relating to the condition of the knife are essentially the same as those encountered in all forms of microtomy. The reader is referred to discussions by Hallén[37, 38] and by David[27] for a detailed treatment of the whole problem. It is necessary to mention a few points here.

Some of the geometrical aspects of section cutting are represented diagrammatically in Fig. 3.3 (p. 29).

For a section to be cut at all, it is necessary for the knife to be bevelled, and to have a clearance angle—or 'bite' into the tissue. As the knife 'bites', it compresses the tissue against the direction of cutting and against the direction of advance of the specimen-carrier. The section is therefore always compressed (or foreshortened) and thicker than is indicated by the setting of the microtome. Theory and practice indicate that the greater the rake angle (see Fig. 3.3), the less the resulting compression and, incidentally, the less the friction between the section and the knife. The clearance angle must be between 6 and 10° if the tissue-block is to pass clear of the knife during the upstroke; therefore, the rake angle can only be increased if the bevel angle, β, is kept as small as possible. One approach would be to use trapezoidal-shaped knives, similar to lathe tools, and bevelled only in the upper facet; however, these knives would be exceedingly difficult to sharpen. Hallén[37] has shown that, in practice, the bevel angle can only be decreased so far; beyond a certain point the edge becomes so fragile that it cannot be sharpened, or the facets polished. This leads us to a further consideration: friction in relation to the quality of the facets. Obviously, in cryostat work, friction is of great importance, and it will vary proportionately with the depth of the facet (dimension 'A' in Fig. 3.3) and inversely with the surface finish of the facets. The cutting facets of a plane wedge steel microtome knife can be lapped to optical flatness (no irregularity greater than ± 50 Å) only if the bevel angle is at least of the order of 30–35°. A subterfuge can be used to obtain at the same time very narrow and flat facets. Using Hallén's[37] lapping machine, straight facets of moderate finish are first cut at a bevel angle of 25°; then, the knife is finished on the glass lap without abrasives, with the knife-holder adjusted to produce a *second bevel*, of 35°. The secondary facets generated in this way are usually less than 10 μ deep, and are optically flat. We have used this method routinely for many years, and it can be strongly recommended. It should be added that the knives are always examined interferometrically before use with a very simple micro-interferometer after Krug and Lau,[49] manufactured in England (W. Watson, Ltd, Barnet, Herts).

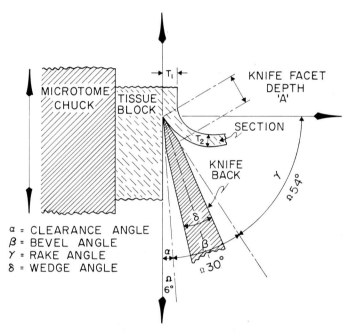

Fig. 3.3 Diagrammatic representation of the geometrical relationships between the knife and the tissue in microtomy.

Knives used in cryostats must *never* be stropped. Stropping curves the facets and their effective bevel angle then becomes that defined by the tangents of the resulting curves, with a corresponding reduction in the rake angle.

A cautionary word ought to be said about the deterioration of even perfect knives in the cryostat. Polished facets are stressed amorphous layers in which plastic flow of the metal has occurred: the so-called Beilby layers.[12] These are easily contaminated; they must be cleaned with cold acetone frequently during cutting, and they are best refinished after each cutting session (2 or 3 min at the glass lap of the Hallén machine are adequate).

The anti-roll plate and the mounting of sections

The single most important mechanical device used in conjunction with microtomes in cryostats is the anti-roll plate, and failure to obtain satisfactory sections can more often be ascribed to the incorrect use of the anti-roll plate, than to any other cause.[3, 67] When a frozen tissue is cut without an anti-roll plate, the section frequently curls upon itself, forming a tight roll on the edge of the knife, because of the physical processes represented diagrammatically in Figs. 3.2 and 3.3. The anti-roll plate was designed to minimize this effect. The plate is a rectangle of glass or plastic (usually about 25 mm wide and 50 mm long), held parallel to the upper cutting facet of the knife and some 50–100 μ above it, protruding somewhat beyond the knife edge, so that the edge of the plate almost touches the leading edge of the tissue-block as this strikes the knife. Obviously, the edge of the plate must be exactly parallel to the edge of the knife. The object of this arrangement is to provide a narrow rectangular channel, having the length of the knife facet (dimension A in Fig. 3.3) as wide as the width of the plate and the depth of the spacer between the plate and the knife, through which the section *must* pass. The severe initial curling of sections is prevented by the first few microns of the plate; the rest of the plate merely prevents the section from being accidentally dislodged.

Several consequences follow from these considerations: (i) Friction at the lower surface of the plate must be reduced to a minimum. In the Coons type of cryostat, the plate is usually made of plain glass and, in the open-top models, of Perspex. These materials have high coefficients of friction, and for this reason Pearse-type instruments are fitted with plates made of pure polytetrafluoroethylene (PTFE). PTFE reduces friction well, but unfortunately it is easily bent and becomes quickly deformed under operational conditions. For this reason, we use a glass plate, made of a suitably cut microslide, coated with PTFE (DuPont 'fluoroglide')[82]. This is frequently cleaned with cold acetone during use, and it is replaced or re-coated as soon as friction noticeably increases. (ii) The vertical spacing between the plate and the knife facet is critical. For sections less than 25 μ thick, this should be of the order of 50–60 μ. We routinely use two short lengths of an adhesive polyester tape, 57 μ thick, attached to the edges of the undersurface of the plate. The spacers must extend to the edge of the plate or they will not rest evenly on the upper facet of the knife. (iii) If the plate protrudes too far beyond the knife edge, it will be struck by the tissue-block, with disastrous results; if it does not protrude far enough, the section will not be guided into the slot. The tolerance of this adjustment is only a few microns. (iv) The plate will not rest squarely on the knife facet if it is not angled in relation to the perpendicular to match the sum of the clearance and the bevel angles of the knife. These adjustments are critical, and they should be carefully carried out. All commercially available cryostats have anti-roll plates with a suitable range of adjustments. (v) A further source of difficulty with anti-roll plates is this: the plate must always be as cold as the rest of the cryostat when this is used under homoeothermic conditions, or as cold as the knife when heterothermic cutting is preferred. This is very difficult to achieve unless the plate can be actuated from outside the cryostat, as in the Pearse-type instruments. (The provision of a remote control for swinging out the anti-roll plate and for increasing its pressure is strongly recommended.)

At the completion of the cutting process a flat section should be lying between the knife and the anti-roll plate (Fig. 3.1A) ready for further processing. For the majority of purposes, it is best to mount the section immediately on a coverglass. The advantages of using coverglasses rather than slides are considerable: (i) it is only with coverglass-mounted sections that the highest possible resolution of the optical microscope can be obtained; (ii) in enzymatic work[7], coverglasses can be treated simultaneously in a Columbia dish (A. H. Thomas, Inc., Philadelphia, U.S.A.), using only 8 ml of solution.

The mounting of cryostat sections on coverglasses is a simple procedure. The coverglass is picked up by the suction device indicated by 'b' in Fig. 3.1 (p. 24), and held parallel to the section as it is brought near it. At a critical distance, dependent upon section-thickness, temperature, and humidity, the section jumps to the glass through its static electrical charge, thereby becoming firmly mounted. This technique is illustrated in Fig. 3.1, B. In the case of unfixed sections, adhesives are not necessary and the section becomes, for all practical purposes, permanently bonded to the glass. Sections of fixed tissues do not adhere quite so firmly.

Freezing-drying facilities

It should be clearly realized that, when currently available cryostats are used, particularly with unfixed tissues, the resulting sections are partly dehydrated before they can be removed from the cryostat.

Some of the unbound water of the tissue-block is lost by evaporation and sublimation in the extremely dry atmosphere of a good cryostat, even before sectioning begins. Perhaps as much as half of the remaining unbound water evaporates from the section before it is removed from the cryostat and used. Since so much water is lost anyway, there is much to be said for extending and controlling this process of cryo-dehydration and deliberately lyophilizing the section. This is easily achieved. A small container kept in a Dewar within the cryostat is connected to an air-ballasted two-stage rotary vacuum pump of some 50 l./min displacement. A vacuum of 10^{-2} mm/Hg will dehydrate coverglass-mounted sections at $-40°C$ (ethyl oxalate/solid CO_2 slush) in under 1 hr. It is not necessary to use a Dewar if one can arrange the expansion coils of the refrigerating system to surround a small container—as was done by Hydén more than ten years ago.[46] Another alternative is to use a thermoelectric heat-sink, our preferred technique. Although Pearse[67] suggests that lyophillized sections are required only for quantitative microchemical work, we recommend them whenever unfixed tissues are cut, both to reduce the variability between sections and as protection against the extraction of tissue-constituents by subsequent treatments.

On cutting ultra-thin cryostat sections

Though the importance of the ultra-microtomy of frozen tissues can hardly be exaggerated, very little work has been done along these lines. Pearse[3] and the authors[32] have both succeeded in cutting sections that were occasionally thin enough for electron microscopy, with slightly modified ordinary cryostats, used under heterothermic conditions. More work along these lines is in progress now, at Princeton and elsewhere. Though the first results are not wholly discouraging it must be admitted that our knowledge is fragmentary and our ignorance is vast. Of the recent attempts at cryostat ultra-microtomy, the most ambitious has been that of Fernández-Morán[35] and his colleagues in Chicago. They installed a Morán ultra-microtome in a cryostat cooled by vaporizing liquid nitrogen, and postulated that it should be possible to cut cross-sections of hydrated DNA molecules at liquid gas temperatures. It is not clear to us, however, from conversations with the author and from studying the papers most carefully whether this promising instrument has yet been used successfully. A somewhat less esoteric approach was that of Stumpf and Roth,[77] who used an open-top special cryostat at $-85°C$ under homoeothermic conditions, with an International Minot microtome. The ease with which they obtained $0.5\ \mu$ sections of unfixed tissues suggests that they might have cut sections adequate for electron microscopy had they used an ultra-microtome in the cryostat chamber. By far the most encouraging results have been those of Bernhard and Nancy[14] at Villejuif. They sectioned glutaraldehyde-fixed and gelatine-embedded tissues with a standard Porter-Blum ultra-microtome, fitted with glass knives and operated in a cryostat with an air-temperature of $-35°C$. The machine was used heterothermically, the tissue and knife being cooled to $-50°C$ with solid CO_2. An anti-roll plate was not used, and difficulty was experienced in flattening the sections. The quality of their electron micrographs is comparable with the best that can be done by freeze-substitution followed by epoxy embedding.[69]

In considering further work in this field, we are of the opinion that the most promising approach would be to begin by attempting to improve Bernhard and Nancy's[14] technique, in the light of what is known about general cryostat methodology. The first improvement would be to provide refrigeration equipment capable of maintaining accurately, in different parts of the cryostat, temperatures in the range of -50 to $-100°C$. Secondly, diamond or corundum knives should be used instead of glass, largely in order to provide a sufficiently long bevelled facet of adequate hardness. But the most significant improvement might well prove to be the use of different methods of flattening and collecting the sections. If they are to be cut dry (as is done for thicker sections), then an anti-roll plate *must* be used; sections could then be picked up and mounted electrostatically, as was done by Koller and Bernhard[48] in another system. The alternative is to use an inert liquid, such as Freon 12, in the knife-trough.

METHODS OF HANDLING UNFIXED TISSUES

Initial steps

Fresh tissues to be sectioned in a cryostat must first be frozen. Every step of the preparative procedure of freezing and handling, from the moment the tissue is exposed in the animal, has a profound effect on the kind of microscopical image subsequently obtained. Every step can be carried out in many different ways, each of which is open to several objections. However, the initial steps: (i) the dissection of the piece of tissue and (ii) its freezing with the least possible delay, and so as to cause least possible disturbance in the tissue-constituents, should not arouse much controversy.

The technique of dissection is straightforward. It is exactly the same as that needed for making any careful cytological preparation. Tissues are kept moist at all times, with a balanced salt solution of appropriate ionic composition containing some glucose or other substrate. Fine, very sharp tools are used; tearing and compression are avoided. It is necessary to work fast, but accuracy is more important than speed. The size of the tissue-block to be cut is limited only by the behaviour of the freezing media subsequently used. It is best to have at least one dimension of the tissue of the order of 2 mm or less. Some workers, as we do ourselves, add to the balanced salt solution a variable amount of one of several 'protective' substances (glycerol, dimethyl sulphoxide, polyvinyl pyrrolidone, polyvinyl alcohol, or albumin). The balanced salt solution is then known as a 'protective medium'. The proper use of protective media is important. It is best discussed in relation to freezing and thawing (see below, pp. 38–39).

Embedding

Some tissues can be frozen for sectioning without preparation other than dissection. This is particularly true of pieces of large organs of chordates. Other tissues, such as testis of chordates, and the vast majority of tissues of invertebrates, should be supported by an 'embedding' medium. This is a different process from the embedding of fixed tissues, since it is neither desirable nor necessary to infiltrate the tissue with the medium: a little support is all that is needed. Gelatine[78] has been used for this purpose, but we much prefer to embed in suitable *tissues*. The medium and the tissue must have the same cutting properties in the cryostat. The tissue used as an embedding medium may serve as a 'built-in control' for whatever reaction is to be carried out with the experimental material. We have found cross-sections of cat or rabbit spinal cord to be very useful media for embedding nervous and glandular tissues of sessile ascidian tunicates,[52] tentacles of pulmonate gastropods,[50] circumoesophageal ganglia of *Helix pomatia* and *H. aspersa*,[28] and a variety of other tissues of awkward shape and size. Kitten liver, on the other hand, proved to be an invaluable medium for supporting peripheral nerves of various species.[32] Testis is best embedded in cross-sections of intestines of the right calibre.[32] The method of mounting the specimen in an intestine is obvious; with spinal cord, the central canal is enlarged with fine scissors and the specimen is mounted in the resulting cavity; with liver, a suitably shaped groove or recess is carved out. It should perhaps be added that the same care should be devoted to the dissection of the embedding medium as is lavished on the experimental tissue. For this reason, it is best always to work with an assistant. The media suggested above will no doubt be complemented by many more in the future. The choice of a suitable medium for a particular tissue should perhaps be considered a challenge to the imagination.

Freezing

The purposes of quenching

This part of cryostat preparative technique is somewhat contentious. It is necessary to consider carefully the purposes of freezing the tissue, the consequences of freezing (and subsequent thawing), and the ways in which the effects of what is commonly described as 'freezing damage' can be alleviated, if not prevented. In cryostat technology, it is customary to bring the tissue to a temperature considerably lower than that of the cryostat, as soon as practicable after the tissue is obtained. This procedure is known as 'quenching'. The purpose of quenching is *not* to bring the tissue to a consistency suitable for sectioning. Most histochemists would agree with Pearse[67] that the most important functions of quenching are: (i) to arrest the metabolism of the tissue as rapidly as possible, and leave it substantially as it was at the time of its excision, thereby avoiding at least the most severe autolytic changes; (ii) to reduce the diffusion of solutes from the tissue (so long as the tissue remains frozen). On the basis of our practical experience with tissues, reinforced by recent theoretical considerations,[53,54] we suggest that there is a further purpose (although this may on occasion be considered instead an unfortunate, if unavoidable, effect of the quenching process), this is (iii) partially to *fix* the tissue (through partial dehydration and disulphide-bonding), thereby reducing the loss of otherwise soluble tissue-constituents during subsequent handling (that is to say, after thawing), but without *greatly* altering the availability of reactive groups in the tissue constituents. It should be noted that the term *fixation* is used here in precisely the sense of Baker's[8,9] definition. Whether freezing *greatly* alters the availability of reactive groups in the tissue-constituents, is open to question.

There are essentially three methods of quenching tissues for microtomy. These reflect the preoccupations of their practitioners: (i) Pathologists (and some others) usually freeze tissues fairly slowly with solid or liquid CO_2. The general approach here is that the precise conditions of freezing do not greatly affect the appearance of the finished preparation (particularly when viewed at low power). (ii) Most histochemists follow Pearse[67] in assuming that, for careful work and optimum preservation of tissue-constituents, it is necessary to quench the tissues as rapidly as possible, usually in a thermally conductive fluid cooled to the temperature of liquid air. (iii) Recently, Cunningham and his associates at the Royal College of Surgeons[25] have advocated a much slower rate of freezing. This technique has gained some adherents amongst students of the distribution of enzymes, particularly in the United States.[60] The first method need not concern us here. Which of the other two is the method of choice, fast or slow quenching?

Physical considerations of quenching

This question is best approached by considering, briefly, the physical process of freezing. More detailed discussions may be found in Pearse's book,[67] in more recent studies by Mazur,[59] and Meryman,[61] and in the *Proceedings* of recent conferences on cryobiology.[75] As tissues are frozen, cells are damaged in various ways, collectively described as 'frost injury'. Mazur[59] recognized three forms of injury: thermal shock, injury due to the growth of intracellular ice, and injury due to the increasing concentration of solutes as water freezes out. To this list should be added the oxidation of sulphydryl bonds to disulphides, described by Levitt.[53,54] Thermal shock is usually defined as: complex effects of low temperatures *above freezing point* and, as such, need not concern us here. The remaining forms of injury *are* important to microtomists, and the method of choice must be that which minimizes their effects. Let us consider first the joint problem of avoiding the effects of increasing concentration of solutes, and intracellular ice. Levitt's damage,[53,54] as will be seen later, can be turned into an advantage in cryostat microtomy.

Tissues, in the present context, must be considered as complex solutions of salts of various ionic species, and organic compounds, complicated by the admixture of various colloidal dispersions, and modified by the presence of membranes. If the block of tissue could be cooled *instantaneously* to a temperature below the (unknown) eutectic point of the most complex solution in the tissue, the tissue might vitrify or freeze homogeneously, without the formation of ice-crystals. Since this is a physical impossibility, freezing appears to proceed as follows: (i) As the block of tissue is cooled below 0°C, its unbound water begins to crystallize: pure ice separates out, and therefore the solutes in the still fluid phase of the system become more and more concentrated. On the basis of Raoult's law, and the Clausius-Clapeyron equations,[57] it can be predicted that ideal non-polar solutions would be concentrated by 1 molal unit per 2°C fall in temperature below freezing point, and solutions of univalent salts by 1 molal unit per 4°C. Eventually, a freezing solution becomes saturated, and the remaining solute precipitates completely, upon a further drop in temperature. If several species of solutes are present, they will freeze out independently at their own eutectic points, and some fluid water will remain present in the system, until the temperature drops below that of the lowest eutectic point of the solutes present. Tissues are not, of course, ideal solutions, and it is likely that some fluid water remains in tissues at temperatures of the order of −40°C. The physical model given above is applicable in the first instance to the extracellular phase of the tissue, in so far as the concentration and species of solutes within cells are different from those without. A similar process, however, can be presumed to occur within cells. It might therefore be taken

4—C.S.I.

as a working hypothesis that, during the freezing process, cells are bathed in a fluid that has a concentration of solutes many times in excess of that obtained under ordinary living conditions, and that organelles, in their turn become surrounded by an abnormally concentrated milieu. (ii) Mazur [59] has demonstrated that, as *cells* are cooled below the freezing point of water, there exists a temperature range in which the intracellular water is supercooled; protoplasm may remain at its normal viscosity even though the cell is embedded in frozen extracellular fluid. Under these conditions, the vapour pressure inside the cell becomes higher than that outside. This being a thermodynamically unstable state, equilibrium is reached either by the supercooled water flowing out of the cell and freezing externally, or by freezing intracellularly. With decreasing temperature, the supercooling increases, and with it the vapour-pressure imbalance. With an increase in vapour-pressure imbalance, the rate at which the supercooled water leaves the cell increases (for any given permeability constant), pushing the system toward equilibrium. When the cooling velocity exceeds a certain value (variable over a range of 5000:1 in the systems studied by Mazur), supercooled water cannot diffuse out sufficiently before freezing. The characteristics of the resulting intracellular ice depend, in turn, on the rate of cooling of the system: the greater the rate of cooling, the greater the number of ice-crystal nuclei seeded, but the smaller and the less perfect the final ice-crystals produced. (iii) Levitt[53] has postulated that, when cells (or indeed any protein solution) are frozen, their gradual dehydration, brought about by factors (i) and (ii), brings structural proteins into close apposition. When this happens, adjacent protein chains become linked by the conversion of matching —SH HS— bonds in the two chains into —S—S— linkages. If these disulphide bonds are stronger than the presumably altered hydrogen bonds and hydrophobe orientations within protein chains, a reasonably permanent denaturation of the proteins could be expected. Levitt confirmed this hypothesis with a model experiment, using a thiolated gel containing a known number of correctly oriented SH groups.[54]

Fast freezing or slow?

The biological consequences of the physical considerations discussed above are complex. There can be no doubt that the exposure of cells to high concentrations of electrolytes during the freezing process is harmful; the extraction of lipids from membranes, large changes in pH, and osmotic effects immediately come to mind. On these grounds, it would seem desirable to use the highest cooling velocity obtainable, and so reduce to a minimum the time of exposure of the cells to these abnormal conditions. But if the cooling velocity is very high, the imbalance in vapour pressure between the external and internal milieux of the cell may result in a varying degree of intracellular ice-formation, with consequent disruption of fine structure. On this basis, the slower the cooling velocity, the better. But, if the time of exposure to a high concentration of electrolytes is critical, then cooling velocities that would undoubtedly produce intracellular ice crystals cannot be avoided. In this case, immensely rapid cooling would be preferable, so as to keep the ice-crystals of a size small enough to produce little structural damage. Quantitative information about measured velocities of cooling, and their balancing effects, is available, so far, only for model systems and some types of isolated cells, suspended in large amounts of extracellular medium. We cannot place the recommendations for microtomists on a sound scientific basis until this information is available for organized tissues. There is little prospect of obtaining this kind of information with existing techniques. Even the rate of cooling of tissues immersed in different quenching media is a matter for speculation (according to Hallett,[39] a thermocouple inserted in a piece of tissue will at best measure the temperature of the thermocouple, and often not even this).

All the available evidence indicates that without the use of protective substances, intracellular ice-formation is unavoidable, unless cooling velocities as low as, perhaps, 1°C/min are used. The crystals may be as small as 500 Å in diameter, when cooling velocities of the order of 2000°C/sec are used.[36, 69] Crystals of this magnitude might be acceptable, if they were unavoidable, in view of the other and more severe forms of damage resulting from very slow freezing. (Methods of avoiding them are discussed in connection with protective media, pp. 38–39.) It should be carefully noted here that the logical argument concerning fast and slow freezing has resolved itself into the relative merits of the fastest possible freezing (e.g. the quenching of minute pieces of tissue in Freon 22 cooled with liquid nitrogen, cooling velocity

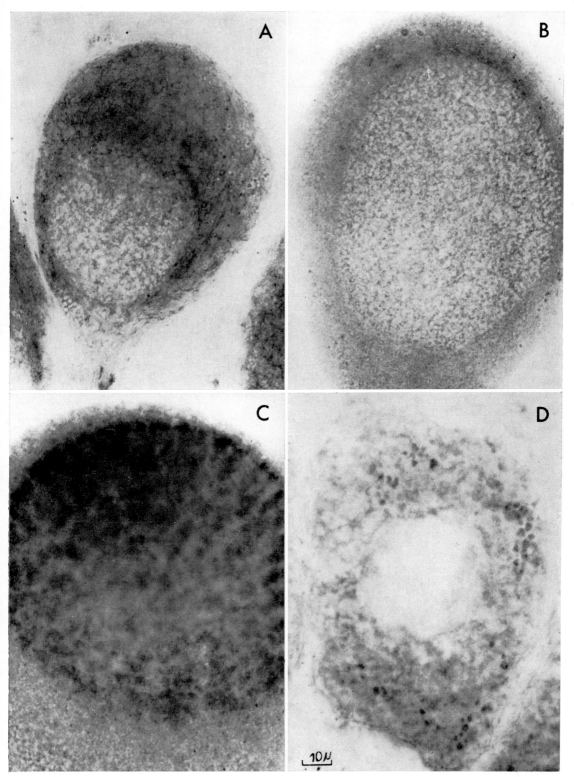

Fig. 3.5 Mesocerebral neurones of *H. pomatia*, treated by tetrazolium methods to demonstrate specific oxidoreductases. A, B, and C are living cells; D is a 3 μ unfixed cryostat section. Substrates: A, isocitrate; B, β-hydroxybutyrate; C and D, lactate.

denatured by the sulphydryl-disulphide transformation[53] that occurs during quenching. If the tissue was frozen after treatment with protective media, these must now protect the tissue for considerably longer periods. However, short of using more powerful refrigeration machinery, and refining the heterothermic technique, to enable tissues to be cut routinely at temperatures of the order of $-65°C$, or lower, there is literally nothing that the microtomist can do to mitigate the effects of this partial thawing.

There is a second and more complete thaw when sections are mounted. If the method recommended here is used—electrostatic mounting on warm and dry coverglasses—the thawing is almost instantaneous (perhaps as fast as $1000°C/sec$) and it is accompanied by dehydration through evaporation. If other methods of mounting were used, much slower thawing could be obtained. Lyophillization after mounting the section merely removes the last traces of water, producing 'permanent' preparations.

What are the consequences of this kind of thawing? The most important is that the intracellular crystal nuclei, seeded during quenching, can now grow, and, if the tissue remains sufficiently long in the critical temperature range, may cause considerable mechanical damage. Further, the tissue once again may be exposed to an excessive concentration of solutes, as during quenching. At the level of fine structure, the effects of thawing can be studied by comparing electron micrographs of frozen cells subsequently embedded without thawing[36,69] with those of frozen cells thawed before embedding.[76,81] In the thawed cells, no trace of the minute holes produced by intracellular ice-crystals remains. Instead, considerable distortion can be seen, along with some direct damage to membranous structures.

These considerations suggest that tissues should be cut at the lowest possible temperature, and they reinforce the argument in favour of sectioning under heterothermic conditions. Further, it might be inferred that sections ought to be thawed as quickly as possible: the technique recommended here is a suitable one for this purpose. The desirability of rapid thawing is also suggested by the fact that the viability of frozen tissues and cells is directly proportional to the velocity of thawing.[33,44,59,61]

Protective media

Purposes and definitions

Protective media may contain glycerol, dimethyl sulphoxide (DMSO), sucrose, polyvinyl pyrrolidone (PVP), polyvinyl alcohol (PVA), or albumin, dissolved either in extracellular fluid or in a suitable balanced salt solution (BSS). The active ingredients can either work as extracellular additives (such as PVP or albumin), or they can penetrate cells: the 'endocellular cryophylactic agents', such as DMSO.[44] Protective media can be used before the tissue is frozen; as storage solutions for collecting frozen sections; or mixed with the incubating or staining solutions subsequently used. Their purpose is to minimize frost damage during and after freezing, and their mode of operation is only partly understood. Glycerol[68] and DMSO[45] were first used as pre-freezing additives in 1949 and 1950, as means of increasing the viability of frozen spermatozoa and blood cells. The histochemical use of protective media is more recent. Scarpelli and Pearse[71] stored, and then stained, cryostat sections in media containing 7·5 per cent PVP in order to avoid what they called 'osmotic damage'. We began dissecting tissues into a 10 per cent solution of crystallized bovine plasma albumin in a Krebs/Ringer saline buffered to pH 7·2 some eight years ago, because this was known to be a harmless medium for living cells, and because this step helped to keep isolated mammalian cells in good condition. The same medium was also used for storing frozen sections, and as an adjuvant in staining,[31,32] though as little as 1 per cent albumin is adequate for these purposes;[79] PVA has also been used recently in storage and staining solutions.[2] Unfixed frozen sections lose as much as 70 per cent of their nitrogenous material (hence proteins) when incubated at $37°C$ in enzymatic reagents;[47] this loss is eliminated when protective media are used.

Presumed mode of action

How do protective media work? The osmolarity of even a 10 per cent albumin solution (about $1·5 \times 10^{-3}$ M) is not significant, and osmosis is not a relevant concept to use in connection with media that penetrate into cells as readily as DMSO does. Most of what is known about the mechanism of the action of protective media is based either on model experiments, or on the effects of varying the composition of the media on the rate of haemolysis of red blood corpuscles in frozen blood.[44] More is known about the behaviour of endocellular agents than about the extracellular additives. Essentially, protective agents reduce the proportion of the water in the tissue that freezes out as pure ice. Even then, the water freezes out as an ice/protective agent complex. If less water freezes out of solution, the increase in concentration of solutes during the transitional phase between the thawed and frozen states is reduced to more acceptable levels. In a sense, this is a form of dehydration.[44] Again, if less freezable water is available within the cell, both the rate of seeding of intracellular ice-crystal nuclei during freezing, and their subsequent growth in the thawing phase, would be correspondingly reduced. In this way, it is not difficult to explain how the protective agents (particularly the endocellular ones) help prevent both types of frost damage during freezing and thawing. It might be remarked in passing that the endocellular protective agents might be expected to be particularly effective during thawing. This hypothesis does not quite explain satisfactorily how protective media prevent the extraction of otherwise soluble tissue-constituents *after* thawing has occurred. It is known empirically that the endocellular agents are much less effective than the extracellular additives in this respect.[32] It is possible that two processes may be involved in the protection of sections in the storage media, and perhaps there are further reactions still during incubation in enzymatic reagents.

The first process in the storage media is probably one of dehydration, in the sense of preventing the hydration of tissue constituents to the point at which macromolecules can become dispersed. Quite reasonable protection is afforded by such unpromising media as 0·88 M sucrose and 1 M NaCl.[32] But this is not all that is involved. Media such as PVP, PVA, and albumin (and presumably other non-dialysable substances) are, mol per mol, thousands of times as effective as sucrose or salts. At temperatures of the order of 37°C, they greatly reduce the rate of penetration into sections of solutes of fairly low molecular weight[32,79], though not in the cold.[32] This suggests that the macromolecules of these media actually function as surrogates of the various cellular membranes damaged during sectioning.[32]

Recommended procedures

On the basis of the considerations discussed above, we recommend three types of protective media: one to be used before freezing the tissue, a storage solution for the frozen sections, and an additive for the solutions subsequently used.

Since protective media cannot be effective if they are unable to reach the cells, and, indeed, cannot prevent intracellular icing if they cannot penetrate into the cells, media such as albumin alone are not very useful for pieces of compact tissue, before freezing. We recommend the following: 20 per cent DMSO, with 1 per cent crystallized bovine plasma albumin, in Krebs/Ringer solution buffered to pH 7·2. This might be called the 'pre-freezing protective medium'. It is used at 0–4°C for 15 min on pieces of tissue not over 1 mm thick. The concentration of DMSO, duration of treatment, and temperature, are those found to be optimum by Rebhun.[69]

The storage solution for frozen sections omits the DMSO. Sections are thin enough for albumin to exert its protective effect, and there is no question of preventing intracellular icing. The storage solution is used at 0–4°C for as short a time as possible.

The additive to enzymatic reaction-media is simply 1 per cent crystallized bovine plasma albumin. Care should be taken to choose a sample of albumin that is free of enzymatic activity of its own. In our experience, commercial samples are often contaminated with lactate dehydrogenase.

Fixation and mounting

Thin cryostat sections of unfixed tissues adhere sufficiently well (particularly if lyophillized and/or if protective media are used) to withstand the subsequent manipulations required for the histochemical demonstration of many substances; this is particularly true of enzymes. Pearse[67] gives a list of many non-enzymatic substances that can be demonstrated successfully in this way, and to this list one might perhaps add lipids by Berg's fluorescence method[13] and primary amines by an aldehyde condensation method.[40]

Eventually, all cryostat sections must be fixed and attached to slides, so as to make semipermanent preparations. Generally, the coverglasses bearing the mounted sections are immersed in formaldehyde-calcium[4] or glutaraldehyde[70] for a few minutes, rinsed again, and mounted with glycerol-jelly or Farrant's medium. Nothing is gained by the drastic procedure of complete dehydration and mounting in resins.

Post-fixation

This is an altogether different technique from the preceding one; the coverglass-mounted cryostat sections of unfixed tissues are subjected to a brief chemical fixation *before* the histochemical or staining reaction is carried out. Post-fixation was introduced by Coons and his colleagues,[24] and it has recently been used to demonstrate certain hydrolases.[10,51]

Post-fixation deserves to be known better, for a different set of reasons. Mounted dry cryostat sections are not shrunken or seriously distorted by any fixatives subsequently used, and they are not likely to be fixed unevenly by even the slowest fixation mixture.[32] Further, since cryostat sections are already partly fixed by the inevitable denaturation of proteins, chemical fixatives subsequently used do not extract nearly so much material from the tissue as they do when used in the conventional way.

A particularly striking example of this phenomenon is the preservation of nuclear constituents by post-fixation. This effect is illustrated in Fig. 3.6. These are photomicrographs of mesocerebral neurones of *Helix pomatia*, fixed in FAM (formaldehyde:acetic acid:methanol), stained in cresyl violet monomer, and photographed at 546 n/m;[26] the photographic density is proportional to the concentration of nucleic acids (both RNA and DNA). A was sectioned unfixed in the cryostat at 3 μ and then post-fixed; B was pre-fixed, double-embedded in collodion and paraffin, and then sectioned at 7 μ. The histochemical specificity of the reactions illustrated was tested by ribonuclease and deoxyribonuclease extractions, and by parallel staining in Einarson's[34] gallocyanine-chromalum. There can be no doubt as to which method produces a more lifelike image or preserves nuclear nucleic acids better. In our laboratories, post-fixation methods also gave results superior to pre-fixation for other histochemical reactions, including the acid haematein test for phosphatides,[5] the Sakaguchi reaction for arginine,[6] and the Hg/nitrite method for tyrosine.[7] The morphological methods described in Baker's *Cytological technique*[9] could also be used satisfactorily, with the added advantage that optimum times could be determined easily for every piece of tissue studied.[32]

Evaluation of cryostat methods in enzymology

Cryostat sections of unfixed tissues are often used for the histochemical localization of sites of given forms of enzymatic activity, particularly those of oxidoreductases. How adequate are present methods for this purpose? Chayen, Wells, and Bitensky[22] suggested two criteria of adequacy: the intensity of the enzymatic reaction in the frozen sections should accurately reflect the level of enzymatic activity in the cells when they were functioning in the animal; and it should be possible, by means of suitable activation, to demonstrate the maximum potential activity of the cells. Another criterion[28,31] is that the localization of the activity within the frozen sections should not vary from that within the living cell. It may be useful to consider separately how well these criteria are met in the case of the oxidoreductases and hydrolases.

Oxidoreductases

Three views are currently held: (i) That oxidoreductases are best studied in unfixed cryostat sections, in order to avoid the inactivation of the enzymes by chemical fixation, even if lyo-enzymes are usually lost during incubation.[67] (ii) That a brief aldehyde pre-fixation does not seriously inactivate oxidoreductases and results in far better morphological preservation than would be possible with cryostat material.[63-65] (iii) That oxidoreductases are best studied in living cells (both in tissue culture and after microdissection), because only thus can soluble factors be kept from diffusing, and morphology be preserved.[28,31] These apparently contradictory views represent efforts by experienced histochemists to draw attention to certain very real pitfalls inherent in the use of each type of preparation.

A measure of objectivity in the assessment of the accuracy of localization of enzymatic methods has

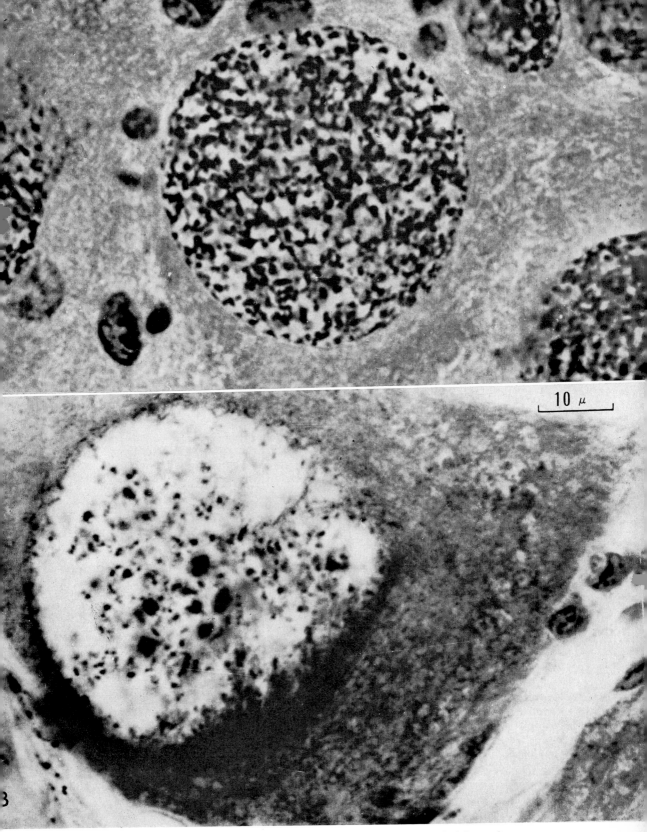

10 μ

Fig. 3.6 Mesocerebral neurones of *H. pomatia*, fixed in FAM and stained in cresyl violet to demonstrate nucleic acids. A is a 3 μ post-fixed cryostat section; B is a 7 μ pre-fixed paraffin section.

become possible now that tetrazolium reduction methods have been adapted for use with living cells in our laboratories.[28,31,32] It is best, when beginning the study of a particular cell, to compare the results for each particular enzyme in living cells, and unfixed, post-fixed, and pre-fixed frozen sections. When the results are qualitatively the same with all four kinds of preparation, differing only in intensity, the most convenient one, or the one providing the best morphological preservation of the tissue can be chosen. Considerable confidence may then be placed in the validity of the results. When the results differ, independent criteria sometimes can be applied; when they cannot, personal preferences become obtrusive, and less confidence can be placed in the validity of the findings.

Some of the results of this type of work have been published.[28,31,50] More will form the basis of a series of papers to be published elsewhere. It is only necessary to mention a few points here.

In the majority of cases, living cells exhibited more enzymatic activity than other kinds of preparations, in the sense that a given level of tetrazolium reduction was obtained in a shorter time (the ratio was usually of the order of $5:1$). Both pre- and post-fixed sections were distinctly less active than those in which fixation was delayed until the completion of the enzymatic process. Fixed sections therefore, cannot be recommended if fresh material is obtainable. In unfixed cryostat sections, some inactivation of oxidoreductases is unavoidable. This happens even when losses of enzyme through extraction are carefully avoided, by the use of protective media. Freezing and thawing by themselves partially inactivate enzymes, presumably through the sulphydryl-disulphide conversion.[53-54] Since many oxidoreductases contain sulphydryl groups in their active centres, this effect can be serious. If enzymes are frozen and thawed rapidly, this effect is less serious (by a factor of 3) than if they are frozen and thawed slowly.[23] A related phenomenon is the 'hybridization' of those enzymes that are partitioned into isozymes, such as lactate dehydrogenase.[23] Some enzymes are more sensitive to freezing and thawing than others; glutamate dehydrogenase may well be the most sensitive.[23] In conclusion, it would seem most unwise, at the present stage, to attempt the histochemical *quantification* of enzymatic activities in cryostat sections, even when the *localization* is identical with that observed in living cells.

As regards precision of localization: when the best possible cryostat sections are compared with living cells, and with the findings of the biochemical literature, the agreement is sometimes very close. Let us take as an example, $NADH_2$:cytochrome c oxidoreductase 1.6.2.1. In living isolated neurones of the spinal cord of the cat, tetrazolium methods indicate that the activity of this enzyme is predominantly localized in the endoplasmic reticulum (Nissl complex), followed by a fair degree of mitochondrial activity, and definite traces of nuclear activity (particularly in the karyoplasm).[28,31,32] In 'protected' unfixed frozen sections of the same tissue (Fig. 3.4, A), the results are similar, except that now the mitochondrial activity is as strong as that of the endoplasmic reticulum. End-feet are seen as clearly as in silvered preparations, presumably because of the mitochondria that they contain. When protective media are omitted, or the preparation is less carefully made, activity is primarily mitochondrial. In pre-fixed material, *only* the mitochondria are at all active. In homogenates of rat liver, manometric methods indicate that 40 per cent of the total enzymatic activity is microsomal, 32 per cent mitochondrial, 10 per cent nuclear, and the balance in the various 'washings'.[65] $NADH_2$:cytochrome c oxidoreductase 1.6.21, can be called a 'good' enzyme for cryostat work (the incubation solution contained, at a concentration of 10^{-3} M, $NADH_2$, as the substrate, and 2,2'-di-p-nitrophenyl-5,5'-diphenyl-3,3'-(3,3'-dimethoxy-4,4'-biphenylene) ditetrazolium chloride[62]—Nitro BT for short—as final acceptor; 1×10^{-4} M cytochrome c as intermediate acceptor, in a protective medium).

Let us now turn to the 'bad' enzymes, and take a cell system that is particularly favourable for histochemical localization work,[28] the mesocerebral neurones of *Helix pomatia* (Fig. 3.5, p. 37). The tetrazolium method of demonstration is similar for all oxidoreductases, here in Fig. 3.5, A, L_s-isocitrate was used as the substrate, D-hydroxybutyrate in Fig. 3.5, B, and L-lactate in Fig. 3.5, C and D; NAD was used as a coenzyme in the three cases. The neurons in A, B, and C were isolated from the living ganglion by microdissection; that in D was frozen and sectioned at 3μ in the cryostat. In the living snail neurone, the activity of L_s-isocitrate oxidoreductase is predominantly mitochondrial; that of D-hydroxybutyrate is confined to some cytoplasmic globules; that of L-lactate is almost entirely nuclear. Since different substrates yield different localizations, the electrons are being captured before the flavoprotein. In frozen sections, the three substrates produce essentially similar results: various membrane-bound cytoplasmic structures display a variable amount of enzymatic activity. In the case of L-lactate oxidoreductase, no trace of nuclear activity remains. For reasons too involved to discuss here, we would tend to accept the living cell as the more valid image, but this might well prove to be a contentious point in the future.

Differences in the localization of those oxidoreductases that were studied in living and frozen unfixed cells are summarized in Table 3.1.

Table 3.1 Histochemical localization of oxidoreductase activity in living cells and in cryostat sections of unfixed tissues

I.U.B. No.	Oxidoreductase	Localization identical in living and frozen	Localization different in living and frozen
1.1.1.1	Alcohol:NAD	×	
1.1.1.2	Alcohol:NADP	×	
1.1.1.8	L-glycerol-3-phosphate:NAD		×
1.1.1.27	L-lactate:NAD		×
1.1.1.30	D-hydroxybutyrate:NAD		×
1.1.1.37	L-malate:NAD		×
1.1.1.40	L-malate:NADP	×	
1.1.1.41	Ls-isocitrate:NAD	×	
1.1.1.42	Ls-isocitrate:NADP	×	
1.1.1.44	6-phospho-D-gluconate:NADP		×
1.1.1.49	D-glucose-6-phosphate:NADP		×
1.1.2.1	L-glycerol-3-phosphate:cytochrome c	×	
1.3.99.1	Succinate:(acceptor)	×	
1.4.1.3	L-glutamate:NADP		×
1.4.3.4	Monoamine:O_2	×	
1.6.2.1	$NADH_2$:cytochrome c	×	(×)
1.6.2.3	$NADPH_2$:cytochrome c	×	(×)
1.6.5.2	$NAD(P)H_2$:2-methyl-1,4-naphthoquinone		×
1.9.3.1	Cytochrome c:O_2	×	

Notes: (×)—Localization different when protective media were not used. Enzymes 1.1.1.1 to 1.6.5.2 demonstrated by tetrazolium methods;[31,32] 1.9.3.1, by a N-phenyl-p-phenylene diamine/8-amino-1,2,3,4-tetrahydroquinoline coupling method.[18]

Hydrolases

The case of the hydrolases is at once simpler and more complicated than that of the oxidoreductases. Simpler, because cold aldehyde fixation does not destroy these enzymes, which can therefore be adequately demonstrated in pre-fixed material.[41-43,66,67] More complicated, because they cannot be demonstrated in living cells with quite the ease of the oxidoreductases,[32] and because somewhat anomalous substrates have to be used in their identification. There are two schools of thought that must be considered: followers of Holt[41-43] and Novikoff,[66] who advocate aldehyde pre-fixation; and followers of Bitensky,[15,16] who recommend the use of unfixed cryostat material, frozen slowly.

Since the results of both schools of thought are basically comparable, the controversy seems to be rather academic. In our experience, several kinds of hydrolase activity (particularly those that can be demonstrated by diazo-coupling and indigogenic methods, i.e. some phosphatases and carboxylic esterases) can be demonstrated in both living and unfixed frozen cells, as well as in pre-fixed ones. But in cells that have not been chemically fixed, the final picture may be varied at will by changing the duration and temperature of the incubation. Recent work on protists and tissue-culture cells indicates that little activity is seen under physiological conditions of incubation.[32] It seems likely that hydrolase activity, to borrow the phraseology of an earlier paper, is one of the few 'forms of physiological activity that are as well carried out by dead cells as by living ones'.[28] Evaluated in this light, *both* the Bitensky and the Holt-Novikoff approaches seem sensible. In the former, no activity is seen in the unfixed material until sufficient damage is inflicted during incubation; in the latter, the activating damage is standardized *ab initio* so that reliable comparisons between tissues can be made. Hydrolase histochemistry, in our opinion, has not yet reached the standards that characterize the best modern work on oxidoreductases.

METHODS OF HANDLING FIXED TISSUES

From what was said earlier in this paper, it should be clear that there is no reason why a cryostat should not be used as an ordinary freezing microtome. In this way excellent thin frozen serial sections of any fixed tissue can be cut. Examples of what can be done in this way were published more than seven years ago;[29] another example is given in Fig. 3.7. This represents three sections of cat spinal cord, fixed and protected by Holt's method,[43] cut in the cryostat at $1\cdot8$ μ (A), $3\cdot7$ μ (B), and $0\cdot7$ μ (C). The sections were then incubated for the demonstration of the enzyme thiamine pyrophosphatase.[66] A and B were photographed with a flat-field apochromatic objective of N.A. $1\cdot32$, and C with a special amplitude-contrast apochromatic objective of the same aperture. Fine detail of the order of $0\cdot2$ μ is sharply delineated, and no obvious or gross artifact is to be seen.

Concerning the preparative technique, a few points ought to be remembered. Working with a cryostat cannot undo the damage done during fixation. For many purposes, brief fixation in cold formaldehyde[4, 43] or glutaraldehyde[70] is satisfactory. But, since intracellular ice-crystal formation, and the noxious effects of the concentration of solutes, can affect fixed tissues as well as fresh ones, care is needed in the subsequent steps. Protective media have been found necessary in most laboratories; Holt and Hicks'[43] simple extracellular additive medium has been found to be satisfactory. Gelatine-embedding is seldom if ever necessary. The method of freezing is also critical, and we unhesitatingly recommend the liquid nitrogen technique given on p. 35.

To conclude this section, we should like to communicate a simple method of 'embedding' pre-fixed tissues that we have found invaluable: embedding in cat liver sausage. Particularly in comparative enzymological work, it is often desirable to have thin sections of composite blocks containing several tissues. A supporting medium is necessary, and this must have cutting properties similar to the tissue. We take the liver of an adult cat, and thoroughly homogenize it in an equal volume of the fixative of choice, at $0\text{-}2°C$, using a high-speed pressure homogenizer at 25–50,000 r.p.m. (Ultraturrax homogenizer from the Hudes Merchandising Co., London). No cells or organelles must be left intact; additional sonication is used if necessary. The homogenate is then centrifuged and resuspended in fresh fixative several times. It is next treated with several changes of $0\cdot88$ M sucrose, and then heated to $90°C$ for 15 min to inactivate any surviving enzymes (other than ribonuclease). Just before use, the homogenate (now of the consistency of *Leberwurst*) is centrifuged down, and the supernatant is poured off. A small aluminium-foil box is filled with the 'sausage'; the fragments of tissue to be frozen are inserted into cavities scooped out of the sausage. The box is then ready to be frozen in liquid nitrogen and everted on to the microtome chuck.

ENVOI

This paper may be objected to on the grounds that there are not enough recipes in it, and that such recipes as are given are not stated directly enough, or in sufficient detail, for practical microtomists to follow them in experimental work. But it was not our intention to write a cook-book. In view of the still nebulous art of cryostat microtechnique, our aim was rather to discuss in plain English some of the physical and biological principles on which the current techniques are based. It is our hope that, by so doing, we might help microtomists to gain a partial understanding of their art, and stimulate them to apply this understanding to the development of special techniques required by their own problems. We hope the reader will write his own cook-book.

There is a warning implicit in every page of this paper. Whilst it is true that a cryostat can do, far

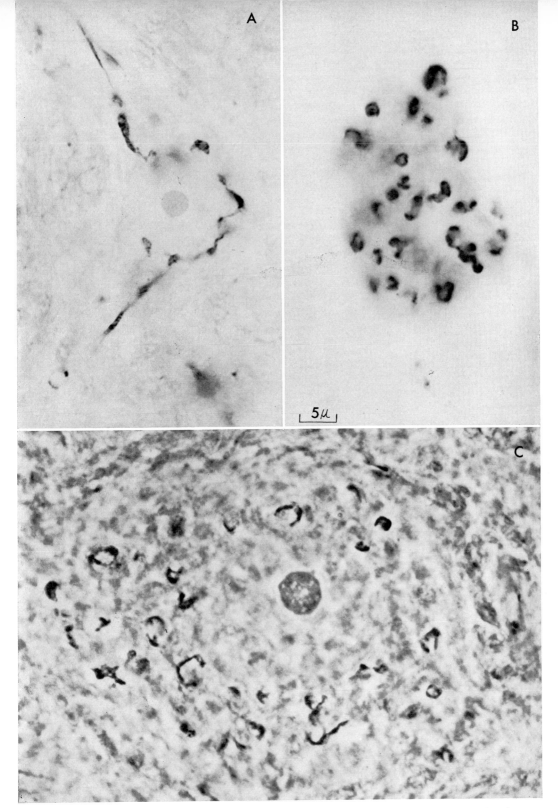

Fig. 3.7 Pre-fixed cryostat sections of the cat spinal cord, demonstrating the activity of thiamine pyrophosphatase. A and B direct microscopy; C, amplitude contrast. A, 1·8 μ thick; B, 3·7 μ; C, 0·7 μ.

more efficiently, anything that a routine freezing microtome can do, and not a few things that it cannot, cryostat sectioning is not a panacea. The deceptive ease with which aesthetically satisfying sections of unfixed frozen tissues can be obtained, makes it more necessary than ever to compare the reactions of fixed tissues with those of living cells. Let us not allow the development of cryostat methods to retard the growth of vital cytology as much as the invention of paraffin sectioning did, ninety years ago.

SUMMARY

Cryostats are automatic microtomes operated within thermostatically controlled containers at temperatures substantially below the freezing point of water. Any work that can be done with an ordinary freezing microtome is more efficiently done with a cryostat. Commercially available cryostats can be used to produce morphologically adequate frozen sections of unfixed or fixed tissues, as thin as 0·3 μ (or as thick as one wishes). Unfixed frozen sections can usefully be dried *in vacuo*, or fixed chemically, for a variety of histochemical and cytological purposes. Unfixed frozen sections may yield equivocal results when used in studies of the distribution of enzymes (especially oxidoreductases); these studies are often better carried out with living cells. Unusual embedding media are described as suitable for cryostat use: fresh mammalian nervous system for unfixed tissues of invertebrates; cat liver sausage for pre-fixed tissues of mammals.

ACKNOWLEDGEMENTS

The work described in this paper was carried out in the Neuropsychiatric Research Unit of the M.R.C., the Department of Biophysics of the Universidade do Brasil, Rio de Janeiro, and the Department of Biology, Princeton University. We are grateful to Dr. D. Richter, Professor Carlos Chagas and Dr. R. D. Allen for the facilities put at our disposal and Dr. L. I. Rebhun for helpful suggestions. The work carried out in Princeton was generously supported by a research grant, RG-08691, from the National Institutes of Health. We are grateful to Miss I. G. Evans, Mr. B. Pickering, Mrs. S. R. Abrams, Mr. J. H. Gomany, Mr. H. E. Schrader, and Mrs. M. K. Nunziato, for their able assistance.

REFERENCES

1. ADAMSTONE, F. B. and TAYLOR, A. B. 1948. The rapid preparation of frozen tissue sections. *Stain Tech.* **23**, 109–16
2. ALTMANN, F. P., BUTCHER, R. G., and CHAYEN, J. 1965. The retention of cell components and structure in unfixed sections during histochemical investigations. *J. R. micr. Soc.* **84**, 400
3. ANONYMOUS. 1961. *Instructions on the use of the Pearse/Slee cryostat*. London: privately printed; distributed by the South London Electrical Engineering Co., Lanier Works, Hither Green Lane, London, S.E.13
4. BAKER, J. R. 1944. The structure and chemical composition of the Golgi element. *Quart. J. micr. Sci.* **85**, 1–71
5. ——. 1946. The histochemical recognition of lipine. *Quart. J. micr. Sci.* **87**, 441–7
6. ——. 1947. The histochemical recognition of certain guanidine derivatives. *Quart. J. micr. Sci.* **88**, 115–22
7. ——. 1956. The histochemical recognition of phenols, especially tyrosine. *Quart. J. micr. Sci.* **97**, 161–4
8. ——. 1958. *Principles of Biological Micro-technique*. London: Methuen
9. ——. 1960. *Cytological Technique*, 4th edition. London: Methuen
10. BARDEN, H. and LAZARUS, S. S. 1961. Demonstration of mitochondrial adenosine triphosphatase. *J. Histochem. Cytochem.* **9**, 626
11. BARER, R. and JOSEPH, S. 1955. Refractometry of living cells. II: the immersion medium *Quart. J. micr. Sci.* **96**, 1–26
12. BEILBY, G. 1921. *Aggregation and Flow of Solids*. London: Macmillan
13. BERG, N. 1951. A histological study of masked lipids: stainability, distribution, and functional variations. *Acta path. microbiol. Scand.* suppl. 90
14. BERNHARD, W. and NANCY, M.-T. 1964. Coupes à congélation ultra-fines de tissu inclus dans la gélatine. *J. micr.* (Paris) **3**, 579–588

15. BITENSKY, L. 1962. The demonstration of lysosomes by the controlled temperature freezing-sectioning method. *Quart. J. micr. Sci.* **103**, 205–9

16. —— and COHEN, S. 1965. The histochemical demonstration of alkaline phosphatase in unfixed frozen sections. *Quart. J. micr. Sci.* **106**, 193–6

17. ——, LYNCH, R., SILCOX, A. A., and CHAYEN, J. 1965. Studies on chilling and sectioning tissue for histochemistry. *J. R. micr. Soc.* **84**, 397–8

18. BURSTONE, M. S. 1961. Modifications of histochemical techniques for the demonstration of cytochrome oxidase. *J. Histochem. Cytochem.* **9**, 59–65

19. CHANG, J. P. and HORI, S. H. 1960. A frozen section freeze substitution technique and an improved cryostat. *J. Histochem. Cytochem.* **8**, 310

20. ——, RUSSELL, W. O., and MOORE, E. B. 1961. An improved open-top cryostat. *J. Histochem. Cytochem.* **9**, 208

21. ——, ——, ——, and SINCLAIR, W. K. 1961. A new cryostat for frozen section technique. *Amer. J. Clin. Path.* **35**, 14–19

22. CHAYEN, J. WELLS, P. A., and BITENSKY, L. 1965. The meaning of a 'good' histochemical reaction in relation to the problem of latent enzyme activity. *J. R. micr. Soc.* **84**, 400–1

23. CHILSON, O. P., COSTELLO, L. A., and KAPLAN, N. O. 1965. Effects of freezing on enzymes. *Fed. Proc.* **24**, Suppl. 15, 55–65

24. COONS, A. H., LEDUC, E. H., and KAPLAN, M. H. 1951. Localization of antigen in tissue cells. *J. exp. Med.* **93**, 173–88

25. CUNNINGHAM, G. J., BITENSKY, L., CHAYEN, J. and SILCOX, A. A. 1962. The preservation of cytological and histochemical detail by a controlled temperature freezing and sectioning technique. *Ann. Histochim.* **7**, 433–5

26. DAVID, G. B. 1955. The effect of eliminating shrinkage artifacts on degenerative changes seen in CNS material. *Exc. med. Neurol.* (Amsterdam) **8**, 777–8

27. ——. 1959. *Quantitative Histology of the Central Nervous System.* Chapter 7, pp. 321–94. Oxford D.Phil. thesis

28. ——. 1964. Cytoplasmic networks in neurones: a study in comparative biophysics. In *Comparative Neurochemistry* (edited by D. Richter), pp. 59–98. Oxford: Pergamon Press

29. —— and BROWN, A. W. 1961b. The histochemical recognition of lipid in the cytoplasmic network of neurones of vertebrates. *Quart. J. micr. Sci.* **102**, 391–7

30. —— and STEEL, R. C. 1959. An electronically-controlled constant velocity microtome. *J. physiol.* London **147**, 33–34P

31. ——, DE ALMEIDA, D. F., CASTRO, G. DE O., and BROWN, A. W. 1962. The fine localization of certain hydrolytic oxidative and hydrogen-transfer enzymes within living nerve-cells of vertebrates. *Acta Neurol. Scand.* **38**, Suppl. 1, 43–4

32. ——, BROWN, A. W., CASTRO, G. DE O., DE ALMEIDA, D. F., and ABRAMS, S. R. Results not previously published.

33. DOLAN, M. F. 1965. The present status of organ freezing and possible future recourses. *Cryobiology* **1**, 199–204

34. EINARSON, L. 1951. On the theory of gallocyanine-chromalum staining and its application for quantitative estimation of basophilia. A selective staining of exquisite progressivity. *Acta Path. microbiol. Scand.* **28**, 82–102

35. FERNÁNDEZ-MORÁN, H., 1964. *Annual progress report, integrated research and training program in molecular biology (ultrastructure and electron microscopy).* Committee on Biophysics, University of Chicago (mimeographed)

36. GERSH, I. 1965. Freeze-substitution. *Fed. Proc.* **24**, Suppl. 15, 233–4

37. HALLÉN, O. 1956. On the cutting and thickness determination of microtome sections. *Acta anat.* **26**, suppl. 25, 1–43

38. ——. 1961. Microtomy. In *Encyclopedia of Microscopy* (edited by G. L. Clark), pp. 385–9. New York: Reinhold

39. HALLETT, J. 1965. Discussion on effect of freezing rates on histochemistry. *Fed. Proc.* **24**, Suppl. 15, 268

40. HAMBERGER, B. and NORBERG, K. A. 1964. Histochemical demonstration of catecholamines in fresh frozen sections. *J. Histochem. Cytochem.* **12**, 48–9

41. HOLT, S. J. 1958. Indigogenic methods for esterases. In *General Cytochemical Methods* (edited by J. F. Danielli) i, 375–98. London: Academic Press

42. ——. 1959. Factors governing the validity of staining methods for enzymes and their bearing upon the Gömori acid phosphatase technique. *Exptl. Cell Research*, Suppl. 7, 1–27

43. —— and HICKS, R. A. L. 1961. Studies on formalin fixation for electron microscopy and cytochemical staining purposes. *J. biophys. biochem. Cytol.* **11**, 31–46

44. HUGGINS, C. E. 1965. Preservation of organized tissues by freezing. *Fed. Proc.* **24**, Suppl. 15, 190–5

45. HUGGINS, M. L., TAPLEY, D. F., and JENSEN, E. V. 1951. Sulphydryl-disulphide relationships in the induction of gels in proteins by urea. *Nature Lond.* **167**, 592–3

46. HYDÉN, H. 1955. The chemistry of single neurones: a study with new methods. In *Biochemistry of the Developing Nervous System* (edited by H. Waelsch), pp. 358–69. New York: Academic Press

47. JONES, G. R. N. 1965. Losses of nitrogenous material occurring from fresh frozen section during incubation. *J. R. micr. Soc.* **84**, 399

48. KOLLER, T. and BERNHARD, W. 1964. Séchage de tissus en protoxyde d'azote (N₂O) et coupe ultrafine sans matière d'inclusion. *J. micr.* (Paris) **3**, 589–606

49. KRUG, W. and LAU, E. 1951. Ein Interferenzmikroskop für Durch- und Auflichtbeobachtungen. *Ann. d. Phys.* **8**, 329

50. LANE, N. J. 1964. Localization of enzymes in certain secretory cells of Helix tentacles. *Quart. J. micr. Sci.* **105**, 49–60

51. LAZARUS, S. S. and BARDEN, H. 1962. Histochemistry and electron microscopy of mitochondrial adenosinetriphosphatase. *J. Histochem. Cytochem.* **10**, 285–93

52. LEE, R. S. 1962. Personal communication.

53. LEVITT, J. 1962. A sulphydryl-disulphide hypothesis of frost injury and resistance in plants. *J. Theor. Biol.* **3**, 355–91

54. ——. 1965. Thiogel, a model system for demonstrating intermolecular disulphide bond formation on freezing. *Cryobiology* **1**, 312–16

55. LINDERSTRØM-LANG, K. and MOGENSEN, K. R. 1938. Studies on enzymatic histochemistry. XXXI. Histological control of histochemical investigations. *C. R. trav. Lab.* Carlsberg, sér chim. **23**, 27–35

56. LOWRY, O. H. 1953. Quantitative histochemistry of the brain. Histological sampling. *J. Histochem. Cytochem.* **1**, 420–8

57. ——, ROBERTS, N. R., WU, M. L., HIXON, W. S., and CRAWFORD, E. J. 1954. The quantitative histochemistry of brain. II. Enzyme measurements. *J. biol. Chem.* **207**, 19–37

58. ——, ——, LEINER, K. Y., WU, M. L., and FARR, A. L. 1954. The quantitative histochemistry of brain. I. Chemical methods. *J. biol. Chem.* **207**, 1–17

59. MAZUR, P. 1965. Causes of injury in frozen and thawed cells. *Fed. Proc.* **24**, Suppl. 15, 175–82

60. MELNICK, P. J. 1965. Effect of freezing rates on the histochemical identification of enzyme activity. *Fed. Proc.* **24**, Suppl. 15, 259–67

61. MERYMAN, H. T. (edited by). 1965. *Low Temperature Research in Biology.* London: Academic Press

62. NACHLAS, M. M., TSOU, K. C., DE SOUZA, E., CHENG, C. S., and SELIGMAN, A. M. 1957. Chemical demonstration of succinic dehydrogenase by the use of a new p-nitrophenyl substituted ditetrazole. *J. Histochem. Cytochem.* **5**, 420–36

63. NOVIKOFF, A. B. 1959. Enzyme cytochemistry: pitfalls in the current use of tetrazolium techniques. *J. Histochem. Cytochem.* **7**, 301–2

64. ——. 1961. Mitochondria (Chondriosomes). In *The Cell* (edited by J. Brachet and A. E. Mirsky) **2**, 299–421. New York: Academic Press

65. ——. 1963. Electron transport enzymes: biochemical and tetrazolium staining studies. In *Histochemistry and Cytochemistry* (edited by R. Wegmann), pp. 465–81. Oxford: Pergamon Press

66. —— and GOLDFISCHER, S. 1961. Nucleosidediphosphatase activity in the Golgi apparatus and its usefulness for cytological studies. *Proc. Nat. Acad. Sci.* **47**, 802–10

67. PEARSE. A. G. E. 1960. *Histochemistry, Theoretical and Applied,* 2nd edition. London: Churchill

68. POLGE, C., SMITH, A. N., and PARKES, A. S. 1949. Revival of spermatozoa after vitrification and dehydration at low temperatures. *Nature, Lond.* **164**, 666–7

69. REBHUN, L. I. 1965. Freeze-substitution: fine structure as a function of water concentration in cells. *Fed. Proc.* **24**, Suppl. 15, 217–32

70. SABATINI, D. D., BENSCH, K., and BARRNETT, R. J. 1963. Cytochemistry and electron microscopy. The preservation of cellular ultrastructure and enzymatic activity by aldehyde fixation. *J. Cell Biol.* **17**, 19–58

71. SCARPELLI, D. G. and PEARSE, A. G. E. 1958. Physical and chemical protection of cell constituents and the precise localization of enzymes. *J. Histochem. Cytochem.* **6**, 369–75

72. SCHULTZ-BRAUNS, O. 1931. Eine neue Methode des Gefrierschneidens für histologische Schnelluntersuchungen. *Klin. Woch.* **10**, 113–16

73. ——. 1932. Verbesserungen und Erfahrungen bei Anwendung der Methode des Gefrierschneidens unterfixierter Gewebe. *Zentralbl. allg. Path.* **54**, 225–34

74. SHIMIZU, N., KUBO, Z., and MORIKAWA, N. 1956. Some improvements in the preparation of fresh frozen sections. *Stain Tech.* **31**, 105–9

75. STOWELL, R. E. (edited by). 1965. Cryobiology. *Fed. Proc.* **24**, Suppl. 15.

76. ——, YOUNG, D. E., ARNOLD, E. A., and TRUMP, B. F. 1965. Structural, chemical, physical, and functional alterations in the mammalian nucleus following different conditions of freezing, storage, and thawing. *Fed. Proc.* **24**, Suppl. 15, 115–41

77. STUMPF, W. E. and ROTH, L. J. 1965. Frozen sectioning below −60°C with a refrigerated microtome. *Cryobiology* **1**, 227–32

78. TAYLOR, J. D. 1965. Gelatin embedding on the tissue carrier for thin sections in the cryostat. *Stain Tech.* **40**, 29–32

79. THOMAS, E. and PEARSE, A. G. E. 1961. The fine localization of dehydrogenases in the nervous system. *Histochemie* **2**, 266–82

80. THORNBURG, W. and MENGERS, P. E. 1957. An analysis of frozen section techniques. 1. Sectioning of fresh-frozen tissue. *J. Histochem. Cytochem.* **5**, 47–52

81. TRUMP, B. F., YOUNG, O. E., ARNOLD, E. A., and STOWELL, R. E. 1965. The effects of freezing and thawing on the structure, chemical constitution and function of cytoplasmic structures. *Fed. Proc.* **24**, Suppl. 15, 144–68

82. WEST, W. T. 1962. The Pearse-Slee cryostat: improvement in the anti-roll plate. *Stain Tech.* **37**, 5

NOTE ADDED IN PROOF

The manuscript of this paper was completed in 1965. Subsequently a short paper by Pearse and Bancroft was published (*J. Roy. micr. Soc.* 1966, **85**, 385–90 Controlled temperature cold microtomy.). The authors carried out experiments similar to those described on page 27 *ff.*, and reached qualitatively similar conclusions. The temperature ranges suggested as ideal are slightly different, but this may be because of differences in the thickness of the sections cut, and in the tissues studied during the experiments. Essentially, the findings of Pearse and Bancroft provide supporting evidence for the conclusions stated here.

Chapter 4

Sea Water and Osmium Tetroxide Fixation of Marine Animals

S. H. CHUANG

INTRODUCTION

Osmium tetroxide has long been known to give a lifelike preservation of cells.[18] The history of its introduction into cytological technique, its properties and its actions on cell constituents and structures have been described.[4] Its wide use in electron microscopy stimulated further studies of its reactions with the various biologically important cell structures and constituents revealed by the electron microscope.

The fixation of tissues by osmium tetroxide is preceded by a drop in pH values.[13] This decrease was presumed to be due to the elimination of basic groups such as tryptophane, the reduction of osmium tetroxide to an acid residue, and the oxidation of certain groups.[1] It was suggested that its reactions with proteins depended on their tryptophane, cysteine, and histidine content.[1]

Since the claim that the pH of fixative solutions modified the fine structure of cells under the electron microscope,[13] the buffering of osmium tetroxide solution to a pH slightly above neutrality is much in vogue. It has also been claimed that variations in the tonicity of sodium chloride solutions used as vehicles for this fixative produced different images of metaphase chromosomes.[16]

However, satisfactory preservation of the fine structure of a great variety of cells has been achieved with unbuffered aqueous solutions of osmium tetroxide.[5,7–10,12]

Osmium tetroxide penetrates very slowly and fixes a block of tissue unevenly.[4] Peripheral cells in slices of liver were better preserved than those more centrally placed.[13] However, addition of sucrose to buffered osmium tetroxide solution has been claimed to produce satisfactory fixation at all depths of *small* tissue cubes and to cause much less coagulation of the cytoplasmic matrix.[6] Prolonged fixation in osmium tetroxide was found to cause diffusion of the amorphous matrix from the endothelial cell of the rabbit,[14] and also progressive destruction of tissue constituents, as shown by their increased extractability in water.[2,7]

The following experiments were carried out to find out if the salts of sea water have any effect on the fixation of marine animals and tissues.

MATERIALS

In the choice of a suitable test tissue three criteria were used, namely: superficial position, uniformity of structure, and ease of identification. After some preliminary experiments with the tentacles of several coelenterates, and the branchial filaments of some polychaetes, under the light and electron microscopes, the outer epithelium of the branchial filaments of the polychaete *Terebella lapidaria* (Linnaeus), and its ectoparasitic peritrich ciliate *Paravorticella terebellae* (Fauré-Fremiet) were chosen. The branchial filaments are hollow tubes arranged in three branched clusters on each side of the anterior region of the body. A simple columnar epithelium forms the outermost sheet of cells on the filament, and is not provided with connective tissue. Apart from occasional mucous cells, this epithelium comprises ordinary epithelial cells with microvilli at their free ends. The cells are easily recognized. Their superficial position makes them readily accessible to the effects of the solvent used for osmium tetroxide. Their arrangement in one single layer ensures that they would be instantaneously and simultaneously exposed to the fixative, and less subject to the effects of slow penetration and uneven fixation. The ciliate provides additional organelles characteristic of the Protozoa for comparison.

METHOD

Several branchial filaments from the same cluster in one single specimen of *Terebella lapidaria* were ligatured at the peak of inspiratory dilation in their respiratory cycle; snipped off with fine scissors at the region proximal to the ligatures; and separated into two portions. One portion was immediately dropped into a cold unbuffered 1 per cent aqueous solution of osmium tetroxide of about pH 5·9. The other portion was dropped into a cold 1 per cent solution of osmium tetroxide in sea water of about pH 8·3. This was prepared by mixing one part of 2 per cent aqueous solution of osmium tetroxide with one part of sea water which had been boiled down to half its original volume. The filaments were left in their respective fixatives, in the refrigerator, at a temperature of about 2°C, for 1 hr. After fixation the ligatures were removed during the dehydration of the tissues in ethanol. The tissues were dehydrated in ethanol: 10 min each in 70 per cent and 80 per cent; 13 hr in 1 per cent solution of uranyl nitrate in 96 per cent ethanol; 10 min in absolute ethanol. They were next left in two changes of propylene oxide for 15 min each, and then soaked for about 6 hr in a mixture of one part of propylene oxide and two parts of Araldite with accelerator. The tissues were finally transferred to Araldite and allowed to polymerize in the oven for about 12 hr each at temperatures of 37°, 45°, and 60°C. Thin sections were cut with glass knives, floated on to a water surface, expanded with xylene vapour, and collected on carbon-coated grids. The sections were further stained on the grid for 20 min at 20°C with lead citrate.[15] They were examined in an Akashi TRS 50 electron microscope.

RESULTS

(A) Branchial filament epithelium

In *Terebella lapidaria* the columnar cell in the outer epithelium of the branchial filament has a large nucleus with usually a single nucleolus, a Golgi apparatus, mitochondria with cristae, ergastoplasm with irregular cisternae, sparse ribosomes, and a few lipid globules. The free end of the cell bears a large number of convoluted microvilli with stout bases. These microvilli are embedded in a moderately electron-dense matrix, with some of the distal portions lying free external to this. In the matrix lie a large number of electron-dense fine filaments. In the greater part of the area of contact between the adjoining epithelial cells, the plasma membranes are in close apposition. In certain regions, however, the membranes separate to form tubular intercellular spaces of various, somewhat meandering, shapes and diameters.

The micrographs showed that the general appearance of the cytoplasm with its abundance of wide ergastoplasmic cisternae and scarcity of ribosomes was similar after both fixatives. In the microvillar layer the fine filaments were equally well preserved (Fig. 4.1, A and B). The plasma membrane, lining the microvilli and the rest of the cell, was similar in appearance, with intercellular spaces between neighbouring epithelial cells equally frequent, and equally large. In longitudinal sections of the epithelial cells the Golgi apparatus appeared as stacks of slightly to strongly curved parallel Golgi saccules, with a large cluster of Golgi vesicles at each end. These saccules and vesicles appeared of equal density with either type of osmium fixation (Fig. 4.1, C and D). Similarly, there was no difference in the structure of the mitochondria, which appeared as sausage-shaped outlines of irregular diameter enclosing evenly distributed cristae of more or less uniform diameter (Fig. 4.1, C and D). The appearance of the nucleoplasm varied in cells fixed with the same solution, due to slight individual variation in the degree of aggregation. This was equally evident in materials fixed in either way. There was no consistent association of a greater degree of aggregation with either fixative (Fig. 4.1, C and D). In the nuclear envelope there was no discernible difference in the size of the electron transparent space between the inner and the outer nuclear membranes.

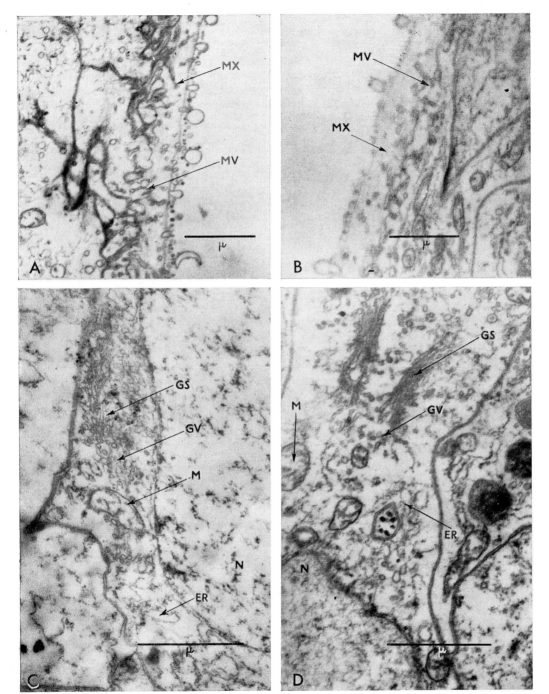

Fig. 4.1 Electron micrographs of longitudinal sections through the outer epithelial cells of the branchial filament in the polychaete, *Terebella lapidaria.*
A (× 24,000) **and C** (× 36,000)   Distal and middle part of cell fixed by OsO₄ dissolved in distilled water.
B (× 24,000) **and D** (× 36,000)   Distal and middle part of cell fixed by Oso₄ dissolved in sea water.

ER, ergastoplasm;   MX, matrix with fine filaments;
GS, Golgi saccule;   MV, microvillus;
GV, Golgi vesicle;   N, nucleus.
M, mitochondria;

(B) Peritrich ciliate

Paravorticella terebellae is a solitary peritrich ciliate with a contractile vacuole. It lives attached to the distal end of the outer epithelium of the branchial filaments of *Terebella lapidaria*. When the branchial filaments were dropped into the osmium tetroxide solutions, these ectoparasites remained attached and were fixed in a partially contracted condition with the oral disk and oral cilia covered over by a fold of the distal part of the body. Blocks of transparent Araldite were trimmed under a binocular microscope so as to orientate the contained ectoparasites for transverse or longitudinal sectioning. The fine structure produced in *Paravorticella terebellae* by the plain aqueous osmium tetroxide solution was hardly distinguishable from that of the osmium in sea water. There was no difference in the general appearance of the cytoplasmic matrix, which was packed with a great number of vacuoles and vesicles, among the more electron-dense organelles. In the periphery of the contractile vacuole the lumen of the tubules that occurred in stacks did not appear to differ in diameter (Fig. 4.2, A and B). The vacuoles lining the wall of the contractile vacuole varied both in size and number in the same section from the same animal: however, they did not seem to differ after the two different osmium fixations. The mitochondria showed a similar wavy outline, with slender and evenly distributed tubular cristae. No discernible swelling of the mitochondria or their cristae was noted after either fixation (Fig. 4.2, A and B). The ergastoplasm with narrow lumen also occurred in animals preserved in the plain aqueous solution of osmium tetroxide (Fig. 4.2, D). The meganuclei sectioned were not in comparable phases, and were not therefore compared. The fine structure of the pellicle was similar. The degree of separation of the outer from the inner pellicle membrane depended on the degree of stretching of the pellicle, and appeared similar in comparable states of stretch, when fixed in either solution. The thickness and density of the layer of tubules lying just below the inner pellicular membrane were comparable (Fig. 4.2, C and D). The vacuolar spaces under the aforesaid layer of tubules differed greatly in both size and number in the same section; the effects on them of the two solutions of osmium tetroxide seemed identical. The cilia appeared equally well preserved (Fig. 4.2, C and D).

DISCUSSION

The fine structure produced by an aqueous solution of osmium tetroxide on the peritrich ciliate *Paravorticella terebellae* is no different from that produced by osmium tetroxide dissolved in sea water. No difference is found in the ultrastructure of the outer epithelium of the branchial filament of the polychaete *Terebella lapidaria*, even though the branchial filament was fixed at the peak of inspiration, when it was distended with body fluid and its volume was at a maximum. Considering the swelling, the increase in weight, and other changes which the tissues underwent during the fixation period in the osmium tetroxide solutions,[3,19] and the shrinkage sustained in the subsequent dehydration in ethanols,[3,11,19] the similarities in the fine structure after the use of osmium tetroxide dissolved in two widely different vehicles are striking. The two solutions differed widely not only in ionic content, but also in pH. The simple 1 per cent solution of osmium tetroxide of about pH 5·9 is hypotonic, but produced a fine structure as good as that produced by osmium tetroxide in sea water with a pH of about 8·3.

Satisfactory fixation with unbuffered aqueous solution of osmium tetroxide in this study confirms its previous successful application to other tissues.[5,7-10,12] These results do not support the claim that fixative solutions should be isotonic with the body-fluid of the animal.[17] Nor do they support the claim that fixative solutions should be buffered for better preservation of the cells.[13] The buffering of fixative solutions now in vogue is based on the assumption that cell permeability is seriously altered just before or during fixation, and that buffers in the fixative would maintain the cell content at a pH close to its normal value during fixation.[13] However, for amphibian erythrocytes a simple 1 per cent osmium tetroxide has been found to be superior to the buffered solution, even if this is supplemented by sucrose,[6] due to the extraction of haemoglobin at higher pH values. This can be prevented, however, by addition of calcium ions.[19] For metaphase chromosomes tonicity, due to sodium chloride, and pH do not appear to be of sole importance for their lifelike preservation, which, however, depends on the type of cations and the dielectric constant of the fixative vehicle.[16] It appears that a particular intracellular constituent

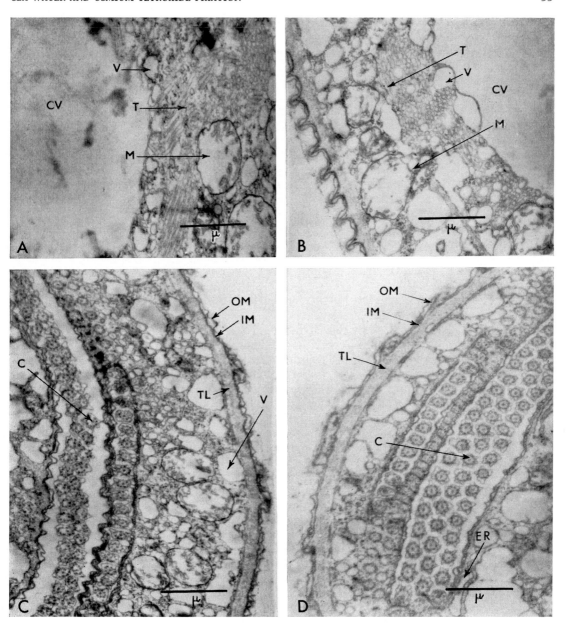

Fig. 4.2 Electron micrographs (× 22,000) of longitudinal sections through the distal end of the peritrich ciliate, *Paravorticella terebellae*.
A and B Fixed by OsO₄ dissolved in sea water.
C and D Fixed by OsO₄ dissolved in distilled water.

C, cilia;	OM, outer pollicular membrane;
CV, contractile vacuole;	T, tubule;
ER, ergastoplasm;	TL, tubular layer;
IM, inner pellicular membrane;	V, vacuole.
M, mitochondria;	

may require a certain combination of factors for its optimum preservation. But since knowledge of the lifelike structure is based mainly on the evidence of light microscopy, such as phase contrast microscopy, the 'best' image of the ultrastructure of cells at the higher magnification and resolution of the electron microscope remains to be decided.

The addition of indifferent salts to fixatives for the sake of their osmotic effects during fixation has been discussed.[4] The osmoregulatory exchanges between the cells and the salt-containing vehicle of the fixative have been assumed to be of importance in fixation. Liver slices show a marked and rapid swelling, an increase in weight, and a slow rise in specific gravity, during immersion in osmium tetroxide solution. These changes have been found to occur not only in aqueous solutions of osmium tetroxide, but also when this is dissolved in saline, balanced Tyrode's solution, etc.[3] The similarity of the fine structure of the protozoon, and of the polychaete tissue, fixed with and without sea water, seems to suggest that the salts of sea water do not appear to be necessary for a satisfactory preservation of the ultrastructure under investigation. Speculation points to two processes occurring at different rates during fixation, namely: the immobilization of the macromolecules of the cells by the fixing agent, and the osmoregulatory interaction between the cell and the indifferent salts in the vehicle of the fixing agent. However, it remains to be demonstrated how important to good fixation are the osmoregulatory exchanges that occur between the cell and the indifferent salts, just before the membrane systems and the macromolecules are transformed by the fixing agent.

SUMMARY

Osmium tetroxide is known to give lifelike preservation of cells. To find out if the salts of sea water have any effect on the fixation of marine animals and tissues, the outer epithelium of the branchial filaments of the polychaete *Terebella lapidaria* Linnaeus and its ectoparasitic peritrich ciliate *Paravorticella terebellae* (Fauré-Fremiet) were fixed in either a cold unbuffered aqueous solution of osmium tetroxide, or a solution in sea water, for 1 hr. The tissues were embedded in Araldite after treatment with ethanol and propylene oxide. Sections were stained on the grid with lead citrate, and examined with an electron microscope.

Similar results were obtained with both fixative solutions on the superficial epithelium of the branchial filament. The general appearance of the cytoplasm, the fine filaments of the microvillous layer, the stacks of Golgi saccules and the clusters of Golgi vesicles, the mitochondria and their cristae were equally well preserved. Slight variation in the degree of aggregation of the nucleoplasm was evident in individual cells, but there was no consistent association of a greater degree of aggregation with either fixation.

After fixation, *Paravorticella terebellae* remained attached, in a partially contracted condition, to the branchial filament. The fine structure of the organism produced by the plain aqueous osmium tetroxide solution was hardly distinguishable from that produced by osmium tetroxide in sea water. The cytoplasmic matrix, the tubules and vacuoles in the vicinity of the contractile vacuole, the mitochondria with their wavy outlines and evenly distributed cristae, the fine structure of the pellicle and of the cilia, were comparable in appearance.

These results are discussed. It is concluded that the salts of sea water do not appear to be necessary for satisfactory preservation of the ultrastructure under investigation.

ACKNOWLEDGEMENTS

The author is indebted to Dr. J. R. Baker, F.R.S., for permission to work in the Cytological Laboratory and also for his constant advice and encouragement, to Professor J. W. S. Pringle, F.R.S., for accommodation in the Department, and to Mr. J. M. McCrae and Miss E. G. M. Collins (Mrs. Williams) for technical assistance. The Akashi electron microscope and the Edwards vacuum evaporator for this work were provided by the Wellcome Trust; the Huxley ultramicrotome, by the Royal Society; and the current expenses by the Department of Scientific and Industrial Research.

REFERENCES

1. BAHR, G. F. 1954. Osmium tetroxide and ruthenium tetroxide and their reactions with biologically important substances. *Exp. Cell Res.* **7**, 457–79
2. ——. 1955. Continued studies about the fixation with osmium tetroxide. *Exp. Cell Res.* **9**, 277–85
3. ——, BLOOM, G., and FRIBERG, U. 1957. Volume changes of tissue in physiological fluids during fixation in osmium tetroxide or formaldehyde and during subsequent treatment. *Exp. Cell Res.* **12**, 342–55
4. BAKER, J. R. 1958. *Principles of Biological Microtechnique*, pp. 81, 120, 123. London: Methuen & Co.
5. ——. 1965. The fine structure produced in cells by fixatives. *J. R. micr. Soc.* **84**, 115–31
6. CAULFIELD, J. B. 1957. Effects of varying the vehicle for osmium tetroxide in tissue fixation. *J. biophys. biochem. Cytol.* **3**, 827–9
7. CLAUDE, A. 1961a. Problems of fixation for electron microscopy. Results of fixation with osmium tetroxide in acid and alkaline media. *Pathol. et Biol.* **9**, 933–47
8. ——. 1961b. Morphologie et organisation de constituants nucléaires dans le cas d'un carcinome rénal de la souris. *C. R. Acad. Sci.* **252**, 4186–8
9. DROCHMANS, P. 1960a. Electron microscope studies of epidermal melanocytes, and the fine structure of melanin granules. *J. biophys. biochem. Cytol.* **8**, 165–80
10. ——. 1960b. Mise en évidence du glycogène dans la cellule hépatique par microscopie électronique. *J. biophys. biochem. Cytol.* **8**, 553–8
11. FINEAN, J. B. 1958. X-ray diffraction studies of the myelin sheath in peripheral and central nerve fibres. *Exp. Cell Res.* Suppl. **5**, 18–32
12. MALHOTRA, S. K. 1962. Experiments on fixation for electron microscopy. I. Unbuffered osmium tetroxide. *Quart. J. micr. Sci.* **103**, 5–15
13. PALADE, G. E. 1952. A study of fixation for electron microscopy. *J. exp. Med.* **95**, 285–98
14. PORTER, K. R. and KALLMAN, F. 1953. The properties and effects of osmium tetroxide as a tissue fixative with special reference to its use for electron microscopy. *Exp. Cell Res.* **4**, 127–41
15. REYNOLDS, E. S. 1962. The use of lead citrate at high pH as an electron opaque stain in electron microscopy. *J. Cell Biol.* **17**, 208–12
16. ROBBINS, E. 1961. Some theoretical aspects of osmium tetroxide fixation with special reference to the metaphase chromosomes of cell cultures. *J. biophys. biochem. Cytol.* **11**, 449–55
17. SJÖSTRAND, F. S. 1956. The ultrastructure of cells as revealed by the electron microscope. *Int. Rev. Cytol.* **5**, 455–533
18. STRANGEWAYS, T. S. P. and CANTI, R. G. 1927. The living cell *in vitro* as shown by darkground illumination and the changes induced in such cells by fixing reagents. *Quart. J. micr. Sci.* **71**, 1–14
19. TOOZE, J. 1964. Measurements of some cellular changes during the fixation of amphibian erythrocytes with osmium tetroxide solutions. *J. Cell Biol.* **22**, 551–63

Chapter 5

Some Cytologically Relevant Aspects of the Physical Chemistry of Dyes

R. B. McKAY

PROPERTIES OF DISSOLVED AND ADSORBED DYES

The colouring of tissues with dyes is essentially an adsorption process in which the rate-determining step is the diffusion of the dye through the pores of the tissues. The diffusing dye is subjected to many types of forces acting either separately or simultaneously, e.g. long-range ionic attraction or repulsion, short-range hydrogen bonding, and various other forms of short-range forces collectively referred to here as van der Waals forces (dipole–dipole interactions, dispersion forces, etc.). Often the dye is retained primarily by ionic interaction with tissue components, e.g. the colouring of nucleic acids by basic dyes, but this is not always so. For example, the nucleic acid components of the chromatin in the spermatogenetic cells of the mouse are coloured strongly by the acid dye, methyl blue;[31] cellulose is basiphil, yet most direct cotton dyes are acid dyes. Further, the colour of dyes attached to tissue components is not always the same as that in solution, e.g. the matrix of cartilage is coloured red by toluidine blue.[4] Clearly, to understand fully the mechanism by which dyes are retained by tissue constituents it is essential to know first the state of the dye in solution (i.e. the extent to which the dye is dissociated into ions, the extent to which it is undissociated, and the extent to which it is aggregated, or in other words associated into clusters containing more than one dye ion), and second the state when attached to tissue constituents. Also, before the light-absorption properties of dyes attached to tissues can be interpreted satisfactorily, it is necessary first to understand the light-absorption properties of dyes in solution, which is a much simpler system. Unfortunately, these requirements have not yet been fulfilled. The present account therefore is an attempt to present briefly the chief facts that are known, and to outline their interpretation. The section on metachromasy, in particular, deals with several controversial aspects, and it must be emphasized that many of the opinions expressed are not generally accepted.

Dyes as electrolytes

Many of the dyes important in cytology are electrolytes, and as such tend to dissociate into ions when dissolved in water. The coloured ion is termed the dye ion, the other (or others) the counter-ion(s) or gegenion(en), e.g. Na^+, Cl^-. Dye ions may be anionic (negatively charged) or cationic (positively charged), and for convenience are referred to here as acid and basic, respectively.

The extent to which dyes dissociate into ions in water is not fully understood, chiefly because the established theories of electrolytic dissociation apply only to simple ions, and cannot be extrapolated satisfactorily to such large and complex ions as dye ions, which are often of comparatively low solubility and tend to aggregate. The most powerful method of studying the dissociation of electrolytes is to measure the ability of their solutions to conduct electricity. The method is simple in principle, but extremely difficult in practice, since an extraordinarily high degree of purity of both the water and the electrolyte are essential for quantitative study. The lowest electrolyte concentration that can be studied depends on how well these conditions are satisfied, and often the measurements in the most dilute solutions are the most critical in the interpretation of the data. Dyes are particularly difficult in that it is often impracticable to purify them to the required standards. In addition, basic dyes tend to fade rapidly at low concentrations in water, and a significant fraction of the dye may also be lost from solution by adsorption to glassware.

Apart from experimental difficulties, there are difficulties in the interpretation of conductivity data. The main one arises from the fact that the conductivity is affected not only by interaction of dye ions and counter-ions, but also by aggregation of the dye. The effect of the aggregation of organic electrolytes on their electrical conductivity is complex and has been dealt with at length elsewhere.[20, 37] Let it suffice to say here that aggregation in water of organic electrolytes other than dyes (e.g. soaps[20, 30, 37]) and indeed of simple electrolytes in solvents of low dielectric constant,[16] is associated with uptake of counter-ions to reduce electrostatic repulsion between the like-charged ions. There is no apparent reason why dyes should behave differently, and indeed there is some evidence that counter-ions are included in dye aggregates.[22, 38]

The best conductivity data available for dyes are those given in reference 42 (an extension of the work in references 38 and 41); eight sulphonated azo dyes have been studied. Briefly, methyl orange and another monosulphonated monoazo dye behave as strong electrolytes, for they are fully dissociated up to about 9×10^{-3} M, the highest concentration studied. The disulphonated bisazo dyes, Congo red, benzopurpurine 4B, and metabenzopurpurine, show evidence of aggregation right down to about 3×10^{-5} M, the lowest concentration studied, yet earlier data[38] indicate that the fraction of counter-ions included in the aggregates is less than 10 per cent at concentrations below about 1×10^{-3} M. Bordeaux extra appears to be slightly associated, but only at concentrations above 2×10^{-3} M. The data on a monosulphonated trisazo dye resemble those of a weak electrolyte, for this dye is incompletely dissociated and highly aggregated even at concentrations below 1×10^{-5} M.

Methylene blue is the only basic dye for which reliable conductivity data are available. This dye appears to be aggregated right down to about 7×10^{-5} M—the lowest concentration studied[41], but unfortunately not low enough to permit satisfactory interpretation of the data.

Aggregation of dyes in aqueous solution

Many methods have been developed in the past to estimate the average number of molecules in dye aggregates: the aggregation number (unaggregated molecules are included in this average, and are considered to have an aggregation number of one). They have been discussed elsewhere.[46] Collectively, the results have proved beyond reasonable doubt that dyes in general do aggregate in water, but they have yielded little information about the mechanism by which the aggregates are formed and their structure. Studies have been confined mainly to azo dyes[1, 25, 41, 45] which are of great technical importance. Some of these form large aggregates that are relatively easy to detect. Anthraquinone dyes have recently been shown to form small aggregates, except in a few specialized cases.[9] Basic dyes have been neglected in the past, although basic cyanine dyes, of little interest to biologists, have received some attention from photographic chemists.[12, 43]

The most useful type of method that has been developed in the past depends on the measurement of rates of diffusion of dyes. For this purpose an excess of salt must be present in the test solutions to overcome the accelerating effect of the diffusion potential, which is an artifact caused by the more rapid diffusion of the smaller counter-ions.[21, 25] A recently developed polarographic method has enabled the aggregation of a variety of dyes to be measured readily as a function of dye concentration.[23] Although electrochemical in character, the method depends solely on the rates of diffusion of dyes through solution to the surface of a dropping-mercury cathode, provided that a sufficient excess of salt is present.

Aggregation numbers estimated by these and other methods are so approximate that they merely indicate whether the aggregates are large or small. In view of the gross assumptions made, and of the inherent sources of experimental error in each method, it is hardly surprising that in the few cases where direct comparison is possible, aggregation numbers estimated by different methods are in poor numerical agreement. Nevertheless, there are areas of broad general agreement and these justify the following conclusions. Unless otherwise stated the data refer to aqueous solutions containing an excess of salt.

There is evidence that at low concentrations dyes (like other organic electrolytes such as soaps[11]) are essentially unaggregated. The polarographic experiments[23,32] suggest that the acid dyes, Congo Red and Solway Ultra Blue B (an anthraquinone derivative), and the basic dye, crystal violet, are monodisperse below a certain critical dye concentration around 1×10^{-4} M, but different for each dye; diffusion experiments suggest that two acid (monoazo) dyes are unaggregated below 1×10^{-3} M;[45] the conductivity data described above show that Bordeaux extra, in the total absence of salt, behaves as a simple strong electrolyte,[38,42] and therefore must be unaggregated in this concentration range. The polarographic data[23,32] suggest that above the critical concentration the aggregation numbers increase with increase in concentration until precipitation occurs. Crystal violet and Solway ultra blue B form only small aggregates in saturated solution, whereas Congo red and a cyanine dye form much larger aggregates. In fact those of Congo red appear to be of colloidal dimensions and there is ample independent evidence to corroborate this finding. The basic dye, methylene blue, appears to be aggregated at concentrations above 3×10^{-5} M;[23] the conductivity data show that the dye is partially aggregated at 7×10^{-5} M in the complete absence of salt.[41] The aggregation number of this dye increases from unity to a small value of about 2 or 3 near 1×10^{-4} M, but shows no further appreciable change over a wide range of concentration even until precipitation occurs.[23]

The data are too few and too inaccurate to show any general relationship between the chemical constitution of dyes and their ability to aggregate. Aggregation probably occurs primarily because of the low affinity of water for the large organic residues in the dye, but there is evidence that short-range attractive forces between adjacent dye molecules play an important part. Four examples illustrate this latter point. The molecules of azo dyes which form large aggregates can assume a planar conformation, thereby enabling large areas of adjacent molecules to come into close contact. In fact benzopurpurine 4B has a greater tendency to aggregate than its isomer, metabenzopurpurine, the molecules of which cannot assume a planar conformation for structural reasons.[25,46] Likewise, the symmetrical 1,1'-diethyl-2,2'-cyanine bromide has a greater tendency to aggregate than its unsymmetrical -2,4'-isomer, the molecules of which are less able to stack compactly.[32] Further, sulphonated anthraquinone dyes appear to form mixed aggregates with non-ionic surface-active agents by interaction of hydrophobic residues in the dye molecules and in the surface-active agents.[9] Finally, there is some evidence that powerful hydrogen-bonding agents can break down the large aggregates of azo dyes in which intermolecular hydrogen bonding is possible, but not of those in which it is improbable.[1]

Aggregation of adsorbed dyes

Studies of the adsorption of dyes to non-porous inorganic powders have shown that often more dye is adsorbed than can possibly be accommodated in a single close-packed layer of dye molecules at the surface of the powder. This has been observed with crystal violet and orange II on both titania and zinc oxide,[13] with a number of dyes on alumina[17] and other substrates,[18] and with basic cyanine dyes on powdered silver halides.[39] Opinions may differ as to how this phenomenon arises, but nevertheless the dye appears to be effectively aggregated, in part at least, in the adsorbed state. The phenomenon appears to be general with basic dyes on negatively charged substrates, but there are exceptions, notably Rhodamine B.[17]

There is also evidence that basic dyes are aggregated when attached to basiphilic regions in formalin-fixed yeast cells.[19]

METACHROMASY

Many dyes when they are used to colour certain tissue constituents (chromotropes) absorb light of shorter wavelength than they do in solution. This phenomenon is known as metachromasy. Metachromasy is also exhibited in aqueous solution when the dyes interact with soluble chromotropes, e.g. nucleic acids,[36] polyphosphates,[7] and gelatin.[8] Metachromasy, however, may be induced in aqueous solution in the absence of chromotropes either by increasing the dye concentration or by adding salt. The present discussion is concerned chiefly with metchromasy of dyes in solution.

The visible absorption spectra of solutions of a dye at different concentrations are conveniently compared by plotting them in terms of the molar extinction coefficient, ε, which is the optical density of a given solution, divided by the concentration of the dye in gram molecules per litre, and by the length of the light-path of the spectrophotometer cell. Experiment has shown that although the shapes of the spectra of all dyes thus plotted depend to some extent on the dye concentration, metachromasy is a larger effect through which the shapes of the spectra are radically altered. The main absorption band of the dye at low concentrations (M-band) is gradually replaced by a band at shorter wavelengths (H- cr hypsochromic band) as the dye concentration is increased (see Fig. 5.1). At high dye concentrations the H-band also diminishes in intensity and broadens towards shorter wavelengths.[34]

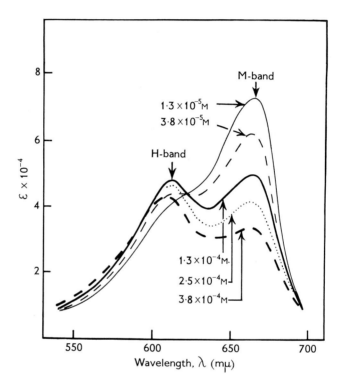

Fig. 5.1   Effect of concentration (indicated) on the visible absorption spectrum of methylene blue in water.

The phenomenon is exhibited only by those dyes in which the ionic charge is distributed by resonance throughout the chromophoric system in the dye molecule, i.e. the common basic dyes and a few acid dyes such as the fluorescein derivatives, e.g. eosin.[15] Sulphonated azo dyes, e.g. Congo red, and sulphonated anthraquinone dyes, e.g. Solway ultra blue B, do not exhibit metachromasy and in their molecules the ionic charges are on the sulphonate groups, which are not integral parts of the chromophoric system. Some acid dyes, however, are really sulphonated basic dyes, e.g. the thiacarbocyanine dye studied in reference 34, and these may exhibit metachromasy.

The H-band is diminished and the M-band is intensified by increase in temperature[27,40,44] and this, together with other factors beyond the scope of the present discussion, suggests that the bands should be attributed to absorption by distinct species of the dye.

Current hypotheses

It is widely held that the thermally unstable species responsible for the H-band is a dimer consisting of two dye ions, and that the M-band is due to absorption by monomeric dye.[15,27,40,47] This follows from the observation that the equilibrium between the M- and H-band conforms approximately[44] to a simple law of mass action relationship of the type $D^+ + D^+ \rightleftharpoons D_2^{2+}$. The ions of the dimer are considered to be oriented parallel to each other, and interaction between their respective electronic systems is considered to result in absorption of light of wavelength shorter than that absorbed by monomeric dye.[47] Unfortunately there is as yet no direct experimental evidence to correlate dye–dye interaction measured by an independent method with the appearance of H-bands. Some difficulties of the hypothesis have been pointed out.[32]

Recently it has been suggested that metachromasy in solution should be attributed primarily to interactions between dye ions and their counter-ions.[34] The close approach of a counter-ion to a dye ion is considered to alter and reduce the distribution of the charge on the dye ion and thereby to shift the absorption band of the dye to shorter wavelengths. This hypothesis provides a rough qualitative explanation of the metachromasy exhibited by dyes in organic solvents. Thus, since the strength of the electrical forces between ions in solution increases with decrease in the dielectric constant, D, of the solvent, it follows that, at low dye concentrations, metachromasy should be perceptible only in solvents of low dielectric constant. In fact marked metachromasy of basic dyes in the concentration range from 10^{-5} M to 10^{-3} M has been demonstrated in the solvents of low dielectric constant, chloroform $(D=5)$[14,35] and anisole $(D=5)$,[14,35] and in binary solvent mixtures of dielectric constant below 3.[3,28,35] In ethanol of moderate dielectric constant $(D=24)$, metachromasy has been detected at high dye concentrations of around 10^{-2} M;[5,26] in solvents of still higher dielectric constant, e.g. acetonitrile $(D=38)$ and formamide $(D=109)$, no metachromasy has been detected even at concentrations well above 10^{-3} M.[2] Indeed, there is evidence[14] that of the molecules of a triarylmethane dye in chloroform, the fraction which absorbs in the H-band, corresponds to the fraction which is associated with counter-ions as determined from conductivity measurements. Moreover, metachromasy can be induced in dye solutions of fixed concentration in binary solvent mixtures by varying the volume ratio of the solvents, and thereby lowering the dielectric constant.[3,34,35] The spectral changes become marked at about the same dielectric constant as that at which the electrostatic interactions between the ions become marked.[35]

Water, however, is quite anomalous, for despite its high dielectric constant $(D=79)$ metachromasy is marked in the concentration range from 10^{-5} M to 10^{-4} M with many basic dyes. Significant H-band formation does not occur in any other solvent of high dielectric constant, except in glycerol[3] but only at concentrations greater than 10^{-3} M. Water, however, is a unique compound; each molecule is triatomic, yet capable of participating in four hydrogen bonds. This leads to abnormally strong intermolecular attractions which give rise to an extensive three-dimensional structure at room temperature and various well-known anomalous physical properties. The effect of the anomalous properties of water on the properties of aqueous dye solutions has not been dealt with in the literature, yet it is tempting to link the anomaly of water in the context of metachromasy with its generally anomalous character. Perhaps the inability of structured water to solvate the dye ions properly leads to dye–dye association (aggregation) which is certainly an important factor in metachromasy in aqueous solution. This latter point is demonstrated in the following paragraph.

Metachromasy can be induced in an aqueous dye solution by adding salt[32,34,36,40] (i.e. increasing the counter-ion concentration), and the spectral changes are identical to those induced by adding dye. Much more salt than dye, however, is required to increase by the same amount the fraction of dye absorbing in the H-band.[34] This implies that dye–dye association to form aggregates is an important factor. It has been proposed that counter-ions are included in the dye aggregates to reduce electrostatic repulsion between the like-charged dye ions and that the action of these counter-ions on the electronic system of the dye ions is responsible for the spectral changes.[34] This hypothesis is incorrect, for recent evidence (J. F. Padday, unpublished results) shows that the counter-ions of several basic dyes in water are not associated with dye ions to any appreciable extent even though the H-band is prominent.

The weight of evidence favours the dimerization hypothesis of metachromasy in aqueous solution in which the spectral changes are considered to be induced by interaction between the electronic systems of two dye ions oriented parallel to each other. Nevertheless, it has not yet been explained how two like-charged and thereby mutually repelling dye ions can exist side-by-side in the absence of counter-ions.

Red forms of Methylene Blue

Red forms of methylene blue have been observed in solution. The dye is pink and rapidly precipitates from solution in solvent mixtures of dielectric constant below about 3.[29, 34] Particles of undissociated dye are responsible for the red colour, and indeed a red form has been detected in crystals of the dye.[10, 36] It has been suggested that the phenomenon is due to strong interaction between dye ions and counter-ions that are in intimate contact and not separated by solvent molecules.[34] In this way the charge distribution in the dye ion is greatly reduced, thereby producing the large colour change from blue to red. It is probably significant that the perchlorate of the dye (precipitated from aqueous solution by adding potassium perchlorate) is red and can be shown by elementary analysis to be anhydrous, unlike the normal dye chloride which is hygroscopic and extremely difficult to obtain in an anhydrous form.

Methylene blue also exists in a red form in organic solvents that contain strongly basic anions, e.g. an ethanolic solution of sodium ethoxide.[33] Further, the dye exists entirely in a red form when dissolved in simple aliphatic amines which are strong Lewis bases.[33] The mechanism by which this red species is produced has not been established, but it is likely that the species is a complex of the dye and the amine.

Metachromasy induced by chromotropic substances

Many basic dyes show colour changes when they interact with chromotropic substances either in solution or in tissues. The colour changes have been attributed to various causes, including the formation of ill-defined aggregates of dye at the active sites in the chromotrope, although the mechanism by which the colour is changed has been left vague. The problem is extremely complex and many factors are undoubtedly involved (see, for example, ref. 8 and other references quoted therein).

It is indeed likely that interaction between the electronic systems of dye ions oriented parallel to each other at the active sites on the chromotrope is an important factor responsible in part for colour changes,[6] but this type of mechanism cannot explain satisfactorily all aspects of the problem. One other factor worthy of consideration, particularly with large colour changes, is the possible effect on the electronic system of the dye ions of electrostatic interaction with charged groups in the chromotrope. In support of this view is the fact that basic dyes exhibit metachromasy when they interact with negatively-charged substrates[4,8,36] and the fact that the colour change is often more marked, the more basiphil the active groups (sulphate > phosphate > carboxylate).[4] Further, in solution the metachromasy of basic cyanine dyes induced by interaction with gelatin is observed only at a pH value above the iso-ionic point of gelatin,[8] i.e. only when the gelatin is negatively charged.

SUMMARY

The dissociation of dyes into ions, and the aggregation of dyes in aqueous solution, are related properties. The limitations of methods of investigating the latter have restricted greatly the study of both properties, and few data are available. The lack of knowledge of the state of dyes in solution has, in turn, hindered progress towards the complete understanding of more complex problems, such as the colouring of tissue constituents and metachromasy. Some aspects of metachromasy are discussed.

REFERENCES

1. ALEXANDER, P. and STACEY, K. A. 1952. The colloidal behaviour of dyes. *Proc. Roy. Soc. (London)* **212A**, 274
2. ARVAN, K. L. and ZAITSEVA, N. E. 1961. Spectral investigation of the influence of the solvent on the aggregation of organic dyes. *Optics and Spectroscopy* (English trans.) **10**, 137
3. —— and ——. 1961. Spectral investigation of the aggregation of dyes in solvents of low polarity. *ibid.* **11**, 38

4. BAKER, J. R. 1958. *Principles of Biological Microtechnique*, Chapter 13, p. 243. London: Methuen; and New York: Wiley
5. BARANOVA, E. G. 1962. Study of the association of rhodamine 6G in ethanol and glycerol solutions. *Optics and Spectroscopy* (English trans.) **13**, 452
6. BERGMANN, K. and O'KONSKI, C. T. 1963. A spectroscopic study of methylene blue monomer, dimer and complexes with montmorillonite. *J. Phys. Chem.* **67**, 2169
7. BIDEGAREY, J. P. and VIOVY, R. 1964. Relations entre les phénomènes d'association en solution et la métachromasie: cas de bleu de toluidine. *J. Chim. Phys.* **61**, 1391
8. BOYER, S., PICHEN, L., and DEGROVE, O. 1965. Interactions en solution: gelatine et colourants sensibilisateurs. *J. Chim. Phys.* **62**, 301
9. CRAVEN, B. R. and DATYNER, A. 1963. The interaction between some acid wool dyes and nonylphenolethylenoxide derivatives. *J. Soc. Dyers Col.* **79**, 515
10. DÄHNE, S. 1962. Die rote Form des Methylenblaus. *Z. phys. Chem.* **220**, 187
11. DEBYE, P. 1949. Light scattering in soap solutions. *J. Phys. Chem.* **53**, 1
12. DICKINSON, H. O. 1947. The aggregation of cyanine dyes in aqueous solution. *Trans. Faraday Soc.* **43**, 486
13. EWING, W. W. and LIU, F. W. J. 1953. Adsorption of dyes from aqueous solution on pigments. *J. Colloid Sci.* **8**, 204
14. FEICHTMAYR, F. and SCHLAG, J. 1964. Einfluss von Lösungsmittel und Konzentration auf die Spektren von Triphenyl-methanfarbstoffen. *Ber. Bunsenges. Phys. Chem.* **68**, 95
15. FÖSTER, T. and KÖNIG, E. 1957. Absorptionsspektren und Fluoreszenzeigenschaften konzentrierter Lösungen organischer Farbstoffe. *Z. Electrochem.* **61**, 344
16. FUOSS, R. M. and ACCASCINA, F. 1959. *Electrolytic Conductance*, Chapter 18, p. 249. New York and London: Interscience
17. GILES, C. H., EASTON, I. A., and MCKAY, R. B. 1964. Mechanism of adsorption of cationic dyes by alumina, and a note on heat changes in solution adsorption. *J. Chem. Soc.* 4495
18. ——, ——, ——, PATEL, C. C., SHAH, N. B., and SMITH, D. 1966. Association of absorbed aromatic solutes. *Trans. Faraday Soc.* **62**, 1963
19. —— and MCKAY, R. B. 1965. Adsorption of cationic (basic) dyes by fixed yeast cells. *J. Bacteriol.* **89**, 390
20. HARTLEY, G. S. 1935. The application of the Debye-Hückel theory to colloidal electrolytes. *Trans. Faraday Soc.* **31**, 31
21. —— and ROBINSON, C. 1931. The diffusion of colloidal electrolytes and other charged colloids. *Proc. Roy. Soc. (London)* **134A**, 20
22. HOLMES, F. H. and STANDING, H. A. 1945. The dyeing of cellulose with direct dyes. Part IV. The electrolytic conductance of aqueous solutions of direct dyes. *Trans. Faraday Soc.* **41**, 568
23. HILLSON, P. J. and MCKAY, R. B. 1965. Aggregation of dye molecules in aqueous solution. A polarographic study. *Trans. Faraday Soc.* **61**, 374
24. LEMIN, D. R. and VICKERSTAFF, T. 1947. The aggregation of direct dyes and of methylene blue 2B in aqueous solution. *Trans. Faraday Soc.* **43**, 491
25. LENHER, S. and SMITH, J. E. 1935. The dyeing of cotton: particle size and substantivity. *J. Amer. Chem. Soc.* **57**, 497 and 504
26. LEVSHIN, V. L. and BARANOVA, E. G. 1959. A study of the concentration extinction of the luminescence of dyes in various solutions and isolation of various types of extinction. *Optics and Spectroscopy* (English Trans.) **6**, 31
27. —— and GORSHKOV, V. K. 1961. Study of the nature of the bonding forces of associated dye molecules in concentrated solutions. *ibid.* **10**, 401
28. —— and LONSKAYA, I. S. 1961. Dependence of the association of rhodamines on the structure of their molecules and the nature of the solvent. *ibid.* **11**, 148
29. LEWIS, G. N., GOLDSCHMIT, O., MAGEL, T. T., and BIGELEISEN, J. 1943. Dimeric and other forms of methylene blue: absorption and fluorescence of the pure monomer. *J. Amer. Chem. Soc.* **65**, 1150
30. MCBAIN, J. W. 1944. Solutions of soaps and detergents as colloidal electrolytes. In *Colloid Chemistry* (edited by J. Alexander), Volume 5, Chapter 3, p. 102. New York: Reinhold
31. MCKAY, R. B. 1962. An investigation of the anomalous staining of chromatin by the acid dyes methyl blue and aniline blue. *Quart. J. micr. Sci.* **103**, 519
32. ——. 1965. Visible absorption spectra of some cationic dyes in aqueous solution. Effect of aggregation. *Trans. Faraday Soc.* **61**, 1787
33. ——. 1966. Red forms of methylene blue. *Nature, Lond.* **210**, 296
34. —— and HILLSON, P. J. 1965. Metachromasy of dyes in solution. Interpretation on the basis of interactions between dye ions and counter-ions. *Trans. Faraday Soc.* **61**, 1800
35. —— and ——. 1966. Metachromatic behaviour of dyes in solvents of low dielectric constant. *Trans. Faraday Soc.* **62**, 1439
36. MICHAELIS, L. and GRANICK, S. 1945. Metachromatic behaviour of basic dyestuffs. *J. Amer. Chem. Soc.* **67**, 1212

37. MOILLIET, J. L., COLLIE, B., ROBINSON, C., and HARTLEY, G. S. 1935. The significance and determination of mobilities in the study of colloidal electrolytes. *Trans. Faraday Soc.* **31**, 120
38. —— and ROBINSON, C. 1934. The aggregation of colloidal electrolytes from transport number and conductivity measurements: some benzidine dyes. *Proc. Roy. Soc. (London)* **143A**, 630
39. PADDAY, J. F. and WHICKHAM, R. S. 1966. Adsorption of cyanine dyes at silver halide surfaces, Part 2: Spectral properties and coverage. *Trans. Faraday Soc.* **62**, 1283
40. RABINOWITCH, E. and EPSTEIN, L. F. 1941. Polymerization of dyestuffs in solution. Thionine and methylene blue. *J. Amer. Chem. Soc.* **63**, 69
41. ROBINSON, C. 1935. The nature of the aqueous solutions of dyes. *Trans. Faraday Soc.* **31**, 245
42. —— and GARRETT. 1939. The degree of aggregation of dyes in dilute solution. *ibid.* **35**, 771
43. SCHEIBE, G. 1938. Reversible Polymerisation als Ursache neuartiger Absorptionsbanden von Farbstoffen. *Kolloid-Z.* **82**, 1
44. SHEPPARD, S. E. and GEDDES, A. L. 1944. Effects of solvents upon the absorption spectra of dyes. v. Water as solvent: quantitative examination of the dimerization hypothesis. *J. Amer. Chem. Soc.* **66**, 2003
45. VALKÓ, E. 1935. Measurements of the diffusion of dyestuffs. *Trans. Faraday Soc.* **31**, 230
46. VICKERSTAFF, T. 1954. *The Physical Chemistry of Dyeing*, 2nd edn., Chapter 3, p. 59. London: Oliver and Boyd
47. WEST, W. and PEARCE, S. 1965. The dimeric state of cyanine dyes. *J. Phys. Chem.* **69**, 1894

Chapter 6

On the Affinity of Dyes for Histological Substrates

D. J. GOLDSTEIN

Differences in the intensity of staining of fixed and sectioned tissue components, on which histology largely depends, are often loosely attributed to differences in 'affinity' for the dye. In the present article affinity will be discussed in more precise terms, and methods by which it may be measured will be outlined.

Knowledge of the affinity of the dye–substrate interaction, by itself, may be of very little value, and in no way obviates the need for studies of the chemical or physical processes involved in staining; but such knowledge is highly desirable if histological and histochemical procedures are to be applied rationally, and interpreted correctly. For example, knowing the affinity and other thermodynamic parameters such as the heat, entropy, rate, and activation energy of staining (not discussed here), may give one an insight into the mechanism of the process. It may also enable one to predict the conditions which will give a desired intensity of staining of a given substrate, or optimal differential staining of a number of substrates. Comparison of the affinities of a dye for different substrates may justify raising an empirical, selective staining method to the status of a specific histochemical test. Comparison of the affinities of a number of dyes for a given tissue component, may enable one to deduce the structural features of the dye molecule on which affinity depends. This could lead to a rational design or selection of dye molecules for a particular purpose.

Unfortunately, there is insufficient information on the affinity of histological substrates for dyes, for such interpretations, or generalizations, to be usefully attempted at the present time. This article will therefore be restricted to, (a) an outline of the theory and procedures involved in the measurement of affinity, and (b) a survey of some data currently available.

THE DEFINITION OF AFFINITY

A definition of affinity is best based on thermodynamic considerations. The reader is referred to Vickerstaff[31] for the derivation of the mathematical relationships, and a full discussion of the principles involved, and it suffices to say that the *affinity* is the tendency of a dye molecule to move from a defined 'standard state' in solution to a defined 'standard state' in the tissue, this tendency being related to the equilibrium constant of the reaction and the free energy change involved. The *standard affinity* (the standard free energy change with the sign changed) is given by the expression

$$-\Delta G^\circ = RT \ln K \tag{6.1}$$

$$= RT \ln \frac{a_t}{a_b} \tag{6.2}$$

where R is the gas constant, T the absolute temperature, K the equilibrium constant of the reaction, and a_t and a_b are respectively the activities of dye in the stained tissue and in a dyebath with which the tissue is in equilibrium.

The *activity* of a substance is related to, but is not generally identical with, its measurable concentration, except in an ideal or very dilute solution; the precise relationship will depend on the mechanism of dyeing which is postulated or discovered, and on the somewhat arbitrary definition of the standard states.

These points will become clearer during the following discussion of two cases of particular relevance to histological dyeing.

Solution of dye in the substrate—the Freundlich isotherm

Witt (1891, cited in[20]) thought staining was in general simply a question of the dye forming a 'solid solution' in the substrate. If this were so, the distribution of dye could be described by the ratio

$$\frac{a_t}{a_b} = K \tag{6.3}$$

where the equilibrium constant K is the familiar distribution coefficient which applies to the partitioning of a solute between two immiscible solvents. The standard states may here be defined as being 1 mole of dye per litre of dye solution or substrate. If the dye forms an ideal solution in both phases, the activities may be replaced by the concentrations, and the standard affinity is

$$-\Delta G^\circ = RT \ln \frac{D_t}{D_b} \tag{6.4}$$

where D_t and D_b are respectively the concentrations of dye in the tissue and the dyebath at equilibrium. The same relationship is valid even if the dye does not form an ideal solution in both phases, provided that the deviation from ideality is similar in both solutions: in this case the activity coefficient (by which the concentration must be multiplied to give the activity) in the numerator and denominator will cancel out.

In practice such a simple relationship is unfortunately seldom if ever observed, even in fat staining (see later). Mostly there is a considerable deviation of the dye solutions from ideality, sometimes due to aggregation of the dye molecules. Let us suppose that the dye aggregates both in the dyebath and in the substrate, with an aggregation number (i.e. a mean number of dye molecules per aggregate) of n in the former and m in the latter. The reversible reaction of staining will then be

$$m(\text{Dye formula})_n \rightleftharpoons n(\text{Dye formula})_m \tag{6.5}$$
$$\text{(Dyebath)} \qquad\qquad \text{(Substrate)}$$

and by the law of mass action

$$K = \frac{(D_t)^n}{(D_b)^m} \tag{6.6}$$

where D_t and D_b are as before the concentrations of dye at equilibrium in substrate and dyebath. Taking logarithms and rearranging,

$$\log D_t = \frac{1}{n} \log K + \frac{m}{n} \log D_b \tag{6.7}$$

whence it will be seen that $\log D_t$ should be linearly related to $\log D_b$. From a log–log plot of pairs of values of D_t and D_b it is possible to evaluate m/n and $(\log K)/n$. One might in this way compare the uptake of a single dye by a number of substrates, or the uptake of a number of dyes by one substrate. Unless, however, one has an independent estimate of the aggregation number n of the dye in the dyebath, it will not be possible to calculate K or the standard affinity.

Equation (6.6) is mathematically equivalent to the so-called Freundlich adsorption isotherm (reference [31], p. 97),

$$D_t = k(D_b)^x \tag{6.8}$$

except that k and x in equation (6.8) correspond to $K^{1/n}$ and m/n respectively in equation (6.6). The Freundlich isotherm is a purely empirical relation which has, however, been found to apply to a number of dyeing situations. Thus Georgievics (1894, cited in reference [20]) found the uptake of indigo-carmine by silk to follow approximately the relation

$$\frac{D_b^{\frac{1}{2}}}{D_t} = \text{constant} \tag{6.9}$$

and he proposed that most or all staining could be represented by the expression

$$\frac{D_b^{1/y}}{D_t} = \text{constant} \tag{6.10}$$

the numerical value of y indicating the affinity of the dye for the fibre. It will be seen that y in equation (6.10) corresponds to n/m in equation (6.6), and not to affinity in the sense in which we have been using the term.

Obedience of data to a Freundlich isotherm does not of course prove that the dye forms a true solution in both bath and substrate, with aggregation occurring in one or both of the phases. Other interpretations may well be valid in certain cases—(reference [31], p. 109).

Adsorption of dye onto a finite number of sites—the Langmuir isotherm

In histological dyeing, the dye does not usually simply form a solution in the substrate, but dye particles are bound to a finite number of discrete sites in the tissue. Thus the coloured cations of basic dyes are bound to acidic groups (sulphates, phosphates, or carboxyls) in the tissue, and the number of dye ions which the tissue is able to take up is limited by the number of such groups available.

From the law of mass action, Langmuir (cited in reference [31], p. 98) showed that if the binding sites are all similar but independent, and can each bind only one dye molecule, and if (in ignorance of the activity of dyes in solution) one replaces the activity of dye in the dyebath, a_b, by the concentration, D_b,

$$D_t = KSD_b/(1+KD_b) \tag{6.11}$$

Here S is the saturation value of dye in the substrate (i.e. the value of D_t when all available sites are occupied), and the other symbols have the same significance as before. A major difference between the Freundlich and Langmuir isotherms is that only in the latter is a plateau of dye uptake approached in concentrated dyebaths.

Rearranging equation (6.11),

$$K = \frac{D_t}{(S-D_t)D_b} \tag{6.12}$$

which by comparison with equations (1) and (2) shows that $D_t/(S-D_t)$ is in fact the activity of dye in the substrate.

This activity could alternatively be expressed as $\theta/(1-\theta)$, where θ equals D_t/S (i.e. the fraction of available sites occupied by dye). If the substrate is half-saturated with dye, $\theta=0\cdot5$ and $\theta/(1-\theta)=1$, so that

$$K = 1/D_b \tag{6.13}$$

i.e. the equilibrium constant is numerically equal to that molar concentration of dyebath in equilibrium with a half-saturated substrate. The standard state of dye in the substrate is here implicitly defined as half-saturation, where the activity equals one. The standard state in the dyebath is, as before, 1 mole per litre.

Provided a Langmuir isotherm can be shown to apply to a real dyeing system, and provided it appears that, from all the available evidence, the necessary assumptions are physically reasonable, the problem of measuring the affinity of the dye for the substrate is essentially one of evaluating K in equation (6.12). If two pairs of values of D_t and D_b are available, equation (6.12) can be solved algebraically. If one has several pairs of measurements, it is preferable to plot the data graphically in such a way as theoretically to give a straight line, and derive K and S from a line fitted to the experimental points by eye or by the method of least squares. Rearranging equation (6.11)

$$1/D_t = 1/S + 1/KSD_b \qquad \qquad (6.14)$$

so that a plot of $1/D_t$ against $1/D_b$ should be linear.

Alternatively,

$$D_t = S - D_t/KD_b \qquad \qquad (6.15)$$

so that a plot of D_t against D_t/D_b will give a straight line of slope $-1/K$, intercepting the D_t axis at a value S. This method, due to Scatchard (cited in reference [11]), gives a better distribution of the experimental points than does the simpler reciprocal plot.

Random scatter of the experimental points about the best straight line will be a function of experimental error, while a systematic deviation may indicate a deficiency in the theory. Thus quite commonly it is found[15] that adsorption of a solute onto a solid follows a Langmuir isotherm in weak solutions (i.e. gives a straight line on a Scatchard plot), but that an additional uptake of solute occurs from concentrated solutions. This could be due to such factors as the presence in the substrate of sites of varying affinity, the weaker sites adsorbing significant quantities of solute only from concentrated solutions. In this case one can report a value for the mean affinity as determined over a stated range of dyebath concentrations.

Most histological dyes are ionic. An experimentally established standard affinity refers to the uptake of dye ion plus inorganic gegenion (usually Na^+ or H^+ in the case of an acid dye, and Cl^- in the case of a basic one), because both cations and anions must be taken up from the dyebath to maintain electrical neutrality in the solid substrate. If both dye ion and gegenion are adsorbed, both will contribute to the measured affinity, although the affinity of the inorganic ion is likely in most cases to be much less than that of the dye ion. By assigning a value to the affinity of the inorganic ion for the substrate, one can calculate the affinity attributable to the dye ion itself. This will not be attempted in the present article.

We shall now examine, in the light of the above theory, some data on histological dyeing.

QUALITATIVE RECOGNITION OF DIFFERENCES IN AFFINITY

As early as 1885 Gierke wrote that dyes attach to tissues because of a physical surface attraction, which might differ quantitatively between different dyes. He placed Methylene blue, Safranin and Bismarck brown in ascending order of firmness of attachment to tissues, but left open the nature of, and the laws governing the attraction.

Although a high affinity between dye and substrate will tend to make the attachment a firm one, the rate of staining or of removal of dye from a substrate will be affected also by other factors. The reader may recall that many classical staining methods depend primarily on differences in the permeability of tissues to dye.[1,12,13,14,20,22] In any study of affinity, however, one must eliminate kinetic factors by ensuring that equilibrium is reached between tissues and dyebath (e.g. by staining for a time long enough for a previously overstained and an unstained section, placed in the same dyebath, to reach the same intensity of staining). It is therefore difficult to interpret Gierke's results or those of Spicer and Meyer,[27] who found that Alcian blue was readily displaced from a carboxyl-containing mucin by Aldehyde fuchsin (or vice versa), but that sulphated mucins tended to retain the dye first applied.

Provided equilibrium is achieved, affinity may be assessed by observing the uptake of dye from dyebaths of varying dye concentration (but constant temperature, pH etc.). The higher the affinity, the greater is the relative uptake from a dilute bath. Thus very dilute methyl green stains only nuclei[3]; and goblet cell mucin from the distal intestine stains in more dilute aqueous toluidine blue solutions than does mucin from the proximal gut[18]. These results are most probably due to true differences in affinity. The intensity of staining is, however, affected not only by the affinity between dye particle and reactive site, but by the number or concentration of such sites in the tissue, and the visual threshold may play a role in the recognition of faint staining.

It may not be out of place here to echo Baker's[2] warning regarding the visual assessment of staining intensity. As he points out, if a given object passes 50 per cent of the light incident upon it, an object four times as thick (or one containing four times as much dye) will pass only one-eighth as much light, so that one forms an exaggerated opinion of the difference in the amount of dye taken up. A corollary is the case of a weakly stained object, which passes (say) 80 per cent of the incident light. An object containing four times as much dye will pass about 41 per cent of the incident light (i.e. about half as much), so that one may get a falsely *low* impression of the relative amount of dye present. It is clear that in the assessment of staining, it is safest either to measure dye concentrations instrumentally, or simply to judge visually whether or not two objects have the same intensity of staining (for which task of matching the eye is excellently adapted).

Dye and other molecules will compete for binding sites in a substrate, and the affinities of different dyes for a single substrate might be compared by studying their competition with each other, or with a third substance (e.g. an inorganic ion). Graumann[18], among others, has studied the inhibition of metachromasia by salts, and more recently Scott and Dorling[25] have observed that staining of polyanions by the basic dye Alcian blue may be inhibited by inorganic salt ($NaCl$ or $MgCl_2$). The concentration of salt necessary depends on the nature of the acid groups involved. In Scott and Dorling's work, dye continued to be bound by sulphate ester groups at a salt concentration five to ten times that which abolished visible staining of carboxylate or phosphate groups. This gives some indication of the relative affinity of salt and dye for a given substrate, and may be valuable in the empirical distinction of various mucins, but throws no light on the absolute affinity of the dye for the substrate or on the relative affinities of the dye for different substrates.

THE QUANTITATIVE MEASUREMENT OF AFFINITY

The most direct way of ascertaining D_t, the dye concentration in a stained substrate, is by microdensitometry. For this to be possible, Beer's law must be obeyed, i.e. the absorbance of monochromatic light by dye in the substrate must be proportional to the amount of dye present. Unequivocal proof of this is rarely possible with histological sections. If Beer's law is valid for all wavelengths of light, similarly shaped absorption spectra will be found with all intensities of staining, but the converse is not necessarily true, so that such spectra are not absolute proof that the absorbance is proportional to the concentration. Even if differently shaped spectra are found at different dye concentrations (as is commonly the case with histological dyes), Beer's law *may* still hold at a particular wavelength. In Fig. 6.1 spectra are given for RNA stained in two concentrations of Azure A, the absorbances in both curves being expressed as a percentage of the peak absorbance. At the higher dye concentration there is a marked widening of the peak, which may be partly due to instrumental factors such as the use of imperfectly monochromatic light. In general the range of wavelength (i.e. the deviation from true monochromaticity) of the light used should be less than the width of the absorption peak to be measured: if this is so Beer's law may be obeyed at the absorption peak, but not on the 'shoulders' of the curve.

For present purposes we shall assume, with some reservations, that the absorbance at the wavelength of maximum absorbance is proportional to the dye concentration, provided that the absorbances compared are not very different.

An indirect approach to the measurement of affinity

As shown previously, provided a Langmuir isotherm is obeyed, the association constant K of the dye–substrate interaction is numerically equal to the reciprocal of that molar concentration of dyebath in equilibrium with a half-saturated dyebath. Now the concentration of dye in a thin section (say 3μ) is about half that in a section which is twice as thick (i.e. 6μ), if the two sections have the same apparent intensity of staining. If the thin section is saturated with dye, which will probably be the case after staining in a concentrated dyebath, the matching thick section will be half-saturated, and from the known dyebath concentration necessary to achieve this, K and hence the affinity can be calculated. In practice, a single heavily stained 3 μ section is compared under a comparison microscope with a series of 6 μ sections stained in varying concentrations of dye. Using this method Goldstein[16] found the following approximate values for $-\Delta G°$ (in kcal/mole) for the staining of formalin-fixed tissue by Azure A (4°C, veronal-acetate buffer pH 4·0): cartilage matrix 3·8 (orthochromasia) and 5·3 (metachromasia); mast-cell granules 4·0 (orthochromasia) and 4·4 (metachromasia); nuclear chromatin and cytoplasmic RNA 3·1; mucin 2·7 to 3·4; thyroid colloid and toad poison-gland secretion 2·3.

A somewhat more elaborate procedure for the evaluation of affinity[17] depends on the fact that (making the same assumptions as before) it can be shown that K equals the reciprocal of that concentration of dyebath in which a thin section (say 3 μ) achieves the same apparent intensity of staining as a

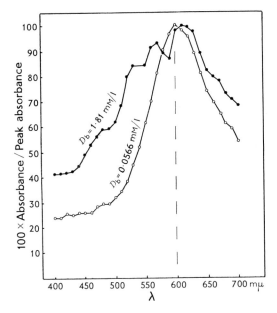

Fig. 6.1 Absorption spectra of cytoplasmic RNA in formalin-fixed pancreas stained in 1·81 or 0·0566 mM/l. Azure A (6 μ sections, 4 days, pH 4·0, 20°C, airdried). The ordinate is the absorbance expressed as a percentage of the peak absorbance. Note the widening of the peak in the case of the more intensely stained section.

thick (6 μ) section stained in a dyebath concentration of $1/(3K)$. Thin and thick sections are stained to equilibrium in a series of dyebaths of varying concentration, and a pair of sections of similar intensity sought, fulfilling this criterion.

The value of K found by the two methods should be the same if a Langmuir isotherm is strictly obeyed, but in fact a higher affinity of Azure A for pancreatic RNA and some other substrates is obtained using the second method, indicating that the apparent affinity tends to decrease with a rise in the dyebath concentration.[17] Both methods have the advantage of requiring no specialized equipment other than a comparison eyepiece, but suffer from the serious drawback that there is no easy way of cutting sections at two different, pre-determined thicknesses. Other theoretical and practical difficulties are discussed in the original articles.

The measurement of affinity by spectrophotometry of sectioned tissues

Most histological objects are both small and non-homogeneous, so that ordinary photometric methods are often not applicable (see references [24] and [29] for a discussion of difficulties inherent in microspectrophotometry). In unpublished work by the author, absorbances of dyes in sectioned tissue components have been measured with an integrating microdensitometer (manufactured by Barr & Stroud Ltd after a design by Deeley[5]), in which distributional error is eliminated by a scanning procedure. The machine records on an arbitrary scale the total relative absorbance of stained material in a defined microscopic field, but has a cut-out so that no absorbance of more than a pre-determined value (of up to 1·0) is recorded. Full details of the experimental procedure and results will be published elsewhere: here only a few results will be given to illustrate the potentialities of the method.

Fig. 6.2 is an adsorption isotherm for the uptake of the basic dye Azure A by nuclei of the rat islet of Langerhans (formalin-fixed 6 μ paraffin sections, stained 4 days at 20°C in varying concentrations of dye in veronal-acetate buffer at pH 4·0 and then either air-dried or dehydrated in ethanol). The abscissa is the dyebath concentration on an arithmetic scale (the highest concentrations being omitted to avoid undue compression of the scale), and the ordinate is the relative absorbance of stained nuclei, each point

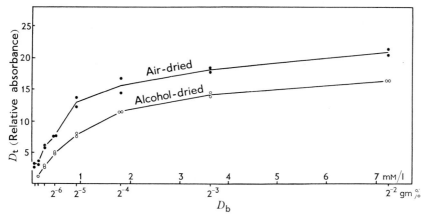

Fig. 6.2 Adsorption isotherm of Azure A (6 μ sections, veronal-acetate buffer pH 4·0, 20°C, 4 days) onto formalin-fixed nuclei of rat islet of Langerhans. Arithmetic plot showing approach to a plateau at high dye concentrations.

representing the mean of readings on at least ten nuclei in a given section. Duplicate sections were used throughout. It will be seen that the dye uptake tends to a limit in strong dyebaths, but the detailed adherence to a Langmuir isotherm cannot readily be assessed.

In Fig. 6.3 the same data are replotted after Scatchard, the relative absorbance being on the ordinate as before, but the abscissa being the absorbance divided by the dyebath concentration. Although there is some random scatter, especially in the air-dried series, the experimental points obtained in the more dilute dyebaths fall satisfactorily onto straight lines, indicating obedience to Langmuir isotherms. In the stronger dyebaths, however, the uptake is greater than would occur if the same Langmuir isotherm were followed throughout. In concentrated solutions the activity of the dye might be expected to increase more rapidly than the concentration,[23] thus invalidating the use of dyebath concentrations in place of activities in the Langmuir equation. This effect could not, however, account for an uptake greater than the saturation value found by extrapolation of the straight-line portion of the graph, and most of the 'extra' uptake is probably attributable to weakly-binding sites which take up dye only from strong baths.

The straight lines in Fig. 6.3 were fitted by the method of least squares to data obtained in dyebaths of up to 7·25 mM/l. (i.e. omitting the highest point in the air-dried series and the highest eight points in the alcohol-dried series), and correspond to affinities of 4·35 and 3·84 kcal/mole for the air-dried and alcohol-dried series respectively. These values are higher than those found by the indirect method, which applies to the uptake of dye from more concentrated baths. The difference in affinity between the alcohol- and air-dried series is curious, as one might have expected alcohol to remove mainly weakly-bound dye from the sections and thus result in a higher measured affinity. It appears, however, that brief

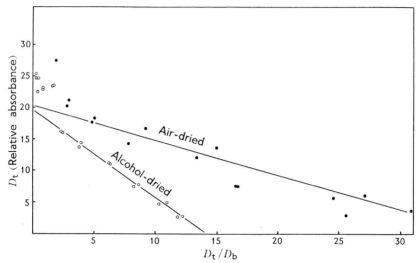

Fig. 6.3   The same data as in Fig. 6.2, in a Scatchard plot. The data fall satisfactorily onto straight lines, except that in concentrated dyebaths there is more uptake than expected (experimental points fall above the straight line).

alcohol dehydration removes relatively more dye from faintly than from heavily stained sections, and one should therefore avoid alcohol dehydration wherever possible in this type of work.

In Fig. 6.4 data are presented on a Scatchard plot for the uptake of Azure A by pancreatic cytoplasmic

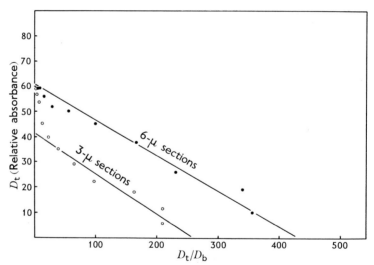

Fig. 6.4   Uptake of Azure A (3 μ or 6 μ sections, 4 days, veronal-acetate buffer pH 4·0, 20°C, airdried) by pancreatic cytoplasmic RNA; Scatchard plot.

RNA (formalin-fixed 3 μ or 6 μ paraffin sections, 20°C, veronal buffer pH 4·0, air-dried). The highest points in the 6 μ series are unreliable, as the absorbance was close to the maximum registrable by the photometer, but the same type of deviation from an ideal Langmuir isotherm is seen as was found for the staining of nuclei, except that here the deviation from the straight line starts in a somewhat weaker dyebath. The best straight lines through data from dyebaths of up to 1·81 mM/l. (i.e. omitting the highest three points in the 6 μ, and the highest four points in the 3 μ series), correspond to affinities of 5·135 and 5·086 kcal/mole for the 6 μ and 3 μ series respectively. This is a very satisfactory agreement, despite the random variation of individual experimental points, and shows that the method is capable of giving reasonably precise and reproducible results.

It is interesting to see that Azure A has, under the defined conditions, a higher affinity for cytoplasmic RNA than for nuclear chromatin, but speculations on the reason for this would be premature.

MISCELLANEOUS DATA RELEVANT TO THE AFFINITY OF DYES FOR HISTO-LOGICAL SUBSTRATES

Although many quantitative investigations of histological staining have been performed in the past, only a few have been carried out in such a way as to throw light on the affinity of dyes for histological substrates. Some of these investigations will now be discussed.

Uptake of lysochromes by fat

Meier[21] measured the partitioning of Sudan III, Sudan black B, and other fat 'stains' ('lysochromes' in Baker's 1958 terminology[1]) between fat and water–alcohol mixtures.

The proportion of dye in the fat phase increased with increasing molecular weight of the dye, decreasing temperature, or decreasing molecular weight or concentration of the alcohol in the mixture, and relatively more dye passed into the fat as the dye concentration was raised. Good straight lines were obtained with a log–log plot of equilibrium concentrations (i.e. a Freundlich isotherm was obeyed). Although Meier invoked surface adsorption phenomena to explain his findings, it seems to the present author that his interesting results could be due simply to a partitioning of dye between the phases, the dye being more aggregated in the dyebath than in the fat. Meier himself showed with light-scattering experiments that dye aggregates, which could increase to colloidal dimensions, existed in the dyebath. In the absence of quantitative information on the aggregation number of dye in the bath, one cannot (as discussed previously) calculate standard affinities from Meier's data.

Affinity of an acid dye for a histological substrate

Deitch[6] studied microphotometrically (with a non-scanning method) the uptake by sections of sea urchin ovary of the acid dye Naphthol yellow S (in 1 per cent acetic acid at pH 2·8, 20°C). Uptake increased with dyebath concentration to a plateau in about 10^{-3} per cent dye, but further uptake occurred in the strongest dyebaths, indicating the presence of additional weaker sites. Considering only the more strongly binding sites, an uptake approximately equal to half the plateau was found in 10^{-6} per cent dye. This corresponds to an affinity of about 12·4 kcal/mole, which is comparable with results for the dyeing of wool by monobasic acid dyes at 60°C.[31] Deitch found that after acetylation (2 hours 5 per cent acetic anhydride) there was less staining in all dyebaths, but the shape of the isotherm was not much altered, showing that while the number of available dyebinding sites was diminished their mean affinity was relatively unaffected.

Uptake of basic dyes by various substrates

Singer and Morrison[26] measured photometrically the absorbance of formalin-fixed fibrin films stained in methylene blue solutions buffered at pH 7·08. Doubling the residual dyebath concentration more than doubled the amount of dye bound, which as they note is contrary to the Langmuir isotherm, but their data can be shown to conform fairly closely to a Freundlich isotherm, giving a reasonably straight line on a log-log plot. This is quite unexpected, but might be due to fibrin dissolving out of their films into the less concentrated dyebaths.[30] In any case, Singer and Morrison's system appears insufficiently understood for affinity calculations on their data to be possible.

Carnes and Forker[4] measured microphotometrically the amount of Toluidine blue or Crystal violet taken up by cartilage from a dilute dyebath and found the concentration of dye in the tissue at equilibrium to be about 5000 times that in the bath. An affinity could be calculated from this result only by assuming, contrary to probability, that the distribution coefficient would remain constant over a range of dyebath concentrations.

Stacey and Wildy[28] estimated colorimetrically dye eluted from formalin-fixed HeLa cells which had been stained for 5 min in Methylene blue (phosphate–citrate buffer, pH 7·0). From their figure it appears that a plateau was approached in strong dyebaths, about 22·5/26 of the plateau value being taken up from a 0·1 per cent dyebath. This corresponds to an affinity of about 4·5 kcal/mole, but this result is unreliable, as Stacey and Wildy themselves found that 5 min was insufficient for equilibrium to be reached.

Szirmai and van der Linde[30] equilibrated unfixed frozen sections of thymus, or isolated thymic nuclei, with varying concentrations of Azure A, and measured the dye remaining in the supernatant after centrifugation. The amount of dye bound was estimated by difference, and was found to follow a Langmuir type of isotherm. With dye concentrations of less than about 0·2 M, however, a soluble dye–substrate complex remained in solution, vitiating the results obtained in weak dyebaths and rendering any calculation of affinity impossible.

Deitch[7] measured microphotometrically Methylene blue uptake by lymphocyte nuclei (air-dried smears, methanol-fixed, post-fixed in formalin, acetylated, aqueous dyebaths). Staining reached a plateau in $10^{-4·5}$ M dye. About half the plateau value was obtained in $10^{-5·25}$ M dye, corresponding to an affinity of approximately 7·05 kcal/mole.

The motility of supravitally stained bull spermatozoa has been measured by van Duijn.[8,9,10] Assuming that a decrease in motility was proportional to the amount of dye adsorbed, he was able to show that the uptake of Toluidine blue, Janus green, Rhodamine B, and Primuline from dilute dyebaths closely followed a Langmuir isotherm. Some data also fitted a Freundlich isotherm, presumably because the substrate was far from saturated. Van Duijn appears to have published his extensive data only in a graphical form from which calculations of standard affinities are almost impossible.

Uptake of an 'electron stain' by nucleic acid and other substances in vitro

Huxley and Zubay[19] measured the weight increase of various substances after exposure to 0·2 to 2·0 per cent (i.e. about 0·005 to 0·05 M) aqueous uranyl acetate. From a Scatchard re-plot of their data, affinities are obtainable of about 4·2, 2·75, and 2·3 kcal/mole for the uptake of the salt by purified DNA, nucleohistone and purified histone respectively, the affinity being somewhat higher in weaker solutions. They note that uranyl chloride or nitrate gave much weaker binding than did uranyl acetate, presumably because the acetate anion binds more firmly to the substrate than do the other anions, and thus contributes more to the total affinity of the salt. Zobel and Beer[32] state (without giving their data in detail) that salmon sperm DNA reacts with uranyl acetate (10^{-4} M to 10^{-3} M, pH 3·5) with an association constant of about 8×10^6. This corresponds to an affinity of about 9·25 kcal/mole, but they also found an additional looser binding at higher pH's or higher concentrations of uranyl ion, which may explain the discrepancy between their results and those of Huxley and Zubay.

CONCLUSION

Both experimental methods and theoretical foundations exist for the measurement of the affinity of dyes or 'electron stains' for histological substrates. Provided different workers express their results in a readily comparable form, or at least present the raw data from which standard affinities can be calculated, there seems to be no obstacle to the rapid accumulation of data which will make attainable some of the objectives outlined in the introduction to this article.

ACKNOWLEDGEMENTS

I wish to thank Dr. P. A. H. Wyatt of the Department of Chemistry, Sheffield University, for criticism of this manuscript. The work reported in this article was assisted by grants to the Department of Human Biology and Anatomy from the Medical Research Council and the Science Research Council.

REFERENCES

1. BAKER, J. R. 1958. *Principles of Biological Microtechnique*. London: Methuen & Co.
2. ——. 1960. *Cytological Techniques*. London: Methuen & Co.
3. —— and WILLIAMS, E. G. M. 1965. The use of methyl green as a histochemical reagent. *Quart. J. Micr. Sci.* **106**, 3–13
4. CARNES, W. H. and FORKER, B. R. 1956. Metachromasy of amyloid. *Lab. Invest.* **5**, 21–43
5. DEELEY, E. M. 1955. An integrating microdensitometer for biological cells. *J. Sci. Instruments* **32**, 263–7
6. DEITCH, A. D. 1955. Microspectrophotometric study of the binding of the anionic dye, Naphthol yellow S, by tissue sections and by purified proteins. *Lab. Invest.* **4**, 324–51
7. ——. 1964. A method for the cytophotometric estimation of nucleic acids using methylene blue. *J. Histochem. Cytochem.* **12**, 451–61
8. DUIJN, C. VAN. 1961. Photodynamic effects of vital staining with diazine green (janus green) on living bull spermatozoa. *Exp. Cell Res.* **25**, 120–30
9. ——. 1962. Toxic and photodynamic effects of toluidine blue on living bull spermatozoa. *Exp. Cell Res.* **26**, 373–81
10. ——. 1964. Effects of Rhodamine B and primuline on bull spermatozoa and their use for fluorimetric determination of live/dead ratios. *Mikroskopie* **19**, 75–87
11. EDSALL, J. T. and WYMAN, J. 1958. *Biophysical Chemistry*, Vol. 1. New York: Academic Press
12. EHRLICH, P. 1879. Über die specifischen Granulationen des Blutes. *Arch. Anat. Phys.* (Physiol. Abt.). Reprinted in *Collected Papers of Paul Ehrlich* (edited by F. Himmelweit). London: Pergamon Press, 1956
13. FISCHER, A. 1899. *Fixirung, Färbung und Bau des Protoplasmas*. Jena: Fischer Verlag
14. GIERKE, H. 1885. Färberei zu mikroskopischen Zwecken. Part 4. *Ztschr. f. wiss. Mikr.* **2**, 164–221
15. GILES, C. H. and MACEWAN, T. H. 1957. Classification of isotherm types for adsorption from solution. *Proc. Second Int. Congress of Surface Activity*, pp. 457–61. London: Butterworths
16. GOLDSTEIN, D. J. 1963. An approach to the thermodynamics of histological dyeing, illustrated by experiments with azure A. *Quart. J. micr. Sci.* **104**, 413–39
17. ——. 1965. A further note on the measurement of the affinity of a dye (Azure A) for histological substrates. *Quart. J. micr. Sci.* **106**, 299–306
18. GRAUMANN, W. 1961. Bestimmung der 'Stärke der Metachromotropie.' *Acta Histochem.* Suppl. **2**, 217–21
19. HUXLEY, H. E. and ZUBAY, G. 1961. Preferential staining of nucleic acid-containing structures for electron microscopy. *J. Biophys. Biochem. Cytol.* **11**, 273–96
20. MANN, G. 1902. *Physiological Histology*. Oxford: Clarendon Press
21. MEIER, W. 1959. Untersuchungen zur Theorie der Fettfärbung. *Z. wiss. Mikrosk.* **64**, 193–208
22. MÖLLENDORFF, W. VON and MÖLLENDORFF, M. VON. 1924. Untersuchungen zur Theorie der Färbung fixierter Präparate. III. *Ergeb. der Anat. u. Entwickl.–Geschichte* **25**, 1–66
23. MOORE, W. J. 1962. *Physical Chemistry*, 4th edition. London: Longmans
24. POLLISTER, A. W. and ORNSTEIN, L. 1959. The photometric chemical analysis of cells. In *Analytical Cytology*, 2nd edition (edited R. C. Mellors). New York: McGraw-Hill Book Co.
25. SCOTT, J. E. and DORLING, J. 1965. Differential staining of acid glycosaminoglycans (mucopolysaccharides) by Alcian Blue in salt solutions. *Histochemie* **5**, 221–33
26. SINGER, M. and MORRISON, P. R. 1948. The influence of pH, dye, and salt concentration on the dye binding of modified and unmodified fibrin. *J. Biol. Chem.* **175**, 133–45
27. SPICER, S. S. and MEYER, D. B. 1960. Histochemical differentiation of acid mucopolysaccharides by means of combined aldehyde fuchsin-alcian blue staining. *Am. J. Clin. Path.* **33**, 453–60
28. STACEY, R. S. and WILDY, P. 1960. Quantitative studies on the absorption and elution of methylene blue. *Exp. Cell Res.* **20**, 98–115
29. SWIFT, H. and RASCH, E. 1956. Microphotometry with visible light. In *Physical Techniques in Biological Research*, Volume 3 (edited G. Oster and A. W. Pollister). New York: Academic Press
30. SZIRMAI, J. A. and LINDE, P. C. VAN DER. 1963. Studies on the dyebinding and metachomasia of thymus nuclei. *Histochemie* **3**, 233–48
31. VICKERSTAFF, T. 1954. *The Physical Chemistry of Dyeing*, 2nd edition. London: Oliver and Boyd
32. ZOBEL, C. R. and BEER, M. 1961. Electron stains. I. Chemical studies on the interaction of DNA with uranyl salts. *J. Biophys. Biochem. Cytol.* **10**, 335–51

Chapter 7

The Biological Activity
of Neutral Red

W. S. MORGAN

In earlier investigations of the exocrine pancreas cells of the mouse,[1] microscopic examination of fresh tissue in normal saline revealed that after the subcutaneous administration of an optimum amount of neutral red, the cytoplasm first became diffusely pink-stained and at about $\frac{1}{2}$ hr discrete dye granules measuring $0.7\ \mu$ appeared in a supra- and para-nuclear position between the zymogen granules and the basement membrane. During the course of the next several hours these dye granules increased in size and appeared as aggregates measuring 2 to 3 μ in diameter. The aggregates reached a maximum development at about 7 hr after which they gradually diminished in size and at approximately 15 hr had disappeared, leaving the cell normal in appearance. This sequence of events was given the name 'neutral red granule cycle' and is depicted in Fig. 7.1, A.

'GOLGI TECHNIQUES' AND NEUTRAL RED GRANULE FORMATION

The application to pancreas of the Aoyama (silver) and the Mann-Kopsch (osmium) techniques (two classical methods of demonstrating the so-called 'Golgi Apparatus'), at similarly-timed intervals after injection of neutral red disclosed some interesting differences from normal tissue.[2] These are shown in Fig. 7.1, B.

Whereas the preparations of pancreas from uninjected animals revealed an argentophil network (Aoyama) or osmiophil bodies (Mann-Kopsch) generally amongst the zymogen granules, those of tissue from animals given neutral red revealed that the metal was deposited on the dye granules at the expense of the Golgi apparatus. Perhaps of even greater interest was the observation that in the later stages of the cycle *all* of the argentophil or osmiophil component of the exocrine cell was confined to the neutral red granules. At the end of the cycle when the dye has disappeared from the cell, the Golgi figures as demonstrated by these classical procedures again assumed their usual appearance.

CELLULAR BASIPHILIA AND CAPACITY FOR NEUTRAL RED GRANULE FORMATION

In related studies[3] it was found that after fixation of tissue containing neutral red granules, in formalin or Helly's solution, paraffin embedding, and staining with hematoxylin and eosin, the dye granules appeared as darkly basiphilic bodies surrounded by a vacuolated zone. Since this vacuolated ring was not present in Aoyama or Sudan black preparations, procedures in which lipid elements are identified or preserved, it was concluded that the neutral red granules were partly lipid and that this component was dissolved out in the course of routine staining methods in which alcohols and other lipid solvents were employed.

The observation that in fixed preparations the neutral red granules were basiphilic suggested the possibility of an association between the dye and ribonucleic acid. In view of the importance of the latter in protein synthesis, experiments were conducted to determine whether when the cell was involved in the process of neutral red granule formation, and much cytoplasmic RNA apparently thus engaged, protein synthesis was diminished. Studies of zymogen granule formation as well as radioactive amino acid incorporation experiments failed to demonstrate any inhibitory effects of neutral red on protein synthetic mechanisms.[4]

The basiphilic nature of the neutral red granule in fixed preparations is as yet unclear but remains one of the most intriguing features of this induced cytoplasmic body. This is particularly true since the

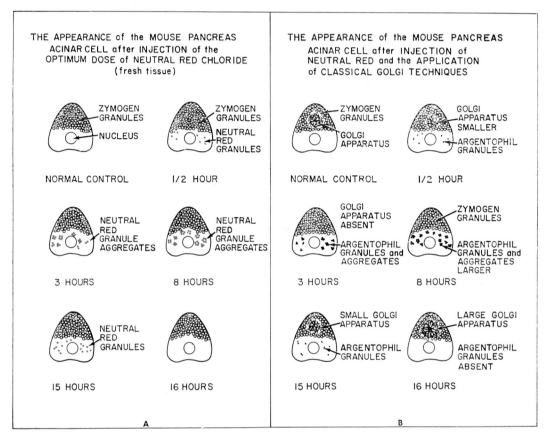

Fig. 7.1 A Shows the 'neutral red granule cycle' with formation of the neutral red granules and aggregates after the subcutaneous administration of the optimum dose of neutral red as visualized in fresh preparations.[1]

B Shows the Golgi apparatus in control tissue as revealed by classical Golgi techniques, the decrease in size of the Golgi apparatus concomitant with the appearance of the argentophil counterparts of the neutral red granules, and finally the reappearance of the Golgi apparatus when the argentophil granules are no longer present.

affinity of a tissue for neutral red is related to its general basiphilia. Thus, the pancreas exocrine cell, probably the most basiphilic cell in the animal body, has the greatest affinity, followed next by the salivary gland, liver, and other tissues in decreasing order of basiphilia.

Along this line, it has been of interest to note that the converse seems to be true, namely that eosinophilic cells have lowered capacity, or lack totally the capacity for neutral red granule formation. This is demonstrated impressively in the salivary gland of the mouse which is composed of mucous, serous, and mixed muco-serous lobules. In the basiphilic, mucus-forming cells neutral red granule formation is marked, while in the adjacent serous glands there is none. A similar comparison can be made in the mouse stomach, where dye granules are seen in the zymogenic cells but not in the acidophilic parietal cells.

CONCERNING THE FATE OF NEUTRAL RED IN THE BODY

Earlier studies disclosed that after the injection of the optimal dose of neutral red in the mouse, about 25 per cent of the dye is excreted in the urine and this occurs during the first 48 hr. More recent spectrophotometric investigations support the view that while temporarily in the form of neutral red granules in the pancreas and many other tissues, the dye molecule is not altered, but that chemical modification associated with elimination of the remaining dye occurs chiefly in the liver and to a less extent in the kidney. Although alcohol-acetic acid extracts of pancreas throughout the course of the neutral red granule cycle continually exhibit the 540 mμ absorption maximum of typical neutral red, similar extracts of liver and kidney late after dye injection reveal a 490 mμ absorption peak of modified neutral red which has apparently been reduced and is now fluorescent. The implication is that while serum levels of the absorbed dye are high, the dye remains unchanged in pancreas, but when serum levels fall, the dye passes from pancreas to the blood to be carried to the liver for final disposition. All of the dye in the intestinal excreta, and some found in the urine late after injection, is in the modified form.

HISTOLOGICAL STUDIES OF THE LIVER

Interest in the metabolic fate of the dye in the animal body has caused us recently to examine the nature of neutral red granule formation in the liver. The methods employed have been largely those used earlier in studies of the pancreas; however, in regard to the demonstration of the Golgi substance we have limited ourselves to the Aoyama procedure which served us well in the earlier investigations. In this study, mice of equal weight were injected with the optimum dose of neutral red and sacrificed at hourly intervals thereafter.

Samples of liver from control mice carried through the Aoyama procedure disclosed a characteristic and reproducible pattern of argentophilia as shown in Fig. 7.2, A and B. Argentophilia was not observed in the hepatocytes but was confined to the Kupffer cells and the sinusoidal lining. The latter may actually represent a further expression of the argentophilia of the Kupffer cells since it is known that the processes of these cells extend for considerable distances along the sinusoids. The Kupffer cells and the argentophilia are largely concentrated in the midzonal region of the lobule with slightly less in the central areas and very little in the portal zones.

In comparable material from animals given neutral red, within 15 to 30 min after injection, small discrete argentophil granules are observed in the hepatocytes close to the sinusoidal membrane. With time there is not much increase in size of the granules but they become more concentrated in the midportion of the cell (Fig. 7.2, D). Their location in the hepatic lobule is primarily in the mid zonal and central areas. Associated with an increase in the number of granules is a progressive decrease in argentophilia of the Kupffer cells (Fig. 7.2, C and D).

ELECTRON MICROSCOPIC STUDIES OF NEUTRAL RED GRANULES

Byrne[5] conducted electron microscope studies of the pancreas of mice given neutral red and concluded that the dye granules were lysosomal in nature. In histochemical investigations, she added further support for this conclusion by demonstrating that the granules gave a positive reaction for acid phosphatase, a common chemical constituent of lysosomes.[6] In this laboratory, Alousi[7] has repeated Byrne's electron microscope studies and has confirmed her findings. Alousi has also studied the dye granules of liver, and observes no significant ultrastructural differences from the granules seen in pancreatic tissue (Fig. 7.3, A and B).

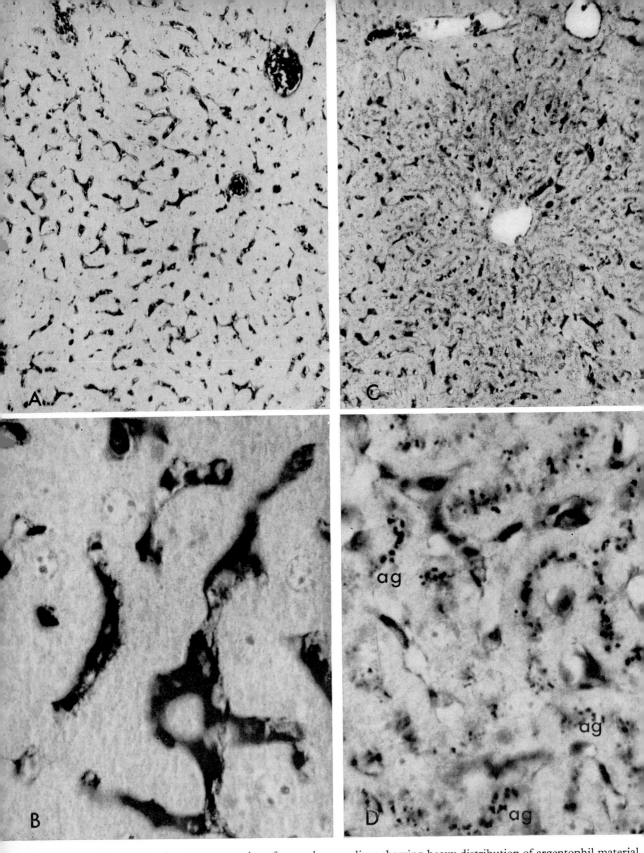

Fig. 7.2 A Aoyama preparation of normal mouse liver showing heavy distribution of argentophil material in Kupffer cells and along sinusoids, predominantly in mid-zonal and central areas of the liver lobule. ×200; B Same preparation. ×400; C Aoyama preparation of liver from mouse injected with neutral red 3 hr earlier. Note the appearance of fine black granules in the hepatic cells. In fresh tissue red dye granules are seen in this same distribution. Note also the decreased argentophilia of the Kupffer cells and sinusoids. ×200. AG = argentophil granule; D Same preparation. ×400.

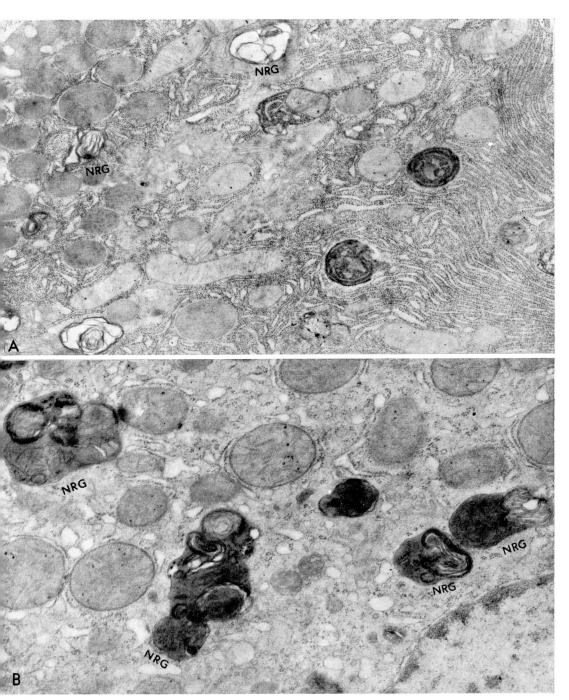

Fig. 7.3 A Portion of pancreatic acinar cell 3 hr after neutral red injection. Interspersed among the zymogen granules and endoplasmic reticulum are autophagic vacuoles revealing partially degraded segments of rough endoplasmic reticulum. One vacuole includes mitochondria. Several myelin figures are seen scattered throughout the field. × 29,000. NRG = neutral red granule; B Portion of hepatic parenchymal cell 1 hr after neutral red injection. Within the cytoplasm there are scattered autophagic vacuoles. The one far to the left illustrates partially degraded mitochondria. Myelin figures are encompassed in several vacuoles. × 23,000. NRG = neutral red granule.

DISCUSSION

The studies described above indicate certain differences between pancreas and liver in the mechanism of handling neutral red. In the pancreas, the substance capable of segregating the dye appears to derive from the Golgi apparatus of the pancreas exocrine cell itself, whereas in the liver the dye is apparently first segregated in droplet form in the Kupffer cells after which the droplets move into the hepatic cells. This apparent movement of the dye granules from one cell to another is certainly of interest. In both the pancreas and liver the formation of the neutral red granules involves an argentophil or osmiophil material as shown by the classical techniques employed to demonstrate the Golgi apparatus. When these histological methods are applied to tissue in the advanced stages of neutral red granule formation, they no longer demonstrate a Golgi apparatus. A discrepancy is encountered here, however, in the fact that in electron micrographs of both pancreas and liver containing neutral red granules, typical Golgi membranes are also visualized in other parts of the cell. This discrepancy can be resolved if one considers that the Golgi membranes, as visualized electron microscopically, may serve as a repository for a lipid material which is discharged from the Golgi apparatus in response to the entrance of neutral red into the cell. Conclusions as to whether or not the Golgi apparatus manufactures this lipid material cannot be drawn from these studies.

The Golgi complex demonstrated electron microscopically in tissue containing neutral red granules, but not shown by the classical Golgi silver or osmium impregnation techniques, may represent Golgi apparatus depleted of the postulated lipid component which accounts normally for the argentophilia or osmiophilia of this body in control tissue. In tissue from animals injected with neutral red, that material has left the Golgi apparatus and is now engaged with neutral red.

Similar changes have been observed under other circumstances. The exocrine pancreas reacts in a similar manner when substances other than neutral red are injected. One of these is acridine orange. There is even a human disease counterpart for this biological process in the condition known as hemochromatosis. Patients with this disease have an abnormally high level of serum iron, and granules containing this metal are frequently deposited in profusion in many tissues, especially the liver and pancreas, to the extent that cellular function may be diminished, or the cell may die. The author has applied the Aoyama technique to liver and pancreas taken from several patients with hemochromatosis. The morphology of these iron-containing granules as demonstrated by this method is strikingly similar to that of the dye granules in the tissues of animals injected with neutral red.

The conclusion to be drawn tentatively from these findings is that pancreas, liver, and most other tissue cells, contain an intracellular material which can be called forth to segregate and thus guard the cell from the potentially harmful effects of foreign substances, which, because of appropriate molecular weight and other physico-chemical properties, can enter the cell. If true, this must serve as an important protective mechanism. One disturbing aspect here, however, is the observation in electron micrographs that the neutral red granules are not purely sacs filled with segregated dye, but that they may also contain mitochondria and bits of endoplasmic reticulum or other cell constituents. These findings suggest that there is much more to this induction of dye granule formation than simple dye segregation and point the way for further work. In what way, if any, lysosome formation may be related to the cell's metabolic handling of the dye is not known at present.

In conclusion, neutral red has long been employed by cytologists in their explorations of cellular physiology, where its colour and relatively harmless effect on cells have been major advantages in its use. These properties are no less valuable to modern investigators, applying the newer approaches of biochemistry and electron microscopy to the study of the cell.

REFERENCES

1. MORGAN, W. S. 1953. Cytological studies of the acinar cells of the pancreas of the mouse. I. The formation of neutral red granules. *Quart. J. Micr. Sci.* **94**, 141–53
2. ——. 1953. Cytological studies on the acinar cells of the pancreas of the mouse. II. The Argentophil, Osmiophil, and Sudanophil Substance. *Quart. J. Micr. Sci.* **94**, 269–79

3. ——. 1958. Affinity of neutral red for cells with basophilic cytoplasm. *Nature Lond.* **181**, 1132–3
4. ——. 1958. A cytoplasmic neutral red-ribonucleic acid component not influencing protein synthesis. *Exper. Cell Res.* **14**, 435–9
5. BYRNE, J. M. 1964. An electron microscopical study of neutral red granules in mouse exocrine pancreas. *Quart. J. Micr. Sci.* **105**, 219–25
6. ——. 1964. Acid phosphatase activity in the neutral red granules of mouse exocrine pancreas cells. *Quart. J. Micr. Sci.* **105**, 343–8
7. ALOUSI, M., STENGER, R. J., and MORGAN, W. S. (unpublished studies)

Chapter 8

Cellular Response to Vital Staining

JENNIFER M. GREGORY (née BYRNE)

Vital staining provides us with an instance of an essentially light microscopical technique whose effect on the cell is best studied by means of the electron microscope. Having established some facts about what structures in the cell at an electron microscopical level are altered during vital staining, and speculated a little about why they are altered, one can then use vital stains with the light microscope with a greater degree of confidence that they are displaying something meaningful than has been possible before.

The first use of a synthetic dye to colour inclusions within the living cell was made in 1887, when Brandt coloured lipid droplets in the cytoplasm of the Protozoon *Actinosphaerium* with Bismarck brown. An independent discovery was made in 1881 by Certes, who showed that mouthless and non-phagocytic Protozoa such as *Opalina* can take up quinoleine blue from solution. He also used Bismarck brown. In the 1880s a great deal of work was done with vital stains, particularly by Pfeffer, who was the first to investigate the vital staining of plant cells, and by Ehrlich on animal cells. Ehrlich introduced the dye neutral red into the field of vital staining.

There are two types of vital colouring, dependent on two rather different phenomena. Certain phagocytic cells will take up coloured particles as part of their ordinary function of eliminating foreign material. The coloured material need not be a dye so long as its particles are sufficiently small to enable it to be taken up by the cell. The only dyes that are suitable for use in this way are those that have a strong tendency to flocculate into particles of colloidal dimensions, and this in general means acid (anionic) dyes, such as the azo-dyes. The coloured particles that one sees in the phagocytic cells are not pre-existing cell inclusions stained by the dye, but ingested dye particles. The dye is effectively outside the cell's cytoplasm and does not affect its metabolism.

The other type of vital staining is the colouring of pre-existing cell inclusions or the formation of new inclusions within a cell by dyes that can penetrate the living cell from dilute solutions. All the dyes that produce this type of colouring are basic (cationic) dyes. It is the response of the cell to the presence within it of this type of dye that concerns us here.

Combined light and electron microscopical studies to see how the cell responds to vital staining have been carried out on the uptake of neutral red by mouse exocrine pancreas cells,[5] and of acridine orange by HeLa cells.[38, 39]

The appearance of neutral red granules in mouse exocrine pancreas cells as seen with the light microscope was studied by Morgan,[27, 28] who worked out an optimal dosage of the dye, which produced the granules without producing any marked pathological changes in the cells. Under these conditions, pancreas cells from mice injected with neutral red show, under the light microscope, a cycle of formation of neutral red granules, followed by their gradual disappearance from the cell. At first[5, 27] the cytoplasm is colourless or very pale pink and no granules can be seen, but an hour after injection of the optimal dose of neutral red, a few separate dark-red granules, 0.5–0.7 μ in diameter, can be seen in the cytoplasm between the nucleus and the zymogen granules. From $1\frac{1}{2}$ to 2 hr after injection the number of granules increases, and the granules tend to aggregate into clusters, although they do not fuse. The aggregates are confined to the supranuclear region of the cell. At this time the aggregates measure up to $1.8\ \mu$ in greatest dimension. The number of granules and the size of the clusters increases up to 8 hr after injection of the dye, and at this time each aggregate measures up to $3.5\ \mu$ in greatest dimension. After this time the neutral red gradually disappears.

What are the neutral red granules? Are they a pre-existing cell inclusion that is stained, swollen and made visible by the dye, or do they arise in the cytoplasm as a result of the action of the dye? Morgan concluded that the neutral red granules were not pre-existing vacuoles that were subsequently stained, but were new formations produced by the presence of neutral red. This view is largely supported by the electron microscopical work.

In electron micrographs of pancreas from mice injected with neutral red, inclusions are found which correspond in size and distribution to the neutral red granules seen with the light microscope. Osmiophil inclusions are seen near the nucleus, often lying between the nucleus and the zymogen granules. The simplest form of these inclusions consists of an ovoid osmiophil body $0.28–0.68\ \mu$ in length, containing one or more electron-lucent vacuoles. The electron-dense material is not homogeneous. This type of inclusion is found 1 hr after injection of the dye. After longer periods of exposure to the dye, the inclusions become much more varied and complicated in structure, and tend to be larger (see Figs. 8.1 and 8.2). They are still confined to the cytoplasm near the luminal end of the nucleus. The complex inclusions commonly consist of an aggregate comprising electron-lucent vacuoles rimmed with electron-dense material, a mass of membraneous material, and often a large vacuole up to $0.7\ \mu$ in diameter (see Fig. 8.1). The whole is enclosed in a membrane and can measure up to $2\ \mu$ in total length. In other cases (see Fig. 8.2) vesicles occur, wholly or partly filled with a mass of electron-dense material, including what appears to be membraneous material. In some cases parallel membranes are seen, each of which can be resolved into dark–light–dark bands. This structure, and the osmiophilia, suggest phospholipid. Morgan, with the light microscope, found that the neutral red granules give a positive reaction with the acid haematein test for phospholipid.[28] Three or more inclusions may occur together, the whole complex measuring up to $4\ \mu$ in greatest dimension.

The inclusions seen under the electron microscope are of about the same size as the neutral red granules of the light microscope, and they occur in precisely the same region of the cell. They have never been found in control tissue, either from normal animals or from animals injected with distilled water. It would thus seem certain that the inclusions seen with the electron microscope are the neutral red granules.

It must be emphasized that no changes in structure of any other cytoplasmic inclusion were seen. The nucleus, mitochondria, zymogen granules, and cytoplasmic membranes appeared to be the same in both experimental and control tissue. The inclusions that represent, under the electron microscope, the neutral red granules of light microscopy do not correspond to anything known in the ultrastructure of normal mouse exocrine pancreas cells. This supports Morgan's view that the neutral red granules do not exist before vital staining takes place. The neutral red granules could be some normal inclusion pathologically altered by the presence of the dye, and it has been suggested, for instance, that they are transformed mitochondria.[43] But in tissue from animals injected with the optimal dose of neutral red, intermediate forms between normal mitochondria and neutral red granules are not seen, so that there seems to be little justification for this view.

Tanaka[41] obtained remarkably similar pictures of neutral red granules in mouse lymph node cells. After subcutaneous injection, the dye becomes segregated in what Tanaka calls 'segresomes'. As seen under the electron microscope these are characterized by a single limiting membrane, a diffuse internal matrix, and also by the occasional appearance of membraneous masses very similar to those found in the neutral red granules of pancreas cells. Tanaka supposes these masses to be lipo-protein.

The intracellular accumulation of acridine orange by HeLa cells, investigated by Robbins and Marcus,[38] follows a very similar pattern to that of neutral red by pancreas cells. HeLa cells grown in dilute solutions of acridine orange accumulate the dye in particles, which cap the nucleus on one side of the long axis of the cell. These particles increase in size and number to a maximum reached after

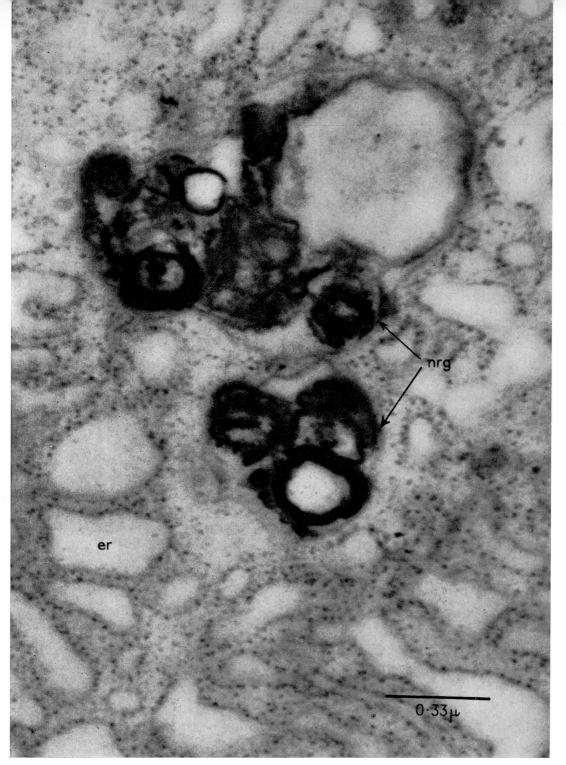

Fig. 8.1 Neutral red granules in pancreas from mouse injected with neutral red (0·0002 g neutral red per gram of mouse body weight, administered as a 1 per cent aqueous solution) and killed after 2 hr. The material was fixed in buffered osmium tetroxide, embedded in Araldite and stained with lead tartrate.
    er, endoplasmic reticulum. nrg, neutral red granule.

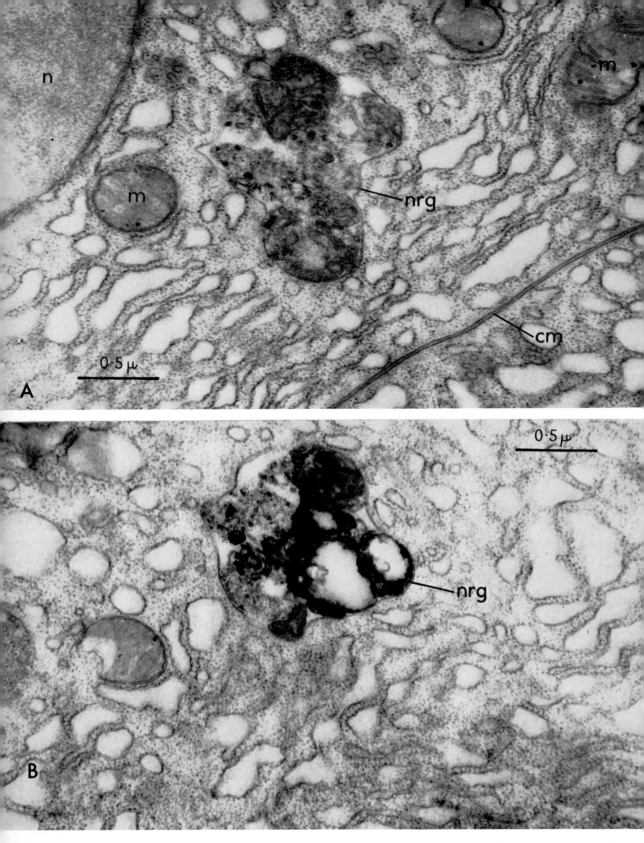

Fig. 8.2 Neutral red granules in pancreas from mice injected with neutral red (0·0002 g neutral red per gram of mouse body weight, administered as a 1 per cent aqueous solution) and killed after 4 hr. The material was fixed in buffered osmium tetroxide, embedded in Araldite and stained with lead tartrate.

cm, cell membrane.   nrg, neutral red granule.
n, nucleus.   m, mitochondrion.

2 days in the dye solution. At this time the particles measure 1–3 μ in diameter. In contrast, HeLa cells grown in high concentrations of the dye show a diffuse reddish stain. Acridine orange particles are not visible in most cells. But if such cells are removed from the dye solution and placed in dye-free culture medium, then, after 2–3 hr most of the dye within the cell is segregated into acridine orange particles.

Electron microscopy of particles in HeLa cells grown for 2 days in dilute acridine orange gives pictures remarkably similar to those of neutral red granules in pancreas.[39] Expanded vesicles surrounded by much thickened osmiophil walls, composed of lamellae with a periodicity of 38 Å are found. It is concluded that these are largely phospholipid. Again, these workers find no recognizable disorganization or degeneration of structure of the nucleus, mitochondria, Golgi vesicles or endoplasmic reticulum in the cells treated with acridine orange, but they equate the acridine orange particles with transformed multi-vesicular bodies. Apparent intermediate forms between the typical multivesicular bodies of the HeLa cell and the fully formed acridine orange particles are seen. Even after 7 days in acridine orange the cytoplasmic inclusions undergo no major additional transformations. The changes that are seen are a matter of degree rather than kind. The walls of the expanded vesicles show increased thickness, and fewer intermediate forms of multivesicular bodies are seen, although unaltered multivesicular bodies can still be found.

Electron micrographs of HeLa cells exposed to high concentrations of dye solution and then allowed to 'recover' in dye-free medium show a different result. Large numbers of electron-lucent vesicles with no thickening of their membraneous walls and little or no internal structure are seen.

In view of the remarkable similarity of structure of neutral red granules and acridine orange particles revealed by the electron microscope it is perhaps significant that Morgan[29] found acridine orange to be the only one of several dyes of similar molecular weight to neutral red to give the same results in the vital staining of pancreas.

The neutral red granules of mouse exocrine pancreas and the acridine orange particles of HeLa cells show considerable morphological resemblance to the lysosomes[14,34] of various cells, particularly those of liver and kidney cells.[11,31,33] The closest resemblance is to the lysosomal derivatives 'autophagic vesicles', 'cytolysomes', 'residual bodies', or 'phagosomes' described in various tissues.[1,12,13,35] But the identification of a lysosome depends on enzymatic and not morphological criteria.[11,14,32] One of the enzymes characteristic of lysosomes, namely acid phosphatase, lends itself to visual identification under the electron microscope by means of the Gomori test, and the presence of acid phosphatase activity is taken to be indicative of a lysosome. Both the neutral red granules of mouse exocrine pancreas[6] and the acridine orange particles of HeLa cells[39] show acid phosphatase activity, suggesting that they are lysosomes or are derived from the lysosomes. Lysosomes are not normally seen in pancreatic tissue, but it may be that, as a response to the presence of the dye, the number within a cell is greatly increased. Novikoff[32] has suggested a developmental or functional inter-relationship between the Golgi apparatus and the lysosomes, and in view of this it is perhaps significant that the neutral red granules of mouse exocrine pancreas are confined to that part of the cell near the Golgi apparatus.

Further instances are known of the presence of acid phosphatase activity in cell inclusions that can be stained in life. The granules in the cytoplasm of cultured astrocytes and fibroblasts, which stain in life with neutral red, methylene blue, toluidine blue, and azure B, show acid and alkaline phosphatase, and esterase activity.[36,37] It is concluded that the granules are lysosomes. In neurons of *Helix aspersa* acid phosphatase activity is found in the cortices of granules known to take up vital stains.[8,22,26] Also, granules which take up neutral red in the tentacular cells of *H. aspersa* show acid phosphatase activity.[23,24] Koenig[15] finds that the glycolipoprotein granules in neural cytoplasm stain with neutral red and methylene blue, and suggests that these granules are lysosomes on the grounds of their enzyme content. Koenig[16] also finds inclusions in kidney, liver, spleen, small intestine, pancreas, heart, and central nervous system cells which colour rapidly in life with neutral red, methylene blue, and azure B. These granules closely resemble in their size, shape, number, and distribution, granules that show acid phosphatase activity. More conclusively, vitally stained rat neurons and renal nephron cells[19] tested for acid phosphatase activity, show such activity in the neutral red droplets. Koenig has also shown that hepatic lysosomes can be stained in life with acridine orange.[17] Again, Brachet[2] suggests that it is likely that lysosomes are identical with vacuoles that can be stained in life with basic dyes. Beaufay (quoted by de Duve[11,12,13]) has shown that in fractionated vitally stained liver cells part of the dye is retained in the particulate fraction, the highest concentration being in the fraction richest in lysosomes. Similarly, Koenig[17,19] finds that lysosomes are stained *in vitro* in particulate fractions by neutral red, methylene blue, azure B, and acridine orange. Neutral red stains the granules of macrophages and polymorpho-nuclear leucocytes,[9] which also shelter under the general heading 'lysosomes' on the grounds of their enzyme complement. Stretching the point somewhat, most Protozoan food vacuoles reveal the properties ascribed as characteristic of lysosomes[30] and they often stain with vital dyes. It is also possible that the neutral red granules of amoebae,[4] the formation of which is induced by the presence of the dye, are identical with granules which have acid phosphatase activity.[30]

Lysosomes are known to act as intracellular phagocytic organelles. Cells that are forced to ingest large amounts of foreign material for the breakdown of which they are not equipped, tend to accumulate such substances in their lysosomes.[13] The lysosomes of the parenchymal cells and Kupffer cells of rat liver[35,42] become enormously enlarged by the detergent Triton WR-1339. Enlargement of hepatic lysosomes is seen in tissue from animals treated with dextran[10] or with poly-vinyl pyrrolidone, obviously due to accumulation of the injected substance.

Vital dyes may be treated in the same way. In their role as intracellular phagocytic organelles the lysosomes may accumulate dye within themselves, becoming enlarged in the process. But the cell's response to the presence within it of a vital stain differs according to the concentration of the dye, and consequently the speed with which segregation of the dye into particles must take place if the cell is not to be killed. The rapid accumulation of acridine orange into particles in HeLa cells after exposure to a relatively high concentration of dye, followed by a period in dye-free medium, which leads to the forma-tion of dilated multivesicular bodies without thickened walls or internal structure is probably the result of a process very similar to that by which detergent is accumulated. The absence of any fine structure to which the dye could bind, and the enlargement of the multivesicular bodies may imply that the dye is free and therefore either transported against a steep concentration gradient or taken up by intracellular phagocytosis. Robbins and others have shown that this type of segregation of acridine orange into particles is energy dependent. Starved HeLa cells stained with high concentrations of acridine orange segregate the dye into particles only if incubated in dye-free medium containing glucose. Omission of the glucose inhibits the formation of acridine orange particles and leads to complete cellular degeneration.

But this is the response of the cell to an overdose of dye. When the HeLa cell is allowed to grow in dilute acridine orange, or when mice are injected with the optimal dose of neutral red, then we see under the electron microscope, not 'empty' dye granules, but granules with a large accumulation of mem-braneous material—presumably phospholipid.

Basic dyes stain phospholipid *in vitro*[3] and it was concluded that the phosphoric acid radicle of the phospholipid would seem to provide a site for such a reaction. There are in the literature a number of coincidences of cell inclusions that stain vitally and are wholly or partly composed of phospholipid that is not necessarily structural, but is free to react. For instance, in the sympathetic neurons of rabbit it is the 'lipochondria' that stain vitally with neutral red[7] and again in the motor neurons of the thoracic ganglia of *Locusta* the 'lipochondria' stain in life with various vital dyes.[40] In the neurons of *H. aspersa* the globules that are easily stained in life by a number of vital stains show a large amount of phospho-lipid[8] and in the oocytes of *Limnea stagnalis* the inclusions which stain in life contain phospholipid and possibly cerebroside.[25] This correlation between 'free' phospholipid content and the ability to stain with vital dyes suggests that, if a dye can penetrate the cell in life, it will stain by combination with the phospholipid. This same conclusion has been reached by Robbins and his colleagues.[39] They find that in a previously unstained HeLa cell the initial segregation of acridine orange into particles is a passive process—occurring rapidly at $3°C$. They suggest that this phase of staining represents the binding of the dye by the phospholipid of the membrane of the multivesicular body. They point out the structural uniqueness suggested by the preferential binding of the dye to it. Koenig also suggests a salt-type linkage between basic vital dyes and glycolipoprotein.[16,19]

As well as basic vital stains, the lysosomes of various tissues accumulate metallic cations,[16] including

Fe^{3+}, Ag^{+}, Pb^{2+}, Cu^{2+} and Hg^{2+}. Other cations such as phenothiazine and reserpine are also taken up by neuronal, glial, perivascular, liver, and kidney lysosomes.[18] Again, two tetrazolium salts, neotetrazolium and nitro-blue tetrazolium, accumulate in the lysosomes of a number of tissues in rat, and undergo reduction to their formazans.[20] These salts do not accumulate in the lysosomes *in vitro*. Such accumulation is accompanied by changes in the number and size of the lysosomes and by changes in their fine structure.

In support of the electrostatic binding of basic dyes, the staining of lysosomes by acridine orange is inhibited by many of the cations taken up by lysosomes. Such staining is inhibited by Ag^{+}, Pb^{2+}, and Cu^{2+}, while chlorpromazine, histone, protamine, Na^{+}, and K^{+} are weakly inhibitory. Lysosomal staining by acridine orange also fails to occur at pH 5·5 or less.[19] These findings strongly suggest that the binding of basic vital dyes depends on electrostatic forces between the cationic dyes and anionic binding sites. As has been suggested earlier, Koenig supposes these binding sites to be the phosphoryl moiety of the phospholipid. In support of this it is found that lysosomes treated with phospholipase C lose their ability to bind basic dyes.[21]

The amount of membraneous material in electron micrographs of neutral red granules and acridine orange particles suggests that the dye not only combines with phospholipid membranes already there, but in some way induces the cell to make available fresh binding sites for the dye. Robbins and co-workers find that the formation of acridine orange particles with large accumulations of phospholipid in HeLa cells is energy dependent.

The segregation of the dye into discrete drops or granules against a concentration gradient either by active transport or by phagocytosis would be helped by the removal of the dye by combination with phospholipid. So it is tempting to suggest that the increase in phospholipid is a cellular response to increase the efficiency of dye segregation, although it must be recognized that the accumulation of phospholipid may be due to the disruption in some way by the dye of the metabolic pathways that normally deal with phospholipid, the resulting accumulations of phospholipid quite coincidentally combining with the cause of the disruption.

The apparently widely occurring staining of lysosomes by vital dyes may restore vital colouring as a useful technique for light microscopical cytology. If vital stains are accumulated in a way that is typical of the cell's response to, and treatment of foreign substances, then, being coloured, they could provide a useful means of labelling lysosomes in living cells for investigating the way in which substances are accumulated in them, and finally expelled from the cell.

REFERENCES

1. BEHNKE, O. 1963. Demonstration of acid phosphatase-containing granules and cytoplasmic bodies in the epithelium of foetal rat duodenum during certain stages of differentiation. *J. Cell Biol.* **18**, 251
2. BRACHET, J. 1957. *Biochemical Cytology*, p. 52. New York: Academic Press
3. BYRNE, J. M. 1962. The uptake of dyes by extracted phospholipids and cerebrosides. *Quart. J. micr. Sci.* **103**, 47
4. ——. 1963. The vital staining of *Amoeba proteus*. *Quart. J. micr. Sci.* **104**, 445
5. ——. 1964. An electron microscopical study of neutral red granules in mouse exocrine pancreas. *Quart. J. micr. Sci.* **105**, 219
6. ——. 1964. Acid phosphatase activity in the neutral red granules of mouse exocrine pancreas cells. *Quart. J. micr. Sci.* **105**, 343
7. CASSELMAN, W. G. B. and BAKER, J. R. 1955. The cytoplasmic inclusions of a mammalian sympathetic neurone: a histochemical study. *Quart. J. micr. Sci.* **96**, 49
8. CHOU, J. T. Y. 1957. The cytoplasmic inclusions of the neurones of *Helix aspersa* and *Limnea stagnalis*. *Quart. J. micr. Sci.* **98**, 47
9. COHN, Z. A., HIRSCH, J. G., and WIENER, E. 1963. The cytoplasmic granules of phagocytic cells and the degradation of bacteria. In Ciba Foundation Symposium, *Lysosomes* (edited by A. V. S. de Reuck and M. P. Cameron). London: Churchill
10. DAEMS, W. T. 1962. Distribution of acid phosphatase in liver tissue of dextran-injected mice, as observed with the electron microscope. In *5th International Congress for Electron Microscopy*, Vol. 2, appendix, VV, 12. New York: Academic Press

11. DUVE, C. DE. 1959. Lysosomes, a new group of cytoplasmic particles. In *Subcellular particles* (edited by T. Hayashi). New York: Ronald Press Co.

12. ——. 1963. The lysosome. *Scientific American*, **208**, no. 5, 64

13. ——. 1963. The lysosome concept. In Ciba Foundation Symposium, *Lysosomes* (edited by A. V. S. de Reuck and M. P. Cameron). London: Churchill

14. ——, PRESSMAN, B. C., GIANETTO, R., WATTIAUX, R., and APPELMANS, F. 1955. Tissue fractionation studies. 6. Intracellular distribution patterns of enzymes in rat liver tissue. *Biochem. J.* **60**, 604

15. KOENIG, H. 1962. Histological distribution of brain gangliosides: lysosomes as glycolipoprotein granules. *Nature, Lond.* **195**, 782

16. ——. 1963. Intravital staining of lysosomes by basic dyes and metallic ions. *J. Histochem. Cytochem.* **11**, 120

17. ——. 1963. Vital staining of lysosomes by acridine orange. *J. Cell Biol.* **19**, 87A

18. ——. 1963. Accumulation of chlorpromazine and reserpine in brain lysosomes. *Neurology* **13**, 365

19. ——. 1965. The staining of lysosomes by basic dyes. *J. Histochem. Cytochem.* **13**, 20

20. ——. 1965. Intravital staining of lysosomes and mast cell granules by tetrazolium salts. *J. Histochem. Cytochem.* **13**, 411

21. —— and GRAY, R. 1964. Action of phospholipase C on lysosomes. *J. Cell Biol.* **23**, 50A

22. LANE, N. J. 1963. Thiamine pyrophosphatase, acid phosphatase and alkaline phosphatase in the neurones of *Helix aspersa*. *Quart. J. micr. Sci.* **104**, 401

23. ——. 1964. Localization of enzymes in certain secretory cells of *Helix* tentacles. *Quart. J. micr. Sci.* **105**, 49

24. ——. 1964. Further observations on the secretory cells in the optic tentacles of *Helix*, with special reference to the results of vital staining. *Quart. J. micr. Sci.* **105**, 61

25. MALHOTRA, S. K. 1961. A comparative histochemical study of the 'Golgi apparatus'. *Quart. J. micr. Sci.* **102**, 83

26. MEEK, G. A. and LANE, N. J. 1964. The ultrastructural localization of phosphatases in the neurones of the snail, *Helix aspersa*. *J. roy. micr. Soc.* **82**, 193

27. MORGAN, W. S. 1953. Cytological studies of the acinar cells of the pancreas of the mouse. Part 1. The formation of neutral red granules. *Quart. J. micr. Sci.* **94**, 141

28. ——. 1953. Cytological studies of the acinar cells of the pancreas of the mouse. Part 2. The argentophil, osmiophil and sudanophil substance. *Quart. J. micr. Sci.* **94**, 269

29. ——. 1958. Affinity of neutral red for cells with basophilic cytoplasm. *Nature, Lond.* **181**, 1132

30. MÜLLER, M., RÖHLICH, P., TÖTH, J., and TÖRÖ, I. 1963. Fine structure and enzymic activity of Protozoan food vacuoles. In Ciba Foundation Symposium, *Lysosomes* (edited by A. V. S. de Reuck and M. P. Cameron). London: Churchill

31. NOVIKOFF, A. B. 1959. In *Subcellular particles* (edited by T. Hayashi). New York: Ronald Press Co.

32. ——. 1961. Lysosomes and related particles. In *The Cell*, Vol. 2 (edited by J. Brachet and A. E. Mirsky). New York: Academic Press

33. ——. 1963. Lysosomes in the physiology and pathology of cells: contributions of staining methods. In Ciba Foundation Symposium, *Lysosomes* (edited by A. V. S. de Reuck and M. P. Cameron). London: Churchill

34. ——, BEAUFAY, H., and DE DUVE, C. 1956. Electron microscopy of 6 lysosome-rich fractions from rat liver. *J. biophys. biochem. Cytol.* suppl. 2, 179

35. —— and ESSNER, E. 1962. Cytolysomes and mitochondrial degeneration. *J. Cell Biol.* **15**, 140

36. OGAWA, K. and OKAMOTO, M. 1960. Cytochemistry of cultured neural tissue. 1. Intracellular granules of astrocytes. *J. Histochem. Cytochem.* **8**, 351

37. ——, MIZUNO, N., and OKAMOTO, M. 1961. Lysosomes in cultured cells. *J. Histochem. Cytochem.* **9**, 202

38. ROBBINS, E. and MARCUS, P. I. 1963. Dynamics of acridine orange-cell interaction. 1. Interrelationships of acridine orange particles and cytoplasmic reddening. *J. Cell. Biol.* **18**, 237

39. ——, MARCUS, P. I., and GONATAS, N. K. 1964. Dynamics of acridine orange-cell interaction. 2. Dye induced ultrastructural changes in multivesicular bodies (acridine orange particles). *J. Cell Biol.* **21**, 49

40. SHAFIQ, S. A. and CASSELMAN, W. G. B. 1954. Cytological studies of the neurones of *Locusta migratoria*. Part 3. Histochemical investigations, with special reference to the lipochondria. *Quart. J. micr. Sci.* **95**, 315

41. TANAKA, H. 1962. Electron microscope studies on vital stain as compared with phagocytosis. A concept of segresomes. In *5th International Congress for Electron Microscopy*. Vol. 2, LL-13. New York: Academic Press

42. WATTIAUX, R., WIBO, M., and BAUDHUIN, P. 1963. Influence of the injection of Triton WR-1339 on the properties of rat-liver lysosomes. In Ciba Foundation Symposium, *Lysosomes* (edited by A. V. S. de Reuck and M. P. Cameron). London: Churchill

43. WEISS, J. M. 1955. Intracellular changes due to neutral red as revealed in the pancreas and kidney of the mouse by the electron microscope. *J. exp. Med.* **101**, 213

Chapter 9

Empiricism—Silver Methods and the Nerve Axon

W. HOLMES

For more than a century solutions of silver compounds have been used to make certain elements of tissues distinguishable. Of these methods, those whose purpose is to display the nerve cell throughout its extent have received the most attention, and one may guess that there have been more variant methods directed to this end, than in any other comparable branch of microtechnique. The reagents used contain silver in solution and the reactions involved bring about precipitation of the silver in a coloured form, usually by reduction. Such processes are fundamentally different from dyeing, and for many years it has been customary to speak of these techniques as 'impregnation' methods. Baker[4] has defined this word as follows: 'The term "impregnation" should not be used unless the process of silvering (or the deposit of some other metal) involves two separate steps; the word is applicable to (the) first step only, during which no visible deposit is made.' It seems proper to refer to the techniques we are considering as staining methods, since a second step is always involved. There is a good precedent for this: Mann[20] refers to 'the ultimate staining effect' of 'the impregnation methods'; it is understood, of course, that 'staining' is not synonymous with 'dyeing'.

John Baker's influence has always been directed against what he has called 'irrational' methods in cytological technique. Such methods are those which are capricious because we do not know all the variables operating, so that the processes are incompletely controlled. He has also urged that investigators should give their reasons for selecting a particular reagent or procedure and an indication of the researches which led them to the choice. Methods are empirical when they are derived only from the accumulated experience of observed uniformities, the uniformities not having been resolved into the laws which underly them.

The silver methods have, from the start, been linked with the development of the technique of photography, and the empirical element was substantial in both of them. The use of metals other than silver, different reducing mixtures, gold toning, and so on, were tried in both fields, and Liesegang, expert in both of them, demonstrated what they have basically in common.[18] Both processes depend on the formation of nuclei of silver atoms, upon which more silver is deposited in the later stages of treatment to produce a visible image. One of the linkages which eventually proved most fruitful was Simarro's method.[29] He deliberately based a silver method on the practice of photography: he soaked tissues in a solution containing a salt of bromine or iodine and then transferred them to a solution of silver nitrate. The material was embedded and sectioned, the processes being carried out in the dark. The sections were illuminated, treated with a photographic reducer, and fixed with sodium thiosulphate.

The importance of Simarro's method lay in its influence on Ramón y Cajal. Ramón was a keen photographer all his life; as a student he manufactured the first rapid bromide plates in Spain, and later wrote many articles on photography (Ramón y Cajal[25]). He was naturally interested in Simarro's method, but found it unsatisfactory; he modified it profoundly, retaining the photographic developer, and produced the first version of his own silver method in 1903.

Students of photography and of silver staining still have something to learn from each other. On the one hand the possibility that light plays a part in the silver methods has been almost completely neglected; on the other the fineness of grain in a successful silver preparation is quite unattainable by the photographic process. This one can see by examining a negative microscopically, or can experience with more emotion by comparing one's beautiful silver preparations with their published photographic image.

Formaldehyde as a fixative was introduced by Blum in 1893 (Baker[2]). This was an event of special importance in the post-mortem room and anatomical laboratory. Specimens of the human nervous system are usually available only after a period of post-mortem change, which is shocking to a cytologist, but such specimens must be studied. The study of pathological change is an essential ancillary to the study of the normal, just as the reverse is true. The high rate of penetration of formaldehyde, the fact that it is a non-precipitant fixative and does not dislocate lipids, gave it pre-eminence in the study of the central nervous system of large animals, particularly of man. The characteristics of Ramón y Cajal's mixed fixatives make them much less suitable.

Max Bielschowsky was a pathologist who appreciated the advantages of formaldehyde for nervous tissue, and from a consideration of its chemistry proposed a new silver method. He pointed out that ammoniacal silver solutions were used by chemists as a test for aldehydes because they are reduced by such compounds, and a deposit of metallic silver is formed. He experimented using such silver solutions with formol-fixed tissues, and having obtained some differential staining of axons proposed the hypothesis that the reducing power of the formaldehyde is lost in most of the tissue because it becomes involved in protein linkages, but it is not so much involved with the substance of the nerve cell and axon so that in those structures it retains its power to reduce ammoniacal silver. To complete his preparations he borrowed from photography the gold toning process, and the use of sodium thiosulphate to remove unreduced silver. His preparations showed neurofibrils in the dendrites, but none in the axons.[7]

In 1903 another reagent containing silver was introduced into cytological technique by Regaud and Dubreuil.[26] This was 'Protargol', which is so different in many properties from silver nitrate and ammoniacal silver nitrate as to establish a new line of development in silver staining methods.

Regaud and Dubreuil were not attempting to stain nerve axons. They were following up some of the earlier 'photographic' methods for the study of the cells of epithelia, the aim of the methods being to demonstrate the boundaries between the individual cells. In these earlier methods fresh pieces of tissue were immersed in solutions of silver in the form of the nitrate or of other inorganic or organic compounds. The results always seemed imperfect because the silver combined with other ions such as chloride in the tissue, and the exposure to light necessary for the completion of the process produced random black precipitates.

Silver salts, like those of other metals, have had many therapeutic uses, but they were often also destructive, and disfiguring when silver or silver oxide were deposited under the influence of light. Regaud and Dubreuil studied a new group of compounds, or mixtures, of uncertain composition which have since been called 'silver proteinates'. Their therapeutic advantage was that they conferred all the benefits of the silver ion without its disadvantages: they did not damage normal tissue and did not form general precipitates even in the presence of chloride and light, but stained the cell boundaries effectively. One of these mixtures, given the trade name 'Protargol' by the Bayer company, was many years later the prime ingredient of Bodian's method for nerve axons, and blackened fingers were no longer diagnostic of the student of nervous histology.

Peters[24] has surveyed the history of the use of these proteinates for silver staining. They are dispersed in water before use and the silver is in the form of particles, not necessarily elementary, of a size of about 0.01μ. The concentration of silver ions is very low; it varies with pH, increasing as the pH falls.

Baker pointed out long ago that when one is studying, for example, the Golgi system, it is not sufficient to reproduce an appearance, the 'Golgi net', which corresponds with what was formerly thought to be a 'good preparation'.[3] So with the silver methods one must know precisely what one wishes, and is able, to make visible.

Clearly the silver methods are not histo-chemical tests. Wolman[32] made a brave attempt in this direction. He aimed to show that all the metallic impregnation methods are based upon a single general principle, which can be deduced from the available information. He considered that the active components of the tissue responsible for nucleus formation were reducing groups which were either free, or formed or made available by preliminary treatments. He concluded that the impregnation of axons is mainly due to a compound which is soluble in hot organic solvents. Sulphydryl and carbonyl groups were thought to be partly responsible. Peters[22] concluded that the reducible silver fraction, that is the silver which lies in the tissue in an unreduced form after impregnation, and which attaches to the silver nuclei on development, is chiefly combined with histidine. This conclusion came from experiments involving the blocking of chemically active protein end-groups.

What one usually sees in a conveniently large mammalian axon, for example in a Bielschowsky or Bodian preparation, is a uniformly coloured object. The uniformity varies only when one is looking through different thicknesses of the longitudinally arranged cylinder or where, in transverse section, one may see a darker line round the circumference which may or may not correspond with the surface membrane of the axon. There is no evidence of internal differentiation, but this might be due to inadequate resolution or to the thickness of the silver deposit. Peters[23] has examined material of this kind, that is, material which had been previously carried through a silver staining routine, with the electron microscope. Some of the axons in his material showed evidence of an internal fibrillar structure. When such fibrils were visible the particles of developed silver seemed to be either inside or on the surface of the fibrils. The particles varied in size between 30 and 700 Å. No silver nuclei were visible in impregnated but undeveloped preparations.

The question whether axons contain numerous fibrillar structures, the neurofibrils, has been with us for more than eighty years. Recently it has been given a new life by electron microscope studies, particularly those in which the process of degeneration in the terminals of divided axons has been compared in silver and electron microscopic preparations.

M. Schultze was the first to introduce osmium tetroxide to microtechnique, in 1864 (Baker[4]), and he was also the first to describe axons as having a neurofibrillar structure. His article on the nerve fibre in Stricker's *Handbook of Histology*[28] documents this view in a variety of situations in the animal kingdom. Cytologists were happy to accept that nerve fibres, with their unique function, should have a unique kind of cytoplasmic structure, and it was assumed that the neurofibrils were the conducting elements of the nerve fibre. The work of Fischer[12] and Hardy[14] which demonstrated that some apparent structures in cytoplasm were artifacts of fixation, did not affect the neurofibril concept, for their experiments never produced long artificial fibres.

S. von Apáthy and A. Bethe used methylene blue and dyes such as toluidine blue to study the neurofibrillar structure of neurons. The nature of their preparations led them to conclude that the neurofibrils of one neuron were continuous with those of all neurons with which it was connected. They thus became opponents of the 'neuron theory', in which it was held that the relation between neurons was one of contact only. The neuron theory was based on the observations of embryologists, and on Ramón y Cajal's studies with the dichromate–silver nitrate technique of Golgi. Golgi's method demonstrates nerve cells as distinct independent units, with axons and dendrites sharply marked off as the extensions of the cell soma. In such preparations the whole neuron is 'solid', with no internal structure visible, nor could Ramón show neurofibrils satisfactorily by the methods of Bethe and his school. He wrote: 'The complicated procedure of A. Bethe was not within reach of everybody. Like that of Apáthy it flourished only in the laboratory of its inventor. . . . I obtained by dint of patience some mediocre and inadequate stains. Attributing the failure to my lack of skill as a preparator, I politely requested from the brilliant originator of the method, a typical preparation to compare with my own. Some weeks later I received, carefully packed, like precious objects two preparations. . . . "These preparations are exceptionally good", the Strassburg professor wrote. "Be careful in handling them, and return them as soon as possible, as we have no others at present." What a disappointment! The jewels of technique, those invaluable preparations, unpacked with emotion and examined with a palpitating heart, were no better than mine.'[25]

Ramón's experience with Bethe and his methods did not lead him to doubt the existence of neurofibrils; it made him conclude that more precise methods must be devised for making them visible, and for reaching a decision whether they are continuous from one neuron to another. By variants of his silver method, particularly in fixation, he succeeded in producing neurofibril techniques, as well as many designed for particular types of neuron. 'Nowhere in cytological technique has empiricism run riot so freely as in the invention of fixative mixtures',[3] and Ramón's repertoire of fixatives for particular purposes contains some remarkable concoctions. It is unfortunate that he, like so many others who have developed silver methods, did not record the events in his experience and in his mind which made him try particular processes, and all those trials which failed. It is unreasonable to suppose that he mixed substances at random: one of his kinds of fixative contains pyridine, and the procedure after fixation ensures that some of the pyridine is carried over into the silver bath. In my own silver method pyridine was added to the buffered silver nitrate because it forms a complex with silver.[16] Blest[8] has tested higher pyridine derivatives and finds them equally effective, and sometimes preferable. Again, several of Ramón's fixatives contain drugs known to affect the nervous system. This seemed absurd in the days when it was thought that silver methods were too crude to detect axon abnormalities resulting from environmental changes such as poisoning, anoxia, or extremes of temperature. We now know better, and it may well be that Ramón was following some such clue as was given by Mann.[20] Mann recorded that he had obtained more perfect Golgi preparations of neurons from the brain of a patient who had died of paraldehyde poisoning than he had ever had before or since from a human brain.

There are obvious advantages in methods of silver staining which can be applied to tissues after they have been embedded and cut into serial sections, and the years from 1920 onwards were marked by the search for such methods. The ammoniacal silver solutions of the Bielschowsky type were so alkaline that they detached the sections from the slide and their high silver concentration often resulted in heavy unwanted deposits on the surface of the sections, however firmly attached. The first significant step towards avoiding these difficulties was made by Bartelmez and Hoerr;[5] they replaced ammoniacal silver by 'Protargol' solutions. They do not tell us why they did so, but they may well have appreciated that the reasons which led Regaud and Dubreuil to choose this reagent in place of silver nitrate applied equally to the problem they had in mind. Their method did not come into general use, and all practitioners of silver staining will agree that the most significant event in the history of silver methods for paraffin sections was the publication of the 'Protargol' method of David Bodian.[9] This involves, among other changes, the addition of metallic copper to the 'Protargol' impregnating bath. Dr. Bodian has very kindly told me the circumstances which led to the formulation of this method, which have never before been published. He has given me permission to quote his words in this essay, and they form an important historical document.

'I am glad to relate the origin of my "Protargol" method, as best as I can recall it. It began in 1934, when I was beginning a study of the optic pathways of the opossum. Results with block and frozen section methods of silver staining were so erratic that I tried some of the paraffin methods, such as Davenport's. They were clumsy, and so I was attracted to Bartelmez' and Hoerr's modification of the Rogers paraffin-Bielschowsky method. Bartelmez had substituted "Protargol" for silver nitrate, after finding that "Protargol" was the only commercial silver proteinate which was usable. Mild silver proteinates such as "Argyrol" and "Neosilvol" were unsatisfactory. The "Protargol"-Rogers method was useful to me, but much too capricious for valuable experimental material.

'In early 1935, I began to use "Protargol" as an impregnating solution, but shifted to photographic developers, following the Cajal procedures. This was surprisingly good for peripheral nerve, but rather inferior for central nervous tissues or ganglia. Results were at least consistent, however, and could be improved with the use of acid fixatives or alcohol. A long series of trials followed, in which a great variety of fixatives and developers were evaluated. Sample sections of an opossum brain fixed by perfusion with 80 per cent alcohol, impregnated with "Protargol", reduced with hydroquinone, and toned with gold, were better than previous material, and I decided to stain half of the set of serial sections with "Protargol". The series had been mounted on 4×5 inch glass slides, and the other half of the series had been stained with toluidine blue simultaneously by means of a brass rack designed for the purpose. Before entrusting the valuable set of paraffin sections to bulk staining with "Protargol" (the optic nerves had been sectioned in this specimen), I tested the possible effect of the brass rack upon the "Protargol" impregnation. I suppose it is possible that the thought occurred that the brass might conceivably improve the stain rather than interfere, but this tantalizing detail escapes my memory.

'The sample sections were placed in "Protargol" in a Coplin staining jar, and a cleaned copper penny was added in order to discover whether the silver would be removed by plating of the copper, and thereby be lost to the sections. Instead, an excellent impregnation resulted, comparable to a successful result with the classical reduced silver methods. The test was repeated with copper wire, in order to make sure that copper, and not another metal in the penny alloy, was responsible. Mercury was also tried, since it was incorporated into the soldered joints of the brass rack, and was also successful but messy. Many tissues and many fixatives were tried, as summarized in my 1937 paper. Incidentally, the opossum brain was impregnated in bulk with the brass rack, with very satisfactory results.

'Some lots of "Protargol" have given superior results with "Protargol" alone, without added copper, especially after fixation with mixtures such as formol-alcohol-acetic, but the consistency of results made possible by the copper is clearly a contribution of Lady Luck. Nevertheless, my debt to Bartelmez is greater than that to the fickle Lady, since it was he who set the stage for my experiments with "Protargol" and with the important effects of fixatives.'

The name 'Protargol' does not define the contents of the mixture that it is, and my own interest in silver staining arose on meeting the consequences of this fact. The 'Protargol' we had been using successfully in Oxford was made by the firm of Bayer in Germany; this became unavailable, and the mixture marketed under the same name by the firm of the same name in the USA was ineffective.[15] 'If cytological technique is to be a scientific subject, the reagent must be known.'[3]

Since Bodian's publication very many workers have attempted to analyse the silver methods, and to formulate, on the basis of their analysis, new and more reliable techniques. There is an underlying feeling that, when all is comprehended, a method of general validity will emerge. But all of us who concoct a new method are aware that we cannot controvert complaints from other practitioners, that the method works very well in a certain part of the nervous system and with certain groups of animals but not equally well in other parts and other groups.

Before we diagnose these difficulties as being due to the empiricism of silver methods we must ask ourselves what elements of axon structure we are trying to show or succeeding in showing by the methods we use. This involves another question: have all axons the same characteristics, so far as our methods can explore them, differing only in dimensions of size?

We are brought back here to the neurofibril concept.

In the days before electron microscopy, many cytologists attacked the question of the reality or otherwise of the neurofibrillar component of the axon. There was great scepticism about the appearances shown by silver methods, and about the preparations and views of the school of Bethe. Observations of the living cell did not give a clear-cut answer one way or the other. Filamentous neurofibrils were sometimes seen, but at other times were quite undetectable. The discovery by Young of the giant axons of the squid made it possible to study the properties of the axoplasm in relatively large quantities; polarized light analysis showed a certain amount of anisotropic organization contributed by crystalloidal micelles of an anisodiametric shape. The birefringence was positive with respect to the long axis of the fibre.[6]

Péterfi[21] suggested that the axon contents are a rodlet sol capable of forming fibrous structures under slight chemical or physical provocation. The review in which he made this suggestion is a landmark in the history of the subject. The review is entitled 'Das Leitende Element' and, since it appeared in a *Handbuch* edited by Bethe, its subject was the neurofibril. But Péterfi was convinced that in the living axon there were only longitudinal micelles which acted as 'nuclei' upon which further material could be deposited, and fuse into longitudinal fibres, when subjected to abnormal conditions. 'Die Neurofibrillen . . . sind im Neuroplasma der lebenden Nervenzelle und Faser in latenter Form, aber auch *nur* in dieser Form vorhanden.'

Investigators using the electron microscope find filaments in the axon, structures around 100 Å thick and of varying length, apparently protein. But in the frog, for example, they form only 0·5 per cent by volume of the axon contents, and they amount to only about 2·6 per cent of the dry weight of extruded squid axoplasm (Hydén, 1960). Peters' observations, already mentioned, on the size and location of the deposited metal in silver preparations suggest that the neurofibrils of light microscopy are not merely formed by an enormous build-up of deposit on the neurofilaments.

It is not necessarily the case that all axons have the same internal structure and properties. The syncitial nature of many invertebrate fibres, for example, in which the axon is a compound of the processes of many neurons, makes a difference not unlikely. And a difference has been established in other cases. Thus Wigglesworth[31] studied the axons of the cockroach; he found that axons of the order of 22 μ in diameter were composed of a great number of uniform fibrils about 0·5 μ thick: 'the whole structure resembles a bundle of cooked spaghetti with raisins in the form of mitochondria lying at intervals between the strands'. These fibrils can be seen, still of the same thickness, in smaller axons down to 3 μ or less in diameter. It is clear that when analysing silver methods, or studying the results of using them, we must bear in mind Baker's remark[3] that subjective ideas of what final result is desirable when we make a microscopic preparation should never be allowed to form in the mind except on the solid ground of comparison with what we believe to exist in the normal cell.

One way in which this necessary correlation can be made is shown by a recent line of work on synaptic structure.

When an axon is cut across, or merely interrupted by pinching, the separated parts are incapable of reuniting, and the distal part, which terminates in a synapse, is separated from its cell body. This distal segment loses its organization and eventually disappears; the process of degeneration passes through stages which are recognizable by particular cytological techniques. Recognizable degenerative changes provide a way of tracing nerve fibre connections in the central nervous system. The question of the location of particular connections can be most precisely answered if one can spot characteristic processes of 'terminal degeneration', that is, changes in the vicinity of the synapse which is the end of the axon. Two silver methods have been developed for tracing terminal degeneration, the first by Glees, and the second by Nauta and Gygax.

The prime necessity for the study of degenerative changes is knowledge of the location and structure of the normal nerve terminals. In the vertebrate central nervous system they were characterized long ago in silver preparations as distinctive bulbs and rings and buttons applied to the surface of the neuron with which they made contact. But silver preparations persistently failed to show these special terminations in considerable areas of the central nervous system, the neocortex for example. Armstrong, Richardson, and Young[1] pointed out that they could perfectly well demonstrate by cytological methods special terminations in areas of the brain in which the silver methods gave negative results. The silver methods behaved consistently, so here is a paradox; it has proved to be a fruitful one.

Boycott, Gray, and Guillery[10] studied parts of the brain of the lizard, and the visual cortex of the rat, by silver methods and by electron microscopy. In the hippocampus of the lizard, for example, there were synaptic endings in the form of argyrophil rings 2–4 μ in diameter. Electron microscopy showed that synaptic endings were more numerous than they were in the silver preparations; all endings contained mitochondria and synaptic vesicles; some of them also contained a bundle of neurofilaments in the form of a ring 2–4 μ in diameter. Each bundle was composed of from 10 to 50 neurofilaments each 50–100 Å thick. In the visual cortex the endings were smaller, about 0·5 μ in diameter, and all were without neurofilaments.

The same authors[11] studied the lizard brain again, in animals exposed to different environmental temperatures. Their work confirmed them in the opinion that some only of the synaptic terminals contain neurofilaments and that the silver methods fail to stain many terminals simply because these methods are specific for neurofilaments. They conclude 'that the neurofilaments of electron microscopy have argyrophil properties and correspond with the neurofibrillae of light microscopy'.

Gray and Hamlyn[13] now looked at the implication of this view for the silver methods specific for the degenerative processes in synapses. The Glees method is known to demonstrate changes after the lapse of a different length of time from that after which the Nauta-Gygax method is positive. Thus the normal optic tectum of the bird contains almost no ring, bulb or button endings. These structures appear, and are demonstrated by the Glees method, after from 7 to 11 days of degeneration. The typical Nauta-Gygax picture of degeneration is not demonstrable till from 28 to 30 days, then the Glees preparations are negative.

The electron microscope was used to follow the terminal changes in the axons after the same periods of degeneration as were used in the silver studies. Eight days after section of the axon, neurofilaments have appeared in the presynaptic cytoplasm, replacing the synaptic vesicles and mitochondria. These filaments are about 100 Å in diameter; they are new formations and it seems unlikely that they can have been newly synthesized, since that part of the axon is separated from its cell body. Gray and Hamlyn suggest that they may be 'solid' protein fibres precipitated from a previously 'soluble' protein in the axoplasm. At 30 days they have disappeared, and so has most of the rest of the terminal, and it is not clear what remnants of the nerve fibre and its termination contribute to the Nauta-Gygax picture.

It seems that we have here good evidence that particular silver techniques specifically demonstrate particular ultra-structural components of the normal and degenerating axon terminal.

Walberg[30] presents the results of his own investigation on this same problem. He emphasizes that we cannot generalize the conclusion that a particular silver method stains only terminals containing neurofilaments to all silver methods. That is, that there is no single quality 'argyrophilia': different fixations and different silvering processes give different results. Such caution is clearly appropriate when we have taken only the first steps on a new line of study. With different material Lund[19] comes to similar conclusions. A modification of one of Ramón's methods is used to demonstrate the normal axons in the central nervous system of the octopus and other cephalopods. It shows a neurofibrillar structure in many of the normal axons, and also shows degenerative changes in cut axons. But Glees's method does not stain the normal axons, as it should if it is specific for neurofilaments, and if neurofibrils are composed of neurofilaments.

The day-to-day use of silver methods is much more for the study of the distribution of nerves, and of their interconnections, than for the investigation of the structural units of the axon. Light microscopy remains essential for building up a picture of a system which is so widely dispersed in three dimensions. A way of using the electron microscope to assist and perfect an investigation by silver methods and light microscopy is shown by Richardson's study of the structure of autonomic nerves in the small intestine.[27] First he developed his own method of silver staining, a variant ammoniacal method, which seemed in the light of everything already known about the visceral nervous system to give the best preparations. He attached particular importance to fixation, and manufactured his own formaldehyde, and kept its properties completely under control. He was insistent that his method should be free of one of the bugbears of autonomic studies: the fact that the reticulin fibres and the strands of Schwann cell cytoplasm may blacken with silver and delude microscopists by appearing as nerve axons. Having obtained from his silver studies a picture of the general organization of his tissue he was well placed to study particular parts of it by electron microscopy. One of his most important conclusions is what has been suspected but never proved: that part of the final autonomic pathway consists of axons of a size which is below the limit of resolution by light microscopy, so that no silver method, however 'perfect', could make them visible by this means. More than half of the axons he saw were of a diameter less than $0 \cdot 1 \, \mu$.

All science begins with empiricism, and the observations which lead to analysis may sometimes be made possible by chance. But the silver methods for axons have never been wholly empirical, and the area of empiricism is progressively diminishing in the hands of practitioners unsatisfied by lack of understanding.

REFERENCES

1. ARMSTRONG, J., RICHARDSON, K. C., and YOUNG, J. Z. 1956. Staining neural end feet and mitochondria after postchroming and carbowax embedding. *Stain Tech.* **31**, 263
2. BAKER, J. R. 1933. *Cytological Technique*. London: Methuen.
3. ——. 1942. In *Cytology and Cell Physiology* (edited G. H. Bourne), 1st edition. Oxford University Press
4. ——. 1958. *Principles of Biological Microtechnique*. London: Methuen
5. BARTELMEZ, G. W. and HOERR, N. L. 1933. The vestibular club endings in *Ameiurus. J. comp. Neurol.* **57**, 401
6. BEAR, R. S., SCHMITT, F. O., and YOUNG, J. Z. 1937. The ultrastructure of nerve axoplasm. *Proc. Roy. Soc.* B **123**, 505
7. BIELSCHOWSKY, M. 1902. Die Silberimprägnation der Axencylinder. *Neurol. Zentbl.* **21**, 579

8. BLEST, A. D. 1961. Some modifications of Holmes's method, for insect central nervous systems. *Quart. J. micr. Sci.* **102**, 413
9. BODIAN, D. 1936. A new method for staining nerve fibers and nerve endings in mounted paraffin sections. *Anat. Rec.* **65**, 89
10. BOYCOTT, B. B., GRAY, E. G., and GUILLERY, R. W. 1960. A theory to account for the absence of boutons in silver preparations of the cerebral cortex. *J. Physiol. Lond.* 152; *Proc. Physiol. Soc.* 3P
11. ——, ——, and ——. 1961. Synaptic structure and its alteration with environmental temperature. *Proc. Roy. Soc.* B **154**, 151
12. FISCHER, A. 1899. *Fixierung, Färbung und Bau des Protoplasmas.* Jena: Fischer Verlag
13. GRAY, E. G. and HAMLYN, L. H. 1962. Electron microscopy of experimental degeneration in the avian optic tectum. *J. Anat., Lond.* **96**, 309
14. HARDY, W. B. 1899. On the structure of cell protoplasm. *J. Physiol. Lond.* **24**, 158
15. HOLMES, W. 1943. Silver staining of nerve axons in paraffin sections. *Anat. Rec.* **86**, 157
16. ——. 1947. The peripheral nerve biopsy. In *Recent Advances in Clinical Pathology* (edited S. C. Dyke), 1st edition. London: Churchill
17. HYDÉN, H. 1960. The neuron. In *The Cell* (edited J. Brachet and A. E. Mirsky), Vol. 4. New York: Academic Press
18. LIESEGANG, R. 1911. Die Kolloidchemie der histologischen Silberfärbung. *Kolloid-chem. Beih.* **3**, 1
19. LUND, R. D. 1965. The staining of degeneration in the nervous system of the octopus by modified silver methods. *Quart. J. micr. Sci.* **106**, 115
20. MANN, G. 1902. *Physiological Histology. Methods and Theory.* Oxford
21. PÉTERFI, T. 1929. Das leitende Element. In *Handbuch der normalen und pathologischen Physiologie* (edited A. Bethe), Bd. **9**, 79
22. PETERS, A. 1955a. Experiments on the mechanism of silver staining. Part I. Impregnation. *Quart. J. micr. Sci.* **96**, 84
23. ——. 1955b. Experiments on the mechanism of silver staining. Part III. Quantitative studies. *Quart. J. micr. Sci.* **96**, 301
24. ——. 1959. Experimental studies on staining nervous tissue with silver proteinates. *J. Anat. Lond.* **93**, 177
25. RAMÓN Y CAJAL, S. 1937. Recollections of my life (translated E. Horne Craigie). *Mem. Am. Phil. Soc.* 8, 1
26. REGAUD, C. and DUBREUIL, G. 1903. Sur un nouveau procédé d'argentation des épithéliums. *C.r. Ass. Anat.* Vième session
27. RICHARDSON, K. C. 1960. Studies on the structure of autonomic nerves in the small intestine. *J. Anat. Lond.* **94**, 457
28. SCHULTZE, M. 1871. In *Handbuch der Lehre von den Geweben des Menschen und der Thiere* (edited S. Stricker). Leipzig
29. SIMARRO, L. 1900. Quoted from abstract in *Zeit. wiss. Mik.* **18** (1901)
30. WALBERG, F. 1964. The early changes in degenerating boutons and the problem of argyrophilia. *J. comp. Neurol.* **122**, 113
31. WIGGLESWORTH, V. B. 1960. Axon structure and the dictyosomes (Golgi bodies) in the neurones of the cockroach, *Periplaneta americana. Quart. J. micr. Sci.* **101**, 381
32. WOLMAN, M. 1955. Studies of the impregnation of nervous tissue elements. *Quart. J. micr. Sci.* **96**, 329

Chapter 10

Intracellular Canals

O. L. THOMAS

In addition to its important role as the regulator of interchange of materials between the cell and its environment, the plasma membrane is usually regarded as a simple retaining sheath serving to restrain the more mobile cytoplasm, thus giving to the cell a definite shape. However, the surface membrane of free living cells is far from static and frequently shows active movement, thrusting out or withdrawing small protrusions. Similarly both ectodermal and endodermal cells from their earliest developmental stages show similar activities for those of their surfaces exposed to the outside world. It is not therefore surprising to find pinocytotic and micropinocytotic activity especially well developed in their definitive counterparts.

The name 'pinocytosis' was suggested by Lewis[20] in 1931 and since that time the phenomenon has been extensively studied in a variety of protozoa and tissue culture cells. These studies show that the plasma membrane is indeed a more dynamic structure morphologically than has been previously believed. Following a local reduction of surface tension the plasma membrane can deeply invaginate the underlying cytoplasm producing intracellular canals lined by the membrane. These canals may wander into the interior of the cell and even lose their connections with the extracellular space by absorption of the proximal parts of the membrane. Is this phenomenon of widespread occurrence in the intact organism; and are canals and spaces produced within cells by this means? Intracellular canals have been described within a variety of cells, and I propose to discuss this question using a few examples from my own experience.

The microdissection studies of Chambers[6] and others have shown that the protoplasmic surface of the cell is remarkably elastic and can be deformed by external forces. Some of the recorded instances of intracellular canals could perhaps be accounted for by such deformation, following artificial pressures caused by the preparation technique. The 'plasma canaliculi' of liver cells, first described in 1895 by J. H. Fraser and E. H. Fraser, possibly fall within this category. These canals were said to be extensive channels which branch and anastomose within the cytoplasm and open into the blood sinusoids which surround the liver cells. As their dimensions are too small to accommodate blood cells they were considered to contain only plasma.

These structures were rediscovered in 1902 by Sir Edward Sharpey-Schafer.[29] Tradition has it that Schafer had given his class for routine study purposes, some sections of liver prepared by the usual carmine–gelatine injection mass procedure. A student observed that the red mass was not only contained in the sinusoids, but was also occupying intercommunicating canaliculi in the interior of every liver cell, even surrounding the nuclei. Schafer was quick to establish the truth of this statement by further injection experiments, and the observation was confirmed independently by Herring and Sutherland Simpson[14] in 1905. The description of the plasma canaliculi was repeated through many editions of Schafer's famous textbook but since 1945 has been dropped from the text, and the canals are now presumably of historic interest only. The author is fortunate in possessing one of Schafer's original slides prepared by carmine injection given him by his esteemed teacher, the late Professor John Malcolm of Otago University, New Zealand (a former demonstrator for Schafer). In this old preparation (Fig.

10.1) the canals are still beautifully clear and match Schafer's description exactly. Are these canals present in the living cell, or are they produced at the moment of injection due to the pressure of the syringe? If indeed they be injection artifacts one would expect the cell membrane to appear ruptured and the mass to have escaped into the cell in the form of irregular shaped blobs having a haphazard distribution. This is not so. They appear as slender tubular structures with smooth contours, and some divide and give rise to side branches. In the absence over the years of further confirmation we are more or less forced to agree that the canaliculi were artifacts of technique. Perhaps the carmine forces its way into the cell by opening up *potential canals and spaces* communicating with the sinusoids. They may also form by a process of simple but extensive invagination of the plasma membrane. As the mass advances into the cell it could carry before it a finger-like pouch of membrane. This would contain the mass within the tubelike structures visible in the preparation. This interesting problem should repay further investigation by modern methods.

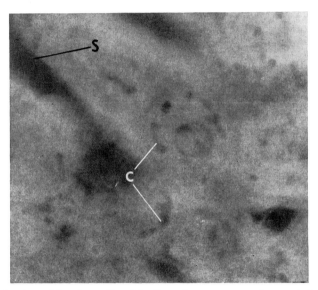

Fig. 10.1 Carmine injection of the sinusoids of the liver from a preparation made by Sir Edward Sharpey-Schafer. Sinusoids 's' are filled with the mass. Discrete intracellular canals are visible within the liver cells 'c'. They form a branching and anastamosing system of channels which communicate with the sinusoids. Note smooth outlines to the canals, 1000 × photomicrograph. The plates in early editions of Schafer's *Essentials of Histology* should also be consulted.

The appearance of 'intrinsic capillaries' within the thyroid follicles of animals administered with thiourea[31] provides us with another example where the plasma membrane of cells is deeply invaginated, but in this case it happens during the life of the animal, and in response to physiological conditions.

Due to the action of the drug the thyroid cells cease their production of thyroxin. Compensatory thyrotrophic stimulation then causes an intense hyperaemia of the gland. The normal interfollicular network of capillary vessels dilate enormously to form sinus-like spaces and these are supplemented by the development of an ancillary system of minute intra-epithelial vessels which invade the thyroid cells, and in doing so deeply invaginate their plasma membranes. This change is shown to be reversible. After discontinuing the drug the capillaries retreat from their intracellular position and resume a normal extracellular distribution. These facts indicate that the plasma membrane of thyroid cells possesses considerable elasticity and potential for deformation.

Intracellular canals are recorded in pancreas cells by numerous investigations over many years. More recently Lacy[19] redirected attention to them, claiming them to be 'Golgi canals'. Saguchi[28] believed that the canals form part of a functional intracellular duct system. The author[33] has attempted to inject the canals from the pancreatic ducts by a procedure similar to that used by Schafer and Herring[14] in the liver. Carmine masses failed to work, but the injection of a 'mass' consisting of laked blood displayed by iron haematoxylin staining gave dramatic results. In the sections, both intercellular and intracellular canals are seen to communicate with the alveolar cavity. These intercellular canals correspond to the secretory capillaries described previously by Zimmerman (Fig. 10.2). They are best regarded as simple extensions of the alveolar cavity insinuated between the cells. Zimmerman's reconstructions of them emphasize that these capillaries have no wall of their own but are formed by groove-like excavations or invaginations of adjoining cells. Due to the injection forces lateral branches of these capillaries could form at various points by further and deeper invaginations of the same plasma membrane (Fig. 10.2 'c'). In this way a system of intracellular canals, such as I have demonstrated, could have their origin. Local saccular spaces interposed on the course of the intracellular canals are also clearly demonstrated by the laked blood injection method and these correspond with the secretion vacuoles of Kupffer, also reported by Müller[24] and Saguchi.[28] My injection experiments support Saguchi's claim that a functional duct system of exogenous origin occurs within the acinar cells of the pancreas. It is also likely that the metallic impregnation techniques colour this identical system.

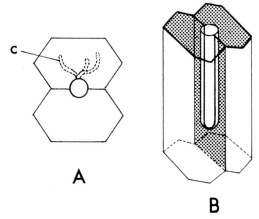

Fig. 10.2   Diagram of two adjacent glandular cells with a secretory capillary between them. A. View in cross section. B. Side view.
        Redrawn after Zimmerman. Note that the capillary does not possess a separate wall of its own but is formed as a groove produced by indentation of the plasma membrane of two adjacent cells. Further invagination of these membranes could produce the intracellular canals 'c'.

The discovery of the 'juice canals' in nerve cells by Holmgren in 1899 started a controversy which extends over the years. In a second paper in 1902 Holmgren[15] described, in the spinal ganglion cells of the rabbit, canals of fairly uniform calibre anastomosing freely to form a dense network extending round the nucleus. Here and there he found these canals communicating with the pericellular space. Holmgren later described similar canals in a wide range of material including not only nerve cells but cells from various epithelia and other sources. Confirmations of his findings were made by a large number of workers (Kolster, Fragnato, Lugaro, Sjövall, and many other famous names). In plant cells, too, similar observations were made. Bensley[3] observed in living onion root-tip cells a system of canals which he homologized with the Holmgren canals of animals. In nerve cells Holmgren believed his canals were filled by the processes of capsular or glial cells and this finding has recently been supported by the electron microscope studies of the prawn (*Leander serratus*) neuron by Malhotra and Meek.[21] These studies show the canals are narrow finger-like invaginations of the cell membrane. They are bounded by an electron-dense membrane continuous with the cell membrane and thus belong to the category of B-cytomembranes of Sjöstrand.[30] Some of the invaginations are extremely long and almost reach the nuclear membrane. Occasionally rounded swellings occur in the course of the canals especially near the surface of the cell.

Prior to the appearance of Malhotra and Meek's paper most authorities would have placed Holmgren's canals, together with the plasma canals of liver cells, as observations best relegated to the limbo. The position today would seem that they are worthy of further investigation by modern means. Historically Holmgren's papers appeared shortly after Golgi's original description of the 'Apparato Reticulo Interno' and workers throughout the world were busy attempting to confirm and extend both observations. Cajal,[5] perhaps the most prolific of these workers, regarded the Golgi apparatus and Holgren's Canals as one and the same thing and applied the name 'conduits de Golgi-Holmgren' to them. He regarded the appearances seen in his bichromate–silver preparations as due to the presence of canals filled with coagulable substance which has an affinity for colloidal silver. Sanchez, one of Cajal's pupils, went so far as to describe communications between the conduits and the tracheal system in insects.

It is important to recall the history of the bichromate–silver technique. It was discovered in 1873 by Camillo Golgi of Pavia through the favour of chance. The famous Italian neurologist happened to place some nervous tissue, which had previously been hardened in potassium bichromate, into a bath of silver nitrate. He noticed that the resultant precipitate of silver bichromate, which forms right within the thickness of the tissue, possesses the valuable property of staining black the cell body and processes of the neurones, normally so refractory to artificial colouration. They appear brilliantly outlined against a clear yellowish background. Furthermore the method has the valuable attribute of staining a few only of the neurones present in any one section. This renders easier the task of untangling the details of a veritable thicket of cells and processes. Golgi's bichromate–silver reaction quickly became one of the most valuable analytical techniques available to the neurohistologist. So successful was the method in the discoverer's hands, that Golgi was rewarded with a Nobel Prize in 1905.

However, the method has always been a very capricious one. Sometimes instead of revealing nerve cells and processes only blood vessels are stained. Such preparations are usually regarded as failures and fresh attempts are made with further specimens of tissue.

In 1891 by means of a slightly modified bichromate–silver procedure, Golgi observed for the first time the intracellular network of communicating strands which now bears his name—the Golgi apparatus. This second important discovery again attracted world attention and many modifications of the technique were devised, especially combinations of the reagents with osmic acid, but the original bichromate-silver reaction of Golgi remained the essential basis of all the procedures.

By 1894 a third important use for Golgi's reaction had been found. Müller,[24] Golgi, Heidenhain, Kölliker, and others employed it to explore the finest ramifications of the ducts or secretory capillaries of exocrine glands, and the intimate connections of the biliary canaliculi of the liver. As with the case of the nerve cell and its processes, these delicate tubular passages give a positive reaction which enables the investigator to follow their course with great precision, for they appear as black strands which turn and twist among the almost colourless gland cells.

The bichromate–silver technique had by now become one of the great tools of the general histologist. Many attempts have since been made to give a chemical rationale for this important reaction but all have failed. It has remained an essentially empirical procedure to this day. We are left to ask why should nerve cells, blood vessels, bile capillaries, and the finest ramifications of ducts of exocrine glands, together with the curious and enigmatic Golgi apparatus, all blacken by what is essentially the same method? Does the Golgi reaction indicate a physiological relationship between the structures stained—a pathway or mechanism as yet not understood?

The author believes significant progress toward answering these questions stems from the histo-chemical researches of Bourne[4] and later Novikoff[26] and his associates. Essentially the methods used by these workers are a modification of the Gomori[12] procedure for phosphatases, using nucleoside mono-, di-, and tri-phosphates, as well as thiamine pyrophosphate, as substrates. By these means preparations

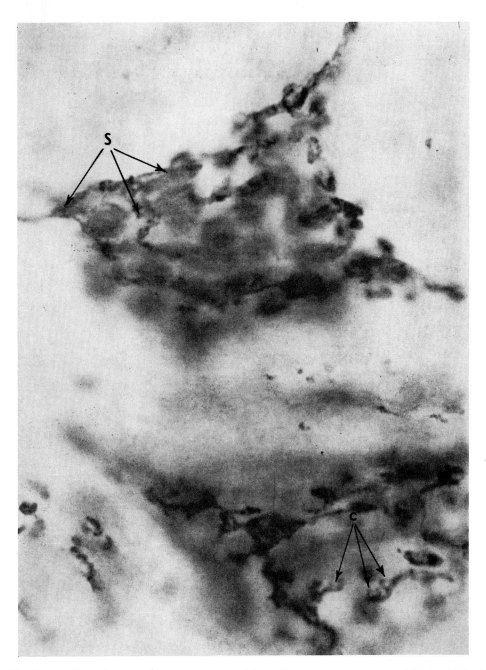

Fig. 10.3 Preparation of an anterior horn neuron of the guinea-pig. Incubated at 37°C in TPP substrate for 30 min. Note the extensive intracellular strands 's'. The claw-like appendages 'c' are typical of a 'good' Golgi apparatus as seen with silver methods. 3000 × photomicrograph.

are achieved which exactly duplicate the results obtained by the Golgi bichromate–silver reaction in all the situations so far examined, nervous tissues and glands alike. Fig. 10.3 is a preparation of an anterior horn cell from the guinea-pig spinal cord. The material was quickly excised from the killed animal and fixed in cold 4°C formal–calcium solution (Baker [2]) for 24 hr. Later 20μ frozen sections were prepared and incubated at 37°C for 30 min. in a buffered substrate pH 7·2 containing thiamine pyrophosphate (TPP), manganese activator, and lead ions to trap the phosphate liberated from the substrate by enzymatic hydrolysis. The lead deposits were blackened by hydrogen sulphide. These deposits localize the enzyme thiamine pyrophosphatase (TPPase) upon or within a delicate network of strands (S) which branch and anastomose with one another. This preparation should be compared with Golgi's[11] original drawings and with his description of the internal apparatus. All the salient features are faithfully reproduced including 'de minces plaquettes ou de petits disques'. The extent of the reaction is remarkable, and well illustrates the fact that the neuron possesses the largest and most elaborate Golgi apparatus of any known cell.

If the enzyme had not been so well localized by the chilled fixative, diffusion from its location in the network prior to incubation could produce a generalized blackening of the cell. (Something akin to this

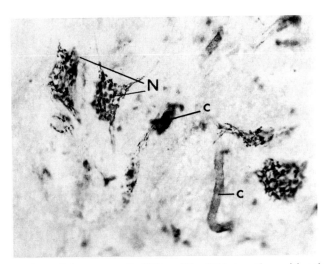

Fig. 10.4 Lower power view of the same cells 'N' as Fig. 10.3, together with others similarly showing internal reticulum. Capillaries 'c' in close relation to the neurons show localization of TPPase in their walls. 800 × photomicrograph.

must happen in the original Golgi neurological methods.) This photomicrograph should also be compared with the illustration of the classical Golgi apparatus of the kitten neurone in the author's earlier paper.[35]

Fig. 10.4 is a lower power view of the same cell together with others similarly showing the internal apparatus. Capillaries in close relation with these nerve cells show localizations of TPPase in their endothelial walls. This association of stained neurones with similarly stained blood vessels is a constant feature of preparations made by the bichromate–silver reactions.

Invertebrate nervous tissues also show this striking relationship between nerve cell and associated vessels. Fig. 10.5 is a TPPase preparation of the cerebral ganglion of *Lumbricus terrestris*. Two nerve cells (N) show localizations of enzyme in the Golgi apparatus (compare illustrations Figs. 4 and 7 from the author's earlier paper,[34] and close by a small capillary (c), cut in transverse section, shows a strongly positive reaction. In gastropods Meek and Lane[22] have shown localizations of TPPase in the cerebral neurones. In the absence of true capillaries within the ganglion mass the TPPase is found in a curious series of spaces within the periganglionic connective tissue.[36]

If the TPPase method is applied to exocrine glands again results are obtained which precisely duplicate the Golgi reaction. The alveolar ducts and their delicate ramifications, the secretory capillaries, are
clearly outlined. Fig. 10.6 shows a section of a pancreatic alveolus prepared by the TPPase technique.
The tortuous secretory capillaries show a positive reaction and are blackened. They are insinuated

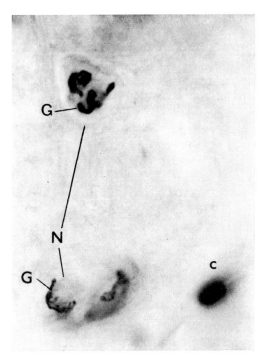

Fig. 10.5   Cerebral ganglia of *Lumbricus terrestris*, TPPase preparation. Two nerve cells 'N' show localizations of the enzyme in the Golgi apparatus 'G'. A capillary 'c' nearby shows a strongly positive reaction.
1200 × photomicrograph.

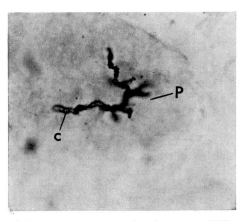

Fig. 10.6   Secretory alveolus of the mouse pancreas. Resting state. TPPase method. Incubation 30 min.
Tortuous secretory capillaries 'c' show positive localization of the enzyme. Note the extensions of the
capillaries between the cells and the delicate protrusions 'P' into the cells themselves. 1000 × photomicrograph.

between the individual cells and end as blind sacs. In some places the lumen of the capillary is visible and the enzyme is seen located in its wall. Small lateral extensions or protrusions of the capillary appear to indent the plasma membrane of the secretory cells and in some instances extend as delicate canals within the cytoplasm. Compare these photomicrographs with the illustrations from the earlier communication of the author[33] where the same tissue has been injected through the duct system with laked blood. Comparison should also be made with a standard Golgi bichromate–silver preparation of the pancreas (Fig. 10.7). *There can be no doubt that identical structure is revealed by all three methods of preparation.*

If mice are first injected with pilocarpine and their pancreatic tissues subsequently subjected to the TPPase method the localization of the enzyme is considerably altered. For a short period, 15–30 min after injection, the secretory capillaries and alveolar cavity appear to lose all trace of the enzyme. Instead, delicate intracellular tubular structures appear within the depths of the cytoplasm itself, and are coloured black due to the presence of the associated enzyme. The impression is gained that these structures are fine channels which tend to run together and course between the zymogen granules (Figs. 10.8, A and

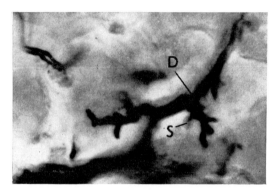

Fig. 10.7   Section of similar tissue to that in Fig. 10.6 but prepared by the original Golgi bichromate–silver method. Again note secretory capillaries 'S'. Note that they drain to a small branch of the main pancreatic duct 'D'. These preparations are thick sections mounted without coverglass in a drop of canada balsam. High-power photomicrography is difficult, but further detail similar to Fig. 10.6 can be seen in the original slide by repeated focusing. 800 × photomicrograph.

10.9). As time elapses these intracellular locations reach a peak in their development and then become fewer and less extensive and finally after 1 hr or more the enzyme is found only in the secretory capillaries. For a while connections between the intracellular and extracellular passages are visible (Figs. 10.8, C and 10.6) and then the preparations gradually resume the resting state (Fig. 10.8, A) but portions of the intracellular system may remain as the finger-like expansions of the capillaries described above.

It is important to stress that the histochemical methods for TPPase used in these studies were originally recommended by Novikoff[26] as a new and reliable technique for the demonstration of the Golgi apparatus in cytological studies at the light microscope level. However, in the above series of pancreatic cells doubt arises in the identification of the Golgi apparatus. Clearly the intracellular passages (Figs. 10.8, B and 10.9) could be described as such but as the scene shifts to the later stages in the secretory cycle (Fig. 10.8, C) identification is more difficult. What is Golgi apparatus and what is in fact an extension of the duct system into the cell? The two components seem to run into one another! We may well ask—Is the Golgi apparatus a functional component of the duct system?

Elucidation of the role of the nucleoside phosphatases and TPPase has been carried a step further by the application of the histochemical technique to electron microscopical studies. Following the pioneering work of Kaplan and Novikoff,[18] and also Holt and Hicks,[16] frozen sections may now be utilized, and precise localizations of the enzymes below the level of light microscopical resolution are possible. The papers of Novikoff[27] and his associates should be consulted for full details of this remarkable achievement, but in general it appears that the plasma membrane localizes the nucleoside phosphatases especially in those cells where infoldings are demonstrable by the electron microscope—for example the B-cytomembranes of kidney cells, the transitional epithelium of the bladder, and many situations where microvilli are present. As would be expected, some localization is present on the granular endoplasmic reticulum whose ribosomes are a possible site for production of the enzyme.

More energetic reactions are found on the γ-cytomembranes (Golgi saccules) in a variety of both animal and plant cells. It is significant that a histochemical method which produces such an exact replica of the Golgi apparatus at a light microscope level should show such precise localization of the reaction in the smooth surfaced γ cytomembranes. There now seems no doubt that the classical Golgi apparatus has its counterpart essentially in the γ cytomembranes of the electron micrograph.

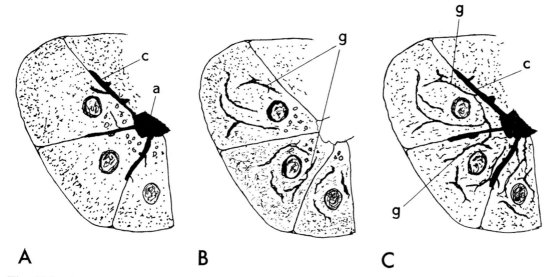

A   B   C

Fig. 10.8   Appearances seen in pancreatic alveoli when TPPase method is applied following pilocarpine injection.
   A. The resting state. Alveolar cavity 'a' and secretory capillaries 'c' show positive localization of enzyme. Note small protrusions from the secretory capillaries extending into the cells.
   B. 15–30 min after pilocarpine injection. The alveolar cavity and secretory capillaries no longer give a positive reaction. Instead delicate intracellular canals or strands 'g' previously invisible are blackened by the reaction.
   C. After 1 hr the secretory capillaries once more give a positive reaction. For a short period the intracellular canals or strands appear to join the secretory capillaries. Later the secretory capillaries and alveolar cavity alone will remain blackened and the preparations then resume the resting stage A.

What is the explanation of this localization ? More exact knowledge is needed. Novikoff has suggested that the enzymes concerned are related to the transport energy requirements of the cell. Also TPPase may influence indirectly the amount of acetyl coenzyme A utilized in the production of phospholipid (long known to be associated with the Golgi apparatus).

Perhaps more surprisingly, why should this precise histochemical method reproduce so faithfully the original Golgi reaction, which, as we have seen, occurs within a seemingly unrelated series or tissues and cells ? Are the enzymes themselves, or their substrates, or both, responsible for the bichromate–silver reaction ?

Many new questions await answers. For example: is there a significant relationship between nerve cell and accompanying vessels, when both give positive reactions ? Such a relationship must certainly pertain between the secretory cell and its duct system in similar circumstances. Why do only some cells in a section of nervous tissue blacken by the Golgi reaction, and others remain unstained? In adjacent cells in a nerve ganglion, the presence of the Golgi apparatus may be shown in one, while its neighbour fails to reveal it. Why, on frequent occasions, do we get vessels stained in these preparations, and not cells at all? Perhaps there is a hidden significance in the well-known capriciousness of the Golgi reaction, and a deeper meaning, at present not understood, to the reaction itself. An experimental approach to this problem may have interesing possibilities.

Many old problems can also be seen in a new light. Is the Golgi apparatus a canalicular system? Many cytologists have argued that it is. We recall especially the papers of Gatenby[8, 9, 10] and his associates, where, in both nerve cells and liver cells, powerful claims were made for the acceptance of the Golgi as a system of canals possessing argyrophilic or osmiophilic walls.

The biliary canaliculi have also been described as giving off diverticula which penetrate the liver

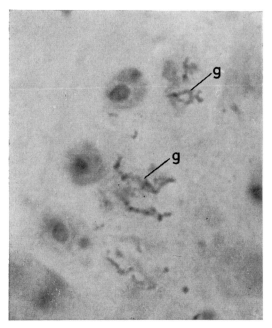

Fig. 10.9 Pancreas of mouse. 30 min after pilocarpine injection. Incubated with TPPase for 30 min at 37°C. Secretory capillaries are not visible, cf. Fig. 10.6, but delicate intracellular canals or threads 'g' course between the zymogen granules. 1500 × photomicrograph.

cells (Cajal, Heidenhain, Gatenby, and Moussa[8]) and the recognition of these intracellular 'flexuose tubules' as the Golgi apparatus of the liver cell, has formed one of the major cytological controversies of the past. In this connection it could be said that wherever the presence of intracellular canals is well established, the existence of a separate Golgi apparatus within the same cells is held in doubt. A good example is the parietal cell of the mammalian gastric glands. Certainly this cell exhibits the best-known example of an intracellular duct system. It is readily demonstrated by the bichromate–silver method (Fig. 10.10) but many authorities, including Golgi himself, have reported the classical apparatus absent. Recently Hally,[13] in an electron study of the mouse parietal cell, failed to locate γ cytomembranes, small granular vesicles or large vacuoles. He believed that spurious resemblance to the Golgi complex may arise where the intracellular canaliculus is cut transversely and some of its microvilli, cut longitudinally, appear as smooth double membranes. Others, cut transversely, resemble Golgi vesicles. Moussa and

Khattab[23] in a recent light microscope study of this cell also found difficulty in distinguishing between branches of the intracellular duct and the Golgi apparatus. From preliminary studies by the author[37] it appears that the parietal cell intracellular duct and its microvilli localize TPPase in a similar manner to the secretory capillaries of the pancreas, and furthermore this cell exhibits a similar secretory rhythm in respect to this localization.

Over the years Baker,[1] and other members of the Oxford School of cytology, have shown instances where seemingly unrelated objects within cells are blackened by the Golgi reaction and hence have been classified together as a single cell organelle. For example, the author[32, 34] has recorded examples both from light and electron studies where mitochondria are responsible in part for the formation of the classical Golgi apparatus. The nucleoside phosphatase techniques explain quite well how this could arise. It is now known that di- and tri-phosphatases present in the mitochondria will at favourable pH values

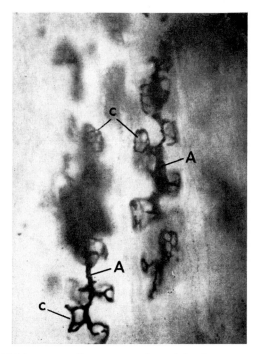

Fig. 10.10   Standard Golgi bichromate–silver preparation of the human gastric glands. A long tortuous secretory acinus 'A' communicates with the intracellular canaliculi within the parietal cells 'c'.

precipitate phosphate from a variety of substrates. In view of the close analogy shown to exist, the silver methods would in all probability precipitate silver at the same sites, and as a result the mitochondria would be incorporated into the 'black reaction'.

Finally, mention must be made of the interesting observations concerning the endoplasmic reticulum of mycobacteria. In what must be one of the simplest vacuolar and membranous systems present within a living cell, Imaeda and Ogura[17] have shown how this is derived from infoldings of both layers of the plasma membrane. They illustrate a scheme showing how a more complex membranous system could be derived from a primary simple invagination of the plasma membrane which receives secondary invaginations into the first, and so on repeatedly. The evidence at present available from a variety of sources: investigations with living cells, histochemistry, light and electron microscope studies, suggests the possibility of a similar process occurring in higher forms, leading to the formation of both the endo-plasmic reticulum and γ cytomembranes. Thus, in a sense, the entire vacuolar membranous system of the cell could tentatively be regarded as a specialized infolding of the plasma membrane producing an intricate intracellular canal system.

In conclusion, as this article is part of a volume in honour of John R. Baker, F.R.S., I would like to pay tribute to my former chief by recommending to the reader one of the classics of cytology, 'The structure and chemical composition of the Golgi element'.[1] Although written in 1944, his fresh and stimulating approach to the complicated problem of the Golgi apparatus, as he then saw it, is pertinent today. The history of cytology abounds with examples of controversies which have raged over the years. This history teaches us that matters which have appeared settled have a habit of becoming unsettled once more as fresh information comes to light. Let us follow Baker's example, and test and challenge the accepted view, for science is, surely, no static discipline.

REFERENCES

1. BAKER, J. R. 1944. *Quart. J. Mic. Sci.* **85**, 1–71
2. —— 1950. *Cytological Technique.* London: Methuen
3. BENSLEY, R. R. 1910. *Biol. Bull., Woods Hole*, **19**, 179–94
4. BOURNE, G. H. 1943. *Quart. J. Exp. Physiol.* **32**, 1
5. CAJAL, S. R. 1889. *Gaceta Medica Catalana*, 1–8
6. CHAMBERS, R. and CHAMBERS, E. L. 1961. *Explorations into the Living Cell*, Cambridge Mass. Harvard University Press, and London, Oxford University Press.
7. DA FANO, C. 1926. Obituary of Golgi, *J. Path. Bact.* **29**, 500–14
8. GATENBY, J. BRONTE and MOUSSA, T. A. A. 1949. *J. Roy. Mic. Soc.* **69**, 185–99
9. —— and —— 1950. *La Cellule*, **54**, 51–64
10. —— 1951. *Nature*, **167**, 185
11. GOLGI, C. 1891. *Arch. Ital. de Biol.* **30**, 62
12. GOMORI, G. 1964. *Microscopic Histochemistry*, pp. 172–87. Chicago, Ill., Univ. of Chicago Press
13. HALLY, A. D. 1959. *J. Anat.* **93**, 217
14. HERRING, P. T. and SIMPSON, S. 1905. *J. Physiol.* **33**, *Proceedings*, xviii
15. HOLMGREN, E. 1902. *Ergebnisse der Anatomie, Wiesb.* **11**, 274–329
16. HOLT, S. J. and HICKS, R. M. 1961. *J. Biophys. Biochem. Cytol.* **11**, 47–66
17. IMAEDA, T. and OGURA, M. 1963. *J. Bact.* **85**, 150–63
18. KAPLAN, S. and NOVIKOFF, A. B. 1959. *J. Histochem. Cytochem.* **7**, 295
19. LACY, D. 1954. *Nature, Lond.* **173**, 1235
20. LEWIS, W. H. 1931. *Bull. Johns Hop. Hospt.* **49**, 17–27
21. MALHOTRA, S. K. and MEEK, G. A. 1960. *J. Roy. Mic. Soc.* **80**, 1–8
22. MEEK, G. A. and LANE, N. J. 1964. *J. Roy. Mic. Soc.* **82**, 193–204
23. MOUSSA, T. A. and KHATTAB, F. I. 1956. *La Cellule* **63**, 137–49
24. MULLER, E. 1894. *Om inter och intracellulare Kortelgangar.* Stockholm: Sanson och Wallin
25. NOVIKOFF, A. B. and NOE, E. F. 1955. *J. Morph.* **96**, 189–222
26. —— and GOLDFISCHER, S. 1961. *Proc. Nat. Acad. Sci.* **47**, 802–10
27. —— ESSNER, E., GOLDFISCHER, S., and HEUS, M. 1961. *Interpretation of Ultrastructure*, pp. 150–92 New York: Academic Press
28. SAGUCHI, S. 1918. *Amer. J. Anat.* **26**, 347
29. SHARPEY-SCHAFER, E. 1938. *Essentials of Histology*, p. 388. London: Longmans
30. SJÖSTRAND, F. S. 1956. *Internat. Rev. Cytol.* **5**, 455
31. THOMAS, O. L. 1945. *Anat. Rec.* **93**, 23–45
32. —— 1954. *La Cellule* **56**, 229–41
33. —— 1955. *Nature, Lond.* **176**, 978–79
34. —— 1961. *La Cellule* **61**, 295–312
35. —— 1963. *Quart. J. Mic. Sci.* **104**, 75–79
36. —— 1963. *Nature, Lond.* **199**, 89–90
37. —— 1965. Unpublished work
38. ZIMMERMAN, K. 1898. *Arch. Mic. Anat. v. Entwick* 611–40

Chapter 11

Image and Artifact—Comments and Experiments on the Meaning of the Image in the Electron Microscope

S. M. McGEE-RUSSELL AND W. C. DE BRUIJN

Man sieht nur, was man weiss—GOETHE

INTRODUCTION

The treatments that must be applied to living material to permit observation in an electron microscope are often rather complicated. The complexity of these procedures compels one to copy them with the utmost care as they are recorded in the literature, even in the finest details, to obtain results consistent with published data. As a corollary, one should be interested in, and, whenever possible, collect information about the contribution of each step in a procedure to the final character of the electron microscope image, and perform experiments to check the influence of any deviation in technique.

In a given procedure one must make the critical decision either (1) to accept a certain final reaction product as representing a 'true image' of the major, or essentially interesting part of the living object, or (2) to reject it as an undesired artifact, although practically no information, at the necessary level of resolution, is available about the real life structure. The 'artifact concept' is consequently personal, and transitional, although it oscillates about a generally accepted, but continuously evolving mean, which is 'accepted scientific fact'. This is essentially the same topic as that discussed by Pantin at the beginning of this book. It is a concept and a problem which commonly occurs in communication, and in the development of understanding between disciplines which are based on different premises, or observers with different viewpoints. An artifact is often a lack of understanding. Today's artifact can often be tomorrow's datum; and, as emphasized by Pantin, the artifacts appropriate at one level of analysis may be explained at a different level, with the achievement of *full* understanding only through comparative study at several levels.

The nature of the image in the electron microscope limits the formation of suitable end products to reactions that create different degrees of electron scattering in the object finally placed in the microscope. Since living structures show minimal intrinsic electron scattering power, the visibility of a given structure is largely determined by:

(1) its ability to form a *stable* reaction product with a chemical that has electron scattering power, and
(2) the electron scattering power of that chemical.

As a rule, one wishes to observe a certain structure in relation to its surroundings and background (natural or artificial). For this, one must have a condition such that the structure has a different electron scattering power to that of the surrounding material. This gives differential contrast, and we must now briefly consider the general concept of contrast, in electron microscopy.

9—c.s.i.

Contrast may be defined as the difference in electron scattering power between a certain structure and its directly adjacent surround that is detected by the human eye in the image on a fluorescent screen or photographic plate, as a shade of grey, or black. Theoretically there are three main modes of contrast:

(1)   Positive Contrast   – object darker than surroundings
(2)   Equal Contrast   – object equal to surroundings
(3)   Negative Contrast – object brighter than surroundings

These definitions correspond to the usages normal in light microscopy. The exact nature of the object must be very clearly specified, at the different appropriate levels of discussion, in order to indicate clearly what constitutes the surroundings, and which is the comparison involved, since in electron microscopy, for example, one may wish to consider 'mitochondrion', 'mitochondrial matrix', or 'mitochondrial membrane', or 'polar part of membrane'. The visibility of the object is generally best if the contrast is at its highest, although in electron microscopy this does *not* necessarily correspond to optimum focus.

The moment in a procedure when greater electron scattering capacity is conferred on, or coupled to the structure, or its surrounding, is of surprisingly little importance in biological tissues. There can be, apparently, no *optimum* choice; any moment would appear to have its advantages and disadvantages. Thus pieces of fresh or fixed tissue, or plastic sections of tissue may be treated with osmium tetroxide or uranyl acetate, or phosphotungstic acid (PTA). Each treatment may produce a somewhat different result; none of the results is without interest.

In the development and application of a processing procedure, the following questions must be asked, and, if possible, answered, in relation to the stage when it is considered that contrast should be introduced:

(1)   How much of the substance of interest was present in the object in life?
(2)   How much of this substance was lost or modified during the steps preceding the moment chosen as important to the experiment? What is the nature of the modification, if any?
(3)   Are chemical reactions available which form stable electron scattering reaction products with *specific* ligands (SH, COOH, NH_2, NH, PO_4) of the structure?
(4)   Are such reactions applicable at the chosen moment, as part of the type of processing?
(5)   If reactions are applicable successfully, what is the stability of the reaction product, and how much is sacrificed in the subsequent steps of the procedure?
(6)   How much stable reaction product is formed in the object, and how much in the immediate surroundings of the object? (This question directly relates to the concept of contrast as discussed in this paper, and the selectivity of the reaction—Q.3.)

At the present time it is difficult to give examples of experimental procedures and data which provide answers to all of these logical questions, simply because too little is known about the processes involved in fixation, dehydration, embedding, and electron contrast 'staining'. However, we have carried out numerous experiments which have relevance to many of the questions, and some significant ideas seem to be emerging.

MATERIALS AND METHODS

Experimental Series I—on dehydration and extraction

Experimental Series I involved sixteen different treatments of tissue of the same origin, divided into four different groups of four treatments each. Small 1 mm cubes of tissue from the lactating mammary gland of a mouse were fixed in 6·5 per cent glutaraldehyde in 0·14 M cacodylate buffer of pH 7·3, with 0·22 M sucrose, for 4 hr at 0–4°C. After fixation, the cubes were stored in a storage buffer of the same composition as the fixative, but without the glutaraldehyde, for 18 hr at 0–4°C.

After storage, the cubes were divided into different groups, and treated as follows: (1) the basic comparison of dehydration sequences involved four different dehydrating agents: acetone, alcohol (95 per cent ethanol plus 5 per cent methanol), Durcupan A (X 133/2097), and glycol methacrylate, which were used in graded series (20, 30, 50, 70, 80, 90, 100 per cent—2 × 10 min each). The cubes of group 1 were therefore fixed only in glutaraldehyde, dehydrated in the four reagents as alternative procedures, and then passed through 1,2,-epoxy propane (2 × 10 min) embedded in Epon C (5A plus 5B) and slowly polymerized (18 hr at 37°C, plus 2 × 24 hr at 60°C). Hence, in group 1, any differences in the final images could be attributed to the different actions of the dehydrating agents on glutaraldehyde-fixed tissue.

(2) In the second group, the cubes were removed from the storage buffer, and 'double-fixed' with a 1 per cent osmium tetroxide solution, in a buffer of the same composition as the storage buffer, before being passed through the same four sequences as the cubes in group 1. Here differences could be attributed to the interaction of fixation and dehydration, by comparing the images obtained with those obtained for group 1.

(3) In the third group, the cubes were first dehydrated through the same four dehydrating sequences, and then brought back to storage buffer and allowed to re-hydrate overnight, *before* being double-fixed with osmium tetroxide of the same composition as was used in the second group. The cubes were then dehydrated again, and carried through the same procedure as before to plastic. In this group, the final images obtained might give some indication of the amount of 'osmium-reacting material' extracted from glutaraldehyde-fixed tissue by dehydrating reagents, and a direct indication of the amount left, if positive membrane contrast could be recovered.

(4) In the fourth group of cubes, the basic sequence was modified only by the omission of the 1,2,-epoxy propane antemedium between the dehydrating agent and the Epon C embedding medium.

Sections were cut from the experimental blocks on LKB ultratomes, with glass knives, and studied in a Phillips 200 electron microscope. They were studied both unstained, and stained by various standard methods, in order to obtain information upon inherent contrast, and apparent reactivity of the tissue with contrasting reagents. Comparisons were made at standard accelerating voltages, with standardized photographic techniques.

Experimental Series II—on fixation and contrasting

Small 1 mm cubes of fresh mouse liver were fixed in a 1 per cent solution of osmium tetroxide in cacodylate buffer with the addition of 0·1 M calcium chloride and 0·1 M potassium ferricyanide, for 1 hr at 0–4°C. This solution had a final pH of 7·0. After fixation the cubes were thoroughly rinsed in dilute cacodylate buffer (2 × 20 min), dehydrated in a graded acetone series (20, 30, 50, 70, 80, 90, 100 per cent, 2 × 10 min each) and passed through a mixture of acetone and Epon C (1:1) (instead of epoxy-propane) for 2 × 10 min, before embedding in Epon C as before.

Blocks of the same material were also double-fixed in glutaraldehyde followed by osmium tetroxide/ferricyanide solution in the following way: 6·5 per cent glutaraldehyde in cacodylate buffer (without sucrose) at pH 7·3 for 18 hr at 0–4°C; followed by storage for several days in cacodylate buffer (without sucrose) but with the addition of 0·1 M calcium chloride (anhydrous compound) and 0·1 M potassium ferricyanide: post-fixation was carried out in the same osmium/ferricyanide solution described above, in the same way.

DEHYDRATION AND EXTRACTION

General discussion

In order to be able to analyse the influence of dehydration upon the final electron microscopical image it is necessary to have some standard reference point, or generally accepted, easily recognized image. The mitochondrion is one of the most significant cell organelles, because of its fascinating morphological variation, its widely analysed physiological functions, and its high phospholipid content. We have taken the appearance of the mitochondrial membranes in the electron microscopical images, as our 'test object' in the experiments, so as to have such a standard reference. Happily, the mitochondrion seems to be the most easily recognized cell organelle after even highly diverse experimental treatments.

We may, therefore, now consider the logical questions that we posed in the introduction, in relation to current knowledge of this particular cell organelle, and the nature of the procedures normal to electron microscopical technique.

Question 1*—How much of the substance of interest is present ?

Phospholipids are most significant in the 'structural activity' of biological tissues (see the discussion by Chayen in this volume), and in the interaction between the tissue and fixatives, or processing treatments. Baker[1] was amongst the very first to emphasize the importance of phospholipids in considering cell structure and activity. Let us agree that phospholipids are a 'substance of interest', for this experimental approach, and discussion.

There can be no doubt that phospholipids are one of the major components of mitochondrial membranes. Recent analyses by Fleischer and colleagues,[9] of beef heart mitochondria, found that about 27 per cent of the isolated mitochondrial material was of lipid nature. Of these lipids, 95 per cent were phospholipid. Table 11.1 is based upon the components they determined, plus data from a recent

Table 11.1

	%	C = C/P	P⁻	N⁺
Phosphatidyl inositol	10	—	10	—
choline	37	2·7	37	37
ethanolamine	31	3·7	31	31
Cardiolipin	16	3·5	32	—
Residue	5	—	—	—
			110	68

review by Green and Fleischer.[11] We have added to the table estimates of the main ligands present as positive and negative charges. These estimates are helpful in considering the mechanisms of fixation, and contrast enhancement. It is plainly dangerous to use beef heart mitochondria as a universal model for mitochondria, but, until such time as the particular mitochondria in a given study are as fully analysed, the data serve as an excellent guide. Hence, for the mitochondrion, question 1 may be answered with some degree of assurance. A high proportion of phospholipid (25 per cent) is present, as is well known.

Question 2*—How much substance is lost in initial steps ?

On this subject there is surprisingly little information available in the literature, for most substances, including the phospholipids. Schidlovsky[20] provided some evidence that considerable amounts of phospholipids were dislocated by ethanol, if ethanol was applied to unfixed spinach leaves prior to fixation with osmium vapour. We mentioned in a preliminary report,[4] and illustrate here more fully, that differences in the images of the same glutaraldehyde-fixed material are detectable in the electron microscope after dehydration in different dehydrating agents, but can be explained only partly by the solubility of phospholipids in the dehydrating fluids. It is necessary to emphasize that the final embedding medium is not without effect upon the tissue and its final image, as a factor additional to the dehydrating agent. A great deal of extraction may take place in the early steps of a procedure involving solvents like ethanol and glycol methacrylate, but a further change may occur during infiltration and curing of the final medium. The well-known differences in the images of tissues embedded in epoxy resin and in methacrylate furnish us with a clear indication of this important fact. However, an initial significant extraction of the tissues is achieved by some of the standard solvents used in electron microscopy. This may be shown by direct estimation, and some observations of this kind are discussed below.

    * See p. 116.

Discussion of direct estimations of the extraction of lipids by dehydrating media

In collaboration with Dr. H. J. Ybema, direct estimations were made of the phospholipid content of media which had reacted with known amounts of fixed tissue cubes (Tables 11.2 and 11.3). Measurements were also made of the phospholipid content of the residual tissue, after treatment with the dehydrating agents, and of the original fresh, unfixed tissue. Five media were studied, in accordance with the previous discussion: acetone, alcohol (95 per cent ethanol plus 5 per cent methanol), Durcupan A (X133/2097), glycol methacrylate, and 1,2,-epoxy propane.

Table 11.2

Qualitative distribution of the phospholipids

Dehydrating media	In the dehydrating media				Remaining in the tissue			
	cepha-lines	leci-thines	tri-glycerides	chol-esterol	cepha-lines	leci-thines	tri-glycerides	chol-esterol
Alcohol (95% ethanol) (5% methanol)	−	+	+	+	+	+	−	−
Ethanol (pure)	+	+	+	+	+	+	−	−
Methanol (pure)	−	+	−	+	+	+	−	−
Acetone	−	+	+	+	+	+	−	−
1,2,-epoxy propane	−	+	+	+	+	+	−	−
Glycol Methacrylate	N.O.	N.O.	+	−	+	+	−	−
X133/2097: Durcupan A	N.O.	N.O.	+	−	+	+	−	−

Symbols: N.O. = Observation impossible through technical limitations.
  + = Observed present.
  − = Not present.

Table 11.3

Quantitative distribution of the extracted phospholipids

Dehydrating media	Total amount of phospholipids estimated (columns 1+2+3)	(1) Phospholipids in the various dehydrants	(2) Phospholipids in the subsequent 1,2,-epoxy propane	(3) Remaining phospholipids in the tissue cubes
	µg P/100 mg dry tissue (≡100%)	%	%	%
Alcohol (95% ethanol) (5% methanol)	209·5	70·4	4·0	26·2
Ethanol	293·0	62·0	2·7	35·5
Methanol	347·0	76·1	4·3	20·0
Acetone	241·3	22·6	3·3	74·0
1,2,-epoxy propane	214·5	45·4	6·5	48·0
Glycol methacrylate	161·1	0·4	15·5	84·1
X133/2097: Durcupan A	194·0	0·6	34·5	64·5

In order to be able to estimate the phospholipid contents of the fluids in practice, the sequence of events normal to the electron microscopical embedding procedure had to be modified slightly. The presence of cacodylate buffer in the dehydrating agents disturbs any measurement of phospholipid content, so the tissue cubes were rinsed in several changes of distilled water in order to remove the storage buffer completely. Furthermore, instead of dehydrating the tissue cubes in a graded series of reagents, about 200 mg of fixed, stored, and rinsed tissue cubes were dehydrated directly in 10 ml of each of the five agents at 100 per cent concentration, for $2\frac{1}{2}$ hr, which is the total time covered in the normal procedure of graded steps. After this treatment, the tissue cubes were in all cases exposed to 1,2, epoxy propane for another $2\frac{1}{2}$ hr. One group therefore received a double treatment with the epoxy-propane.

The phospholipid contents of the dehydrating agents, the epoxy-propane solutions, and the residual cubes of tissue were analysed qualitatively by thin-layer chromatography on silica-gel (Merck) with a mixture of di-n-butyl ether:acetic acid:water (40:35:5) as a chromatographic solvent phase according to Marinetti and colleagues.[15] In order to measure the phospholipid contents of the dehydrating agents, the fluids were evaporated at a temperature below 40°C under a nitrogen stream. The residue obtained was redissolved in a chloroform:methanol mixture (2:1), washed with water, and clarified with methanol according to Folch-Pi and colleagues.[10] The tissue cubes were treated with acetone magnesium chloride, followed by a chloroform:methanol mixture (2:1). This mixture was washed, and clarified with methanol. The full details of the procedure are given in another paper (Ybema and Leiznse[25]). Quantitative estimations of the phospholipid content of the dehydrating agents, the epoxy-propane solutions, and the tissue cubes, were performed on the chloroform/methanol solutions of each sample using the micro-method for phosphorus determination of King.[14]

The dehydrating agents glycol methacrylate (GMA) and Durcupan A (X 133/2097) could not be evaporated in the way stated above. Therefore we extracted the experimental samples of these agents with petroleum-ether, which was then evaporated *in vacuo* at low temperature. The resulting residue was redissolved in the chloroform:methanol mixture. As a consequence of this procedure, some of the phospholipids apparently escape measurement, because of incomplete extraction of the dehydrating agents by the petroleum-ether.

The phospholipid content of fresh tissue was determined using the same methods. The qualitative and quantitative distributions of phospholipids extracted, are given in Tables 11.2 and 11.3. From the measurements it seems clear that:

(1) during the fixation and storage procedures, about 80 per cent of the phospholipid in the fresh tissue was lost;

(2) the phospholipid content of 200 mg samples of fixed tissue varied considerably, although the samples were taken from a common 'pool' of tissue cubes from one mouse;

(3) the different dehydrating agents extracted different amounts of phospholipids. The cholesterol was totally extracted by all the agents.

(4) Of the 'classical' dehydrating agents, acetone seems to extract the smallest amount of phospholipid from fixed tissue cubes. This observation is apparently in accordance with our experiments on the osmium reaction, and our observations of the final electron microscopical image.

(5) The experimental dehydrating agents, glycol methacrylate and X 133/2097, extract very little phospholipid (see comment 6).

(6) Epoxy-propane has a considerable effect upon the phospholipid content which remains in tissue cubes after treatment with either glycol methacrylate or X 133/2097.

(7) Epoxy-propane has comparatively little effect upon the small phospholipid content which remains in tissue cubes after dehydration by the 'classical' dehydrating agents.

Fig. 11.1   In all the micrographs m = mitochondrion, n = nucleus.
    A. (Symbolized treatment (see text): G/Alc/EP/Epon/Ur Ac)   Mouse; mammary gland. Mitochondria show equal contrast; membranes have the same electron scattering capacity as the matrix, after uranyl acetate staining (1 per cent aq. Ur Ac pH 4·3, 30 min at 37°C). Virtual absence of membrane contrast throughout preparation.
    B. (G/Ac/EP/Epon/Ur Ac)   Mouse; mammary gland. The mitochondrial membranes now appear in negative contrast. After uranyl acetate staining (same conditions) the mitochondrial matrix has greater electron scattering capacity than the membranes. Lipid vacuole also appears negative in contrast.

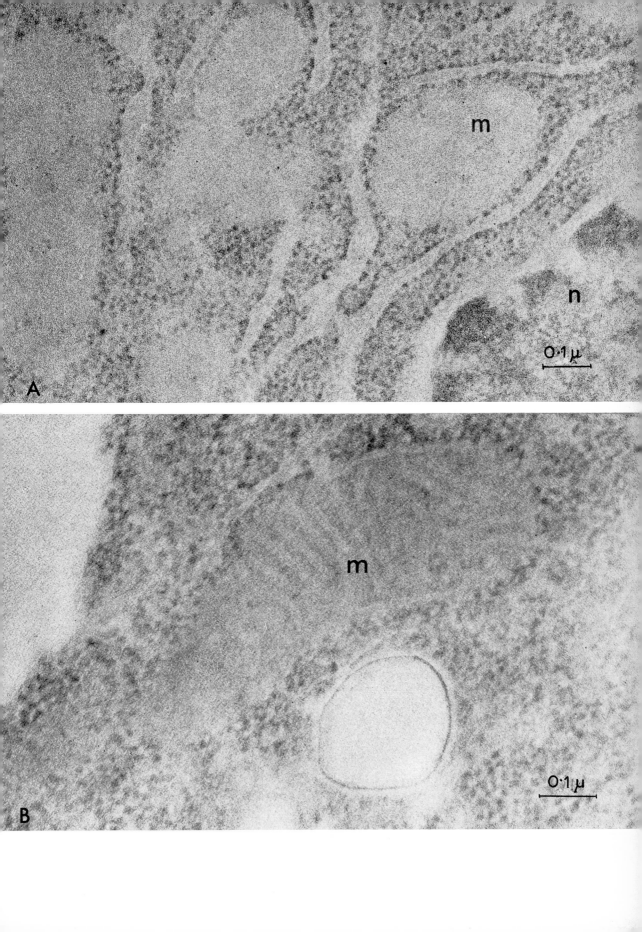

Since epoxy-propane has become a standard antemedium used before embedding in Epon, the final statement (7) is a somewhat reassuring conclusion. However, of course, the data confirm that dehydration, as a procedure, normally extracts a very significant quantity of phospholipid. Our demonstration of the apparently lesser extractive effect of acetone deserves to be emphasized. Acetone is fully miscible with Epon C, so that the epoxy-propane may be omitted from a schedule involving acetone. Such an omission does not lead to any technical difficulties during infiltration, polymerization, or sectioning. Further comparative experiments using acetone and alcohol as alternative dehydrating agents should prove instructive.

Discussion of Experimental Series I on dehydration and extraction

The influence of dehydration on the image of tissues fixed in glutaraldehyde

It is worth while to take advantage of the microscopist's opportunity to inspect the end products of diverse histological procedures in terms of the tissue structure, and compare the chemical measurements of extraction with the final electron microscopical image. The results of such examinations are illustrated in Figs. 11.1 and 11.2 (experimental series I).

After alcohol (Fig. 11.1, A) and acetone (Fig. 11.1, B) dehydrations, the images are significantly different, with fixation only in glutaraldehyde (G fixation). The mitochondria in the tissue dehydrated in alcohol show *equal contrast*, as defined above (Fig. 11.1, A), and are therefore difficult to distinguish. The mitochondria in the tissue dehydrated in glycol methacrylate (GMA) also have equal contrast (the latter are not illustrated but their contrast is similar). After both acetone and Durcupan the mitochondria show *negative contrast* in the membranes. Glutaraldehyde fixed tissue has virtually no intrinsic contrast in the electron microscope. In the cases illustrated in the figures, the electron scattering power was conferred as the final step of processing of sections, in a standard manner (1 per cent uranyl acetate solution pH 4·3 for 30 min at 37°C), and the visibility of the structure in the section is due to the formation of reaction products between the uranyl ions and the ligands in the remaining substance of the tissue (see question 3, p. 116).

Treatment of G-fixed tissue with the solvents, *before* a secondary post-fixation with osmium tetroxide, which involves a dehydration, and a rehydration, is illustrated, for acetone, in Fig. 11.2, A. Here, after acetone, the mitochondrial membranes appear barely visible, in *positive contrast*, of very low order. After treatment of this kind with the other dehydration media, even this degree of contrast was not achieved by the osmium post-stabilization, and low order *equal contrast* was seen in the mitochondria, with no differentiation of membranes (not illustrated). Control samples of tissue double-fixed in the normal way (GO, see methods tabulation), gave a 'normal' picture of the mitochondria. Membranes were then in distinct positive contrast after acetone, alcohol, and Durcupan A dehydration, but were much less distinct, although still positive in contrast, after GMA dehydration.

The results show that, in the particular case of acetone, enough material is present in the tissue after complete dehydration, to form a stable reaction product with aqueous osmium solutions after rehydration of the acetone treated tissue. However, the inability of osmium tetroxide, and 'staining' solutions of heavy metals, to develop positive contrast in sections of G-fixed tissue dehydrated with alcohol, Durcupan, and GMA, would certainly seem likely to be related to the loss of phospholipids during the dehydration (compare with Table 11.3). The comparatively lesser influence of the final antemedium of epoxy-propane is illustrated by the comparison of Figs. 11.1, B, and 11.2, B, where, in Fig. 11.2, B, omission of the epoxy-propane from the schedule did not alter the relative contrast levels. In both figures

Fig. 11.2   A. (G/Ac/Os/Ac/EP/Epon/Ur Ac)   Mouse; mammary gland. After this treatment, the uranyl acetate staining produces positive contrast in the mitochondrial membranes which is just detectable. Hence, an osmium reacting component is retained through acetone dehydration, and rehydration.

B. (G/Ac/—/Epon/Ur Ac)   Mouse; mammary gland. This image shows little difference from Fig. 11.1, B. Omission of the epoxy propane treatment does not modify the negative contrast obtained in the mitochondrial membranes after fixation only in glutaraldehyde, and acetone dehydration.

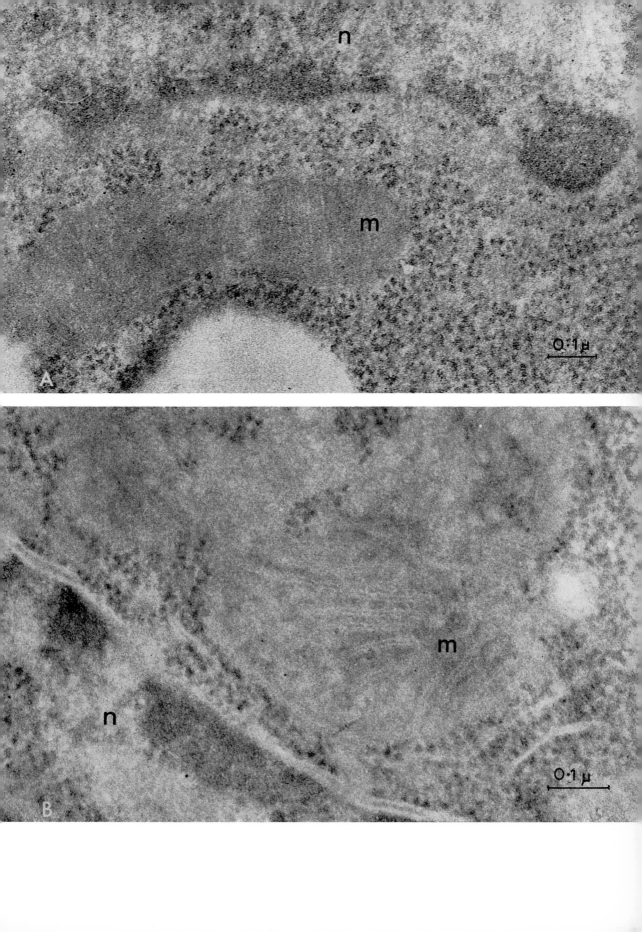

(11.1, B, and 11.2, B), the mitochondrial membranes are shown in negative contrast, and about the same density of contrast is shown in the mitochondrial matrix. This correlates reasonably well with the chemical estimates of extraction, which suggest that normal sequences of dehydration remove most of the phospholipids before reaching the epoxy-propane which then has little effect, quantitatively. Unusual dehydration sequences with the new compounds available (GMA and Durcupan) seem to retain phospholipids, which are then later extensively extracted by the epoxy-propane.

The negative contrast of the G-fixed tissue dehydrated with acetone, despite the apparent demonstration of the retention of more material, after acetone, may be explained in at least three possible ways. Either the Epon C removes the last amounts of phospholipid during infiltration, or the polymerization reaction modifies the phospholipids to such an extent that they lose their reactivity, or, finally, the usual staining reactions, with uranyl acetate, lead salts, or osmium tetroxide, do not possess enough 'selectivity' when applied to the sections, to differentiate the retained phospholipids from other material present in the immediate surroundings, even though phospholipids are still present and unmodified. If the remaining phospholipids are very evenly distributed in all cell elements, the effect would be the same as a lack of selectivity in the contrasting reactions. The positive membrane contrast that may be shown after rehydration, and stabilization with osmium, does not suggest that this is likely. It seems more probable that the last amounts of phospholipid are affected by the treatment with Epon C.

The influence of dehydration on the image of tissues fixed in osmium

Figure 11.3, A and B, show two very different images, which were obtained after dehydration in acetone (Fig. 11.3, A), and GMA (Fig. 11.3, B), with schedules which in other respects were identical, beginning with a fixation in osmium tetroxide (treatments: osmium/acetone/epoxy-propane/Epon/uranyl acetate, compared with osmium/GMA/epoxy-propane/Epon/uranyl acetate). Mitochondrial membranes are well preserved by acetone, as they are, also, by alcohol and Durcupan A. In GMA-dehydrated tissue the membrane contrast is very irregular, but positive. The cytoplasmic and mitochondrial matrix materials both appear rather 'patchy'. From these images, and similar comparisons, it is clear that the character of the final image is controlled not by the fixation alone, but also, to some extent, by the way the reaction products are stabilized or de-stabilized during post-treatments (see question 5). The same fixation may be differently affected by different post-treatments. The effects are compound effects dependent upon the sequence of steps, and the individual reagents at each step, up to and including the final stage, polymerization of the embedding medium. Destabilization of reaction products by embedding media may be most clearly demonstrated by polymerizing G-fixed tissue in methacrylate at 60°C. The solution and dissipation of the tissue achieved by this procedure is remarkable. It may be avoided, however, by polymerization at low temperature with ultra-violet light (McGee-Russell and De Bruijn[16]). The value of the emphasis which electron microscopists always place upon the details of technical treatment, is significantly confirmed by experimental comparisons of this kind. Interaction between the tissue, the fixative, the dehydrating agents, and the components of the final embedding medium may influence even the final stage of micrography in the electron microscope, through an effect upon the stability, under irradiation, in vacuum, of the final section. Sabatini and colleagues[21] (page 29) found that methacrylate sections were noticeably less stable to the electron beam when the included tissue was doubly fixed in aldehyde followed by osmium, rather than fixed only in either aldehyde or osmium.

Fig. 11.3 A. (Os/Ac/EP/Epon/Ur Ac)   Mouse; mammary gland. Here, after osmium fixation and acetone dehydration, a normal positive contrast image is obtained by the standard staining with uranyl acetate.
B. (Os/GMA/EP/Epon/Ur Ac—see text)   Mouse; mammary gland. The stabilization achieved by a primary osmium fixation is less resistant to dehydration in glycol methacrylate, than it is to acetone. After GMA, the mitochondrial membranes show positive contrast, but the image is somewhat disorganized. The endoplasmic reticulum shows a 'patchiness' reminiscent of polymerization damage.

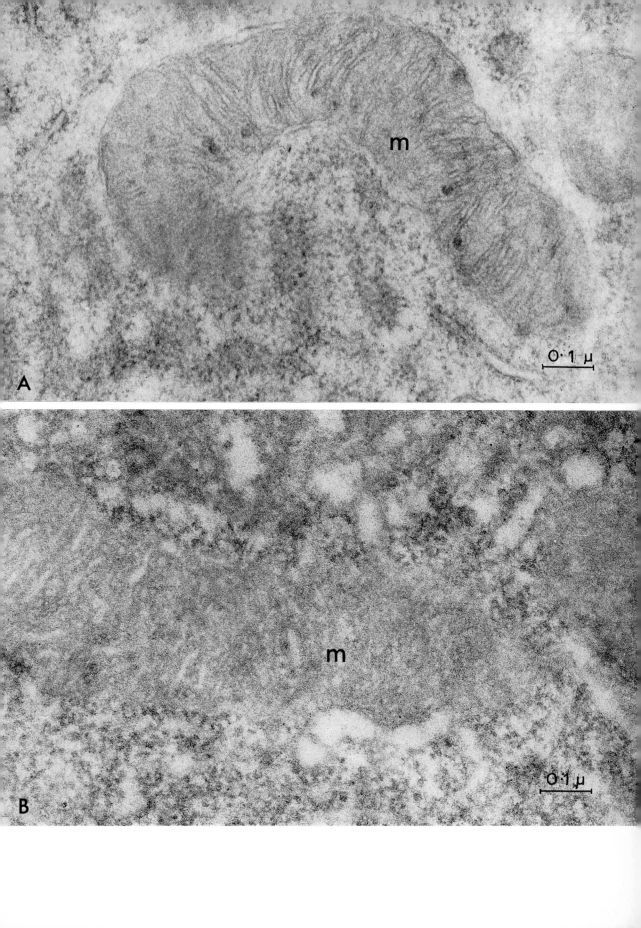

FIXATION AND CONTRASTING

General discussion

The direct extraction of components by solvents in early steps of procedure, involves a real loss of material, which may by measured, as we have shown. Fixation is often, in a sense, a 'constructive' step, since, as pointed out by Baker,[1] one purpose of fixation is to stabilize the tissue against the subsequent steps in processing. However, we must accept that even the most 'constructive' fixative may also remove cell material, and that the constituents retained are certain to be modified in chemical activity (this relates directly to the second part of question 2: how much substance is modified, and what is the modification?). The chemistry of fixation, and the nature of the changes involved in fixation are the subject of active work in many laboratories, at the present time, for the good reason that there remains a great deal to be discovered, and understood.

Hake[12] recently considered the destructive action of osmium tetroxide on amino acids, polypeptides, and proteins. Stoeckenius and Mahr[22] have studied the action of osmium tetroxide on phospholipids and other related lipids. They give examples of the infra-red analysis of the changes induced in various pure lipids by osmium tetroxide. The results of these two studies show clearly that the chemical structure of lipids is modified; but it remains difficult to explain, in chemical terms, why the tissue is 'fixed' in terms of Baker's definition. The mode of action of osmium tetroxide in cross-linking cellular material is not unequivocally established (see the reactions shown in Table 11.4). Criegee and Richter[5] suggested that

Table 11.4

$$
\underset{\text{C}}{\overset{\text{C}}{\|}} + OsO_4 \longrightarrow
\begin{array}{c} C{-}O \\ \backslash \\ OsO_2 \\ / \\ C{-}O \end{array}
\longrightarrow
\begin{array}{c} C{-}OH \\ \\ C{-}OH \end{array}
+ H_2OsO_4
$$

(after Stoeckenius)

$$
\begin{array}{c} R \\ | \\ H{-}C{-}NH_2 \\ | \\ O \\ \| \\ C{-}OH \end{array}
\xrightarrow[-NH_3]{+\frac12 O_2}
\begin{array}{c} R \\ | \\ C{=}O \\ | \\ O \\ \| \\ C{-}OH \end{array}
\xrightarrow[-CO_2]{+\frac12 O_2}
\begin{array}{c} R\ \ O \\ | \ \ \| \\ C{-}OH \end{array}
$$

(after Hake)

the primary cyclic osmium dioxide ester and the diol together form a secondary stable ester (the 'secondary Criegee ester'), but there is some doubt as to the presence of this reaction product in tissues. Becker[2] and Wigglesworth[23] have suggested alternative systems of cross-linking the primary cyclic osmium dioxide esters. It is generally accepted that some method of cross-linking is a necessary part of 'fixation'.

Present knowledge of the reactions between glutaraldehyde and phospholipids is also slight, although there is no evidence, as yet, to show that it will deviate much from the known action of formalin (e.g. Deierkauf and Heslinga[6]). As an example of a significant reaction, one may mention the report of Woolman and Greco[24] on formaldehyde, outlining a reaction between aldehydes and double bonds. Sabatini and colleagues[21] reported that glutaraldehyde/osmium double fixed tissue (GO) was significantly lower in contrast than tissue fixed only in osmium, when there was no post-aldehyde wash procedure. After a prolonged post-aldehyde wash in 0·2 M sucrose solution buffered to pH 7·4 with 0·1 M phosphate or cacodylate, they state that the contrast obtained by post-osmication was then greater than that obtained in tissues fixed in osmium tetroxide only, and embedded in Epon (Sabatini,[21] p. 29). This might be related to the possible reaction between the aldehyde and the double bonds, which, if like that discussed by Woolman and Greco[24] for formaldehyde, could have a significant effect upon the contrast obtained by subsequent osmication. Other chemical changes could also have significant effects upon electron contrast producing reactions, and upon the reactivity of the tissue towards light microscopical procedures. Analysis including information from both approaches may be helpful.

Question 3*—Are there chemical reactions for specific ligands, which give electron contrast?

In response to this question one may point to recent work which is suggestive of possible approaches to the problem. Riemersma and Booij[18] and Riemersma[19] have considered the interaction of a tri-complex-compound involving anionic dye molecules (Brilliant Scarlet 3 R), uranyl ions, and lecithin, with osmium tetroxide. They consider that the evidence at present tends to suggest the formation of a stable osmium dioxide-complex at the polar group of lecithin. With similar theoretical postulates, Elbers and colleagues[7] showed, recently, that other combinations of positive and negative ions with electron scattering capacity were suitable for the stabilization of lecithin and other amphoteric phospholipids against the action of post-treatments. Known chemical reactions for phospholipids which may be relevant include: the quantitative reaction for phosphorus content reviewed by Hoogwinkel and van Niekerk,[13] based on the method of Fiske and Subbarow;[8] and the cisaconite anhydride (CAZA) method for quarternary ammonium compounds which may be applied to N-containing phospholipids, reviewed by Boelsma-van Houte.[3] These techniques may contain useful clues for the study of biological tissues in the electron microscope.

Question 4*—Are such reactions applicable to electron microscopical techniques?

This question follows naturally upon question 3, and it may not be answered in the affirmative for methods of chemical or biochemical origin, without considerable experimental work to achieve mutually appropriate procedures. The histologist has nowadays, however, the considerable advantage of direct observational control of the nature of his material, through use of the electron microscope, at a very fine level, approaching molecular dimensions, as well as the 'built-in controls' (of complexity, texture, coherence, structural relationships, and so on) which are familiar to any cellular morphologist. Cellular organelles are also likely objects in which to expect a concentration of particular ligands, suitable for differential study. Nevertheless, despite this expectation, many ligands are fairly evenly distributed over the whole cell, and then, even if specific reactions are applicable, no marked differential display of cell elements may occur. Unfortunately, the ligands themselves do not have widely different physico-chemical properties. Therefore, there are limited possibilities for differentiating them. If the ionic character of a chemical is involved, in the reaction producing electron contrast, there are only two possibilities for selectivity, correlated with the ligand having either a negative or a positive charge. To the histologist, who is used to studying tissues through the differential use of acid and basic dyes, this does not seem an entirely unpromising situation. In the final experimental series considered in this article (see below), one approach to the problem is illustrated and discussed.

Question 5*—What is the stability of the reaction product in subsequent steps? Question 6*—What is the relative quantity of reaction product in object and surrounding?

On these questions, little more than guesses may be offered for any given procedure, at this time. Schidlovsky[20] showed that certain dehydrating solutions disturb the stability of osmium reaction products. Elbers and colleagues[7] suggested that the dielectric constant of the dehydration medium might play a role in stabilizing the somewhat weak London–van der Waals forces that hold tri-complex systems together. It is difficult to rule out an effect from *any* step in a processing procedure. What that effect is likely to be, and what factors it is dependent upon, must be determined by further investigation of the experimental system, and if necessary, by basic measurements of some of the physical and chemical constants. For example, it is doubtful whether the dielectric constants of some of the new materials in common use in electron microscopy have ever been measured! If relatively specific methods of contrast enhancement are devised, densitometric measurements appear to offer the most likely means of establishing answers to question 6.

* See p. 116.

Discussion of Experimental Series II on fixation and contrasting

The possible influence of positive and negative ions during osmium tetroxide fixation

It is generally accepted, and was demonstrated in a critical tissue, the brain, by Palay and his colleagues,[17] that the addition of calcium ions to the osmium fixative gives better membrane preservation. Theoretically, this could be explained on the assumption that the bivalent calcium ions neutralize all the negatively charged phosphate groups of the phospholipids, and with them form a tri-complex, together with one of the lower osmium oxides. This would be comparable to the situation discussed by Riemersma for lecithin, where the tri-complex includes uranyl.

In this experimental series, we imposed conditions such that the osmium tetroxide had to react with the tissue more or less in competition with a tri-complex combination, calcium ferricyanide (Ca^{2+} $FeCN_6^{3-}$), which is known to stabilize lecithin at neutral pH (Elbers[7]).

Figure 11.4, A and B, illustrates the images obtained of 'unstained' and heavily 'stained' material. The treatment of the tissue, and of the sections was similar, apart from the contrast enhancement procedure applied (uranyl acetate followed by Karnovsky's lead). Hence the symbolized comparative treatments were: Os/CaFeCN₆/acetone/Epon compared to Os/CaFeCN₆/acetone/Epon/uranyl acetate plus lead. In the 'unstained' section (Fig. 11.4, A), all membrane components of the cell are displayed, with rather high electron scattering capacity. Most of the other organelles have a low electron scattering capacity, or appear absent. Fig. 11.4, B shows that the attempt to add additional electron scattering power to the section enhanced only the membrane contrast, and did not reveal any further detail except in the nucleus, where a fine fibrous element was enhanced in contrast. This result must be compared with the result illustrated in Fig. 11.5, A and B, where staining revealed considerable further detail, in material first fixed with glutaraldehyde, and subsequently treated with the osmium/calcium ferricyanide tri-complex combination.

Tissue fixed directly with the osmium/calcium ferricyanide solution resembles, at first sight, potassium permanganate fixed material. At places in the section where orientation is favourable, the unit membrane structure may be resolved. Our impression is, that by adding tri-complex ions to the osmium tetroxide fixative, the selectivity of the osmium reaction is altered, and it becomes more 'specific' for membranes, and possibly, for phospholipids. This increase in 'specificity' is obtained, apparently, at the expense of the non-membraneous (or non-phospholipid) part of the cell, which appears 'poorly fixed'. It is rather difficult to explain this result completely at the present time, and it will have to be discussed again in the light of future experiments. Apart from the theoretical interest of the phenomena observed, some technical profit may result from the increased 'specificity' and selective contrast obtained by the addition of tri-complex ions to the osmium fixative in the double fixation procedure of glutaraldehyde followed by osmium/calcium ferricyanide. Figure 11.5, A and B, shows the two images (one 'unstained' and one 'stained'), which were obtained when glutaraldehyde-fixed liver cubes were stored in a storage buffer containing 0·1 M calcium chloride (anhydrous) and 0·1 M potassium ferricyanide for several days, and then further fixed with a 1 per cent solution of osmium tetroxide in cacodylate buffer with the addition of the same ions, for 1 hr at room temperature.

Figure 11.5, A, shows an unstained section of material of this type, in which, again, like the tissue fixed directly in the osmium/calcium ferricyanide fixative, the membranes show up against an even grey background. In this case too, the unit membrane structure is visible in certain places. If such a double-fixed section is stained densely, with uranyl acetate followed by Karnovsky's lead solution, then,

Fig. 11.4   A. (Os CaFeCn₆/Ac/—/Epon/unstained)   Mouse; liver parenchyma cell. Only membraneous elements display a strong positive contrast, well displayed in the mitochondria and nuclear envelope. The nuclear contents and other elements are very low in contrast. The general aspect is reminiscent of permanganate fixation. Contrast is high for unstained tissue.

B. (Os CaFeCN₆/Ac/—/Epon/Ur Ac & Karnovsky)   Heavy staining with uranyl acetate and lead of the same type of material as A, gives only slight increase in electron scattering capacity of the membraneous elements, and does not materially change the contrast of other cytoplasmic elements. Within the nucleus fine fibrous material gains contrast.

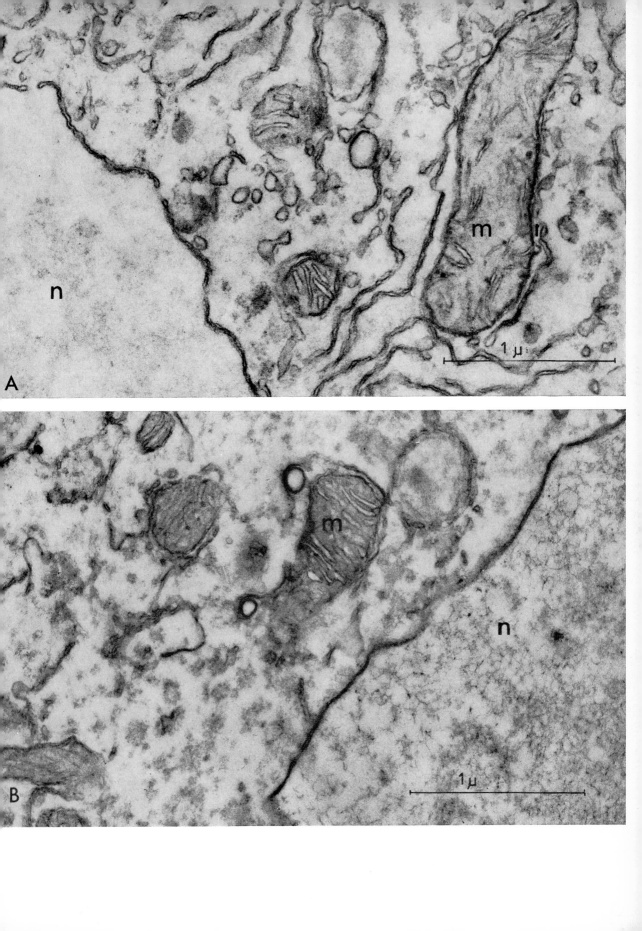

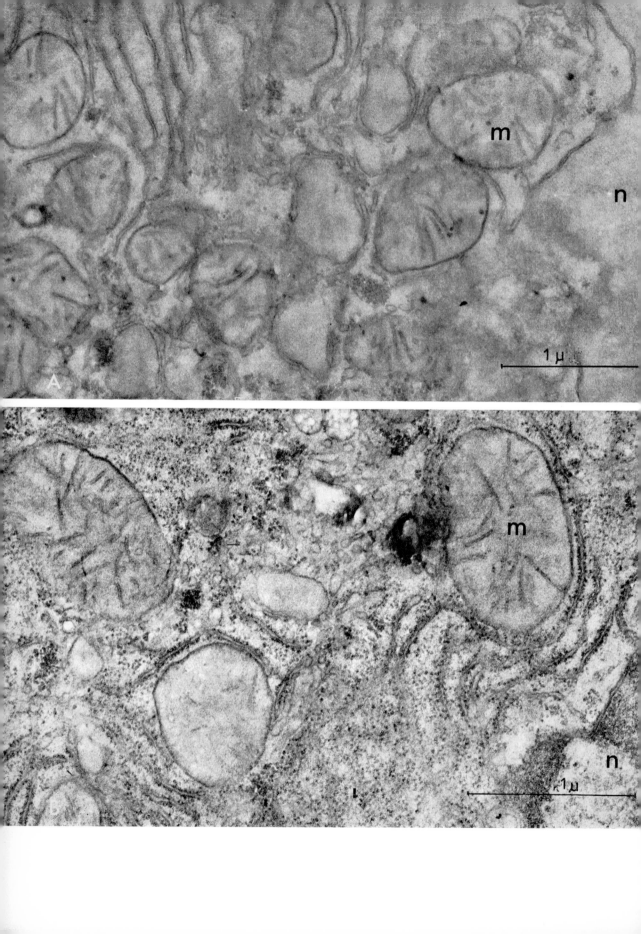

unlike the material treated only with the osmium/tri-complex as primary fixative, the background, which is of even contrast when unstained, is then seen to be occupied by the normal cytoplasmic components, stained in the normal way (Fig. 11.5, B).

From these results, it would appear that the presence of the ferricyanide ions in addition to the calcium ions in the fixative, either in some way prevents the osmium reaction from occurring, or, if it occurs, prevents cross-linking of the proteins. Proteins that are not cross-linked might then be lost during subsequent further processing. If the proteins are first cross-linked by an aldehyde reaction, as in glutaraldehyde fixation, the osmium/ferricyanide complex still produces electron contrast mainly in the membranes, and does not give significant contrast in other organelles (Fig. 11.5, A). Since, however, the other organelles are still present, as they may be shown to be, by staining with uranyl followed by lead, the addition of ferricyanide to the combination of osmium and calcium seems to have somewhat separated the 'membrane contrasting capacity' from the 'non-membrane contrasting capacity' of the osmium solution. This may have useful practical applications.

CONCLUSION

In conclusion, it is suggested that we have shown that the concepts of: 'image as related to truth', and: 'artifact as related to the act of observing', are completely inseparable in cytology and electron microscopy, as they are in other aspects of human endeavour, as denoted by the motto from Goethe at the head of this article. The study of the chemistry of fixation, and the study of the electron microscopical images resulting from the application of chemical procedures to the complexity of living tissues, have both a great way to go before all the puzzles are resolved. Osmium tetroxide remains, apparently, unique, as a substance capable both of fixing and contrasting biological tissues. It is also remarkable for its capacity to preserve more substances, in unsuspected ways, for electron microscopical study, than chemists might, at one time, have been willing to predict. The fixative action of glutaraldehyde, which produces virtually no contrast in its own right, and the negative contrast which is obtained through the intervention of heavy metals, after glutaraldehyde fixation, together demonstrate that 'fixation' and 'contrasting' are potentially separate chemical reactions. Images with positive electron contrast may yet be obtained without the use of osmium tetroxide, but with high selectivity for morphological entities, through further investigations of both the chemistry *and* the visible morphological image, considered together. Such an approach should also yield new facts and new hypotheses relevant to the future development of histochemistry, many aspects of which are discussed elsewhere in this book.

SUMMARY

The final image of a biological tissue in the electron microscope is shown, experimentally, to be affected by each step in a processing procedure. Fixation, dehydration, infiltration, and polymerization of the final embedding medium may all affect the nature of the final image. Each of these steps interacts with the preceding and subsequent steps. A processing procedure must be considered as a whole, with

Fig. 11.5 A (G/Os CaFeCN$_6$/Ac/—/Epon/unstained) Mouse; liver parenchyma cell. Fixation with glutaraldehyde prior to treatment with the osmium/calcium ferricyanide tri-complex does not alter the contrast and appearance of the mitochondrial membranes, and other membranes of the cell (compare Fig. 11.4 A and B), or produce contrast in the nucleus. The level of contrast difference between 'background' and membranes is lower; the non-membraneous elements may be considered to have gained some contrast over that shown in Fig. 11.4.

B. (G/Os CaFeCN$_6$/Ac/—/Epon/Ur Ac & Karnovsky) Mouse; liver parenchyma cell. Heavy staining of the same material as in A produces a much more conventional image; the non-membraneous elements such as ribosomes and nuclear chromatin have gained strong positive contrast, and have 'emerged' from the 'background' of equal contrast material seen in A. The staining stresses the unit membrane structure of the membraneous elements, throughout the image.

many significant parts. Some logical questions concerning the interaction between a biological tissue and a processing procedure are therefore asked. Relevant information is discussed.

Direct chemical estimations of the phospholipids extracted from fixed tissue by dehydrating media (alcohol, acetone, epoxy propane, glycol methacrylate, Durcupan) are tabulated. They show that about 80 per cent of the phospholipid in fresh tissue is lost during fixation and storage procedures; that cholesterol is totally extracted by all the dehydrating agents tested; and that glycol methacrylate and Durcupan extract less phospholipid than the other media tested. Such direct chemical measurements of the material extracted by processing steps may be related to the appearance of the tissue in the electron microscopical image. The mitochondrion is a good 'standard object' for such comparisons. It seems to be the most easily recognized cell organelle after highly diverse experimental treatments. Examples are given of images obtained after experimental treatments of tissues during fixation and dehydration. Some of the problems of the effects of fixation and contrasting on the nature of the final image are discussed.

REFERENCES

1. BAKER, J. R. 1945. *Cytological Technique. The Principles underlying Routine Methods*, 2nd edition. London: Methuen
2. BECKER, R. 1959. Diplom Arbeit Germany, Techn. Hochschule Karlsruhe
3. BOELSMA-VAN HOUTE, E. 1965. *Histochemie van fosfolipide in verband met atherosclerose van de aorta.* Doctor's degree thesis, University of Leiden
4. DE BRUIJN, W. C. and MCGEE-RUSSELL, S. M. 1964. Some experiments on mixed embedding media and dehydration procedures for electron microscopy. *Proc. 3rd European Region. Confer. on Electr. Micr. Prague*
5. CRIEGEE, R. and RICHTER, W. 1936. Osmic acid esters as intermediate products in oxidation. *Ann. Chem.* **522,** 75–96
6. DEIERKAUF, F. A. and HESLINGA, F. J. M. 1962. The action of formaldehyde on rat brain lipids. *J. Histochem. Cytochem.* **10,** 79–82
7. ELBERS. P. F., VERVERGAERT, P. H. J. T., and DEMEL, R. 1965. Tri-complex fixation of phospholipids. *J. Cell Biol.* **24,** 23–30
8. FISKE, C. H. and SUBBAROW, Y. 1925. The colorimetric determination of phosphorus. *J. Biol. Chem.* **lxvi,** 375–400
9. FLEISCHER, S., KLOUWEN, H., and BRIERLEY, G. 1961. Studies of electron transfer systems. XXXVIII. Lipid composition of purified enzyme preparations derived from beef heart mitochondria. *J. Biol. Chem.* **266,** 2936–41
10. FOLCH-PI, J., ASCOLI, M. L., MEATH, J. A., and LE BAROW, F. N. 1951. Preparation of lipid extracts from brain tissue. *J. Biol. Chem.* **191,** 833–41
11. GREEN, D. E. and FLEISCHER, S. 1964. Role of lipid in mitochondrial function. In *Metabolism and Physiological Significance of Lipids*, p. 582. London, New York, Sidney: J. Wiley and Sons Ltd.
12. HAKE, T. 1965. Studies on the reactions of OsO_4 and $KMnO_4$ with amino acids, peptides and proteins. *Lab. Invest.* **14,** 1208–12
13. HOOGWINKEL, G. J. M. and VAN NIEKERK, H. P. G. A. 1966. Method for the micro-determination of phosphorus in biological substances. *Koninkl. Nederl. Akademie van Wetensch. Amsterdam, Proc.* **B63,** 475
14. KING, E. J. 1932. The colorimetric determination of phosphorus. xxxiii. *Biochem. J.* **26,** 292–97
15. MARINETTI, G. V. and STOTZ, E. 1956. Chromatography of phosphatides on silicic acid impregnated paper. *Biochim. et Biophys. Acta* **21,** 168–70
16. MCGEE-RUSSELL, S. M. and DE BRUIJN, W. C. 1964. Experiments on embedding media for electron microscopy. *Quart. J. Micr. Sci.* **105,** 231–44
17. PALAY, S. L., MCGEE-RUSSELL, S. M., GORDON, S., and GRILLO, M. A. 1962. Fixation of neural tissues for electron microscopy by perfusion with solutions of osmium tetroxide. *J. Cell Biol.* **12,** 385–410
18. RIEMERSMA, J. C. and BOOIJ, H. L. 1962. The reaction of osmium tetroxide with lecithin, application of staining procedures. *J. Histochem. and Cytochem.* **10,** 89–95
19. —— (personal communication)
20. SCHIDLOVSKY, G. 1965. Contrast in multilayer systems after various fixations. *Lab. Invest.* **14,** 1213–33
21. SABATINI, D. D., BENSCH, K., and BARRNETT, R. J. 1963. Cytochemistry and electron microscopy. Preservation of cellular ultrastructure and enzymatic activity by aldehyde fixation. *J. Cell Biol.* **17,** 19–58
22. STOECKENIUS, W. and MAHR, S. C. 1965. Studies on the reaction of osmium tetroxide with lipids and related compounds. *Lab. Invest.* **14,** 1196–207

23. WIGGLESWORTH, V. B. 1957. Use of osmium in the fixation and staining of tissues. *Proc. Roy. Soc.* B **147**, 185–99

24. WOOLMAN, M. and GRECO, J. 1952. The effect of formaldehyde on tissue lipids and on histochemical reactions for carbonyl groups. *Stain Techn.* **27**, 317–24

25. YBEMA, H. J. and LEIZNSE, B. 1960. The turnover rates and turnover times of some phospholipids in the grey and white matter of brain tissue of the rat. *Koninkl. Ned. Akad. Wetenschap. Proc. Sci. C.* **63**, 652–63

Chapter 12

The Development of Specific Histochemical Methods for the Localization of Substances with the Electron Microscope

S. BRADBURY

INTRODUCTION

The development within the last twenty years of the techniques of biological electron microscopy has placed a tremendously powerful morphological tool in the hands of the cell biologist. Structures with dimensions of only a few Ångstrom units have been described, and a whole new concept of cell structure has arisen. Although it is not yet possible to attain the full theoretical resolution of the electron microscope, the current achievements represent a hundredfold improvement upon the figures attainable with the optical microscope. It is here that the advantage of the electron microscope lies. Its drawbacks, however, are familiar to all practising electron microscopists. The limitations imposed by the scattering of electrons in the thickness of the section mean that, for practical purposes, the greatest thickness which can, at present, be examined, is of the order of 800 Å; in consequence, as cells may be up to 150 μ in diameter, the end product of electron microscopical examination is a very two-dimensional conception of a structure which exists in three dimensions. Again, the electron microscope cannot be used, at the time of writing, to examine living cells, so that it cannot present us with information about the dynamic processes taking place in the cytoplasmic organelle systems which we are able now to recognize. A third major disadvantage is that the contrast in the image is obtained by differential scattering of electrons in the specimen; this is proportional to the mass per unit volume. If the thickness of the section is maintained at a constant figure, then an increase in the density of the material is essential if the contrast is to be increased. It has been the custom in recent years to obtain this contrast by the use of osmium tetroxide (either as a primary fixative, or following initial fixation of the tissues by other reagents such as aldehydes); the resultant deposition of the heavy metal in the tissue increases the final contrast in the electron microscope. The deposition is not specific in chemical terms, so the densities which are apparent in the electron micrographs cannot be interpreted in terms of the presence of the various chemical groupings which are known to occur in the cytoplasm of the cell.

These disadvantages of the electron microscope are now recognized, and many attempts are being made to overcome them. Methods have been developed for the serial reconstruction of three-dimensional structures from series of micrographs of thin sections,[45] and the adaptation of the techniques of autoradiography to electron microscopy has been successfully accomplished,[34, 46] with valuable results.[11] In combination with techniques whereby antibodies are conjugated with electron-dense markers,[57] progress is being made towards the understanding of some of the dynamics of the cell at the level of resolution of the electron microscope.

The problem of the lack of chemical information in electron micrographs is currently arousing widespread interest. Various physical methods for the differentiation of chemical elements and compounds at this level now exist; the use of electron diffraction techniques is widespread in the study of crystalline materials, and analysis of the characteristic X-ray emission which results from the irradiation of a specimen with a beam of electrons has been exploited in the micro-probe analyser,[12, 16] which is now a valuable research instrument. A review of the possibilities of the methods of physical analysis in the electron microscope has recently been published by Cosslett.[15]

At the optical microscope level, standard histochemical techniques have given us much valuable information, and have enabled a chemical interpretation to be placed on many of the staining reactions in common use. If such specific histochemical techniques can be modified, or new ones invented, for use with the electron microscope, then the localization of various chemical constituents of the cell will be carried a stage further, and our aim of synthesizing the biochemical and morphological approaches* will be assisted. The production of a colour reaction, which is so valuable in optical histochemistry, is useless for electron histochemistry. Instead, one must aim for enhancement of contrast in the final image. Two possible approaches have so far been used: the first is the production of relatively large quantities of insoluble, electron-dense deposits within the cell, for example at the site of a reaction mediated by an enzyme. The second involves treatment of the tissue with some reagent that will react with a particular group in the cell and cause an increase in the electron density of that region. If the process produces a great increase in contrast then it is possible to appreciate the fact by simple visual inspection, but small increases will require the additional use of densitometric methods for their assessment.

Various attempts have been made to modify existing histochemical techniques with the object of producing electron-dense end-products within the specimen. Many of the earlier efforts proved to be unsatisfactory, because the fine structure of the tissue was so severely disrupted that the value of the method could not adequately be assessed. The preservation of a recognizable fine structure is perhaps the most stringent requirement of this new technique, assuming even greater importance than all the other well-known hazards, which are also applicable to histochemistry at the optical level, such as the inactivation of reactant groups. This latter problem also proved particularly troublesome in the early days of electron histochemistry, when the only fixative available for electron microscopy was osmium tetroxide. This compound is very reactive, and combines with many of the end groups present in the tissue,[20,58] thus preventing their reaction with other reagents. This led many of the earlier workers to use unfixed tissues for their histochemical studies, with the result that the morphology was destroyed almost completely. The related problems of (1) preserving the reactivity of the groups in the tissue; and at the same time of (2) retaining a recognizable cellular morphology, are thus of prime importance in electron histochemistry. At the moment it seems a problem which cannot be solved completely and our efforts must be devoted to achieving a satisfactory compromise between the requirements of chemical localization on the one hand, and preservation of a satisfactory electron microscopical morphology on the other. Great progress in this field has recently been achieved by the development of fixatives other than osmium tetroxide; various aldehyde mixtures have been evolved[50,51] which not only preserve the fine structure adequately, but also do not interfere too seriously with histochemical reactions. The aldehydes provide a reasonable compromise between the conflicting demands of the two separate disciplines, and have enabled the development of electron histochemistry to become a practicable proposition.

ENZYMES

By far the greatest effort devoted to the establishment of electron histochemical techniques has been concerned with the study of enzymes. To date this field of work has proved the most fruitful, probably because the Gomori-type of reaction, used in optical histochemistry for the localization of hydrolytic enzymes, has proved very suitable for adaptation to the needs of electron microscopists. In the Gomori methods, the enzyme in the tissue acts on a substrate in the incubation medium to release a product (often phosphoric acid) which immediately combines with the 'capture' reagent (lead or calcium ions) to precipitate an insoluble metal salt at the site of the enzyme (Fig. 12.1).

Early work[9,56] reported the successful localization of acid and alkaline phosphatases in the microvilli, and in the Golgi apparatus, of the intestinal epithelial cell. These early attempts at the application of the Gomori reaction to tissues prepared for the electron microscope were impaired by the poor degree of structural preservation of the tissue, but they pointed out the possibilities and potentialities of the

* The reader is referred to the editors' remarks at the beginning of Part 4 (p. 394).

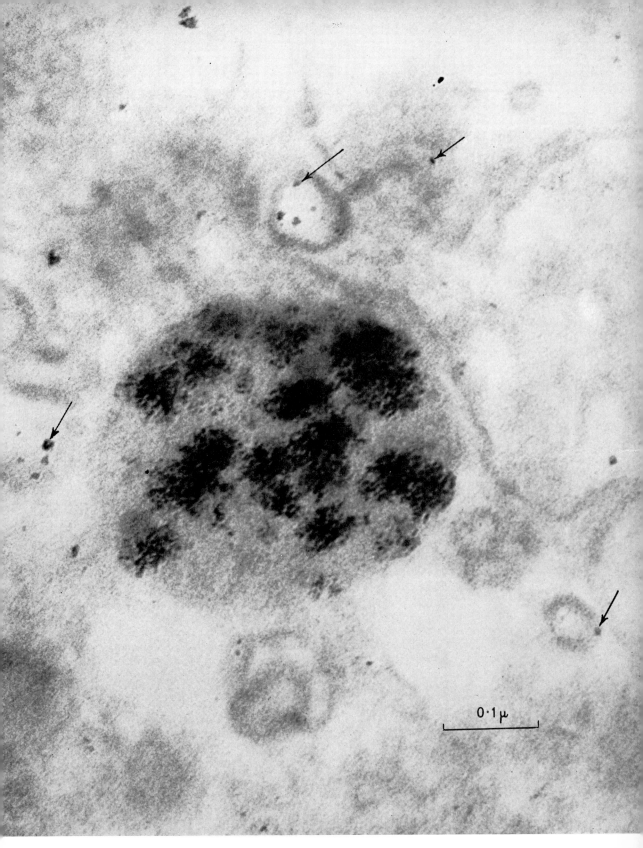

Fig. 12.1 The Gomori reaction for acid phosphatase, applied to a section of bovine adrenal medulla. Note the heavy deposits of lead phosphate in the large lysosome in the centre of the field. Other smaller deposits (some of them indicated by arrows) represent a non-specific attachment of metal to lipid components in the section. These latter deposits are found in control preparations in which the enzyme has been inactivated.

method. Attempts at the localization of phosphatases received a stimulus in 1961, when it was discovered by Essner and Novikoff[18] and by Holt and Hicks[23] that formalin would serve as a satisfactory fixative for the fine structure, and at the same time allow a much greater retention of enzyme activity. With the use of formalin fixation, and a standardized technique, these methods produced essentially repeatable results, especially once it was recognized that the incubation times for the electron microscope were much shorter than for corresponding tissues which were to be examined optically; this reduction of the incubation time had the effect of greatly reducing the non-specific deposition, and so improving the localization. As Pearse points out in his valuable review of enzyme histochemistry at the electron microscope level,[42] the selective powers of the research worker in choosing suitable cells to photograph has undoubtedly played a very important part in the development of these methods. Since the development of the aldehyde fixatives for electron microscopy,[50, 51] the range of enzymes which can be demonstrated by the Gomori-type of metal precipitation technique has increased considerably. The work of Barrnett and his collaborators has added extensively to our knowledge of the nucleoside phosphatases, especially in relationship to the endothelia of small blood vessels,[35] the sarcoplasmic reticulum and myofibrils,[61, 49] and in neurons and neuroglia.[63] The work of Holt and Hicks[23] and of Novikoff and his collaborators[18, 39] has resulted in a mass of valuable information on the localization of acid phosphatase and the characterization of the lysosome. Similarly, other work of Novikoff and his collaborators[40] has resulted in the appreciation that in vertebrates the nucleoside diphosphatases are characteristic constituents of the Golgi apparatus, and may be used as a 'marker' for this cell organelle.

The dehydrogenase system of enzymes has also been successfully studied with the electron microscope. The first attempt at their localization[3] involved the use of potassium tellurite, which can be reduced (with difficulty) by dehydrogenase systems. The final insoluble crystalline product was visible in the electron microscope, in association with the cristae of the mitochondria. An alternative approach was to apply those methods in which the action of the enzyme reduces various tetrazolium compounds, with the production of a formazan. The formazans would be expected to produce a definite increase of contrast in the electron micrographs even though these substances do not contain a heavy metal in their molecule. The first attempts at the production of formazans for electron microscopy[53, 54] suffered from the drawback that the particle size of the diformazan which was deposited in the tissue was between 300–900 Å, so severely restricting the resolution which could be attained. Also these formazans, produced from Nitro BT, were, to some extent, soluble in lipids and lipoproteins, so that the localizations obtained with them were suspect. The application of other tetrazolium salts to these reactions, for instance the use of Tetranitro-blue tetrazolium chloride by Ogawa and Barrnett[41] in the location of succinic dehydrogenase, allowed the demonstration of formazan particles with diameters of only 60–90 Å, and so their localization was shown to be within the osmiophilic layers of the mitochondrial cristae.

Similarly, methods have been developed for the localization of cholinesterase at the fine structural level. Lehrer and Ornstein[33] adapted for the electron microscope a simultaneous diazo-coupling technique, in which α naphthyl acetate was the substrate and 'hexazonium pararosanilin' the coupler. In their paper these authors stress a point which has been overlooked in recent years, namely that it is not necessary to produce a reaction product in which there is incorporated a heavy metal, since useful contrast increments can be achieved with the aid of purely organic reagents. It is likely that, in the near future, with the increasing use of densitometric methods of evaluating the results of electron histochemical tests, such organic reaction products will supersede the present techniques involving the particulate deposition of a heavy metal salt.

Most of the recent attempts to localize cholinesterase have relied upon modifications of the method of Koelle and Friedenwald,[28] in which acetylthiocholine is hydrolysed by the enzyme in the tissues, and the resultant thiocholine is precipitated by combination with copper ions. Miledi[37] used this technique to demonstrate cholinesterase in the synaptic vesicles in the motor end-plate, and in the synaptic cleft at the neuromuscular junction. Karnovsky[26] believes that these compounds form very specific substrates, and that they are hydrolysed by cholinesterases at very fast rates. In his method the thiocholine produced from the substrate reacts with ferricyanide ions and reduces them to ferrocyanide; these latter ions are then captured by copper ions to produce fine, electron-opaque deposits of copper ferrocyanide which sharply mark out the sites of enzyme activity. As a consequence of a very thorough series of control experiments, Karnovsky concludes that his technique gives a reliable picture of the ultrastructural location of cholinesterase.

NUCLEIC ACIDS

Unlike the localization of hydrolytic enzymes, no convenient precipitation technique exists for the nucleic acids, and hence most of the reported studies on these compounds have relied on the increase of contrast which may be produced by treatment with salts of heavy metals. Perhaps the first attempt to increase the contrast of nucleic acids was the work of Bernstein[6] in which iron, in the form of ferric chloride, was used to produce a definite increase in the electron density of the nucleic acid component of T_2 phage, which he used as a test object. Recent unpublished observations by the present author confirm that ferric salts work in the manner suggested by Bernstein, and this method may well be capable of further development into a useful electron-histochemical reaction. Other heavy metals which have been used for increasing the contrast of the nucleic acids include the uranyl salts[24] and trivalent indium.[66] One further interesting metal, which appears to possess some specificity for the phosphate group of the nucleic acids, is bismuth[1]. This substance seems to be very promising as a histochemical reagent, in view of its very low affinity for protein.

The reactions which have been suggested for nucleic acid are entirely empirical, but if they are carried out with the appropriate controls, such as selective extraction by enzymes, then valuable histochemical information may be gained. It seems that very few attempts have been made to apply the standard optical histochemical test for deoxyribonucleic acid (the Feulgen test) at the electron microscope level of resolution. One such early attempt[25] used a hexamine-silver compound to reveal the sites of aldehyde produced by the acid hydrolysis. The resolution was poor, however, and the ultimate resolution would in any case be limited by the size of the reduced silver. Bryan and Brinkley[10] have reported similar results.

PROTEIN GROUPINGS

Only sporadic attempts have been made to localize some of the characteristic amino-acids which are found in proteins. One approach, suggested by Lamb and his co-workers,[30] was the use of dinitrofluorobenzene to couple with the —NH_2, —SH or the phenolic groups of tyrosine. The product was then reduced with chromous sulphate and diazotized. This was followed by coupling with either 'K-acid' or 4 methyl 1,2 mercaptobenzene. In the former case, treatment with silver nitrate would produce an electron-dense silver salt, whilst treatment of the methyl mercaptobenzene with lead nitrate would give rise to the lead derivative, also of considerable electron density. Using bull spermatozoa as test objects, they found a marked increase in opacity to electrons (as determined by densitometric measurements on the electron microscope plates). This technique was not followed up, but with the superior technical methods now available it may well prove worth reinvestigation and development.

A few years later Kendall and Mercer[27] obtained similar increases in the electron density of protein by incorporating lead into a stable chelate with the protein. This work was carried out on tissues fixed in alcohol, and, in consequence, there was considerable disruption of the fine structure. Unfortunately, as in optical histochemistry, many of the most promising techniques for the localization of proteins are not satisfactory after fixation in aldehydes; and hence their successful application, combined with acceptable preservation of fine structure, may well depend upon the introduction of completely different methods of tissue preparation.

Harris and Mazia[22] attempted to use the protein stain mercuric bromophenol blue as an electron microscopical technique. Although they claimed some increase in the electron density of protein structures, in view of the dubious status of the method in standard histochemical practice, it seems that further success is unlikely from this approach.

The coupling of the —SH residue with various organic reagents, especially organic mercurial compounds, has possibilities for electron microscopy. Bahr and Moberger[2] used methyl mercury chloride, a compound with a high degree of specificity for sulphydryl groups, and Barrnett and his co-workers[4] tried the reagent 4 hydroxy-1-naphthol mercuric acetate. Nelson[38] tried the standard histochemical blocking agents for —SH, and claimed that with the electron microscope there was a marked increase in density; this was especially so with 0·0025 M n-ethyl maleimide and with 0·001 M p-chloromercuribenzoic acid. Although the authors of these papers obtained some measure of success, no method has yet been evolved which can demonstrate the —SH group with any degree of certainty; and a fruitful field of study is open here. I recently carried out some promising preliminary experiments (using sections of tissue which had been fixed for a short time in glutaraldehyde) in which another standard sulphydryl reagent, RSR or 1(4-chloromercuriphenylazo 2-naphthol) of Bennett,[5] produced marked increases in electron density of the granules in the cells of the posterior lobe of the bovine pituitary. The preservation of the fine structure was not good, however, and further work on this aspect is at present in progress.

The most promising new development in the field of protein localization has come from the work of Tice and Barrnett.[62] These workers have synthesized several new diazophthalocyanin compounds which contain either lead, copper or manganese in their molecules. All of these compounds will couple with the phenolic groups of proteins which have been fixed in aldehydes, in an analogous reaction to the famous 'coupled tetrazonium' test. The products when examined in the electron microscope show a significant density, especially with the use of mono (3 diazo) lead phthalocyanin at acid pH. This compound caused marked density increases in the components of the nucleus and endoplasmic reticulum, and in pancreatic zymogen granules.

CARBOHYDRATES

The first papers published in this field were devoted to the localization of glycogen in cells, either by the direct use of the periodic acid/Schiff method,[7] by staining with Best's carmine before examination in the electron microscope,[59] or by the use of alkaline solutions of silver salts.[36] More recently, attention has been turned to the possibilities of the localization of acidic mucosubstances, perhaps owing to the fact that glycogen is often fairly easy to detect, because of its characteristic 'rosette' morphology, after staining with alkaline solutions of lead salts. Again electron microscopists have turned to the convenient adaptation of a well-known test used with the optical microscope; this is the Hale reaction, which involves the uptake of colloidal ferric hydroxide in the tissue, and the subsequent location of this salt by producing from it crystals of ferric ferrocyanide or Prussian blue. Gasic and Berwick[19] applied this technique to isolated tumour cells, and succeeded in demonstrating the presence of the small cubic crystals of ferric ferrocyanide at the surface of these cells. By the use of selective enzyme extractions, they obtained evidence which suggested that the substance binding the colloidal iron was in fact a sialomucin. These authors used the complete procedure, maintaining that although it would be possible to detect the particles of colloidal metal in the electron microscope, the small size of the individual metal particles would require very high magnifications for their detection, and it would be very easy to confuse these particles with 'bodies of a similar size of a non-specific nature'.

Their method, successfully modified for use on sections by Bradbury and others,[8] was used to demonstrate the mucoprotein nature of the fibrinoid of the mouse placenta. Independently, work by Curran and colleagues[17] surveyed the use of the colloidal iron method, using sections of rat jejunum as a test object. Both of these studies have emphasized that the reaction is perfectly satisfactory if the first stage alone is carried out, as there is no difficulty in visualizing the very electron-dense particles of the colloidal ferric hydroxide, and the chances of confusing them with any pre-existing tissue constituent are small. Curran and his co-workers applied the reaction to small blocks of tissue, and also attempted to carry out the reaction by floating thin sections upon solutions of the reagents. They concluded that the best results were obtained by carrying out the reaction before the tissue was embedded, particularly if the 'target' cells were near the surface of the slice of tissue. The specificity of this reaction is low, but Curran feels that it may be used with some confidence, because the uptake of metal can be blocked by methylation (a technique which esterifies the carboxyl groups and removes the sulphate groups of the acid mucopolysaccharide). In addition, non-specific uptake on cut surfaces, and in the nucleoproteins, was virtually absent, and there was no binding in the case of tissues which had been fixed in formalin saturated with barium hydroxide, a technique claimed by Conklin[13] to remove mucopolysaccharides.

In recent months I have been engaged in a comprehensive study of colloidal metal techniques. Evidence has been obtained which indicates that the uptake of colloidal metal particles by the muco-substances is simply by a process of electrostatic binding, in which the positively charged particles of the metal are attracted and held by the negative charges on the ionized mucosubstance. It is a happy accident that colloidal ferric hydroxide, the first such reagent to be used by histochemists, possesses particles which have a positive charge. Many other metals and metallic salts can easily be prepared in the form of stable sols, e.g. platinum and gold, but the majority have particles which carry a negative charge. When these reagents were substituted for colloidal iron in the reaction, it was found that, as predicted, there was no specific uptake, only a limited amount of surface absorption (Fig. 12.2, B). When the charge on the particles was reversed, by the addition of ferric chloride solution to the platinum sol (or by the addition of thorium nitrate to the gold sol), the uptake of metal was once more specific (Fig. 12, C). The use of colloidal platinum with a reversed charge has some advantages over colloidal iron, as there is less tendency to aggregation (Curran also noticed the tendency of iron to aggregate) and the particle size is slightly smaller. The work of Curran with respect to blockage by methylation has been confirmed, and it has also been found that incubation of the tissues with the cationic detergent cetyl pyridinium chloride will also prevent the uptake of metal. This is attributed to the negative sites on the mucosubstances being occupied by the detergent molecules. In consequence of this competition there can be no uptake of metal. The action of the enzyme 'Hyalase' also diminishes the uptake of colloidal metal very markedly, both in the microvilli of the intestinal epithelial cell and in the placental fibrinoid, which were chosen as test-objects for this work. A detailed account of these researches will be published elsewhere in due course.

A similar technique for the localization of acid mucopolysaccharides is due to the work of Revel,[48] who floated methacrylate sections on an acid solution of colloidal thorium dioxide (thorotrast). Again he found that the particles of colloidal oxide were absorbed onto the surface of the sections only where the acidic mucosubstances were located. He, also, used methylation, and 'Hyalase' digestion as controls.

One of the well-established techniques for staining acid mucosubstances, is with the copper phthalo-cyanin dye Alcian blue; recent observations[62] indicate that this dye may also prove valuable in electron microscopy for the same purpose, as it produces an increase in electron density in sites known from other evidence to contain acid mucosubstances.

NORADRENALIN AND 5-HYDROXYTRYPTAMINE

Recent work has directed attention to the possibility of locating with the electron microscope, sites which are rich in noradrenalin or 5-hydroxytryptamine. One such technique[64] involves the reaction of glutaraldehyde with the noradrenalin to form a Schiff-base, which is then revealed by the precipitate which it forms with ammoniacal silver. An alternative approach[67] has succeeded in applying the well-known chromaffin reaction to electron microscopy. Treatment of the tissue with potassium dichromate at pH 4·1, following fixation in glutaraldehyde, renders the noradrenalin electron-dense, whereas adrenalin does not react in this way.

These methods are of great interest as, by their aid, interesting physiological problems may be studied. For instance, they have shown that in the ventro-medial nucleus of the rat hypothalamus, the transmitter substance (which resembles noradrenalin) seems to be located in small granules, which are quite distinct from the so-called synaptic vesicles. Wood[68] also claims to be able to modify the method so that it is specific for the substance 5-hydroxytryptamine, which is also of great physiological interest; much more work is, however, needed to establish the chemical specificity of these reactions when applied at the electron microscope level.

LIMITATIONS OF THE PRESENT METHODS

The rapid development of the methods surveyed above may be traced largely to the introduction of effective methods of fixation by aldehydes.[50] These allowed partial satisfaction of the conflict between

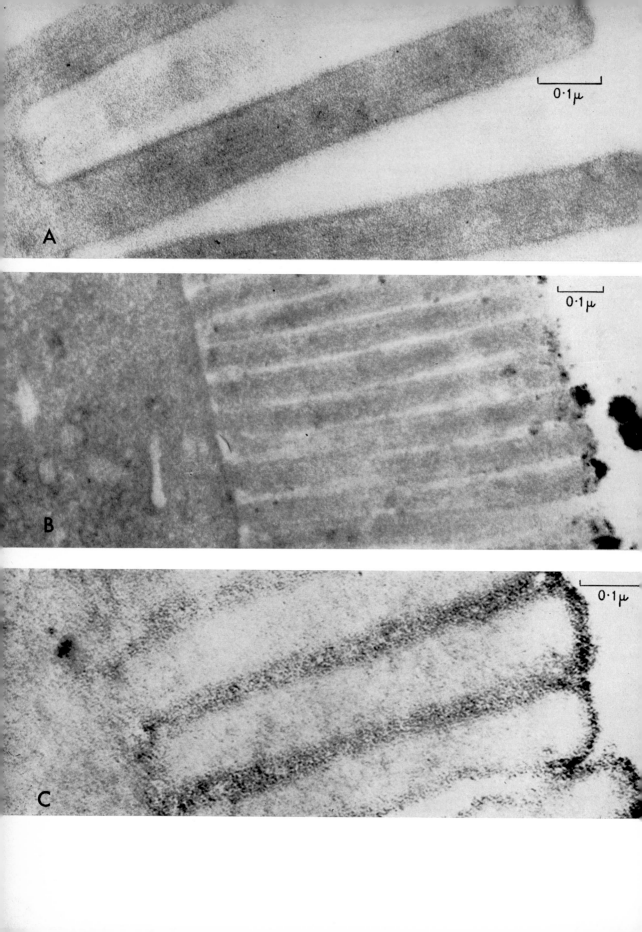

the acceptable preservation of the ultrastructure and the retention of the reactive chemical groupings. Success can only be achieved in this new field if the requirements of both electron microscopy and histochemistry are constantly borne in mind by the research worker, so that the one is not satisfied at the expense of the other.

Most of the present techniques represent direct attempts at adapting existing histochemical techniques to conform to the requirements of electron microscopy, and in order to achieve this it has often been necessary to fall back upon methods of very dubious chemical specificity. There is a danger in translating these methods to the electron microscope level in that, once a satisfactory contrast in the reaction product has been achieved at this level, no further thought is given to the specificity of the reaction. In consequence, large amounts of misleading information may be acquired, and much effort and time subsequently has to be spent in reconciling conflicting views.

It is often not realized that the requirements for the characters of substrates and reaction products, for example, are much more stringent when the technique is required to give valid information at the electron microscope level. Tsou[65] has emphasized this: he stresses that in addition to the accepted histochemical requirements of the substrate, there must also be no reaction with the plastic materials used as embedding media, the reaction product must be stable under the electron beam, and it must cause sufficient electron scattering to allow the detection of its presence. Most important of all, it should not possess any crystalline structure which can be resolved with the electron microscope. The newer formazans, especially that produced by Nitro BT, are satisfactory in this respect, as they deposit in an amorphous form, and their particles are about 35 Å diameter.

One possible source of error, recently pointed out,[47] is reaction of the end-product of the histochemical test with some of the reagents used in the later stages of tissue preparation; in particular, lead phosphate, formed in the Gomori technique may be lost by solution in the osmium tetroxide used for the treatment of the sections after the incubation has been carried out. It is obvious that care must be used to avoid false-negative reactions, from this and similar causes.

The limitations inherent in electron microscopy itself also impose themselves on attempts to develop electron histochemistry. This is particularly true when we attempt to consider the possible changes in the localization of substances during the course of an activity cycle of a cell or organelle. The electron microscope is poorly adapted to provide information about temporal changes in structure, and this also applies when chemical information is being obtained by the same instrument. Again, too, because of the stringent requirements as to the thickness of section which can be examined, the actual amount of tissue which can be surveyed is limited. So the problem is firstly, to obtain tests of sufficient sensitivity, and secondly, to obtain adequate samples, in order to give a representative picture of the localization of any substance in the three-dimensional structure of the cell. It is here that the importance of running parallel optical and electron microscopical studies cannot be over-emphasized, for the technique of optical histochemistry, although lacking the resolution of the electron microscope, enables the sampling to be extended to a much greater bulk of tissue, and to greatly extended time scales. In this way some measure is provided by which the results of the electron microscopy can be judged and extended.

There remains the uncertainty of obtaining adequate penetration by the very large molecules often used in electron histochemistry for the production of electron density at the specific sites in the tissue. In early work with bulk tissue, the comment was often made that there was a reaction in the cells in the outer part of the block, often in a zone which was only 100 μ or so in thickness. The current practice is to carry out the reactions on 40 μ frozen sections, so that the chances of the reagents reaching all the cells

Fig. 12.2 The use of colloidal metals to demonstrate acidic mucosubstances.
A. Control preparation of the microvilli of epithelial cells of the mouse small intestine, fixed in glutaraldehyde.
B. A similar preparation which has been treated with colloidal platinum. This metal has a negative charge; note that there is only irregular non-specific binding.
C. A preparation which has been treated with colloidal platinum which has had its charge reversed. Note that there is now a regular attachment of the particles of colloidal metal to the surface of the microvilli.

are now much improved. It is desirable, however, that this particular problem should be studied in detail in the near future. With the use of colloidal metals, there is also the further uncertainty of whether the large charged particles can actually penetrate the cells, and thus be available for binding onto substances inside the cell. If, as seems likely, there is no penetration of the cell by these colloidal particles, then it will prove a great drawback to the extension of their use in electron histochemistry. If they are limited to the localization of surface deposits of mucosubstances, then other, better methods must be developed.

The development of techniques in this new field has been directed to the production of particulate deposits in the section, either by the use of Gomori-type reactions, which precipitate the salt of a heavy metal, or by the binding of the particulate colloidal metals. Both of these are valid approaches, but by their very nature, it follows that the potential resolution of the microscope cannot be fully exploited, because of the large particle size. With, for example, the Gomori technique, a particle of reaction product is often deposited in the region of a membrane; because of the size of the particulate deposit it is often not possible to say whether the enzyme was located on the outside of the membrane, in its actual substance, or on the inner aspect of its surface. Exactly the same problem is faced by those who use electron autoradiographic techniques, and here, constant efforts are being made to increase the resolution of the method, by the production of smaller silver grains. If the metal-salt deposition techniques are to play a role in the future development of electron histochemistry, then efforts must be directed towards reducing the size of the particulate end-product, so that the resolution may be improved. Such considerations form a very strong argument in favour of developing completely new methods of electron histochemistry, which do not depend on the production of a precipitate, or microcrystalline end-product.

FUTURE PROSPECTS

Although the achievements of electron histochemistry to date have yielded interesting and often exciting information, the field is still at a very rudimentary stage. The way ahead is long and tedious. Nevertheless, the prospects of extending the value of the electron microscope encourage perseverance, and the production of new techniques in this field must inevitably occupy the attention of workers for some time to come. Recent technical advances, however, suggest that these efforts will soon be successful.

In the field of specimen preparation, attempts are being made, following the lead of optical histochemistry, to get away from the necessity of using chemical fixation. Freeze-drying is now practicable; also the technique of freeze substitution;[21] the micrographs which have been published leave little doubt that, very soon, the technical problems still associated with these methods will be solved. Another, completely new, approach is that of 'inert-dehydration',[43] in which there is no chemical fixation, but simply a substitution of the water in the living tissue by a suitable reagent such as ethylene glycol. This is a physical method, which seems to leave the proteins in their native state and hence in a condition which is very suitable for the use of histochemical reactions. A second method evolved by the same author[44] allows the preparation of thin sections in a completely anhydrous environment, and so increases the chances of retaining much more material *in situ*, a point of special importance when one is studying the more labile mucosubstances or carbohydrates.

Other promising developments in this field of specimen preparation include the introduction of new types of embedding media such as the so-called 'water-soluble methacrylates',[31, 32] which not only require much less stringent conditions for the dehydration of the tissues, but also allow the use of enzymes as selective extracting agents. This latter technique has not yet attained its full possibilities as a means of enhancing the specificity of many of the more empirical staining reactions which have been developed for producing contrast in the electron microscope.

Attention must also be paid to methods of revealing chemical groupings with the electron microscope. So far efforts have largely been aimed at devising methods for the production of particulate deposits of electron-dense material in the tissue. In the future we will doubtless move away from this concept, towards utilization of the *measurable* increase of contrast caused by the use of organic reagents of high chemical specificity. This may well be helped by the use of lower accelerating voltages for the electron beam than are common at the present time; as Cosslett[14] has shown, assuming constancy of section thickness, a given degree of contrast enhancement can be produced by a 25 per cent uptake of stain using electrons accelerated at 50 kV, whereas if the voltage is reduced to 10 kV only 5 per cent uptake of the same stain would be needed to achieve the same increase in contrast. It may well be necessary for the manufacturers of biological electron microscopes in future to provide much lower ranges of accelerating voltages than are found on present instruments.

Encouraging progress has already been made in the use of organic compounds;[62] other approaches to the same problem have been made by Seaman and others,[52] and by the synthesis by Seligman and his co-workers,[55] of reagents with reactive —SH in their molecules, which are able to react with osmium tetroxide after uptake into the tissues to form 'osmium black', an electron-dense end-product which is also suitable for optical microscopy. This situation, therefore, allows the correlation of light and electron histochemical observations on material prepared with the same technique.

At the same time completely new approaches to the subject are required. The movement away from the traditional methods and reagents of optical histochemistry is a valuable step forward, especially when the new methods are developed jointly by a histochemist, an electron microscopist, and an organic chemist. Integration of the purely chemical approach with the other available physical techniques, such as micro-incineration combined with electron diffraction for the characterization of the ash deposits,[60] is also a promising approach. Further correlation must also be made with the techniques of electron autoradiography, which are already providing so much valuable information, and with the microprobe analysis techniques, which are now under active development. Determination of the chemical specificity of any new technique is a formidable task when applied at the level of resolution of the electron microscope, but it must be attempted if this new approach is to play its proper part in the development of our knowledge of the composition of the structures revealed by the electron microscope.

SUMMARY

One of the present disadvantages of the electron microscope is that the densities apparent in micrographs cannot be interpreted in terms of the chemical groupings which exist in the cell cytoplasm. The early difficulties surrounding attempts to develop an electron histochemistry are considered, especially the compromises needed to obtain histochemical data and at the same time preserve an acceptable morphology.

Some of the methods (largely adapted from optical histochemistry) now available for the localization of enzymes, protein groupings, nucleic acids, and carbohydrates are summarized. The limitations of these methods, especially with regard to specificity, and imperfect localization or low contrast of the end-product, are discussed.

Future developments in specimen preparation, e.g. the use of freeze-drying and new embedding media are outlined and the need to develop completely new chemical approaches is stressed, together with the desirability of integration between the chemical methods of localization of compounds and the physical approach to analysis at the electron microscope level of resolution.

REFERENCES

1. ALBERSHEIM, P. and KILLIAS, U. 1963. The use of bismuth as an electron stain for nucleic acids. *J. Cell Biol.* **17**, 93–103
2. BAHR, G. F. and MOBERGER, G. 1954. Methyl mercury chloride as a specific reagent for protein bound sulphydryl groups. *Exp. Cell Res.* **6**, 506–18
3. BARRNETT, R. J. and PALADE, G. E. 1957. Histochemical demonstration of the sites of activity of dehydrogenase systems with the electron microscope. *J. biophys. biochem. Cytol.* **3**, 577–88
4. ——, ——, GOLDSTEIN, T. P., and SELIGMAN, A. M. 1958. Histochemical demonstration of protein bound sulphydryl and disulphide groups in epidermis and hair with the electron microscope. *J. Histochem. Cytochem.* **6**, 101–2
5. BENNETT, H. S. 1951. The demonstration of thiol groups in certain tissues by means of a new colored sulfhydryl reagent. *Anat. Rec.* **110**, 231–47
6. BERNSTEIN, M. H. 1956. Iron as a stain for nucleic acids in electron microscopy. *J. biophys. biochem. Cytol.* **2**, 633–4

7. BONDAREFF, W. 1957. Morphology of particulate glycogen in guinea pig liver revealed by electron microscopy after freezing and drying and selective staining 'en bloc'. *Anat. Rec.* **129**, 97–113

8. BRADBURY, S., BILLINGTON, W. D., and KIRBY, D. R. S. 1965. A histochemical and electronmicroscopical study of the fibrinoid of the mouse placenta. *J. Roy. micr. Soc.* **84**, 199–211

9. BRANDES, D., ZETTERQVIST, H., and SHELDON, H. 1956. Histochemical techniques for electron microscopy: alkaline phosphatase. *Nature, Lond.* **177**, 382–3

10. BRYAN, J. H. D. and BRINKLEY, B. R. 1964. A silver-aldehyde reaction for studies of chromosome ultra-structure. *Quart. J. micr. Sci.* **105**, 367–74

11. CARO, L. G. and PALADE, G. E. 1964. Protein synthesis, storage and discharge in the pancreatic exocrine cell. An autoradiographic study. *J. Cell Biol.* **20**, 473–95

12. CASTAING, R. and GUINIER, A. 1950. Applications des sondes électroniques à l'analyse metallographique. *Proc. of the conf. on electron microscopy, Delft, 1949.* 60–3

13. CONKLIN, J. L. 1963. Staining reactions after formalin-containing fixatives. *Stain Techn.* **38**, 56–9

14. COSSLETT, V. E. 1958. A symposium on electron staining. The quantitative aspects of electron staining. *J. Roy. micr. Soc.* **78**, 18–25

15. ——. 1965. Possibilities and limitations for the differentiation of elements in the electron microscope. *Lab. Invest.* **14**, 1009–19

16. —— and SWITSUR, V. R. 1963. Some biological applications of the scanning microanalyser. In *X-ray optics and X-ray microanalysis* (edited by H. H. Putee, V. E. Cosslett, and A. Engstrom), p. 507. New York: Academic Press

17. CURRAN, R. C., CLARK, A. E., and LOVELL, D. 1965. Acid mucopolysaccharides in electron microscopy. The use of the colloidal iron method. *J. Anat.* **99**, 427–34

18. ESSNER, E. and NOVIKOFF, A. B. 1961. Localization of acid phosphatase activity in hepatic lysosomes by means of electron microscopy. *J. biophys. biochem. Cytol.* **9**, 773–84

19. GASIC, G. and BERWICK, L. 1963. Hale stain for sialic acid-containing mucins. Adaptation to electron microscopy. *J. Cell Biol.* **19**, 223–8

20. HAKE, T. 1965. Studies on the reactions of OsO$_4$ and KMnO$_4$ with amino acids peptides and proteins. *Lab. Invest.* **14**, 1208–12

21. HARREVELD, A. VAN and CROWELL, J. 1964. Electron microscopy after rapid freezing on a metal surface and substitution fixation. *Anat. Rec.* **149**, 381–6

22. HARRIS, P. and MAZIA, D. 1959. The use of mercuric bromophenol blue as a stain for electron microscopy. *J. biophys. biochem. Cytol.* **5**, 343–4

23. HOLT, S. J. and HICKS, R. M. 1961. The localization of acid phosphatase activity in rat liver cells as revealed by combined cytochemical staining and electron microscopy. *J. biophys. biochem. Cytol.* **11**, 47–66

24. HUXLEY, H. E. and ZUBAY, G. 1961. Preferential staining of nucleic acid containing structures for electron microscopy. *J. biophys. biochem. Cytol.* **11**, 273–96

25. JURAND, A., DEUTSCH, K., and DUNN, A. E. G. 1958. Application of silver-Feulgen method in electron microscopy. *J. Roy. micr. Soc.* **78**, 46–9

26. KARNOVSKY, M. J. 1964. The localization of cholinesterase activity in rat cardiac muscle by electron microscopy. *J. Cell Biol.* **23**, 217–32

27. KENDALL, P. A. and MERCER, E. H. 1958. A technique for staining intracellular protein. *J. Roy. micr. Soc.* **78**, 40–3

28. KOELLE, G. B. and FRIEDENWALD, J. S. 1949. A histochemical method for localizing cholinesterase activity *Proc. Soc. Exp. Biol. Med.* **70**, 617–22

29. LAMB, W. G. P., STUART-WEBB, J., BELL, L. G., BOVEY, R., and DANIELLI, J. F. 1963. Specific stains for electron microscopy. *Exp. Cell Res.* **4**, 159–63

30. LEDUC, E. H. and BERNHARD, W. 1961. Ultrastructural cytochemistry. Enzyme and acid hydrolysis of nucleic acids and proteins. *J. biophys. biochem. Cytol.* **10**, 437–55

31. ——, MARINOZZI, V., and BERNHARD, W. 1963. The use of water-soluble glycol methacrylate in ultra-structural cytochemistry. *J. Roy. micr. Soc.* **81**, 119–30

32. —— and HOLT, S. J. 1965. Hydroxy propyl methacrylate, a new water-miscible embedding medium for electron microscopy. *J. Cell Biol.* **26**, 137–55

33. LEHRER, G. M. and ORNSTEIN, L. 1959. A diazo coupling method for the electron microscopic localization of cholinesterase. *J. biophys. biochem. Cytol.* **6**, 399–404

34. LIQUIER-MILWARD, J. 1956. Electron microscopy and radioautography as coupled techniques in tracer experiments. *Nature, Lond.* **177**, 619

35. MARCHESI, V. T. and BARRNETT, R. J. 1964. The localization of nucleoside diphosphatase activity in different types of small blood vessels. *J. Ultr. Res.* **10**, 103–15

36. MARINOZZI, V. 1961. Cytochimie ultrastructurale. Fixations et colorations. *Experientia* **17**, 429

37. MILEDI, R. 1964. Electron microscopical localization of products from histochemical reactions used to detect cholinesterase in muscle. *Nature, Lond.* **204**, 293–5

38. NELSON, L. 1960. Cytochemical studies with the electron microscope III. Sulphydryl groups of rat spermatozoa. *J. Ultr. Res.* **4**, 182–90
39. NOVIKOFF, A. B. and ESSNER, E. 1960. The liver cell. Some new approaches to its study. *Amer. J. Med.* **29**, 102–31
40. —— and GOLDFISCHER, S. 1961. Nucleosidediphosphatase activity in the Golgi apparatus and its usefulness for cytological studies. *Proc. Nat. Acad. Sci.* **47**, 802–10
41. OGAWA, K. and BARRNETT, R. J. 1965. Electron cytochemical studies of succinic dehydrogenase and dihydronicotinamide-adenine dinucleotide diaphorase studies. *J. Ultr. Res.* **12**, 488–508
42. PEARSE, A. G. E. 1963. Some aspects of the localization of enzyme activity with the electron microscope. *J. Roy. micr. Soc.* **81**, 107–17
43. PEASE, D. C. 1966. The preservation of unfixed cytological detail by dehydration with 'inert' agents. *J. Ultr. Res.* **14**, 356–78
44. ——. 1966. Anhydrous ultrathin sectioning and staining for electron microscopy. *J. Ultr. Res.* **14**, 379–90
45. PEDLER, C. M. H. 1965. The serial reconstruction of a complex receptor synapse. *Proceedings of the International Congress of Anatomy, Wiesbaden,* Eye structure, IInd Symposium, edited by J. W Rohen, **29**; 53
46. PELC, S. R. 1963. Theory of electron autoradiography. *J. Roy. micr. Soc.* **81**, 131–9
47. REALE, E. and LUCIANO, L. 1964. A probable source of errors in electron histochemistry. *J. Histochem. Cytochem.* **12**, 713–15
48. REVEL, J.-P. 1964. A stain for the ultrastructural localization of acid mucopolysaccharides. *J. Microscopie* **3**, 535–44
49. ROSTGAARD, J. and BEHNKE, O. 1965. Fine structural localization of adenine nucleosidephosphatase activity in the sarcoplasmic reticulum and the T system of rat myocardium. *J. Ultr. Res.* **12**, 579–91
50. SABATINI, D. D., BENSCH, K., and BARRNETT, R. J. 1963. Cytochemistry and electron microscopy. The preservation of cellular ultrastructure and enzymatic activity by aldehyde fixation. *J. Cell Biol.* **17**, 19–58
51. ——, MILLER, F., and BARRNETT, R. J. 1964. Aldehyde fixation for morphological and enzyme histochemical studies with the electron microscope. *J. Histochem. Cytochem.* **12**, 57–71
52. SEAMAN, A. R., HAWKER, J. S., and SELIGMAN, A. M. 1961. Electron microscopic demonstration of cytochemical functional groups of macromolecules by the use of new electron opaque reagents. *J. Histochem. Cytochem.* **9**, 596–7
53. SEDAR, A. W. and ROSA, C. G. 1958. Cytochemical demonstration of succinic dehydrogenase with the electron microscope, using nitro blue tetrazolium (NBT) as an indicator dye. *Anat. Rec.* **130**, 371
54. —— and ——. 1961. Cytochemical demonstration of the succinic dehydrogenase system with the electron microscope using nitro-blue tetrazolium. *J. Ultr. Res.* **5**, 226–43
55. SELIGMAN, A. M. 1964. Some recent trends and advances in enzyme histochemistry. *Proc. 2nd Int. Congress of Histo- and Cytochemistry. Frankfurt, 1964,* 9–21
56. SHELDON, H., ZETTERQVIST, H., and BRANDES, D. 1955. Histochemical reactions for electron microscopy: acid phosphatase. *Exp. Cell Res.* **9**, 592–6
57. SINGER, S. J. 1959. Preparation of an electron-dense antibody conjugate. *Nature, Lond.* **183**, 1523–4
58. STOECKENIUS, W. and MAHR, S. C. 1965. Studies on the reaction of osmium tetroxide with lipids and related compounds. *Lab. Invest.* **14**, 1196–1207
59. THEMANN, H. 1960. Zur elektronenmikroskopischen Darstellung von Glycogen mit Best's Carmin. *J. Ultr. Res.* **4**, 401–12
60. THOMAS, R. S. 1964. Ultrastructural localization of mineral matter in bacterial spores by microincineration. *J. Cell Biol.* **23**, 113–33
61. TICE, L. W. and BARRNETT, R. J. 1962. Fine structural localization of adenosine triphosphatase activity in heart muscle myofibrils. *J. Cell Biol.* **15**, 401–16
62. —— and ——. 1965. Diazophthalocyanins as reagents for fine structural cytochemistry. *J. Cell Biol.* **25**, 23–42
63. TORACK, R. M. and BARRNETT, R. J. 1963. Nucleoside phosphatase activity in membranous fine structures of neurons and glia. *J. Histochem. Cytochem.* **11**, 763–72
64. TRAMEZZANI, J. H., CHIOCCHIO, S., and WASSERMAN, G. F. 1964. A technique for light and electron microscopic identification of adrenalin and nor-adrenalin storing cells. *J. Histochem. Cytochem.* **12**, 890–9
65. TSOU, K. C. 1964. Application of histochemical enzyme substrates for electron microscopy. *Proc. 2nd Int. Congress of Histo- and Cytochemistry. Frankfurt, 1964,* 165
66. WATSON, M. L. and ALDRIDGE, W. G. 1961. Methods for the use of indium as an electron stain for nucleic acids. *J. biophys. biochem. Cytol.* **11**, 257–72
67. WOOD, J. G. and BARRNETT, R. J. 1964. The histochemical demonstration of nor-epinephrine at a fine structural level. *J. Histochem. Cytochem.* **12**, 197–209
68. ——. 1965. Electron microscope localization of 5 hydroxytryptamine (5HT). *Texas Reports of Biol. and Medicine* **23**, 828–37

Chapter 13

The Histochemistry of Phospholipids and its Significance in the Interpretation of the Structure of Cells

J. CHAYEN

At the time when Dr. John R. Baker was laying a firm basis for the histochemistry of lipids, few workers would deign to concern themselves with 'fats'. Everyone knew that cells were made of protein or of nucleoprotein; 'fats' were not fashionable to work with. To some extent this was understandable because the biochemistry of lipids and of the various lipid–protein and carbohydrate–lipid complexes was relatively unknown before the elaboration of the new methods, and the detailed investigations of Folch (e.g. references 1, 2, 3) and then of Klenk (see, e.g. Lovern[4]). However, because of the decisive simplicity of Dr. Baker's test experiments, the essentials on which the present histochemical understanding of lipids rests, had already been prepared by him even before the biochemistry was fully known. Thus his study of the lipids is not only of importance in histochemistry, but is a lesson in the value of true scientific inquiry, without concern for current fashions or 'band-wagons'.

THE DEMONSTRATION OF PHOSPHOLIPIDS

At present there are two main types of procedure for demonstrating the presence of phospholipids in sections or smears. The first depends on the physical nature of these lipids: since they are more or less fatty they will concentrate lipid colourants, namely coloured or fluorescent compounds which dissolve very readily in lipids, from aqueous or alcoholic solutions. Free fats, like fat droplets (especially if they present a large fatty surface), will concentrate such lipophilic substances very readily. The colouring of fat droplets by Sudan compounds or by oil red O, dissolved in 60 or 70 per cent alcohol, is well known. This colouring depends on the different partition coefficients of the lipophilic substance in the free fat, and in the partially aqueous solvent. The process is analogous to the extraction of a hydrocarbon in alcoholic dispersion by shaking it in a separating funnel with ether and water. The hydrocarbon will concentrate in the ethereal layer.

There are certain disadvantages about this type of procedure, as it is used generally. The first is that the colourant is dissolved in an alcoholic solution, and this may affect the tissue lipids. The second is that, for most of these compounds, a relatively large volume of fat is required to ensure that sufficient of the colourant will be concentrated into the tissue. Hence, in the main, only free fats will be demonstrated. The advantage of Sudan black is that it is concentrated into some protein-bound lipids, but it still suffers from the drawback that it must be dissolved in 70 per cent alcohol. This means not only that the alcohol may affect the tissue lipids, but also that the colourant is very soluble in the solvent, so reducing its efficiency in partitioning between the solvent (70 per cent alcohol) and the tissue lipid. The use of the fluorochrome 3, 4–benzpyrene[5] overcomes both these difficulties. It is dispersed in a hydrotropic solution, that is to say it is suspended in an aqueous medium in which it is insoluble, and therefore it can partition itself more readily into the tissue lipids. Moreover, since it is in aqueous dispersion, it can be added to a tissue culture medium or to a Ringer's solution and so can be applied to living cells. Rhodamine B has been used in a similar manner (e.g. by Strugger[6]), but seems to be less lipophilic. Phosphine 3R[7] has also been of some value but, to the present author, it seems a less powerful lipid colourant than is benzpyrene.

Thus it is possible to assess the degree of fattiness of a part of a tissue section, or even of a living cell, by the use of such colourants. The second type of procedure depends on Baker's[8] acid haematein method. It is indeed remarkable that no other method has yet been found to equal this for the amount of information it can give concerning phospholipids. The aniline-blue and orange G stain of La Cour and Chayen[9] almost certainly depends on the physical nature of phospholipids which confers on them a capability of differentiating between the milling and levelling dyes used in this mixture (for an extremely clear discussion see Baker,[10] the present author is indebted to Dr. Baker for this interpretation). An attempt has been made[11] to stain phospholipids by a specific reaction for choline, but the reaction itself is rather severe, and suitable primarily for extracellular matter. Thus we are left with the original acid haematein method, which differs from other methods, for example the benzpyrene technique, in that it depends not on the physical fattiness of the lipids but on their chemical nature. The evidence deserves to be related in some detail.

BAKER'S ACID HAEMATEIN METHOD

The original study on this technique[8] is a model for scientific research workers. Dr. Baker reviewed the earlier related methods, and then applied his modification to samples of biologically interesting

Table 13.1 The intensity of the acid haematein reaction with various materials tested on cigarette paper (All substances at concentration of 0·1 μmole per drop except where stated)

Substance	1st test[a]	2nd test	
		1st slide	2nd slide
1. Phosphatidyl ethanolamine	+ + +	+ + +	
2. Lecithin	+ +	+ + + +	+ + + +
3. Hydrogenated lecithin	−		
4. Lysolecithin	−	−	−
5. Fatty acids from hydrolysis of lecithin	−		
6. Aqueous hydrolysate from lecithin	−		
7. Mixture of 5 and 6	−		
8. Lysolecithin + oleic acid	±	+	+
9. α-glycerophosphate + oleic acid	−	−	−
10. Oleic acid	− [b]	±	doubtful
11. Phosphatidic acid	+ +		
12. Lysophosphatidyl ethanolamine		+	+
13. Fatty acids from hydrolysate of phosphatidyl ethanolamine		+ + + +	
14. Glycerophosphate + ethanolamine hydrolysate of phosphatidyl ethanolamine		−	
15. 14 + 15		+ +	
16. Phosphatidyl serine		+ + + +	
17. Sphingomyelin		+	+
18. Cerebroside		+	+
19. Sulphatide (only 0·02 μmole)		−	
20. Palmitic acid		−	−
21. Triolein		−	−
22. Tristearin		−	−
23. α-Glycerophosphate		−	−
24. Cholesterol		doubtful	−
25. Cholesterol oleate		+	+

[a] Retention of organic phosphate confirmed by staining method.
[b] Positive with oil red O, showing that it had not been lost from the cigarette paper.

chemicals impregnated into cigarette paper. In spite of the relative impurity of the lipids which could be prepared at that time, he was able to show that his acid haematein method differentiated this class of compound clearly from others. Some substances appeared to interfere with the reaction; in the light of present knowledge such interference was probably due either to their contamination with lipids, or to the presence of lipids which were bound by chemical linkages. The extent to which lipids form stable associations with proteins and with carbohydrates was not known, even biochemically, at that time. In attempting to discover the mechanism of this test, he examined a cerebroside—a galactolipin—which therefore lacked the phosphate moiety, and he suggested that its lack of response might indicate that this acidic group was essential for the reaction.[12] Unfortunately, at that time, it could not be known that such cerebrosides were frequently isolated with fully saturated fatty acids, so that this factor was not considered. This work, and its related more applied studies, together with other work by his students on lipid histochemistry, culminated in the excellent review by Cain[13] which has been a guide for many histochemists.

More recently the cigarette-paper experiments have been emulated, but with lipids isolated by modern biochemical procedures.[14,15] It was found that while phosphatidyl choline and ethanolamine yielded strong stains with the acid haematein procedure, this response was more or less abolished if they were previously hydrogenated chemically. All the response was lost if the unsaturated fatty acid was removed enzymatically to yield a saturated lysolecithin. The fatty acids isolated from the hydrolysed phosphatidyl ethanolamine gave a very strong reaction, while the glycerophosphate and ethanolamine moieties gave no response (Table 13.1). The final conclusion was that the reaction depended primarily on the unsaturated bond of the fatty acids; the phosphate may have some influence if only in allowing a hydrophilic entry for the dichromate to the molecule, or in making a stronger complex with the lake.

It seems possible that the mechanism of the acid haematein reaction may be similar to that involved in the formation of chrome colours in textiles (see Fieser and Fieser[16]). In these the fibre is impregnated with the chromium by boiling in sodium dichromate in the presence of a reducing agent like formic or oxalic, acid. These agents reduce the dichromate to chromium hydroxide $Cr(OH)_3$ or possibly to the hydrated oxide, $Cr_2O_3 . 3H_2O$. The chromium then lakes the azo-dye to the textile:

azo-dye

In the acid haematein reaction, the unsaturated bonds of the fatty acids would take the place of the lactic or oxalic acid reducing agent, and haematein would be substituted for the azo-dye. Whether the lake can be attached to the fatty acid, or whether it preferentially migrates to the phosphate moiety, is not known. (This question of the reaction of chromium with phospholipids has been discussed by Baker.[10])

The significance of the double-bond in relation to the acid haematein method when applied to tissue sections was demonstrated by Bitensky,[17] who subjected fresh, unfixed cryostat sections of rat liver to bromination, and showed that this depressed the stainability of the tissue by the acid haematein reaction. She showed, moreover, that when such sections were treated with methanol-chloroform for relatively short periods, and were then tested by the acid haematein method, they stained much more strongly, not less strongly; prolonged treatment with this fat-solvent did appear to extract the unmasked lipid. That the increased stain was due to lipid which had become unmasked was shown, subsequently, by the fact that such sections also stained very much more intensely with benzpyrene. Thus a paradoxical situation obtained, in which a fat-solvent increased the apparent fattiness and phospholipid content of the tissues. This work was confirmed and extended,[18] particularly in studies on lipids of arterial walls. The explanation would seem to lie in the mechanism by which methanol–chloroform acts with such peculiar efficiency: it seems that the role of the methanol is two-fold, firstly to overcome hydrophilic obstruction to the chloroform, and secondly to split lipid–protein bonds; once the binding has been ruptured the freed lipid is soluble in fat-solvents.

Bitensky and Chayen[19] then made use of this unmasking of phospholipid by methanol–chloroform. Serial sections of rat liver were treated by the acid haematein method:

(a)   without prior fixation (as for substances tested on the cigarette papers): a fair reaction was obtained, typically on mitochondria and similar-sized particles.

(b)   after bromination: even after treatment for 5 hr the subsequent acid haematein reaction was weak, but appreciable. Thus it looked as if bromination did not entirely abolish the reaction when done in sections.

(c)   after methanol–chloroform for short periods: the acid haematein reaction was greatly enhanced, throughout the cell.

(d)   after methanol–chloroform followed by bromination: there was no response whatsoever to the acid haematein reaction. In contrast to (b) it followed, therefore, that bromination can completely block the fully unmasked fatty acids of phospholipids. The only obvious explanation of these apparently conflicting results (b and d) was that the bromination of unfixed sections did block all free phospholipids, but that the acid haematein reaction itself unmasked some more unsaturated groups. When all these groups were fully unmasked by the methanol–chloroform, then bromination could entirely abolish the reaction, since there were no more unsaturated fatty acids for the acid haematein reaction to unmask.

To examine this further, other sections were treated with an aqueous 1 per cent solution of calcium chloride at 60°C for 1 day and for 3 days, and were then tested with the acid haematein method; they showed progressive increase in staining, so confirming that the hot water (and possibly the calcium) could unmask lipids.

This conclusion raised two major doubts: firstly, if the reaction itself does unmask some phospholipids, how much of the structural phospholipids—as contrasted with those associated with free fats— were in an unmasked state in life? Secondly, what was the structural significance of the fact that methanol–chloroform unmasked unsaturated bonds before it rendered the lipid soluble in fat solvents? The answers to these questions are inter-related and, as with most problems in lipid histochemistry, their solution derives from Dr. Baker's earlier studies on the pyridine extraction test.

BOUND LIPIDS

Once he had developed the acid haematein method, Baker[8] was concerned to provide some form of control for it. At that time, the simplest control for fatty material seemed to be to extract it with a fat solvent, but he, like the lipid biochemists of that period, found this surprisingly difficult. He advocated the use of pyridine, acting on sections which had been fixed in a weak Bouin's fluid which did not seriously enhance their resistance to extraction. Pyridine was probably the best solvent then available, and yielded results which were rather confusing, but which, in the light of present knowledge, are not surprising. Thus, while some acid haematein-positive material was extracted by this solvent, other parts of cells became strongly positive to the reaction. Consequently, although he was well aware of the possibility that this latter positive reaction represented some unmasked lipid, Baker[8] emphasized that the only parts which definitely contained lipid were those which gave a positive acid haematein test but a negative pyridine extraction test (the acid haematein procedure done after extraction with hot pyridine). This question has been discussed by Chayen and others,[20] who showed that as much as 90 per cent of the matter extracted by pyridine could be non-lipid material, an appreciable proportion of which might possibly be 'masking' the structural lipid which, when unmasked, gave this disturbing increase in staining after treatment with this lipid–solvent. Consequently the present author favours the following tests:

(a) for the proof of the presence of lipid material: benzpyrene; (b) for evidence that the lipid material is phospholipid: the acid haematein test with bromination as control, possibly after unmasking with methanol–chloroform.

The whole question of the extraction of lipids, and hence of structurally bound lipids had to wait for the biochemical studies of Folch and his co-workers,[1,2,3] and of the many workers who developed chromatographic and other techniques for the study of these substances (some are referred to by Chayen and others;[20] others can be found in Lovern[4]). Two studies may be of particular interest for the histochemical aspect of the problem.

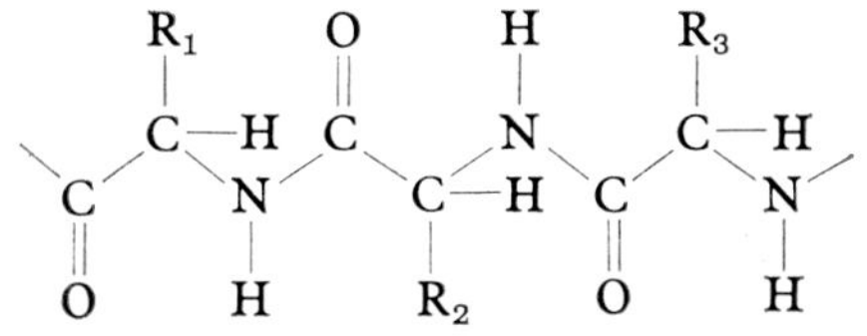

(a) Conventional polypeptide pattern of a protein molecule.

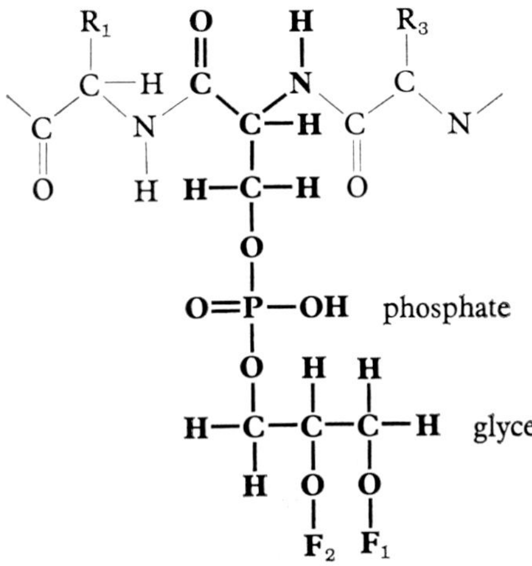

(b) Phosphatidyl serine (in heavy type) inserted in the place of serine ($R_2CH.NH_2.COOH$) in the polypeptide chain.

Fig. 13.1

The first concerned nucleohistone, which was one of the substances which appeared to give a falsely positive response to the acid haematein test. The nucleohistone was prepared in bulk from calf thymus, and it was then degraded by fairly severe acidic or alkaline hydrolysis. The hydrolysate was then analysed for such components of phospholipids as choline and glycerophosphate, none of which should have been present in nucleohistone. To the surprise of the investigators, remarkably high contents of these compounds were found,[20] including sphingomyelin[31] indicating that the acid haematein test had not been misleading. Of particular interest was the fact that much of this lipid material could not be demonstrated until the protein had been totally digested with acid. Consequently, much of this lipid must be of a structural nature, bound tightly to the structural protein of chromosomes or nuclei, very much as indicated by the detailed studies of Lloyd Thomas and his co-workers.[21,22] Hence this lipid would be expected to be 'masked' in most preparations, whereas the formalin-soluble lipid–protein complex which seems to act as a sheath around the chromosomes[23,24] may be less strongly masked.

For the second study,[25] two lipid–protein complexes (A and B) were isolated from formalin which had been used for fixing human lungs. Each complex was extracted separately with successive baths of boiling ethanol–ether. Complex A yielded over 30 per cent of its weight to this solvent; Complex B was hardly affected. They were then subjected to degradative hydrolysis, and the fatty acids liberated by this hydrolysis were collected and weighed. Complex A yielded very little; Complex B now yielded nearly 30 per cent of its weight as fatty acids, in spite of the fact that before its degradation the fat solvent had been able to extract no fat.

LIPID–PROTEIN STRUCTURES

We are faced, therefore, with the task of explaining how fats can be so linked within a lipid–protein complex that they are not extractable by fat solvents, and how the unmasking and freeing of such lipids can be correlated with the freeing of the double bonds of the fatty acids.

The clue to the first question may lie in the fact that the most intractable lipids can be isolated only after the associated protein has been degraded mainly to the constituent amino-acids. This might indicate that in these complexes, the amino-acid of such molecules as phosphatidyl serine, was combined by both its amino and carboxyl group, within the polypeptide chain of the protein. Other phosphatidic

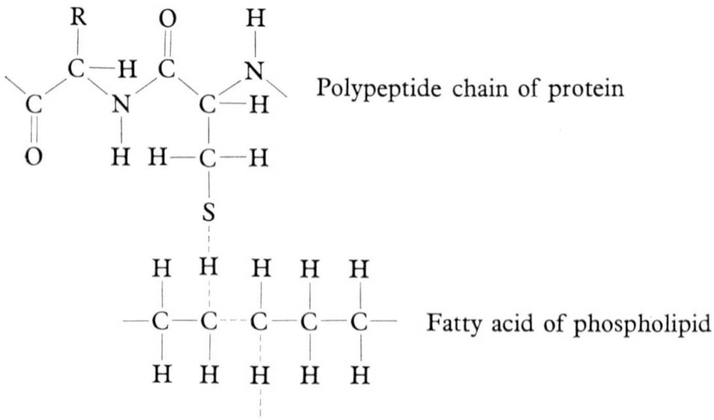

Fig. 13.2 Scheme of the type of mechanism by which fatty acids might be linked through a potentially unsaturated group to a protein. The potentially unsaturated group, becoming —C=C—, is shown by broken valency lines.

acid-amino acid compounds are known, and these might be considered as side-arms to the protein rather than as separate phospholipids loosely connected to the protein (Fig. 13.1).

The second question is open to an even more exciting possibility. When the bromination experiments described above were being done, yet another test was tried. Sections were brominated and then treated with methanol–chloroform. When these were stained with the acid haematein method, they produced almost as much reaction as did those which had not been brominated but had been unmasked with methanol chloroform. Thus the bromine had not been able to penetrate into the double-bonds of the fatty acids. This seemed very surprising: it is difficult enough to believe that unsaturated bonds can be present but not available to the dichromate until they have been magically unmasked—but dichromate is a large ion relative to bromine. It is therefore far more difficult to believe that bromine cannot penetrate to these double-bonds. The more likely answer, however surprising it may be, is that the double-bonds do not exist, as such, in life, but are only potentially unsaturated. Thus the bonds could perhaps be linked structurally as in the sort of configurations shown in Fig. 13.2. Some suggestion that hydrogen-bonding might be involved could be derived from the fact that urea can unmask some acid–haematein-positive material.[18] Moreover, the removal of protein by proteolytic enzymes may enhance the acid haematein response.[18]

THE STRUCTURAL SIGNIFICANCE OF LIPID–PROTEIN COMPLEXES

It is particularly the evidence that some lipids may become extractable only after proteolytic or hydrolytic degradation of 'protein', that leads to the possibility that protoplasm itself may not be protein, but lipid–protein with the lipid tightly conjugated to the protein. It is possible that some of this association is effected through those bonds which, when the association is disrupted, become 'unsaturated' and so become 'unmasked' to the acid–haematein test. The importance of unsaturated fats in reducing the dangers of arteriosclerosis may be connected with the importance of such bonds in the integrity of arterial walls.[18] These bonds certainly become dislocated in dietary protein deficiency[26] and in tissues under various types of stress.[27, 28]

The other vital function of lipid–protein complexes is in maintaining the structure of membranes. Of particular interest to those concerned with metabolic histochemistry, whether from the viewpoint of the action of drugs, of toxic substances, or of disease, is the function of the membranes around intracellular organelles such as lysosomes and mitochondria. This functional structure can be tested by the rate at which intra-mitochondrial or intra-lysosomal enzymes can act on exogenous substrate, since the rate of activity is a function of the speed with which the substrate, and co-factors, can penetrate through the membrane. For such studies, quantitative enzyme histochemistry is necessary and has the advantage over conventional biochemistry that, if done correctly, the permeability of the membranes will not be modified by the isolation, or preparative, procedures.

Finally, mention should be made of those other lipid–protein structures which lie on the surface of, as well as deep down inside the body of, chromosomes. It is possible that it is they, rather than the nucleic acids, which are the site of the primary action of ionizing radiations. This follows from the work of Elbert and others[29] which showed that the breakage per unit dose of X-rays in different gas environments approximated to the effects expected at a lipid–water, not a nucleoprotein–water interface. Certainly, however, the action of radiation on chromosomes is more likely to be on these chromosomal lipid–protein complexes rather than on those of the lysosomal membrane which, by releasing deoxyribonuclease, has been imagined[30] to affect the chromosomes secondarily.

ENVOI

This essay contains much speculation which may not find favour. The object of it, however, is to try to show how the concept of 'lipid' has altered from that of 'fat', regarded mainly as a sort of 'food reserve inclusion' in cells, to one of a vital structural component of living matter. Not only is its molecular composition of importance in the architecture of membranes, and of cellular 'ground substance', but the way in which it is bonded to protein may affect the physiological function of the cell. Yet this appreciation of the all-pervading importance of lipids stems directly from the clear-sighted, delightful studies of Dr. John R. Baker, which were performed without regard to whether such investigations were fashionable or not.

REFERENCES

1. FOLCH, J., ARSOVE, S., and MEATH, J. A. 1951a. *J. biol. Chem.* **191**, 819
2. —— and LEES, M. 1951b. *J. biol. Chem.* **191**, 807
3. —— and LE BARRON, F. N. 1956. *Canad. J. Biochem. Physiol.* **34**, 305
4. LOVERN, J. A. 1957. *The Chemistry of Lipids of Biochemical Significance*, 2nd edition. London: Methuen
5. BERG, N. O. 1951. *Acta Path. Microbiol. Scand.* Suppl. 90
6. STRUGGER, S. 1938. *Protoplasma* **30**, 85
7. VOLK, B. W. and POPPER, H. 1944. *Amer. J. clin. Path.* **14**, 234
8. BAKER, J. R. 1946. *Quart. J. micr. Sci.* **87**, 441

9. LA COUR, L. F. and CHAYEN, J. 1958. *Exp. Cell. Res.* **14**, 462
10. BAKER, J. R. 1958. *Principles of Biological Microtechnique*, pp. 235–6. London: Methuen
11. BOELSMA-VAN HOUTE, E. 1965. *Histochemie van Fosfolipiden in verband met Atherosklerose van de Aorta.* Doctor's degree thesis, University of Leiden: Steufert Kroese
12. BAKER, J. R. 1947. *Quart. J. micr. Sci.* **88**, 463
13. CAIN, A. J. 1950. *Biol. Rev.* **25**, 73
14. CHAYEN, J., BITENSKY, L., and LONG, C. 1964. *Proc. 2nd Int. Congr. Histochem. Cytochem.* p. 148
15. LONG, C. and CHAYEN, J. 1964. *Ann. Rep. Roy. Coll. Surg.* p. 101
16. FIESER, L. F. and FIESER, M. 1953. *Organic Chemistry*, p. 939. London: Harrap
17. BITENSKY, L. 1962. Ph.D. thesis, University of London
18. MAGGI, V., CHAYEN, J., GAHAN, P. B., and BRANDER, W. 1964. *Exp. Mol. Path.* **3**, 413
19. BITENSKY, L. and CHAYEN, J. (unpublished data)
20. CHAYEN, J., GAHAN, P. B., and LA COUR, L. F. 1959. *Quart. J. micr. Sci.* **100**, 325
21. SMITH, J. T., FUNCKES, A. J., BARAK, A. J., and THOMAS, L. E. 1957. *Exp. Cell Res.* **13**, 96
22. BRUEMMER, N. C. and THOMAS, L. E. 1957. *Exp. Cell Res.* **13**, 103
23. LA COUR, L. F., CHAYEN, J., and GAHAN, P. B. 1958. *Exp. Cell Res.* **14**, 469
24. CHAYEN, J., GAHAN, P. B., and LA COUR, L. F. 1959. *Quart. J. micr. Sci.* **100**, 279
25. —— CHAYEN, R., CUNNINGHAM, G. J., and BITENSKY, L. 1961. *Path. et Biol.* **9**, 925
26. GAHAN, P. B., BITENSKY, L., CHAYEN, J., CUNNINGHAM, G. J., and MAGGI, V. 1963. *Quart. J. micr. Sci.* **104**, 39
27. NILES, N. R., BRAIMBRIDGE, M. V., CUNNINGHAM, G. J., BITENSKY, L., and CHAYEN, J. 1964. *Lancet*, May 2, 963
28. BITENSKY, L. 1963. *Brit. Med. Bull.* **19**, 241
29. EBERT, M., HORNSEY, S., and HOWARD, A. 1958. *Nature, Lond.* **181**, 613
30. ALLISON, A. C. and PATON, G. R. 1965. *Nature, Lond.* **207**, 1170
31. CHAYEN, J. 1961. In *The Biochemists Handbook* (edited C. Long), p. 754. London: Spon

Chapter 14

Phospholipids and Nuclear Structure

ALEKSANDRA PRZEŁĘCKA

INTRODUCTION

In the last few years, the rapid development of the morphological and biochemical sciences has revealed many details of cellular structure. The cytoplasmic subunits have become well defined not only in their morphological but also in their physiological and biochemical aspects. As for the nucleus, exhaustive information has been provided about its physiology, and its role in heredity; and much is known about its structural pattern and biochemistry. Nevertheless an exact understanding of the inter-connection of the known facts is still lacking in many cases; many problems are still awaiting a more adequate methodological approach.

The present article contributes a piece of information which, however, is still difficult to interpret from the point of view of cell physiology. It is based both on the data reported in the literature, and on the results of work in this laboratory.

SURVEY OF RESULTS REPORTED IN BIOCHEMICAL LITERATURE

Since phospholipoproteins appear to be important components of all subcellular structures, it could be accepted almost *a priori* that they are found in cell nuclei as well. This suggestion conflicts with the old opinion, based on the authority of Miescher and Schmiedeberg that lipids may be 'foreign to nuclei in general'.[27] However, as proved much later, by analysis of the nuclear fraction from various tissues, lipids appear to be a constant component, though amounting to only a low percentage of nuclear dry mass (Table 14.1).

When compared with the lipid content of other cellular fractions, the amount of lipid in the nuclei forms the lowest proportion. On the other hand, the percentage of phospholipids in the nuclear lipids is very high, while neutral fats and cholesterol are found only in minute quantities.[35] The data reported in the literature which deviate from this general pattern may be ascribed to the specificity of the tissues investigated and, even more, to the degree of purity of the nuclear fraction. As follows from the results of Gurr and his co-workers, the two methods generally applied for isolation (the use of a citric acid or sucrose medium) respectively yield nuclei differing in essentials. Those isolated by the first method, unlike the others, are deprived of their characteristic double membrane. In consequence, their lipid content is much lower. Barnum and his colleagues reported that the phospholipid content in mouse liver nuclei isolated in citric acid medium equals 3·4 per cent of the dry weight, whereas the content of protein, deoxyribonucleic acid, and ribonucleic acid equals 66, 27, and 3·4 per cent respectively.[7] The pattern of nuclear phospholipids is similar to that of the whole cell, except for cardiolipins, which seem to be specific for mitochondria.[20] Nuclear phospholipids in comparison with other cellular phospholipids are rather insoluble in ether[36] and in petroleum ether.[26] However, their fatty acid composition is similar

Table 14.1   Lipid and phospholipid content in cell nuclei fraction

Tissue investigated	The isolation medium	Total lipid. % of the dry weight	Phospholipids. % of total lipid	References
Rat liver	Citric acid, pH 3·4–4	3·2– 7·2	—	16
Rat liver	Citrate, pH 6·0–6·2	7·5–10·8	—	16
Rat liver	Citric acid	3	—	20
Rat liver	2·2 M Sucrose	8	—	20
Rat liver	0·25 M Sucrose	16	93	35
Dog liver	Citrate, pH 6·0–6·2	16·5	95	39

to that found in other subcellular particles.[18,20] In relation to the whole rat liver cell, the nucleus contains about 10 per cent of total cellular protein, 6–7 per cent of total RNA and 2–2·5 per cent of total cellular lipid.[9]

All the above data give no information about the location of phospholipids within the nucleus. Only in the paper of Gurr and his colleagues[20] is it suggested that phospholipids are confined to the nuclear envelope. The amount of phospholipid found by these authors in nuclei isolated in citric acid is, according to their calculation, sufficient to constitute two lipid monolayers which form the single membrane coating the nuclei isolated in this medium. However, the authors remark that, although their suggestion seems probable, the results obtained by them cannot entirely rule out the possibility of another location of phospholipids in the nucleus, e.g. in chromosomes.

Mirsky and Ris succeeded in isolating chromosomes from the lymphocytes of calf thymus. They have not presented any data concerning the occurrence of lipids in this material. They mention, however, that for the estimation of DNA phosphorus, they previously extracted the isolated chromosomes with alcohol and ether to remove lipids. The same procedure was applied for determination of the nucleoprotein content of the 'residual chromosomes' which remain after extraction of the whole chromosomes with M NaCl.[28] This precaution suggests that the lipid phosphorus might interfere in estimations of DNA phosphorus in the investigated structures. According to an earlier work, phospholipids consist of about 2 per cent of the dry weight of the isolated chromosomes.[13]

SURVEY OF RESULTS REPORTED IN CYTOCHEMICAL LITERATURE

The presence of phospholipids in chromosomes has been suggested chiefly by cytochemical investigations. Applying lysochrome, Sudan Black B to paraffin sections of testes of *Triturus, Locusta* as well as of the apical root meristem of *Hyacinthus*, Idelmann[21] revealed lipids in chromosomes of those tissues. In interphasal nuclei only the heterochromatic regions were uniformly positive.

In the same year Chayen and others reported the presence of phospholipids in some plant meristem nuclei and in mouse liver nuclei.[12] The preservation of phospholipids was achieved in paraffin sections only when the tissue was fixed in Levitzki's fluid containing equal parts of 10 per cent formol and 1 per cent chromium trioxide (F-Cr), instead of Baker's formaldehyde/calcium mixture (F-Ca). Formaldehyde, with or without addition of $CaCl_2$, extracted nuclear phospholipids, as confirmed by chromatography of the remaining fixative fluid.[2]

In another paper, the same authors postulated that phospholipids confined to chromosomes are masked, possibly by proteins.[11] It is known from practice with Baker's acid haematein test (AH) for phospholipids, that in some cases cell nuclei give a more distinct reaction in paraffin sections after pyridine extraction than before extraction. Baker's analysis of the test excluded in such case the presence of phospholipids, but indicated the presence of nucleoprotein.[6] However, it appeared later that nucleoprotein (namely calf thymus nucleohistone) contains possibly as much as 6 per cent of very tightly bound sphingomyelin.[10] According to Chayen and his colleagues pyridine would unmask some proteins closely associated with phospholipids in the same way as trichloracetic acid. These phospholipids are then responsible for the positive result of the AH test.[11]

The presence of masked lipids in cell nuclei was also postulated by other authors. Diverse chemical compounds, like some organic acids (formic, acetic, citric, and oxalic acids[1]), α-naphthol,[19] chloroplatinic acid, chromium trioxide, copper acetate, mercuric chloride, osmium tetroxide,[14] and uranyl acetate[15] are claimed as agents capable of unmasking nuclear lipids.

As follows from the above survey, the problem of the presence of lipids in nuclear chromatin has been reconsidered in the last ten years, and remains still in dispute. The experience of our laboratory indicates that there are considerable differences between nuclei of various tissues in respect to their response to the AH test. Many nuclei show a positive AH staining in the nucleoli, which disappears entirely or nearly so after pyridine extraction. In some cases the nuclear membrane has a distinct phospholipid character. But interpretation of the results with nuclear chromatin, obtained with the AH test, involves great difficulties. Moreover, the nuclei differ not only when they originate from various organs, but also from different developmental stages, as in some developing insect ovarioles.

Biezenski and others have found differences in the nuclear phospholipid spectrum in various organs of adult and immature rat.[8] Changes in the appearance of intranuclear lipids are known to accompany certain pathological conditions.[2,4] La Cour, Chayen, and Gahan have shown that cytochemical location of nuclear phospholipids undergoes experimental changes in some plant meristems grown at low temperature.[23] It seemed interesting to observe more carefully whether there are any developmental changes in the appearance of nuclear phospholipids in the follicular vesicle of an insect. Some methodical experiments were also performed to examine the reliability of the AH test in the case of nuclear phospholipids.

PHOSPHOLIPIDS IN THE NUCLEI IN THE FOLLICULAR VESICLE OF
GALLERIA MELLONELLA

Observations were carried out on the ovarioles of *Galleria mellonella*. The ovariole from a newly hatched moth contains a long chain of about 100 follicular vesicles, varying from undifferentiated ones up to fully mature ones ready for oviposition. The meroistic follicular vesicle of *Galleria* consists of an oocyte, linked by a cytoplasmic strand with a nutritive chamber formed by a few trophocytes, all being covered with the follicular epithelium. When it reaches maturity, the oocyte is deprived of its nutritive chamber, which has previously undergone autolysis.[31]

Following the suggestion of Chayen and co-workers,[12] the ovarioles were fixed in Levitzki F-Cr mixture to prevent the nuclear phospholipids from escaping. A positive reaction with AH, suggesting the presence of phospholipids, was observed in nuclei of all three kinds of cells of the follicular vesicle. However, some differences, connected with the developmental stage of the vesicle, could be observed.

From the point of view of the structural differentiation which these cells undergo, the follicular epithelium is the most stable element, at least prior to the formation of the chorion. During all this time, the AH test is positive in chromatin of these cell nuclei. Dark blue spots localized on the granular nuclear structures are observable. They are rather faint in young pre-vitellogenic vesicles, and more distinct in subsequent stages of development (Fig. 14.1, A, E). The small size of these nuclei makes it difficult to distinguish the particular structural components well, which can be done easily with the nucleus of the oocyte. In the nucleus of a very young oocyte, the nuclear membrane remains unstained after AH, and tiny dark blue spots are visible only on chromosomes which at this period are easily distinguishable (Fig. 14.1, B, C). The nucleolus associated closely with one of the chromosomes gives a very faint positive reaction (Fig. 14.1, B). In the course of further development, the reaction in the nucleolus becomes more intense (Fig. 14.1, D, G, I). At the beginning of vitellogenesis the staining does not cover all the nucleolar surface, which gives the impression of a strongly inhomogeneous structure (Fig. 14.1, F). At more advanced stages in the period of yolk synthesis, the AH reaction in the nucleolus is so intense that it masks all its structural differentiation (Fig. 14.1, J). Moreover, in some nuclei strange inclusions connected with nucleoli are observable. They give a positive reaction for phospholipids (Fig. 14.1, K). At this stage the chromosomes are no longer distinguishable. Yet the phospholipid-like spots which were seen among them previously, become gradually more conspicuous and remain distinctly visible. A positive reaction of the nuclear membrane appears with the onset of vitellogenesis. Then the membrane

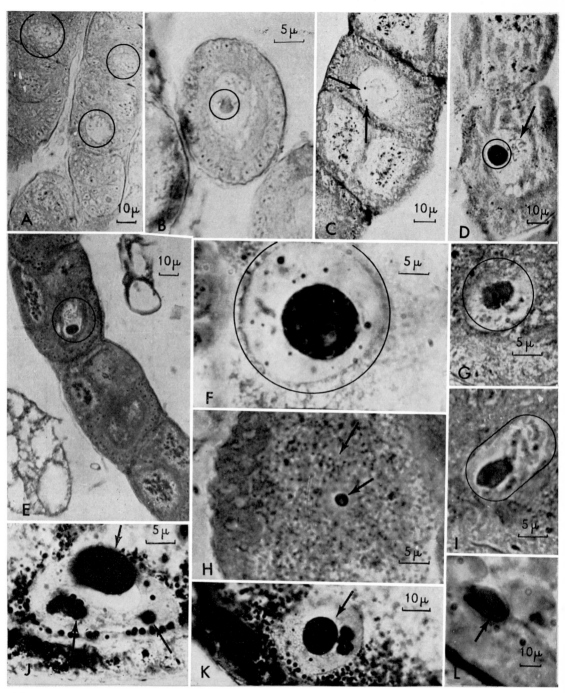

Fig. 14.1 Developmental changes in affinity to acid haematein test in nuclei of *Galleria* follicular vesicle. All Formol–Chromium fixed, paraffin embedded, A–H reaction. A–E, G, I, previtellogenetic vesicles; A, B, very young vesicles, nucleolus of oocyte nucleus almost unstained, faintly stained spots distinguishable in chromosomes; C–E, G, I, stages older than in A, B, nucleolus and single spots visible in chromosomes of oocyte nucleus stained more intensely than in A, B, F, H, J, K, vitellogenetic vesicles: F, oocyte nucleus, nucleolus, and the membrane are heavily stained; numerous stained granulations are distinguishable in the karyoplasm; H, polyploid trophocyte nucleus with stained chromatin and nucleolus; J, oocyte nucleus, AH positive granulations in the karyoplasm, nucleolus stained uniformly; K, oocyte nucleus, AH positive inclusions connected with nucleolus, L, pycnotic trophocyte nucleus, stained uniformly.

becomes strongly undulated, with numerous minute perpendicular processes directed towards the ooplasm. They are distinctly seen even in the light microscope (Fig. 14.1, F).

In the nuclei of trophocytes the positive staining of the membrane is hardly observable. However, numerous nucleoli as well as the chromatin structure become distinctly stained. The staining of the latter is seen in the form of numerous dark blue granules, their size and amount increasing in the course of the vesicle development (compare Fig. 14.1, H, with Fig. 14.1, A). In some trophocytes, the nuclear chromatin becomes differentiated into distinct units, which are probably the prophasic chromosomes in those endomitotically dividing cells (Fig. 14.1, E).

In the autolysing trophocyte chamber, in strongly pycnotic and in disintegrated nuclei a phospholipid-like material could be detected as distributed uniformly (Fig. 14.1, L).

To analyse the results of the AH test, a methodical study was performed.[32] Firstly, the ovarioles were stained, concomitantly to the AH test, with Sudan black B and with Rawitz's basic fuchsin, applied after a prolonged tanning (T-BF);[30] secondly, the acid haematein test was performed on sections after various extraction procedures.

When corresponding sections were stained with the methods mentioned above, a good conformity of results was obtained (Fig. 14.2, A, E). None of the three methods is sufficiently selective for revealing phospholipids in the nuclei. However, when used in combination, they give a better indication of the presence of phospholipids than the AH test alone.

Application of these tests to pure substances reveals differences in their staining affinities from some non-lipid compounds. In contrast to the acid haematein, basic fuchsin, applied after treatment with tannic acid, shows no affinity towards nucleoprotein (Table 14.2). In the light of these observations, the results obtained in the follicular vesicle of *Galleria* suggest that phospholipids are responsible for the observed AH stainability in the investigated nuclei.

Table 14.2   Analysis of the reaction with acid haematein[a] with tannic acid/basic fuchsin,[b] and Sudan black B

Object of staining	Test		
	Acid haematein	Basic fuchsin with previous tannic acid mordant	Sudan Black B
Pure substances:			
Casein	+	−	
Mucin	+	−	
Nucleoprotein	+	−	
Lecithin	+	+	+
Cephalin	+	+	+
Sphingomyelin	+	+	+
Triglycerides	−	−	+
The nuclei of the follicular vesicle of *Galleria m.*, F–Cr fixed, paraffin embedded,			
Nucleolus	+	+	+
Chromatin	+	+	+

[a] Baker.[6]
[b] Przełecka.[30]

After extraction with pyridine by Baker's procedure, the nuclei retain their affinity for AH. Only nuclei of very young follicles are not stained in these conditions, whereas the test is positive at advanced stages. But in this case, the pattern of the dark blue stained elements differs distinctly from that obtained in unextracted sections and is much more extensive (Fig. 14.2, B). This is especially conspicuous in the

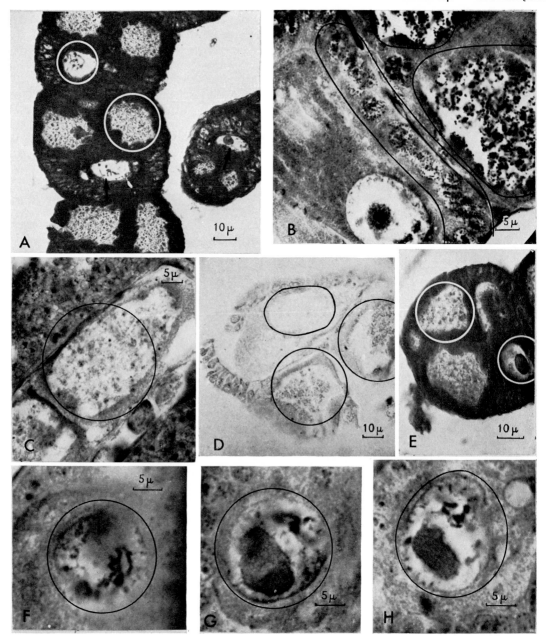

Fig. 14.2 A, E, comparison of results obtained in nuclei of *Galleria* vesicles after application of Sudan black B with those after tannic acid–basic fuchsin; A, vitellogenetic vesicles, Sudan black B, Fo–Cr/ paraffin; E, same stage as in A, after tannic acid–basic fuchsin, Flemming/paraffin. A similar staining of nuclear chromatin and of nucleoli after the two methods as after AH is observable; B–D, F–H, the effect of lipid and/or protein extraction on affinity to acid haematein in *Galleria* vesicles; B, F–H, Bouin–pyridine extraction after Baker, paraffin embedding, AH. B, an intense staining of trophocyte and follicular epithelium nuclei; F–H, oocytes nuclei, decrease of reaction in nucleoli, whole chromosome—threads uniformly stained; C, hot methanol–hot butanol prolonged extraction, trypsin digestion, AH test, considerable decrease of AH affinity in trophocyte nucleus; D, hot methanol–hot butanol prolonged extraction, trypsin followed by pepsin digestion, complete supression of nuclear affinity to AH test.

large nuclei of the oocyte (Fig. 14.2, F–H). According to the above suggestion of Chayen and his colleagues,[11] this may indicate the presence of phospholipids bound to nucleohistone which are unmasked by pyridine. If this hypothesis is correct, the histone-bound lipids should be removed either by the enzymatic digestion of histone with or without subsequent lipid extraction, or they should be eliminated by lipid solvents only, after breaking their link with histone. The first possibility was tested by digesting the sections from Bouin–pyridine-treated ovarioles with trypsin with or without repeated pyridine extraction: The second, by treating the ovarioles with hot methanol, (1 hr at 60°C), followed by a prolonged extraction with hot n-butanol, (40 hr at 60°C), to split the lipo-proteins, and to ensure an extraction of all lipids. The results of the first series of experiments are presented in Table 14.3.

Table 14.3

Procedure applied / Nuclei investigated	Fixation in Bouin's fluid, pyridine extraction, dehydration with graduated alcohols and toluene, 6 μ paraffin sections,		
	No subsequent extraction	Trypsin digestion (0·1 mg/ml phosphate buffer, 1 hr, 37°C)	Trypsin digestion followed by repeated pyridine extraction
	Acid haematein test		
Trophocyte n.	+	+	+
Follic. ep. n.	+	+	+
Oocyte nucleus	+	+	+

The intense dark blue staining characteristic for nuclei in ovarioles after the Bouin–pyridine procedure was obtained equally well after removing histones with or without their hypothetical phospholipid moiety. However, this resistance of the AH stainable compounds may be evoked by some intranuclear changes caused by fixation. Therefore in the second series of experiments, the first step of the whole procedure, namely the fixation of ovarioles in Bouin's fluid, was omitted. Immediately after dissection of the ovarioles, lipid extraction was performed. As described above, hot methanol was followed by a prolonged hot n-butanol treatment. After this procedure, a complete extraction of phospholipids is probably secured, at least of these phospholipid granules which are detectable within the limits of light microscopical resolution.

Sudan black B applied to such material leaves both the cytoplasm and the nuclei unstained. However the AH test gave in the same cell nuclei, exactly the same picture as in Bouin–pyridine treated sections: the nuclei were heavily stained dark blue. Looking for the compounds responsible for this effect, the successive elimination of nucleic acids, histones, and non-histone protein was performed. It appeared that after trypsin digestion, the nuclei, except the oocyte nuclei, were almost completely AH negative (Fig. 14.2, C). In all the nuclei, a *complete* elimination of stainability was achieved *only* by treatment with trypsin followed with pepsin, or with pepsin only (Fig. 14.2, D). The results of these experiments are presented in Table 14.4.

The above results demonstrate that the phospholipid moiety of the presumed lipoprotein complex is *not* a factor responsible for the AH positive reaction in the Bouin–pyridine material. Moreover, the extent of this reaction within the nucleus in Bouin–pyridine tissue is not proportional to the small percentage of total phospholipid content found by biochemical analysis. It seems rather probable that nuclear proteins, both histone and non-histones, are responsible for the intense pyridine-resistant AH staining. Phospholipids, as shown above, seem to be responsible for the much less extensive AH staining of nuclear structure found in unextracted material.

As Chayen and his colleagues have pointed out, pyridine extraction removes not only lipids, but also a part of the nucleohistones.[11] The residual compounds 'unmasked' in that way react with chromium and subsequently form dark blue lakes with acid haematein. They seem, however, as discussed above, to be *not* phospholipids, but proteins.

12—C.S.I.

Table 14.4

Procedure applied / Nuclei investigated	(1) Extraction, methanol, 1 hr, 60°C (2) Extraction, n-butanol, 40 hr, 60°C					
	No subsequent treatment	HCl, 1N 3 hr, 60°C	Trypsin digestion 1 hr, 37°C			Pepsin digestion 3 hr, 37°C
			No further treatment	Pyridine after Baker	Pepsin 3 hr, 37°C	
			Acid haematein test			
Trophocyte n.	+	+	–	–	–	–
Follic. ep. n.	+	+	–	–	–	–
Oocyte nucleus	+	+	+	+	–	–

DISCUSSION

Proteins detected by the AH test in Bouin–pyridine-treated nuclei in the ovariole of *Cynips*, are intensely stained with fast green at pH 8·1, which suggests that they are basic proteins.[22] However, the nuclear proteins of varied tissue cells probably differ in their nature. This fact could involve differences in the affinity of nuclear constituents to AH after pyridine extraction. The physiological state of the cell may also exert an influence on the phospholipid content of the nuclei. Nevertheless the appearance of phospholipids seems generally to be observed in the nuclear membrane, in nucleoli, and even in chromatin (Fig. 14.3). It is interesting that, whereas in interphase nuclei, and in the early meiotic prophase, only single phospholipid spots can be detected in chromatin, in the strongly contracted mitotic chromosomes there is uniform staining.[21, 34, 11, 17]

An increased rate of phospholipid turnover accompanying the mitotic activity of tissues was recently reported. Levin and his associates[25] have found an increased P^{32} uptake into phospholipids, by cell fractions of rat liver regenerating after partial hepatectomy. It is especially interesting that P^{32} turnover in nuclear cephalins and in supernatant sphingomyelins occurs after prophase and prior to, or during, the metaphase; whereas P^{32} turnover in lecithin of all the fractions was detected in the interphase as well as during the mitotic cycle. An electron microscope investigation has shown the accumulation of small electron-opaque bodies in the liver cytoplasm at early intervals after partial hepatectomy. These were assumed to be precursors of fat droplets, which appeared at later postoperative intervals.[38] However, this author failed to notice any changes in the nuclear structure. As shown in the accompanying micrographs, no granules or droplets of the dimensions occurring in cytoplasm were visible within the nucleus. But whether the more delicate or even quite minute bodies of the same electron opacity, which are present, have anything in common with lipids, remains undecided. The necessity of considering the fixatives used for electron microscopy as cytochemical reagents has been pointed out in many discussions. Osmium tetroxide remains inert towards nucleic acids, under the conditions of time and temperature preferable for electron microscopy (i.e. at 0°C for a very short time), as was shown by the extensive study of Bahr,[5] but it reacts with greases and oils, with their constituents like oleic acid and ethanolamine, as well as with some aminoacids, peptides, and proteins. In several cases, the reaction with proteins develops only after a prolonged time. With lipids, the reaction is almost instantaneous, whereas histones become blackened by osmium tetroxide at 0°C only after 2 hr, and protamine after 24 hr. Lehmann has reported that the nucleus of *Amoeba proteus* and nuclei of the *Tubifex* embryo, fixed in buffered osmic acid for 10 min at 0°C, seems to be rich in osmiophilic compounds present in the nuclear membrane, in nucleoli, and in chromonemata. He considers it very probable that these findings indicate the presence of lipid in nuclei.[24] The results of the analyses done by Bahr[5] strongly suggest that these are phospholipids.

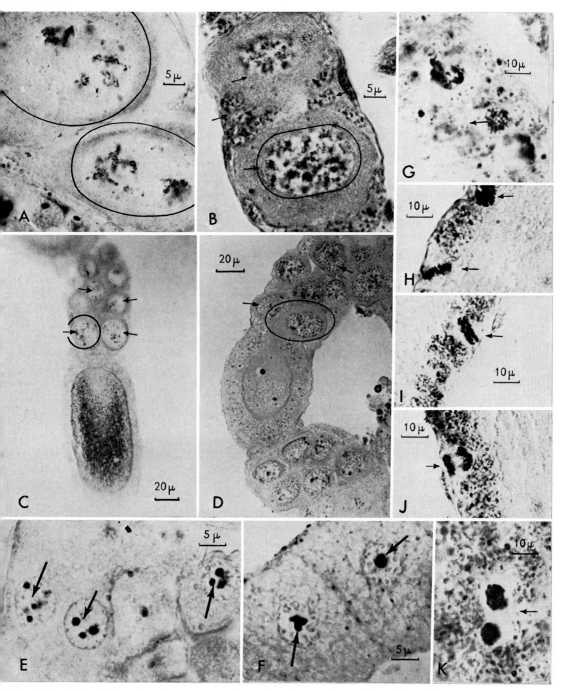

Fig. 14.3 AH staining of nuclear structure in various insect tissues. A, B, *Carausius morosus*, young pre-vitellogenetic vesicles; A, Fo–Cr/paraffin, in the oocyte nucleus: single stained spots arranged regularly on the spiralized meiotic chromosomes are conspicuous; B, Bouin–pyridine/paraffin, the reaction in nuclei much more extensive than in A. C, D, *Cynips folii*, ovarioles; C, Fo–Ca/paraffin, single stained spots in trophocyte nuclei are visible; D, Bouin–pyridine/paraffin, staining of trophocyte nuclei more intense than in C; E, *Apis mellifica*, midgut epithelium of a bee worker Fo–Cr, paraffin positive reaction in nucleoli; F, *Drosophila melanogaster*, salivary glands, Fo–Cr/paraffin, positive reaction in nucleoli; G–K *Carausius morosus*, Fo–Cr/paraffin, successive mitotic phases in follicular epithelial cells, chromosomes stained uniformly; A, B and G–K kindly provided by Dr. A. Dutkowski; C, D, by Dr. A. Kraińska; E, by Dr. A. Hartwig.

Consequently, all the present data, both from the literature, and from our experimental work, seem to indicate that the presence of phospholipids in nuclear structure is highly probable. The problem arises as to whether they are synthesized inside the nuclei or are transported from the cytoplasm. Cytological observations on fixed material, necessarily static in their nature, provide contradictory information. Seite and Chambost reported the extrusion of nucleolar lipoprotein into the cytoplasm of cat nerve cells.[33] Thoenes, after an electron microscope study, described the presence of lipid droplets in the enlarged nucleoli in liver cells of the rat damaged by thioacetamide. According to his observations the lipid droplets are derived from cytoplasm and are taken up into the nucleus by an infolding of the nuclear envelope.[37] In our autoradiographic study, a distinct labelling of nucleolar lipids was found after 60 min incubation of the isolated ovarioles of *Galleria* in a medium containing tritium labelled sodium palmitate. The nucleoli of the ovarioles incubated for shorter periods of time remained relatively unlabelled (unpublished results). This result supports the observation reported by Thoenes, and seems to be also in agreement with our present knowledge based on biochemical investigations, which indicates that synthesis of lipids is restricted to cytoplasmic structures.

However, the essential problem of the physiological role of phospholipids in nuclear structure remains unsolved. The most straightforward speculation may concern the nuclear envelope, where phospholipids probably regulate many compounds crossing this nucleo-cytoplasmic barrier. But the interpretation of the occurrence of phospholipids in chromatin and nucleoli involves difficulty. Pollister assumes that the approximate average nucleoprotein composition of an interphase nucleus in a mammalian liver cell may be as follows: 9 per cent DNA, 1 per cent RNA, 11 per cent histone, 14 per cent residual protein (called chromosomine) and 65 per cent other non-histone proteins.[29] Phospholipids are probably connected with one fraction of the histone, and amount to about 28 per cent of this protein fraction.[10] However, the possible place of phospholipids in any of the newly suggested molecular models of chromosomes presents great difficulties. Presumably, phospholipids are associated with the structure of chromosomes, rather than with their physiological activity. This is in accordance with their abundant occurrence on mitotic chromosomes, which, from the point of view of metabolic activity, are inert in comparison with the interphase chromosomes. On the other hand, nucleoproteins may differ in different cells.[29] The intranuclear phospholipid content seems also to be conditioned by the actual physiological state of the cell. This is indicated by our observations on the developmental changes in *Galleria* ovarioles, and by the data concerning injured or pathological cells. These changes in nuclear phospholipid features, conditioned by the actual physiological state of the cell, are most distinct in nucleoli. But whether this phenomenon is connected with the developmental alteration of structure, or with the physiological activity of the nucleolus, remains obscure. However, it is possibly of importance that it is just in the follicular vesicles of *Galleria mellonella*, an organism with a high lipid metabolism, that the nucleoli are particularly rich in phospholipids. Much further work is, however, needed in order to elucidate these problems.

REFERENCES

1. ACKERMANN, G. A. 1952. A modification of the Sudan Black B technique for the possible cytochemical demonstration of masked lipids. *Science* **115**, 629–31
2. ALTMANN, H. W. 1952. Über den Funktion Formwechsel des Kernes im exokrinen Gewebe des Pankreas. *Z. Krebsforsch.* **58**, 632–45
3. ———, HAMPTON, S. E., and CHAYEN, R. 1964. Phospholipids in formalin extracts of nuclei isolated from sheep lungs. *Nature, Lond.* **202**, 1215–16
4. AMICK, C. J. and STENGER, J. R. 1963. Ultrastructural alterations in experimental acute hepatic fatty metamorphosis. *Lab. Invest.* **12**, 859–60
5. BAHR, G. F. 1954. Osmium tetroxide and ruthenium tetroxide and their reactions with biologically important substances. *Expl. Cell Res.* **7**, 467–79
6. BAKER, J. R. 1946. The histochemical recognition of lipine. *Quart. J. micr. Sci.* **87**, 441–65
7. BARNUM, C. P., NASH, C. W., JENNINGS, E., NYGAARD, O., and VERNUND, H. 1950. The separation of pentose and desoxypentose nucleic acids from isolated mouse liver cell nuclei. *Arch. Biochem.* **25**, 376–83
8. BIEZENSKI, J. J., SPEAT, T. H., and GORDON, A. L. 1963. Phospholipids patterns in subcellular fractions of adult- and immature-rat organs. *Biochim. biophys. Acta.* **70**, 75–84

9. CHAVEAU, J., MOULE, Y., and ROUILLER, CH. 1956. Isolation of pure and unaltered liver nuclei. Morphology and biochemical composition. *Expl. Cell Res.* **11**, 317–21
10. CHAYEN, J. and GAHAN, P. B. 1958. Lipid components in nucleohistone. *Biochem. J.* **69**, 49 p.–50 p.
11. ——, ——, and LA COUR, L. F. 1959. The mature of a chromosomal phoospholipid. *Quart. J. micr. Sci.* **100**, 279–84
12. ——, LA COUR, L. F., and GAHAN, P. B. 1957. Uptake of benzpyren by a chromosomal phospholipid. *Nature, Lond.* **180**, 652–3
13. CLAUDE, A. and POTTER, J. S. 1943. Quoted by Dounce, A. L. 1955. The isolation and composition of cell nuclei and nucleoli. In *The Nucleic Acids* (edited E. Chargaff and J. N. Davidson), Vol. II, p. 122. London: Academic Press
14. CLAYTON, B. P. 1959. The action of fixatives on the unmasking of lipids. *Quart. J. micr. Sci.* **100**, 269–74
15. —— (personal information)
16. DOUNCE, A. L. 1943. Further studies on isolated cell nuclei of normal rat liver. *J. biol. Chem.* **151**, 221–32
17. DUTKOWSKI, A. (in preparation)
18. GETZ, G. S. and BARTLEY, W. 1961. The intracellular distribution of fatty acids in rat liver. The fatty acids of intracellular compartments. *Biochem. J.* **78**, 307–12
19. GOLDMANN, J. 1929. Über die Lipoidfärbung mit Sudan (Scharlach R)-α-Naphthol. *Zbl. Path.* **46**, 289–90; quoted by Romeis, B. 1948. *Mikroskopische Technik*
20. GURR, M. I., FINEAU, J. B., and HAWTHORNE, J. N. 1963. The phospholipids of liver-cell fractions. I. The phospholipids composition of the liver-cell nucleus. *Biochim. biophys. Acta* **70**, 406–16
21. IDELMAN, S. 1957. L'existance d'un complexe lipides-nucléoprotéines à groupements sulfhydrilés au niveau du chromosome. *C.r. hebd. Seanc. Acad. Sci. Paris*, **244**, 1827–8
22. KRAINSKA, M. 1966. Distribution of phospholipids in oogenesis of Cynips folii. *Folia Histochem. Cytochem.* **4**, 389–97.
23. LA COUR, L. F., CHAYEN, J. and GAHAN, P. B. 1958. Evidence for lipid material in chromosomes. *Expl. Cell Res.* **14**, 469–85
24. LEHMANN, F. E. 1958. Functional aspects of submicroscopic nuclear structures in *Amoeba proteus*, and of the mitotic apparatus of *Tubifex* embryos. *Expl. Cell Res.* suppl. **6**, 1–16
25. LEVIN, E., JOHSON, R. M., and ALBERT, S. Phospholipids metabolism in cell fractions of regenerating liver. *Arch. Biochem. Biophys.* **73**, 247–54
26. —— and CHARAGAFF, E. 1952. Phospholipide composition in different liver cell fractions. *Expl. Cell Res.* **1**, 154–2
27. MIESCHER, F. and SCHMIEDEBERG, O. 1898. Quoted by Gurr, *et al.* 1963. *Biochim. biophys. Acta* **70**, 406–16
28. MIRSKY, A. E. and RIS, H. 1947. The chemical composition of isolated chromosomes. *J. gen. Physiol.* **31**, 7–18
29. POLLISTER, A. W. 1952. Nucleoproteins of the nucleus. *Expl. Cell Res.* suppl. **2**, 59–74
30. PRZEŁECKA, A. 1959. A. Rawitz's inversion staining. *Quart. J. micr. Sci.* **100**, 231–9
31. ——. Nucleic acid metabolism and cell interaction in the ovariole of *Galleria mellonella*. *Folia Histochem. Cytochem.* (in press)
32. —— and DUTKOWSKI, A. (in preparation)
33. SEITE, R. and CHAMBOST, G. 1958. Élaboration figurées de la cellule nerveuse: étude du ganglion rachidien chez le chat. *Z. Zellforsch.* **47**, 498–506
34. SERRA, J. A. and SEIXAS, M. P. 1962. Distribution of lipids in the chromosomes during mitosis and meiosis. *Rev. portug. Zool. Biol. Geral.* **3**, 247–54
35. SPIRO, M. J. and MCKIBBEN, J. M. 1956. The lipides of rat liver cell fractions. *J. biol. Chem.* **219**, 643–51
36. STONEBURG, C. A. 1939. Lipids of cell nuclei. *J. biol. Chem.* **129**, 189–96
37. THOENES, W. 1964. Fat in nucleolus. *J. Ultrastruct. Res.* **10**, 194–206
38. TROTTER, NANCY L. 1964. A fine structure study of lipide in mouse liver regenerating after partial hepatoectomy. *J. Cell Biol.* **21**, 233–44
39. WILLIAMS, H. H., KAUCHER, M., RICHARDS, A. J., MOYER, E. Z., and SHARPLESS, G. R. 1945. The lipid partition of isolated cell nuclei of dog and rat liver. *J. biol. Chem.* **160**, 227–32

Chapter 15

Lipochondria, Neutral Red Granules and Lysosomes: Synonymous Terms?

NANCY J. LANE

INTRODUCTION

In 1963 Baker,[5] in discussing invertebrate neurons, wrote that 'future research will show whether the lipochondrion may be regarded as an overgrown and modified lysosome'. Thirteen years earlier he had described spherical or subspherical lipoid particles in the cytoplasm of cells for which he coined the term 'lipochondria'.[3] These organelles were visible in living cells and exhibited an affinity for vital dyes such as neutral red. Their lipid nature was manifested in fixed preparations by their positive response to the acid haematein test for phospholipids and their sudanophilia after staining with Sudan black B.

Lipochondria are common in the cells of invertebrates, especially in their neurons, and it is in these cells that they have been most thoroughly investigated. However, they also appear to be present in vertebrate neurons and other cells of vertebrates. Baker[5] has recently extended the definition of lipo-chondria to include characteristic ultrastructural details; in the electron microscope they are limited by a unit membrane and contain electron-opaque inclusions in addition to the membranes considered to be phospholipid (Figs. 15.1, A; 15.3, A; 15.3, B; and 15.4, C).

Cytologists from the Oxford School, under the guidance of J. R. Baker, have conducted many of the light microscopical studies on invertebrate neurons. These studies have included examinations of cyto-logical details by phase and interference microscopy, observations of the cellular sites of uptake of vital dyes and investigations into the cytochemical features of the cells. Refinement of cytochemical procedures permitting localization of sites of enzymatic activity at the electron microscopical level[52, 63] has made possible the demonstration of acid phosphatase activity in lipochondria. This, with the presence of a delimiting unit membrane, satisfies the morphological criteria[49] and permits their tentative identification as lysosomes, first defined biochemically by De Duve and co-workers in 1955.[19]

LIPOCHONDRIA AND LYSOSOMES IN INVERTEBRATE NERVE CELLS

Neurons of orthopteran insects

Over the years there has been considerable controversy concerning the cytological details of orthop-teran neurons. Do they contain lipochondria or Golgi bodies, or both? The neurons from the cerebral, thoracic or abdominal ganglia of locusts, grasshoppers, and cockroaches have been extensively investi-gated at the light microscopical level; however, the limitation in resolving power severely restricted these studies. Some investigators claimed that both the curved sudanophilic dictyosomes* and the smaller

---

* The term 'dictyosome' has been used by light microscopists to identify the crescentic or ring-shaped Golgi complexes present in the cytoplasm of cells, particularly invertebrate neurons, after certain pre-parative procedures. The term implies that the Golgi bodies in these cells are separate entities rather than an interconnecting network like the Golgi in most vertebrate neurons. Although the discontinuous nature of the dictyosomes in invertebrate neurones has not been proved in serial sections with the electron micro-scope, the term is a convenient one for indicating that each region of stacked Golgi saccules *appears* to be separate from every other region.

granules were the Golgi 'element'.[6,24] Other workers believed that the classical osmiophilic or argentophilic 'Golgi' reaction occurred on the surface of the lipochondria and that both these duplex spheroids, *and* the homogeneous bodies, comprised the lipochondria, with no Golgi bodies, *per se*, existing.[38,39,66,67] A third interpretation was that the sudanophilic lipochondria were separate from the Golgi, but that the two bore a topographical relationship to one another.[47,58]

Early ultrastructural examinations showed that the neurons, in addition to the other cellular constituents, contain two morphologically distinct organelles.[2,7] Ashhurst and Chapman describe these as dense spheroidal bodies containing phospholipid lamellae, and lamellar–vacuolar fields; however, they refer to both organelles as lipochondria. More recent electron microscopical studies provide evidence that these two organelles, the spheroidal lipochondria and the lamellar Golgi, are quite distinct from one another, although they are often closely associated spatially[25,32,34B,71] (Fig. 15.1, A, B, C and D).

It might be expected that a cytochemical study of enzymatic activities would lead to clarification of the uniqueness, and of the relationships of the two organelles. In vertebrate neurons, the Golgi saccules* are characterized by high levels of activity of the enzyme thiamine pyrophosphatase (TPPase) and nucleoside diphosphatase.[51] The 'dense bodies', often containing electron opaque grains and sometimes membrane arrays, are lysosomes, characterized by high levels of acid phosphatase activity.[50] A light microscopical investigation by Lee[36] on the thoracic neurons of the grasshopper *Locusta* showed that acid phosphatase was present only in the Golgi, while TPPase was restricted to the smaller lipochondria. However, by electron microscopy on the grasshopper, *Melanoplus*, neurons, acid phosphatase is found to be present in lipochondria, in some Golgi saccules, and Golgi vesicles (Fig. 15.2, A); TPPase is found in some Golgi saccules and vesicles and, as far as can be ascertained, in all lipochondria[32] (Fig. 15.2, C).

Thus, although the lipochondria of orthopteran neurons are lysosomes, they exhibit a somewhat broader range of substrate specificity than most vertebrate lysosomes. Not only do the lysosomes of *Melanoplus* neurons hydrolyse monophosphates at acid pH, but also some nucleoside diphosphates and triphosphates at neutral pH[32,34B] (Figs. 15.2, B, and 15.2, C).† However, even some vertebrate neurons split nucleoside monophosphates at neutral as well as acid pH; this is particularly evident in certain neurons of rat spinal ganglia where the Golgi is also in the form of dictyosomes. In these cells, cytidine and thymidine monophosphate are hydrolysed at pH 5 and 7 not only by the lysosomes, but also by one or more saccules of the Golgi and an associated region of fenestrated smooth membranes called GERL.[54] Perhaps the similarity in enzyme content between Golgi and lysosomes in both these neuron types is significant in terms of their functional inter-relationship.

It has been adequately demonstrated that the lipochondria in the neurons of orthopterans segregate vital dyes.[38,39,66] Since the lipochondria may be classified as lysosomes, it follows that the sites of vital dye uptake in these cells also correspond to lysosomes. It is perhaps noteworthy that a faint staining also

---

* Although the flattened agranular membranes of the Golgi have been variously referred to as lamellae, cisternae, and saccules, the last of these will be used in the present discussion.

† In the parenchymal cells of insect prothoracic glands, Osinchak[56] finds that the lysosomes display TPPase as well as acid phosphatase activity. The glia cells that ensheath the neurons in *Melanoplus* contain lysosomes that hydrolyse thiamine pyrophosphate and adenosine triphosphate at pH 7·2, as well as monophosphates at pH 5·0 (Fig. 15.2).[32,34A]

Fig. 15.1 Neurons from the thoracic ganglion of the grasshopper, *Melanoplus differentialis*. Tissue fixed in 2 per cent glutaraldehyde-cacodylate for $1\frac{1}{2}$ hours, post-osmicated, and embedded in Araldite.

A, B. The lipochondria (L) are variable in structure; most contain membraneous arrays and granules. Some lie close to the Golgi apparatus (G) which shows typical saccules and vesicles. Also seen are nuclear pores (NP), spiral ribosomal aggregates (polysomes ?) (P), and mitochondria (M). In the inset a lipochondrion is seen at higher magnification; the arrows point to areas where the delimiting unit membrane is cut transversely.

C, D. Golgi complexes in the form of 'dictyosomes'. Note the saccules (GS), and vesicles (GV). Golgi elements often show such crescentic configurations. Note the lipochondria (L) in association with the Golgi apparatus. 'Fenestrated' regions (F) frequently lie close to the Golgi and appear to be part of it. Alveolate or coated vesicles (AV) occur, often in continuity with the fenestrations.

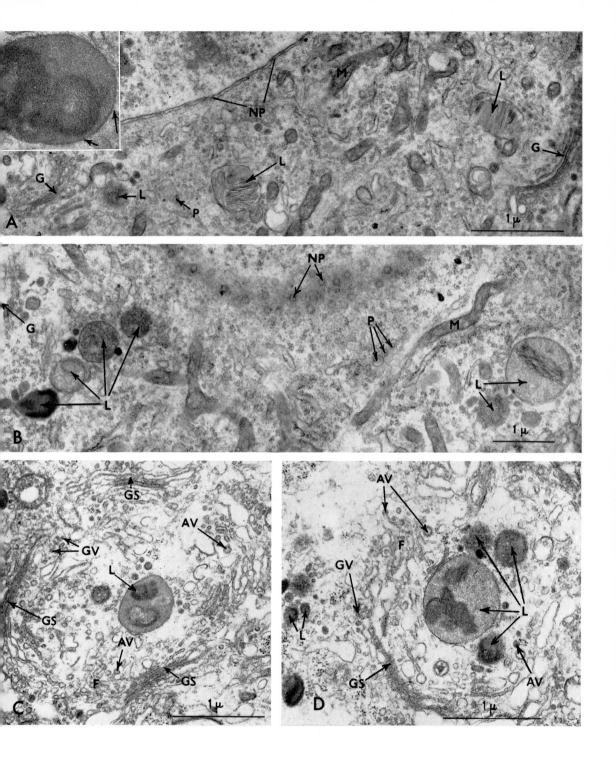

occurs in larger cytoplasmic bodies whose form resembles that of the Golgi dictyosomes. It is possible that this colour is due to dye uptake by the small lipochondria found to be associated with the inner border of the dictyosomes[34B] (Figs. 15.1, C and D). However, electron microscopical studies on vitally stained cells would be required to prove that no uptake at all occurs in the Golgi apparatus. In vitally coloured vertebrate neurons, ultrastructural investigations indicate that dye uptake occurs chiefly in the lysosomes[43] (see discussion on page 178).

Neurons of gastropod molluscs

In a detailed cytochemical study, Chou[14,15] ascertained that the neurons of the snail *Helix* contained three types of lipid globule which he termed triglyceride droplets, 'blue' lipochondria, and yellow lipochondria. He believed that no Golgi apparatus *per se* existed and that the dictyosomes seen in classical impregnation preparations occured by metallic deposition on the distorted surfaces of lipochondria.[14] Such was also the interpretation of many cytologists in the Oxford School.[4] This view was keenly contested by Gatenby and his colleagues[22] who thought that the Golgi apparatus, consisting of a canalicular system, existed in the living cell quite separately from the lipid granules and droplets that were also present.

The results of an electron microscopical study by Chou in collaboration with Meek[16] were claimed to substantiate Chou's earlier hypothesis: they believed that lamellar Golgi membranes could only be found in neurons after inadequate preservation of the lipochondria, and were therefore an artifact of fixation. However, when further ultrastructural studies were made on the neurons of *Helix*, it was observed that the electron-opaque lipochondria were quite distinct from the lamellar Golgi complexes, irrespective of the fixative used.[18,44,45] (Fig. 15.3 A). Moreover, Dalton found, as more recent workers corroborate,[21] that when classical Golgi preparations are viewed with the electron microscope after prolonged osmication, the reduced osmium is deposited along one of the surfaces of the stacked Golgi saccules, not on the lipochondria.

Enzymatic studies at the light microscopical level, show that the lipochondria in *Helix* correspond to lysosomes in that they contain acid phosphatase.[30] They are also capable of splitting thiamine diphosphate.[30] Ultrastructural studies confirm these observations; the reaction product for both enzymes is deposited over electron-sense bodies with the morphological features of lipochondria.[45] Moreover, the evidence indicates that both phosphatases are also present in what may be the same two or three saccules of the Golgi.[45]

Unlike the situation in most vertebrates, in the neurons of *Helix* as in those of *Melanoplus*, the activity of acid phosphatase is not restricted to the lysosomes or lipochondria, nor is TPPase limited to

Fig. 15.2 Thoracic ganglia of *Melanoplus*; fixed in 2 per cent glutaraldehyde-cacodylate, sectioned, incubated for enzymatic activity, post-osmicated, and embedded in Araldite. Control sections incubated in substrate-free medium showed no lead precipitate.

A. Neuron from section incubated for acid phosphatase by the Gomori procedure, using cytidine monophosphate as substrate.[49] Note the lead phosphate reaction product in a body with the size and shape of a lipochondrion (L) and in the Golgi saccules and vesicles (arrows).

B. In this preparation, the substrate used was adenosine triphosphate (ATP) in the method of Wachstein and Meisel.[70] Note the reaction product for ATPase activity in the lipochondria (L), but not in the Golgi apparatus (G). At the top, the plasma membranes of the glial cell surrounding the neuron also hydrolyse ATP.

C. Neurons from section incubated according to Novikoff and Goldfischer,[51] using thiamine pyrophosphate (TPP) as substrate. The lead reaction product, indicative of TPPase activity, is in the lipochondrion (L) and Golgi saccules and vesicles (G). Note, at lower left, the TPPase reaction product in the membranes of the glial cell ensheathing the neuron.

D. Section of a glial cell incubated for ATPase activity as in Fig. 15.2 B. Note the reaction product in the glial lysosomes (L) and apposed plasma membranes (arrows). The mitochondria are free of lead as are the two neurons (N) between which the glial cell lies.

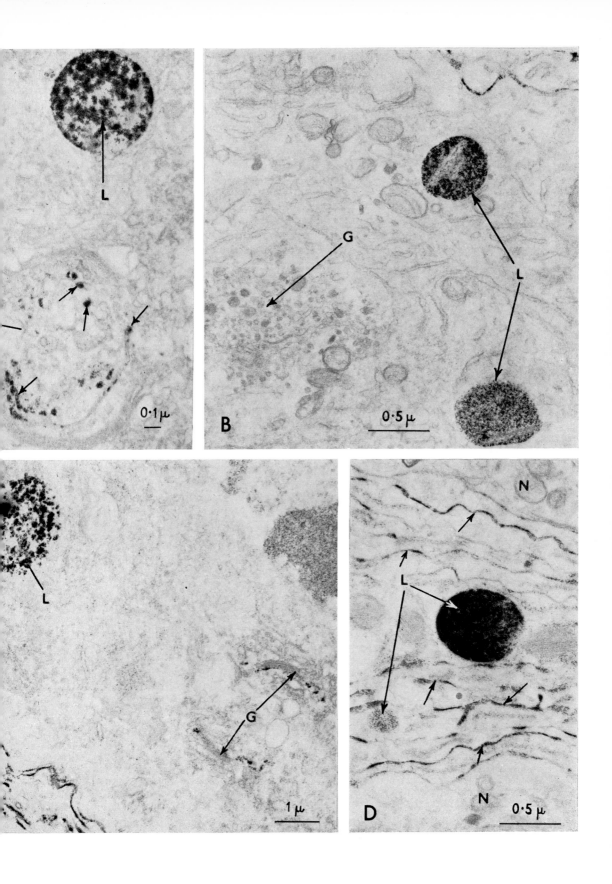

the Golgi. Hence these phosphatases cannot be used as 'markers' for specific organelles. The fact that both lysosomes and Golgi hydrolyse similar substrates suggests that they contain the same enzymes. If biochemical analyses prove this to be so, it may be significant in terms of the developmental or functional inter-relationships between the lysosomes and the Golgi, particularly since the two are so closely associated spatially within the cell.

In the fresh water snail *Planorbis trivolvis*, speculations of this sort are further stimulated by the fact that TPPase, present in the lysosomes and Golgi, is also found in the endoplasmic reticulum in the cerebral neurons[33] (Fig. 15.4, A). In addition, both the lysosomes and some of the Golgi saccules and vesicles, hydrolyse monophosphates at acid pH as well as adenosine triphosphate at neutral pH[33] (Fig. 15.3, C and 15.4 B). The ultrastructure of the phosphatase-rich lysosomes is similar to that of lipochondria in other invertebrate neurons (Fig. 15.3, B) and their cytochemical reactions are those of typical lipochondria.[33] In sections of cells incubated for short periods, the typical lipochondrial ultrastructure is clearly evident as well as the overlying lead reaction product that indicates the presence of phosphatase activity (Fig. 15.3, C, inset).

In the optic tentacles of *Helix*, certain neurosecretory cells are present that contain bodies exhibiting the characteristics of lipochondria.[29] These bodies prove to be the sites of activity of acid phosphatase as well as TPPase.[31]

In conclusion, then, from studies on the cerebral neurons of several genera, in addition to those on tentacular neurosecretory cells, it appears that the lysosomes of gastropod neurons exhibit a similar sort of substrate specificity to those of the orthopteran neurons studied. In relation to this, it might be mentioned that a similar substrate specificity is displayed by the lysosomes in the glial cells surrounding the neurons of *Planorbis*;[33] in this respect they resemble those of the glia of *Melanoplus*.[32] Ultrastructurally, however, they are found to be larger and more complex, containing membrane arrays presumed to be phospholipid and electron-dense clumps which may be digestive residues[33] (Fig. 15.3, C). In this connection, Pipa and co-workers have found cytochemically demonstrable phospholipid in lamellated dense bodies in the glia of cockroach ganglia.[59]

The neurons of some other gastropods have been investigated cytologically and are found to contain bodies with some of the features of lipochondria and lysosomes. However, as yet no enzymatic studies for phosphatase distribution have been made on these cells. These include the neurons of the snail *Lymnaea*[14,15,9] and the neurons of opisthobranch molluscs, especially *Aplysia* (see references in[68]).

Neurons of other invertebrates

Electron microscopical investigations have shown that nerve cells of some other arthropods and certain annelids contain electron-dense bodies with the ultrastructural characteristics of lipochondria or lysosomes. For example, neurons from the horseshoe crab *Limulus*,[8] neuroglandular cells in the cockroach *Leucophaea*,[64] nerve cells in the earthworm *Lumbricus*[61] and in the prawn *Leander*[42] all contain

Fig. 15.3 A.   Neuron from the snail *Helix aspersa*, fixed in buffered osmium tetroxide and embedded in Araldite. Lipochondria of two sorts are present; a phospholipid droplet (pld) and mixed lipid droplets (mld) (Chou and Meek's 'blue' and yellow lipochondria[16]). The Golgi apparatus near the lipochondria consists of Golgi saccules (gl) and Golgi vesicles (gv). Also seen is ribosome-studded endoplasmic reticulum (rnp). Insert shows lamellae, presumably phospholipid (plm), within a lipochondrion. (From Meek and Lane.[45])
  Both these electron micrographs are from the cerebral ganglion of the fresh-water snail, *Planorbis trivolvis*. Ganglion fixed in glutaraldehyde-cacodylate.
  B. Neuron from ganglion that was post-osmicated and embedded in Araldite. Note lipochondria (L) bounded by unit membranes (arrows) and containing ferritin-like granules, amorphous dense clumps, and membrane arrays (PL). Insert shows a lipochondrion composed mainly of membrane arrays.
  C. Section incubated for acid phosphatase by the Gomori lead method using cytidine monophosphate as substrate; post-osmicated and embedded in Araldite. Note reaction product over near-spherical lipochondria (L) and Golgi saccules (GS) and vesicles (GV). Insert shows a cell incubated for ATPase as in Fig. 15.2B for a shorter time enabling one to see the lipochondrial structure as well as the reaction product.

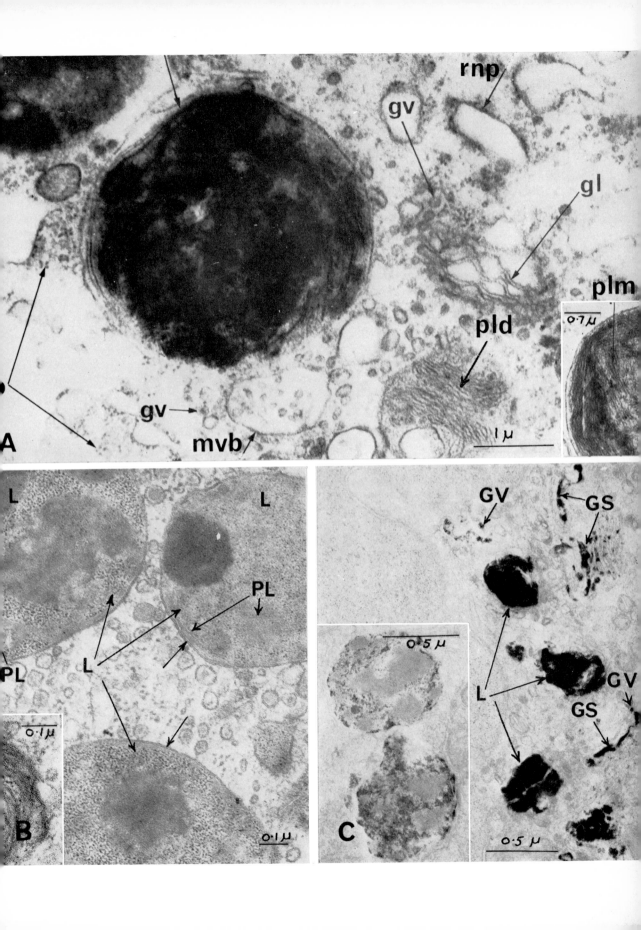

lamellated electron-opaque, membrane-delimited, spheroidal bodies. In the last two instances, light microscopical cytochemical studies have also been made. The bodies in the neurons of *Leander* have been identified as lipochondria,[40] and in *Lumbricus*, they have been shown to contain acid phosphatase.[61] In the absence of examinations for the localization of acid phosphatase it is not possible to state whether the dense bodies in the other nerve cells may accurately be termed lysosomes, but their structural resemblance is pronounced and future enzymatic studies should clarify their nature.

LIPOCHONDRIA AND LYSOSOMES IN VERTEBRATE NEURONS

Casselman and Baker[13] described cytoplasmic bodies in rabbit neurons with the cytochemical reactions of lipochondria. It has since been shown[23] that the nerve cells of rabbits contain lysosomes with cytochemically demonstrable acid phosphatase. Evidence produced by other investigators also suggests that the lipochondria or dense bodies of vertebrate neurons contain lysosomal enzymes. For example, at the electron microscopical level, Novikoff and Essner[50] find that the electron-dense bodies in the neurons of many vertebrates, such as rat and pigeon, are acid phosphatase-rich. Chouinard[17] has observed dense bodies in the neurons of the cat which possess four different lysosomal enzymes; he finds that these have ultrastructural features similar to those of lipochondria. Similar membrane-delimited dense bodies in vertebrate nerve cells have been described frequently in the literature,[1, 48, 57] and some authors[1] suggest that they contain phospholipid. Finally, Koenig[27] has shown that vital dyes are taken up by the glycolipoprotein granules in mammalian nerve cells; these same granules he observes to be acid-phosphatase rich. It appears then, that the dense bodies in the perikarya of vertebrate neurons are both lipochondria and lysosomes.

LIPOCHONDRIA AND LYSOSOMES IN CULTURED ANIMAL CELLS

In cultured vertebrate neural cells, Ogawa and his colleagues[55] have reported the uptake of neutral red by acid phosphatase-rich granules. Robbins, Marcus, and Gonatas[60] in a study of cultured HeLa cells, found that the particles that take up acridine orange also contain acid phosphatase, and at the electron microscopical level, have a content of phospholipid myelin figures. They assume that the dye uptake is associated with the phospholipid. In addition, it has been found in cultured L cells and HeLa cells, that cytoplasmic phospholipids have a distribution corresponding to the sites of acid phosphatase activity.[37] In cultured KB cells, the vital dyes neutral red and acridine orange are taken up by cytoplasmic bodies with the same size and distribution as the granules that display acid phosphatase activity[35] (Fig. 15.4, E–G). In these cells too, then, it seems that the vital dyes are concentrated by the lysosomes. These lysosomes respond positively to the acid haematein test for phospholipids. Further,

Fig. 15.4 A, B. Neuron from *Planorbis trivolvis*. Ganglion fixed in glutaraldehyde-cacodylate post-osmicated and embedded in Araldite.
    A. Treated like that in Fig. 15.2 C. Note reaction product for TPPase in lysosomes (L), Golgi (arrows), and endoplasmic reticulum (ER).
    B. Treated like that in Fig. 15.3 C. Reaction product is in some of the Golgi saccules (GS) and a few vesicles (GV).
    C. Portion of a glial cell ensheathing *Planorbis* neurons, treated as in Fig. 15.3 B. Note the large lipochondria in the cytoplasm, containing clumps, granules and membrane arrays. Insert is an enlargement of part of X, showing the membrane array (probably phospholipid) and the limiting unit membrane (arrow).
    D. Cultured KB cell. Fixed in glutaraldehyde, post-osmicated, and embedded in Epon. Note the lysosomal dense bodies (L). Golgi apparatus G; mitochondria, M.
    E, F, G. Cultured KB cells. E. living cells exposed to 0·04 per cent neutral red. F. living cells exposed to 1×10^{-5} M acridine orange, $\times 950$; and G. glutaraldehyde-fixed cells incubated for acid phosphatase. Note similarity in size and distribution of the cytoplasmic granules (arrows) in the three preparations. (From Lane and Novikoff.[35])

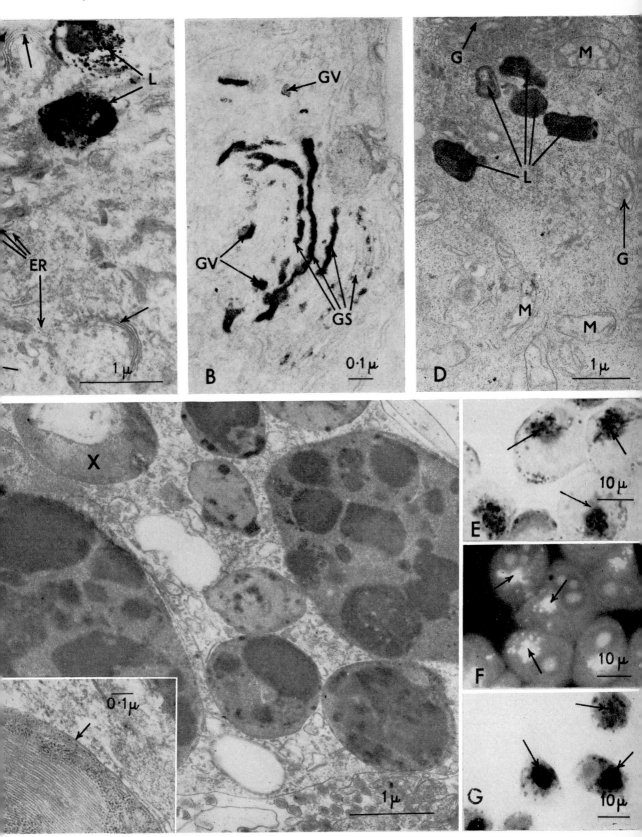

they exhibit ultrastructural characteristics common to many lipochondria; an electron-opaque lamellar or granular matrix delimited by a unit membrane[35] (Fig. 15.4, D). Such results provide further evidence for the supposition that in these cases, the terms *lipochondrion*, *neutral red granule*, and *lysosome* are synonomous.

INTRACELLULAR SITES OF VITAL DYE UPTAKE

There are varied reports in the literature linking vital dye uptake with the lipochondria or the lysosomes. Firstly, a brief recapitulation of the evidence for this presented so far will be made. Cytologists in the Oxford School, working with the light microscope, found that the lipochondria were the sites of uptake of vital dyes in invertebrate cells;[15, 41, 67] these phospholipid-rich lipochondria contain lysosomal acid phosphatase activity.[30, 32, 33, 45] In vertebrate cells, Holt[26] has demonstrated a correspondence between the localization of cytoplasmic phospholipids and the lysosomal enzyme activity; it has also been found in vertebrate tissues that the 'neutral red granules' contain phospholipid.[13, 46] In cultured vertebrate cells, the phospholipid-containing bodies that take up vital dyes also display acid phosphatase activity.[35, 55, 60]

There is further evidence scattered throughout the literature on the uptake of vital dyes by lysosomes in different types of cells. Kovács and Hafiek[28] have shown that the acid phosphatase-rich lysosomes are the sites of uptake of neutral red in male accessory sex glands of mice. Cohn[20] reports that the lysosomal granules of polymorphonuclear leucocytes and of macrophages stain avidly with neutral red. Beaufay[20] has observed that when neutral red is injected into rats, it shows a lysosome-like distribution after centrifugation of liver homogenates. In protozoans, Müller and his colleagues[20] find that the neutral-red-stained bodies have a distribution similar to that of acid phosphatase positive structures; this has also been shown by other investigators.[62] In amoebae, neutral red colours granules with a phospholipid moiety.[10] Schmidt and Tanaka, studying many different cell types, have concluded that the sites of vital dye uptake in the cytoplasm are the 'cytosomes'[65] and the 'segresomes'.[69] These bodies, the authors suggest, correspond to lysosomes. All of these studies lead to the conclusion that the lipochondria, neutral red granules, and lysosomes are actually one and the same organelle; in other words that the three terms are synonymous in the situations described above. It must be noted, however, that this synonymity would not hold true for certain types of lysosome, such as Golgi vesicles or pinocytotic vacuoles. In these cases, the lysosomes could not be considered to be lipochondria.★

More recent studies on the problem of uptake of vital dyes have been made at the ultrastructural level. Koenig and his colleagues have investigated the uptake of such dyes as neutral red or methylene blue in vertebrate neurons; under the electron microscope it can be seen that the dyes induce the formation of lamellar structures within the normally granular, acid phosphatase-rich lysosomes.[43] Byrne,[11, 12] in an ultrastructural study of mouse pancreas, showed that neutral red was taken up by membranous, acid phosphatase-rich bodies. She suggested that the lysosomes accumulate or concentrate dye within them, in the manner of intracellular phagocytes; the dye being precipitated out of the cytoplasm in association with phospholipid.[11] Since the mouse pancreas normally has few lysosomes, the acid phosphatase-rich membranous bodies seem to be induced by the action of the dye; it may therefore be that they are a form of residual body formed from autophagic vacuoles.[11] Both residual bodies and autophagic vacuoles are considered to be a variety of lysosome, since both contain acid phosphatase activity.[49]

Both Koenig and Byrne have carried out investigations on the mode of uptake of vital dyes; Schmidt[65] has made extensive inquiries into the various intracellular pathways of vital dyes when being taken into

★ [*Editorial note:* We believe than an over-rigid application of biochemically 'sound' definitions *at a different level of integration* (the morphological level of cell structure) does not always usefully contribute to clarity (see Pantin, pp. 6–7), and that the terms 'Golgi vesicle' and 'pinocytotic vacuoles' are, in some circumstances, more useful than the terms 'lysosome' or 'lipochondrion'. It remains questionable whether an enzymatic activity alone is a good classificatory feature for cell structure (see also David and Brown, p. 40–44).]

cells. The details of their conclusions will not be considered here, however, as they are not pertinent to the present discussion.

CONCLUDING REMARKS

The proposition set forth in this discussion, namely, that lipochondria and neutral red granules are lysosomes, is based on cytochemical evidence which shows that bodies with the size, shape, structural details, and distribution of lipochondria contain reaction product for acid phosphatase activity. The next step toward a firmer identification of the lipochondria of invertebrates as lysosomes would be the cytochemical demonstration in them of other lysosomal hydrolases, such as β-glucuronidase, aryl sulphatase, or hydrolytic esterase. Proof of the structure-linked latency of the enzymes in the lipo-chondria would also add weight to the argument, since such a latency is an inherent property of lyso-somes as originally defined by de Duve.[19] Clearly, however, for unequivocal evidence that invertebrate lipochondria are lysosomes, biochemical analyses are necessary. At present, there is no supporting data of this kind chiefly because of the paucity of material available from invertebrate sources. Such bio-chemical evidence would be of additional value in ascertaining whether the phosphate-splitting ability of the lipochondria is due to a number of fairly specific enzymes, or one, relatively non-specific phos-phatase, capable of splitting a number of substrates over a considerable pH range.

This raises the question of substrate specificity. It seems that the lysosomes of invertebrate neurons generally display certain differences in substrate specificity from those of vertebrate cells, in that they hydrolyse nucleoside di- and triphosphates at pH 7·2, as well as monophosphates at pH 5·0. It seems questionable, however, whether these differences are very profound. In some cases lysosomes of verte-brates are also capable of hydrolysing more than one substrate over a considerable pH range; the hepatic lysosomes, for example, of several mammalian species split a number of phosphates not only at acid pH, but also at neutral and alkaline pH.[53] A rather similar situation occurs in the neurons of rat spinal ganglia.[54]

In these rat neurons, as in the invertebrate neurons studied, acid phosphatase is not restricted to the lysosomes; it is also present in the saccules of the dictyosome-like Golgi and the adjacent fenestrated region. This similarity in enzyme distribution in these two different neuronal types, leads to speculations concerning the physiology of the dictyosome form of Golgi. Acid phosphatase is not ordinarily found in Golgi saccules when the Golgi is in the form of a network. Could the morphology of the Golgi dictyosome be associated in any way with its content of acid phosphatase activity or vice versa?

In invertebrates, the similarity in enzyme content of the Golgi and lysosomes (both contain acid phosphatase and TPPase and, in *Planorbis*,[33] ATPase), in addition to their intimate spatial association, suggests the possibility of a functional or developmental relationship between them. Since the Golgi vesicles seem to form from the saccules, perhaps the enzymes are transported to the lysosomes via the vesicles.[49] On the other hand, the Golgi saccules may be directly involved in the origin of the lysosomes. The presence of TPPase in the endoplasmic reticulum as well as the Golgi and lysosomes in the neurons of *Planorbis*, suggests that the phosphatases may originally be synthesized in the cisternae of the endo-plasmic reticulum; from there they might be transported to the Golgi and subsequently the lysosomes, or alternatively, directly to the lysosomes.

It is evident that one can only speculate at present as to the possible inter-relationships between these organelles. Although tentative hypotheses may be made,[34,50] the morphological evidence is not as yet sufficient for any definite conclusions concerning the site of origin or the role of the lipochondrion in the invertebrate cell. Hence their dynamic functional inter-relationships with other organelles are even less certain.

In spite of the issues yet to be elucidated, the present situation seems at least tentatively resolvable: the evidence indicates that the lipochondria are the sites of uptake of vital dyes, in other words they correspond to the 'neutral red granules'; they also contain lysosomal acid phosphatase and have a fine structure similar to that of many vertebrate lysosomes. Therefore it may be concluded that in the cells considered here, the lipochondria, and the neutral red granules are also lysosomes.

13—C.S.I.

SUMMARY

A survey of the light microscopical investigations on invertebrate neurons, vertebrate neurons and cultured animal cells indicates that the phospholipid-containing lipochondria are the sites of uptake of vital dyes, corresponding to the 'neutral red granules'. Their fine structure and possession of acid phosphatase activity show them to be lysosomes. Similarities and differences between lipochondria of invertebrates and vertebrates are examined. The slight differences in the substrate specificities of their phosphatases do not detract from their many similarities.

ACKNOWLEDGEMENTS

I would like to thank Dr. Alex Novikoff for many stimulating discussions while I was in his laboratory and for critical review of this manuscript.

This work was supported by a grant from the National Institute of General Medical Sciences (GM 12, 427).

REFERENCES

1. AFZELIUS, B. A. and FRIDBERG, G. 1963. The fine structure of the caudal neurosecretory system in *Raia batis*. *Zeit. Zellforsch.* **59**, 289–308
2. ASHHURST, D. E. and CHAPMAN, J. A. 1962. An electron-microscope study of the cytoplasmic inclusions in the neurons of *Locusta migratoria*. *Quart. J. micr. Sci.* **103**, 147–53
3. BAKER, J. R. 1950. Studies near the limit of vision with the light microscope, with special reference to the so-called Golgi-bodies. *Proc. Linn. Soc.* **162**, 67–72
4. —— 1959. Towards a solution of the Golgi problem: recent developments of cytochemistry and electron microscopy. *J. Roy. micr. Soc.* **77**, 116–29
5. —— 1963. New developments in the Golgi controversy. *J. Roy. micr. Soc.* **82**, 145–57
6. BEAMS, H. W. and KING, R. L. 1932. Cytology of orthopteran nerve cells. *J. Morph.* **53**, 59–95
7. —— SEDAR, A. W., and EVANS, T. C. 1953. Studies on the neurons of the grasshopper with special reference to the Golgi bodies, mitochondria and neurofibrillae. *La Cellule* **55**, 291–304
8. BERN, H. A., NISHIOKA, R. S., and HAGADORN, I. R. 1962. Neurosecretory granules and the organelles of neurosecretory cells. In *Neurosecretion*, Memoirs of the Soc. for Endocrin., No. 12, pp. 21–34. New York: Academic Press
9. BOER, H. H. 1965. A cytological and cytochemical study of neurosecretory cells in Basommatophora, with particular reference to *Lymnaea stagnalis* L. *Arch. Néerl. Zool.* **16**, 313–86
10. BYRNE, J. M. 1963. The vital staining of Amoeba proteus. *Quart. J. micr. Sci.* **104**, 445–58
11. —— 1964. An electron microscopical study of neutral red granules in mouse exocrine pancreas. *Quart. J. micr. Sci.* **105**, 219–25
12. —— 1964. Acid phosphatase activity in the neutral red granules of mouse exocrine pancreas cells. *Quart. J. micr. Sci.* **105**, 343–8
13. CASSELMAN, W. G. B. and BAKER, J. R. 1955. The cytoplasmic inclusions of a mammalian sympathetic neurone: A histochemical study. *Quart. J. micr. Sci.* **96**, 49–56
14. CHOU, J. T. Y. 1957. The cytoplasmic inclusions of the neurones of *Helix aspersa* and *Limnaea stagnalis*. *Quart. J. micr. Sci.* **98**, 47–58
15. —— 1957. The chemical composition of lipid globules in the neurones of *Helix aspersa*. *Quart. J. micr. Sci.* **98**, 59–64
16. —— and MEEK, G. A. 1958. The ultra-fine structure of lipid globules in the neurones of *Helix aspersa* *Quart J. micr. Sci.* **99**, 279–84
17. CHOUINARD, L. A. 1964. Evidence for the lysosomal nature of the PAS-positive granules in the Purkinje cell of the cerebellar cortex in the adult cat. *Canad. J. Zool.* **42**, 103–12
18. DALTON, A. J. 1960. Morphology and Physiology of the Golgi complex. In *Cell physiology of neoplasia*, pp. 161–84. Austin, Texas: The University of Texas Press
19. DE DUVE, C., PRESSMAN, B. C., GIANETTO, R., WATTIAUX, R., and APPELMANS, F. 1955. Tissue fractionation studies. 6. Intracellular distribution patterns of enzymes in rat-liver tissue. *Biochem. J.* **60**, 604
20. DE REUCK, A. V. S. and CAMERON, M. P. 1963. Eds., In Ciba Foundation Symposium on *Lysosomes*. Boston, Mass.: Little, Brown and Co.
21. FRIEND, D. S. and MURRAY, M. J. 1965. Osmium impregnation of the Golgi apparatus, *Amer. J. Anat.* **117**, 135–50

22. GATENBY, J. B. 1951. The Golgi apparatus of liver and nerve cells. *Nature, Lond.* **167**, 185
23. GOLDFISCHER, S. 1964. The Golgi apparatus and the endoplasmic reticulum in neurons of the rabbit. *J. Neuropath. Exp. Neurol.* **23**, 36–45
24. GRESSON, R. A. R., THREADGOLD, L. T., and STINSON, N. E. 1956. The Golgi elements of the neurones of *Helix, Locusta* and *Lumbricus. La Cellule,* **58**, 7–16
25. HESS, A. 1958. The fine structure of nerve cells and fibers, neuroglia, and sheaths of the ganglion chain in the cockroach (*Periplaneta americana*). *J. Biophys. Biochem. Cytol.* **4**, 731–42
26. HOLT, S. J. 1959. Factors governing the validity of staining methods for enzymes, and their bearing upon the Gomori acid phosphatase technique. *Exp. Cell Res.* suppl. **7**, 1–27
27. KOENIG, H. 1962. Histological distribution of brain gangliosides: Lysomes as glycolipoprotein granules. *Nature, Lond.* **195**, 782–4
28. KOVÁCS, J. and HAFIEK, B. 1964. The effect of neutral red on the cells of the male accessory sex glands of mice. *Acta Biologica, Acad. Sci. Hung.* **15**, Suppl. 6, 46
29 LANE, N. J. 1962. Neurosecretory cells in the optic tentacles of certain pulmonates. *Quart. J. micr. Sci.* **103**, 211–26
30. —— 1963. Thiamine pyrophosphatase, acid phosphatase, and alkaline phosphatase in the neurones of *Helix aspersa. Quart. J. micr. Sci.* **104**, 401–12
31. —— 1964. Localization of enzymes in certain secretory cells in the optic tentacles of the snail, *Helix aspersa. Quart. J. micr. Sci.* **105**, 49–60
32. —— 1965. Phosphatase distribution in lysosomes and cytoplasmic membranes in the thoracic ganglionic neurons and glia of the grasshopper (*Melanoplus differentialis*). *J. Cell Biol.* **27**, 56A
33. —— 1966. The fine-structural localization of phosphatase in neurosecretory cells within the ganglia of certain gastropod snails. *Amer. Zool.* **6**, 139
34A. —— 1968. The thoracic ganglia of the grasshopper, *Melanoplus differentialis*: Fine structure of the perineurum and neuroglia with special reference to the intracellular distribution of phosphatases. *Zeit. Zellforsch.* **86**, 293
34B. —— 1968. Distribution of phosphatases in the Golgi region and associated structures of the thoracic ganglionic neurons in the grasshopper, *Melanoplus differentialis. J. Cell Biol.* **37**, 89
35. —— and NOVIKOFF, A. B. 1965. Effects of arginine-deprivation, ultraviolet radiation and X-irradiation on cultured KB cells: a cytochemical and ultrastructural study. *J. Cell Biol.* (in press)
36. LEE, R. S. 1963. Phosphatases in the neurones of *Locusta migratoria. Quart. J. micr. Sci.* **104**, 475–82
37. MAGGI, V. and RIDDLE, P. N. 1965. Histochemistry of tissue culture cells: A study of the effects of freezing and of some fixatives. *J. Histochem. Cytochem.* **13**, 310–17
38. MALHOTRA, S. K. 1955. Golgi bodies in nerve cells of insects. *Nature, Lond.* **176**, 886–7
39. —— 1956. The cytoplasmic inclusions of the neurones of certain insects. *Quart. J. micr. Sci.* **97**, 177–86
40. —— 1960. The cytoplasmic inclusions of the neurones of Crustacea. *Quart. J. micr. Sci.* **101**, 75–93
41. —— 1961. A comparative histochemical study of the 'Golgi apparatus'. *Quart. J. micr. Sci.* **102**, 83–7
42. —— and MEEK, G. A. 1961. An electron microscope study of some cytoplasmic inclusions of the neurones of the prawn, *Leander serratus. J. Roy. micr. Soc.* **80**, 1–8
43. MCDONALD, T. F. and KOENIG, H. 1965. Ultrastructural changes in neuron lysosomes induced by *in vivo* staining with neutral red. *Proc. Am. Assoc. Neuropathol.* **15A**
44. MCGEE-RUSSELL, S. M. 1964. On bridging the gap in cytology between the light and electron microscopes by some combined observations on snail neurones, *Quart. J. micr. Sci.* **105**, 139–62
45. MEEK, G. A. and LANE, N. J. 1964. The ultrastructural localization of phosphatases in the neurones of the snail, *Helix aspersa. J. Roy. micr. Soc.* **82**, 193–204
46. MORGAN, W. S. 1953. Cytological studies of the acinar cell of the pancreas of the mouse. Part II. The argentophil, osmiophil, and sudanophil substance. *Quart. J. micr. Sci.* **94**, 269–79
47. MULIYIL, J. A. 1935. The effect of ultra-centrifuging on the ganglion cells of certain Orthoptera. *Zeit. Zellforsch.* **23**, 627–56
48. MURAKAMI, M. 1962. Elektronenmikroscopische Untersuchung der Neurosekretorischen Zellen im Hypothalamus der Maus. *Zeit. Zellforsch.* **56**, 277–99
49. NOVIKOFF, A. B. 1963. Lysosomes in the physiology and pathology of cells: contributions of staining methods. In: Ciba Foundation Symposium on *Lysosomes.* (edited by A. V. S. de Reuck and M. P. Cameron), pp. 36–77. Boston, Mass.: Little, Brown and Co.
50. —— and ESSNER, E. 1962. Pathological changes in cytoplasmic organelles. *Fed. Proc.* **21**, 1130–42
51. —— and GOLDFISCHER, S. 1961. Nucleoside diphosphatase activity in the Golgi apparatus and its usefulness for cytological studies. *Proc. Nat. Acad. Sci.* **47**, 802–10
52. —— ESSNER, E., GOLDFISCHER, S., and HEUS, M. 1962. Nucleoside phosphatase activities of cytomembranes. In *The Interpretation of Ultrastructure,* Vol. 1, pp. 149–92. New York: Academic Press
53. —— IACIOFANO, P., and VILLAVERDE, H. 1965. Observations on the staining reactions of hepatic lysosomes. *J. Histochem. Cytochem.* **13**, 29A
54. —— QUINTANA, N., VILLAVERDE, H., and FORSCHIRM, R. 1964. The Golgi zone of neurons in rat spinal ganglia. *J. Cell. Biol.* **23**, 68A

55. OGAWA, K., MIYUNO, N., and OKAMOTO, M. 1961. Lysosomes in cultured cells. *J. Histochem. Cytochem.* **9**, 202

56. OSINCHAK, J. 1965. The electron microscopic localization of acid phosphatase and thiamine pyrophosphatase activity in prothoracic glands of the insect *Leucophaea maderae*. *Anat. Rec.* **151**, 395

57. PALAY, S. L. 1960. The fine structure of secretory neurons in the preoptic nucleus of the goldfish (*Carassius auratus*). *Anat. Rec.* **138**, 417

58. PIPA, R. L. 1961. Studies on the hexapod nervous system. IV. A cytological and cytochemical study of neurons and their inclusions in the brain of a cockroach, *Periplaneta americana* (L.) *Biol. Bull.* **121**, 521–34

59. ——, NISHIOKA, R. S., and BERN, H. A. 1962. Studies on the Hexapod nervous system. V. The ultrastructure of cockroach gliosomes. *J. Ult. Res.* **6**, 164–70

60. ROBBINS, E., MARCUS, P. I., and GONATAS, K. 1964. Dynamics of Acridine orange-cell interaction. II. Dye induced ultrastructural changes in multivesicular bodies (Acridine orange particles). *J. Cell Biol.* **21**, 49–62

61. RÖHLICH, P., AROS, B., and VIGH, B. 1962. Elektronenmikroskopische Untersuchung der Neurosekretion im Cerebralganglion des Regenwurmes (*Lumbricus terrestris*). *Zeit. Zellforsch.* **58**, 524–45

62. ROSENBAUM, R. M. and WITTNER, M. 1962. The activity of intracytoplasmic enzymes associated with feeding and digestion in *Paramecium caudatum*. The possible relationship to neutral red granules. *Arch. Protistenk.* **106**, 223–40

63. SABATINI, D. D., BENSCH, K., and BARRNETT, R. J. 1963. Cytochemistry and electron microscopy. The preservation of cellular ultrastructure and enzymatic activity by aldehyde fixation. *J. Cell. Biol.* **17**, 19–58

64. SCHARRER, B. 1963. Neurosecretion. XIII. The ultrastructure of the Corpus Cardiacum of the insect *Leucophaea maderae*. *Zeit. Zellforsch.* **60**, 761–96

65. SCHMIDT, W. 1962. Licht-und Elektronenmikroskopische Untersuchungen über die Intrazelluläre Verarbeiting von Vitalfarbstoffen. *Zeit. Zellforsch.* **58**, 573–637

66. SHAFIQ, S. A. 1953. Cytological studies of the neurones of *Locusta migratoria*. Part I. Cytoplasmic inclusions of the motor neurones of the adult. *Quart. J. micr. Sci.* **94**, 319–28

67. —— and CASSELMAN, W. G. B. 1954. Cytological studies of the neurones of *Locusta migratoria*. Part III. Histochemical investigations, with special reference to the lipochondria. *Quart. J. micr. Sci.* **95**, 315–20

68. SIMPSON, L., BERN, H. A., and NISHIOKA, R. S. 1963. Inclusions in the Neurons of *Aplysia californica* (Cooper, 1863) (Gastropoda Opisthobranchiata). *J. Comp. Neur.* **121**, 237–58

69. TANAKA, H. 1961. 6. Segresome and the cellular uptake of varied substances as revealed in the electron microscopy. *Annual Report* of the Institute for Virus Research, Kyoto University. **4**, 118–52

70. WACHSTEIN, M. and MEISEL, E. 1957. Histochemistry of hepatic phosphatases at a physiologic pH. *Am. J. Clin. Path.* **27**, 13–23

71. WILLEY, R. B. and CHAPMAN, G. B. 1962. Fine structure of neurons within the Pars Intercerebralis of the cockroach, *Blaberus craniifer*. *Gen. Comp. Endocrin.* **2**, 31–43

NOTE ADDED IN PROOF

The author of this chapter has requested the editors to point out that it was written and submitted for publication in December 1965: and that consequently no discussion of the papers pertinent to the subject published after that date has been included.

Chapter 16

The Method of Combined Observations with Light and Electron Microscopes applied to the Study of Histochemical Colourations in Nerve Cells and Oocytes

S. M. McGEE-RUSSELL

INTRODUCTION

The historical development of the discipline of biological electron microscopy, with the rapid advances achieved, the high order of technical skill necessary, and the considerable investment of time involved, has not always produced close co-ordination between results obtained with the light microscope and results obtained with the electron microscope. Co-operation between light microscopists and electron microscopists has, also, not always been productive, unless proper skill has been acquired by scientists in the use of both instruments in their most sophisticated forms. The advantages of the light microscope as a survey tool or a 'rapid scanning system' preliminary to electron microscopy were very early appreciated by Gatenby and Lutfy.[16] However, it was, and it is still, difficult to agree with their conclusion that the light microscope would be reduced to a preliminary, scanning, role alone. It is too highly developed and sophisticated an instrument. The two instruments, light and electron microscopes, are complementary, providing us with different images and different information which, taken together, enable us to achieve a more nearly accurate concept, or interpretation, of cell structure and activities. Recent papers in the literature of electron microscopy, for example Wood and Luft,[58] Huxley and Zubay,[23] McGee-Russell,[30] amply demonstrate the value of a comparative and complementary approach.

It may be accepted as axiomatic that any method of studying cells involves its own special limitations and, therefore, as a direct consequence, its own special interpretations and 'artifacts'. This is true even of the most carefully controlled studies of living cells, since, in a very real sense, every observer is his own artifact. No channel of sensory input is without limitations; no analysis of sense data is inherently accurate without external referents. Baker,[4,5] has always emphasized the value and necessity of interpreting cell structure in terms of the living cell, compared to the cell after processing of whatever kind. The current literature shows, unfortunately, that, once again, rather few cytologists take the time to study fresh cells representative of their experimental material, despite the sophisticated optical and photographic techniques now available to make their task easier. Baker's emphasis on the living cell[4,5] certainly deserves to be re-emphasized here in this book. However, it seems to me that Baker's emphasis on the *comparison* of images obtained by different techniques is as important as his emphasis on the living image. This, therefore, is one of the main points to be made in this article. Physiological cytology must develop as a comparative subject at all levels—at the level of technical procedure, at the level of method of observation, at the level of cell type, tissue type, and type of animal, and finally at the level of interpretation, which must, consequently, be based upon results from as many of the modern methods of study as possible.

The development of techniques for closely correlating images in the light microscope with images in the electron microscope (McGee-Russell;[30] de Bruijn and McGee-Russell;[10] Grimley;[18,19] Potts[45]) leads one to reconsider the value of scanning tissues embedded in plastic with the light microscope. Methods have been devised for the colouring of tissues in plastic sections mainly for morphological orientation and identification, and as a substitute and alternative to phase-contrast microscopy. Some of the recommended methods act as a very general, although excellent, stain, like methylene blue and toluidine blue (with or without borax) (Richardson and colleagues;[52] Meek[37]); others, such as Nile blue A and Sudan black B (McGee-Russell and Smale[32]) have a greater selectivity. As soon as a colouring process displays selectivity, one is faced with a familiar difficulty, embodied in the questions: 'Is this selectivity specific?' and 'Is this differential colouring to be regarded as a histochemical reaction?' At this stage in the development of the subject it does not seem to me necessary to become particularly worried by either of these questions. Having seen repeated in the last twenty years of microscopy and microtechnique, events and difficulties analogous to those of the previous two centuries (see McManus, pp. 209–223), one may discern that, at this juncture, it is more necessary to establish 'selectivity', than it is to engage in speculation or controversy about 'specificity'. Once the selectivity (and consistency) of a method is established, the chemical and physical bases of that selectivity may be investigated; but *first* we must find out, in practice, which methods are truly selective on plastic-embedded tissues. The wealth of information in the past histological literature may be a helpful guide, but it is not a direct guide, nor an entirely reliable one. Staining methods suitable for ordinary paraffin histological sections *may* be applicable to plastic-embedded tissues, after appropriate fixation (see de Bruijn and McGee-Russell[10]), but in each case the methods must be submitted to practical test, and may need considerable modification.

In a given scientific field at any one point in time one may notice the existence of trends or 'fashions'. In electron microscopic histology and cytology (up to this time) 'electron histochemistry' has become equated with procedures derived, often, from 'enzyme histochemistry', and 'substrate incubation procedures'. The cells are fixed, initially, in a manner considered appropriate to the preservation of as much enzyme activity and structure as possible; then reacted upon in an experimental manner; and finally fixed 'as for electron microscopy' and embedded in plastic, by what is regarded as a standard procedure. There is some evidence (Wood and Luft;[58] de Bruijn and McGee-Russell[10]) to show that standard procedures are much less standard, and variations are much more significant in the precise character of the final image, than might be assumed! Another aspect of electron histochemistry involves the direct treatment of plastic sections of tissues, specially fixed and embedded, with reactive solutions such as enzymes, followed by monitoring with the electron microscope after heavy metal staining (Marinozzi[36]). Special water-soluble plastics have been investigated in association with these techniques (Rosenberg, Bartl, and Lesko;[53] Bartl and Bernhard[7]) on the assumption that the special character of the embedding medium would render the tissue more accessible to reagents (Bartl;[6] Bernhard;[8] Monneron[41]). Both types of approach tend to involve the concept that tissues embedded in the conventional embedding media (Epon, Araldite, Methacrylate, and Vestopal) are relatively inert and inaccessible; and tend to cause one to overlook the potential value of techniques producing colour differences visible in the light microscope, which are *compatible* with subsequent electron microscopy of the same sections, in the technique which I call 'combined observations'. Tissues embedded, for example, in Epon, are not, by any means whatever, chemically inert. They are directly accessible to selective colouring procedures and histochemical methods which are appropriate to the light microscope, whilst they remain simultaneously suitable for study in the electron microscope, as I shall illustrate in this article. The recent work of Monneron and Bernhard,[42] which is closely related to this point, is discussed later. The fact that it is possible to analyse the amino acid content of dinosaur bone is a considerable tribute to the indestructibility of protoplasm and its secretory products. We need not despair of determining the chemistry of the tissue (as pointed out by Pantin in this book, p. 9) merely because we have fixed it, and preserved it in a plastic rather than in rock!

Just as Baker[4] (2nd edition, p. 7) pointed out that tissues embedded in paraffin are not impermeable to water, so, today, it may be pointed out that tissues embedded in Epon or other plastics are also not impermeable to water, nor to reagents dissolved in water, nor to other solvents. This is especially true of the cut surface of the section. At the surface, the elements of the tissue are directly accessible to reagents, in a way that elements embedded in the depth of the section, and completely surrounded by plastic, may not be. There are suggestions in the literature that the action of electron-contrasting reagents is dependent upon the coincidence of cut surface and structural element, and that successful contrast enhancement may *not* be achieved on, for example, viruses which lie inside the section, and never reach a cut surface (Shalla, Carroll, and de Zoeten[54]). A phenomenon similar to this has been observed in my work, where, in sections (25 μ thick) of whole mouse heads embedded in Epon plastic, after staining the sections with toluidine blue, it could be seen that the two sides of the section were stained, but that a central zone, in the middle of the thickness of the section, was unstained. This resulted in a 'double image' of the tissue, particularly well shown in large blood vessels, which appeared as two overlapping rings. In focusing through the section, a central 'blurred' area, without stain, and with poor image quality through light scatter, was a striking visual effect. The phenomenon of limited penetration by light-microscopical dyestuffs may be eliminated by prolonged staining, and elevation of the staining temperature. It is possible that similar steps might ensure full electron contrasting of all elements in a thin section, with uranyl acetate or lead throughout its thickness (unless extraction or encapsulation phenomena intervene), but the degree of contamination might be excessive.

Elevation of the temperature is a most significant step to ensure interaction between the plastic-embedded tissue and light-microscopical colouring reagents. It is not, however, always necessary, and some reactions may be carried out successfully at room temperature. It is always worth testing a method at room temperature first, as there are minor problems, such as rapid deterioration of reagents, at elevated temperatures. The histochemical methods discussed in this article are successful at room temperature, with Epon-embedded tissues.

Baker[5] defined histochemistry in the following way: 'Histochemistry is the microscopical study of the location of chemically recognizable substances or groups of substances in the tissues of plants and animals'. This definition must nowadays be extended by the insertion of the words 'and biochemically' after 'chemically', so that it includes the biochemically recognizable substances, and the large and important biological molecules, whose specialized physiological activity, usually enzymatic, is so often the object of interest in the field encompassed by histochemistry today. One may now also, I believe, make a real distinction between two sorts of histochemistry. One, which is often based upon well established chemical reagents, and physical or chemical reactions, I would name '*Substantive Histochemistry*' (histochemistry appertaining to the nature of substances). It is concerned, usually, with the identification and localization of the major classes of chemical substances, including those characteristic of living organisms (carbohydrates, fats, proteins, and nucleic acids), as well as the chemical elements and their compounds (Ca, Fe, etc.) '*Extractive Histochemistry*', as developed by Bernhard and his colleagues in particular, on plastic sections, is a sub-category of Substantive Histochemistry. The other sort of histochemistry, which often involves the use of biochemically active biological molecules in complex substrate mixtures, I would call '*Vivicative Histochemistry*' (histochemistry appertaining to life processes). It is concerned, usually, with the identification and localization of special biological activities in tissues, frequently enzymatic activities. Vivicative Histochemistry would include the sort of quantitative histochemical approach discussed by Chayen in this book (p. 270), as well as the field usually described as 'Enzyme Histochemistry'. It would also encompass histochemical techniques applicable directly to the living cell, and dependent upon the metabolic activity of the fully living cell for their success.

In this article I hope to show how Substantive Histochemistry may be applied to plastic sections suitable for study consecutively in both the light and the electron microscopes; and how the information from the two approaches may be fully cross-correlated, to some profit. Studies of the type discussed in this article have been most selective and successful when applied to insect tissues and mollusc tissues. They have not, so far, been as striking when applied to vertebrate tissues, although this results, probably, more from a lack of extensive exploration of suitable vertebrate tissues (which is currently being remedied), than from a lack of selectivity in them, since, for example, liver gives excellent results with the PAS method.

PART I: THE PERIODIC ACID SCHIFF TECHNIQUE APPLIED TO TISSUES IN PLASTIC FOR COMBINED OBSERVATIONS

MATERIALS AND METHODS

Antheraea pernyi ganglion material as a test tissue

Through the courtesy of Dr. P. Faulkner, who maintained a colony of the Assam Silk Moth, *Antheraea pernyi*, in this laboratory, material was collected from the pupae for electron microscopy, in collaboration with Mr. Brian Rothwell, as part of a training project. Each animal was dissected dorsally, and the body cavity was immediately flooded with chilled veronal acetate buffered osmium tetroxide solution (1 per cent solution, pH 7·2) with added calcium chloride as used by Palay and colleagues.[43] The ganglia, and other organs required, were dissected out carefully under the osmium, using iridectomy scissors and watchmakers' forceps; and, by handling only the trachea external to the organ, any traumatic injury to the tissue was avoided. The ganglionic chain was subdivided so as to give one ganglionic mass to a block, by dissection with razor blades on a wax plate, and further fixed by immersion in a pot of the same chilled osmium tetroxide solution, to give a total fixation time of 1½ hr. The blocks were rinsed with veronal buffer, dehydrated through a graded series of methyl alcohol solutions (20, 30, 50, 70, 80, 90 per cent, 2 changes of 10 min each; absolute methyl alcohol, propylene oxide, 2 changes of 20 min each; propylene oxide/fully catalysed epon resin 1:1 for 1 hr; fully catalysed resin (Epon 5A plus 5B) 1 hr; blocked out in number 00 gelatin capsules, and embedded in Epon in accordance with the recommendations of Luft.[27] The ganglia were sectioned on Huxley ultramicrotomes with either glass or du Pont diamond knives. Ultrathin sections were stained conventionally with aqueous uranyl acetate solution (2 per cent) by total submersion on copper grids at 37°C for 1 hr, to improve electron contrast. Semi-thin sections (0·5–2 μ), suitable for light microscopy, were stained with toluidine blue solution, basic fuchsin solution (1 per cent aqueous), Grimley's trichrome method for plastic sections[17] (4 per cent toluidine blue/4 per cent malachite green counterstained and differentiated with 1 per cent aqueous basic fuchsin), Nile blue, or Sudan black B (McGee-Russell and Smale[32]), after attachment to glass slides. The staining was carried out, usually in 1–2 min, on the miniature hot plate previously described by McGee-Russell, Smale, and Banbury[33] as part of the process of microtomy, rather than as a prolonged staining in the oven as was hitherto recommended. For 'combined observations' of *identical* ultra-thin sections with both light and electron microscopes (McGee-Russell[30]), perfectly flawless, wrinkle-free sections were mounted on special 'Finder' grids devised in collaboration with Mr. W. D. Hogben of Graticules Limited (Maxtaform Grids Type H 6 in copper), which have every grid hexagon reference-indexed with letters and 'dots' (Fig. 16.1), and so greatly facilitate the task of comparative study and the location of identical elements in the two instruments. It is necessary to insist upon the perfection of the sections because, interestingly enough, the light microscope proves to be *more* sensitive to sectioning artifact than the electron microscope, and 'chatter' of broad character will completely spoil the optical image although moderately satisfactory electron microscopy may be carried out on the same section. This is a direct consequence of the smaller field of view in the higher power instrument, which tends to look at the object only within one band of chatter at a time, if the chatter is wide, and the lowest normal working magnification is about 1000 to 2000 times. The surface quality of the section must, also, be of a high order, with no sign of 'scrubbing', for scrubbing will give a considerable amount of scattered light, and as 'noise' this will destroy, completely, any useful image of the object, which is at the limit of visibility when it is an ultra-thin section. Use was made of the 'ring and cup' technique, discussed elsewhere (McGee-Russell[31]), which is based on the original idea of Marinozzi,[36] in order to apply histochemical techniques suitable for light microscopy to ultra-thin or semi-thin sections. The sections were mounted on grids only at the end of the histochemical technique, after free floating sections in 'rings' had been passed 'through' the test solutions and any wash or rinse solutions, all held as balls of liquid in polythene 'micro-cups' (Beckman Catalogue Number 159-426).

The periodic acid schiff technique employed

For the purposes of these studies a very simple PAS method was employed. It may therefore be submitted to some criticism on histochemical grounds. For paraffin sections, and frozen section histology and histochemistry, the difficulties and subtleties of the PAS method have been discussed in the literature at great length. However, as a beginning, it was pleasing to confirm that a simple, well-established technique had the desired selectivity when applied straightforwardly to osmium-fixed, plastic-embedded tissue, in Epon plastic. The fact that the PAS method works on material prepared for electron microscopy

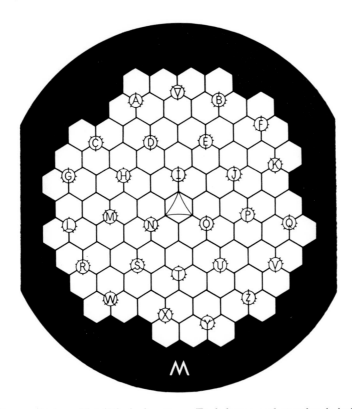

Fig. 16.1 Maxtaform grid type H 6 'Finder' pattern. Each hexagonal opening is indexed alphabetically (A^1, A^2, A^3) by letters and bumps built into the grid bars. Grid centre is distinctive. Alphabetical readout in normal left/right linear pattern is easy under electron microscopical magnifications of about 1500 or less, and the location of sections on the grid may be recorded during ultra-microtomy, by inspection under the binocular microscope. The pattern is a true finder pattern, and gives site recognition, directional orientation for the next search traverse with the electron microscope stage controls, and quick determination of right and wrong sides of the grid. This grid type is a considerable help to combined observations by light and electron microscopy.

has already been emphasized by many workers (Houck and Dempsey[22]—methacrylate; Moe, Behnke and Rostgaard[39,40]—Vestopal; Pease[44]—most embedments, and a particularly interesting discussion; Richardson, Jarrett and Finke[52]—Araldite; and many others). The PAS method employed was based upon the technique published by McManus.[34,35] Yellow or bronze sections (interference colour observed floating in the boat, indicating thicknesses of 90–160 mμ) were cut, for combined observations with light and electron microscopes after mounting on naked Maxtaform H 6 Finder grids. Thicker sections were cut for mounting on glass slides for light microscopy alone, and these sections were treated conventionally by applying the PAS technique on the slide. The yellow or bronze ultra-thin sections were transferred straight from the boat, inside a plastic ring, onto a 'ball' of the first reagent (0·5 per cent aqueous periodic acid) held behind the boat, in a 'cup' mounted in the 'cup-block' previously described.[33] The sections were allowed to react with the reagent for 5 min. The section-loaded ring was then transferred to four successive 'balls' of distilled water for rinse washing, allowing several seconds on each ball; and then, subsequently, transferred to a ball of Schiff's reagent, in a cup, for 15 min. In each case the sections, held within the ring, floated freely at the top of the meniscus of the ball of reagent, in a stable attitude, as previously described.[33] The sections were then treated with three cups of sulphurous acid (2 min each); and then washed and rinsed by passage onto balls of fresh distilled water in four cups. Finally the sections were released from the ring onto the meniscus of a further clean ball of distilled water, and mounted on a grid in the way described in the previous account (McGee-Russell[31]). It was usually convenient to handle only one or two sections within a ring, at a time. In this way, little difficulty was experienced in applying to unmounted ultra-thin sections a light-microscopical histochemical technique which produces visible colour, and involves several treatments, in the complete absence of any grid metal likely to interfere with or confuse the reaction.

Methods of observation: optical microscopy

Direct transmission optical microscopy

The normal colour of a positive PAS reaction is a clear red or pink, with a slight blue tint, when it is viewed in the absence of any counterstain. If counterstains such as haemalum are applied to the nuclei, the tint of the PAS positive may change more towards a purplish red colour. A clear pink or red colouration also characterizes the positive and selective colourations which are obtained upon tissues embedded in Epon, after preparative treatment as for electron microscopy. Direct transmission microscopy with achromatic or apochromatic objectives, of plastic sections after PAS, without counterstaining, should be carried out, as it should for paraffin or frozen sections, with all powers of the optical microscope, including oil-immersion objectives. However, in complex tissues, phase-contrast microscopy should also be applied, in order to give one better orientation within the tissue. This is especially profitable with plastic sections.

Phase-contrast optical microscopy

Phase-contrast microscopy may, most usefully, be applied to plastic sections after treatment with the PAS reaction only, since no counterstain, with its possible obscuring effects, need then be applied, yet the tissue may be studied in full detail, and the position and relation of the PAS positive material to the rest of the specimen fully understood. Most commercial positive phase-contrast objectives (which, according to the usual definition of positive and negative phase-contrast, produce dark contrast images of objects of high refractive index, against a light background) give a further valuable *enhancement* of visibility of the PAS positive result, by the transformation of the red colouration into a brilliant blue colour, through effects which are dependent upon the precise nature of the coatings upon the phase annulus or diffraction plate. This useful effect has been observed and reported by Shanklin and Azzam[55] for paraffin sections of cerebral arteries stained by PAS. On plastic sections, faint positives, with very pale pink colourations, are easily overlooked by direct transmission optical microscopy, but are much more noticeable by positive phase-contrast, through the 'blue enhancement effect'. Even fainter positives, in thinner sections, may be detected by the use of an alternative form of negative phase-contrast, that manufactured by Reichert based on the work of Wilska,[56, 57] known as 'Anoptral-Contrast'. This is a most valuable method for the examination of plastic sections at high resolution in the light microscope.

Anoptral-contrast optical microscopy

The image obtained with anoptral-contrast objectives is an image of outstanding contrast and resolution, since image-degrading light-scatter at the phase annulus has been considerably reduced through the use of multiple coatings (Françon[15]). The image is one to which microscopists must often become acclimatized, if their previous experience has been restricted to dyed specimens viewed by direct light, and positive phase-contrast images. Objects of high refractive index appear as bright objects on a golden brown background, and the brilliance of the image, with the thicker objects, is often disconcerting to the microscopist who is not familiar with the system. Image quality is highest for thin specimens, and this too may be why the virtues of these objectives are not as widely appreciated as those of positive phase-contrast objectives. Anoptral-contrast is particularly valuable for the examination of plastic-embedded tissues, since one has the opportunity to make very thin sections. The background colour of the image formed by anoptral-contrast objectives is produced by the special coatings applied to the phase annulus, and it is this special character which is interesting in relation to the PAS positive result. Under anoptral-contrast objectives, manufactured by Reichert (this must nowadays be specified, since another manufacturer, Metronex, Warsaw, now makes an anoptral system based on Wilska's recommendations[56, 57]), the pink PAS positive colour is transformed, not into the bright blue seen with positive phase-contrast objectives, but into a brilliant orange, of even higher visibility and contrast. Very faint positive colourations are raised to an unarguable level of detection; the faint positives brought to one's notice may usually be confirmed by returning to direct microscopy, but are of such a low order of colour saturation that considerable time and concentration must be devoted to achieve this confirmation. The very faint positives are easily overlooked, and ignored, if the sections are viewed by direct microscopy only.

The value of special microscopical techniques, such as anoptral-contrast, for the optical enhancement of faint differential histochemical colourations achieved on plastic sections, is clear from this one example. It is also clear that this represents a valuable new approach, which should be further developed, since it drives the limits of detection, and the sensitivity of methods, still further towards the biochemical level. Sophisticated optical techniques must be applied to other histochemical methods which may be found to be selective on plastic sections of tissues, particularly if they produce colours different from the PAS positive colouration.

The sensitivity of the anoptral-contrast system, and its value in viewing PAS positive results, is illustrated in Fig. 16.2, A and B. Here, the light micrographs (A and B) are of a bronze Epon section (thickness about 140–150 mμ) which is invisible by direct transmission microscopy. No sign of colour can be detected by direct visual light microscopy. With anoptral-contrast, when viewed with attention, sufficient detail was seen, under the oil immersion objective, for both orientation and photography. The histochemical positive colourations in the section were seen as a faint clear orange image, just sufficient to focus and record by colour photomicrography as well as black and white. Fig. 16.2, A is a direct black-and-white record of the image seen, and the areas marked (pg) are where the PAS positives may be seen as dark contrast shadows in the cytoplasm of a nerve cell from the ganglion of *Antheraea pernyi*. The contrast of this histochemical positive (which is *only* visible under anoptral-contrast in such a thin section) may be improved by appropriate filtration of the transmitted illumination. Interposition of a blue-green, fairly narrow-cut, Ilford 303 gelatine filter into the light path greatly increases the relative darkness and relative contrast of the orange-coloured regions, as shown in Fig. 16.2, B.

Methods of observation: electron microscopy

Using an A.E.I. EM 6 electron microscope at an accelerating voltage of 50 kV, with 25-micron objective apertures, and 400-micron condenser apertures, satisfactory images may be obtained from Epon sections of double-fixed material (glutaraldehyde followed by osmium) which have been treated only with the PAS technique, *without* any other 'post-staining' procedure. The osmication during fixation produces enough contrast for adequate microscopy on the somewhat thicker sections (yellow and bronze interference colours) required for the technique of 'combined observations' with both light and electron microscopes on the same section (McGee-Russell[30]). For higher resolution studies, adjacent thinner sections of silver interference colour may be handled more conventionally, and stained with uranyl acetate. There is little point in using the PAS technique on the thinner sections, which are silver in interference colour, since the visibility of the positive colouration is below the present limits of detection by optical microscopy.

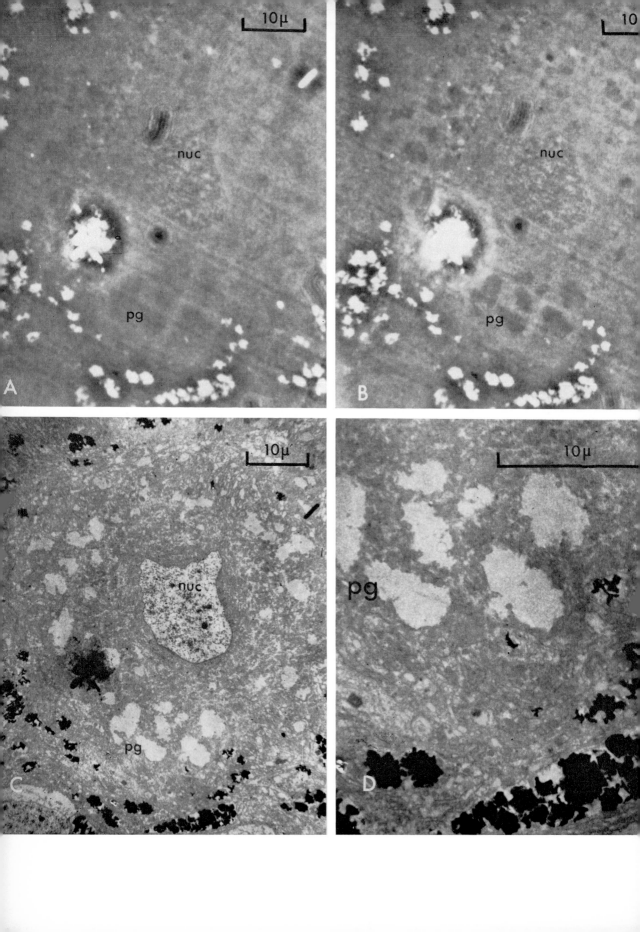

In this way, through combined observations on the *same section*, and high resolution electron micro-scopy of adjacent sections, in the conventional way, an unequivocal correlation may be made between the histochemical properties of objects at the light microscopical level, and their morphological characteristics at the electron microscope level.

The problem of controls—the selectivity and specificity of the PAS reaction on plastic sections

Highly selective colourations are achieved with the PAS reaction when applied to tissues fixed in osmium tetroxide ('O-fixed'), or double fixed in glutaraldehyde followed by osmium ('GO-fixed'). After fixation with glutaraldehyde alone ('G-fixed'), the PAS reaction is not selective. As would be expected from the nature of the fixative, a pink colouration is achieved on the whole tissue, and nearly all cell elements. This is hardly surprising, since it may be assumed that all elements still present in the plastic section have interacted with the aldehyde fixative, and that the PAS reaction is detecting this general interaction and coupling.

Nevertheless, the PAS reaction is highly selective when the G-fixation is followed by a post-fixation in osmium, to give GO-fixed tissue, or if the initial fixation is a conventional osmium fixation procedure, such as veronal acetate buffered osmium tetroxide with added calcium chloride (Palay and colleagues[43]). The sites of the selective colourations obtained are especially interesting, because they appear to parallel closely the 'specificities' displayed by PAS with tissues handled in the manner heretofore conventional for histochemistry. The following question immediately arises: 'What control procedures may be applied, in order to establish histochemical validity for the selective colourations obtained on plastic sections?' The experimental answer is surprising—none of the simple standard methods of obtaining a control preparation (in which the selective colouration is *not* obtained) works. After treatment of sections with saliva, amylase or diastase for periods up to 24 hr at elevated temperatures, positive colourations are still obtained on the selective sites. This confirms the findings of Foster.[14] Sodium periodate may be substituted for the periodic acid, or the periodic acid oxidation may be completely omitted, without altering anything except the intensity of the colouration. The Schiff base still selectively colours the same sites, and is clearly visible under the viewing conditions outlined above. By direct optical microscopy this may not be so obvious, at first glance, but the two other methods of optical observation, discussed above, will reveal the colouration distinctly. It would appear then, that semi-thin plastic sections could be treated with Schiff reagent alone, rinsed with distilled water and allowed to dry, before mounting on a glass slide in Epon, to obtain a highly selective but faint colouration for light microscopy. This is the case. The full, more conventional PAS treatment results in more intense colourations, but they do not differ in location, or selectivity. Hence, at this time, it is not clear what conditions will have to be met to obtain valid controls with the PAS reaction on plastic sections for light microscopical observation. The most likely treatment to investigate is that of bleaching the sections with hydrogen peroxide to oxidize the osmium/tissue complex or reaction product, and render the tissue components more accessible to selective enzymatic extraction. This treatment is known to resuscitate the selective colouring properties of tissues fixed in osmium, and permit the differential staining of double fixed tissues (de Bruijn and McGee-Russell[10]). Different plastic embedding media will almost certainly involve different criteria, and the results obtained with one plastic medium will probably not prove to be suitable for direct extrapolation to another embedding material, without careful consideration, and experimental study. Very considerable increases in the times of incubation, and elevation of the temperature may be needed to allow full penetration of reagents into the thicker sections. Many pointers for the future development of

Fig. 16.2 Combined observations. Four micrographs of an identical area of the same ultra-thin section of *Antheraea* ganglion: A, B, light micrographs; C, D, electron micrographs. The ultra-thin section (of bronze interference colour) is stained only by the PAS technique. A, viewed by anoptral phase-contrast; B. viewed by anoptral phase-contrast plus an Ilford 303 blue-green filter in light path to gain increased detail and contrast; C, low-power electron micrograph to same scale as A and B; D, higher power electron micrograph of area containing PAS positive material; pg, pools of PAS positive glycogen in cytoplasm; nuc, nucleus of ganglion cell.

control procedures for the substantive histochemistry of plastic sections may be found in the publications of the workers associated with Bernhard's laboratory in Villejuif. Monneron,[41] in particular, has recently demonstrated the applicability of enzyme extraction procedures to tissues embedded in Epon, with success rivalling, or surpassing, the success achieved with the hydrosoluble plastics (personal communication; Monneron;[41] Monneron and Bernhard;[42] Bartl;[6] Bartl and Bernhard;[7] Bernhard[8]). The combination of a proteolytic enzyme treatment (pronase), and subsequent treatment with a second enzyme, e.g. a selective nuclease (RNAase) is a promising line of attack, since Monneron has shown such treatments to have a significant effect upon the contrast of ribosomes subsequently obtained by 'double electron-contrasting', using uranyl acetate and Reynolds' lead citrate. In devising control procedures it is clear that full attention must be paid to the method of fixation employed, since Monneron has demonstrated that each of the three methods of fixation, G, GO and O, produces a different subsequent accessibility of the tissue in Epon to extraction procedures. From our experiments, I would expect that the method of dehydration employed may also have a significant effect.

Further investigation must establish the conditions necessary to obtain 'valid histochemical controls' for the selective PAS reactions of plastic embedded tissues, and determine the exact nature of the chemical and physico-chemical factors which govern the selectivity. We can expect soon the rapid development of a highly selective 'Substantive Histochemistry' suitable to plastic embedded tissues, which allows superbly fine localization within remarkably well stabilized structures, which are available for ultrastructural study down to molecular dimensions. Such a substantive histochemistry will be compatible with information from many other approaches, and should prove a useful tool for the study of cell structure and activity.

RESULTS

The microanatomy of the ganglia

The general disposition of the tissues in the ganglia of *Antheraea* pupae is illustrated in Fig. 16.3. The ganglion is enclosed within a capsular layer of smooth homogeneous character (NL), the neural lamella (Ashhurst[1]), when seen by light microscopy. The layer appears to contain at least two elements. It may be subdivided into an outermost layer and an innermost layer which are in continuity with each other, but in places are split apart by the interposition of tracheolar cell cytoplasm surrounding tracheoles. The inner layer shows a clear, but faint, PAS positive reaction in plastic sections. The outer layer is negative, as illustrated in Fig. 16.4. However, scattered throughout the capsular layer, without regard to this general subdivision into two layers, there are occasional fine fibrous elements which are also PAS positive. These may be small denser fibrous aggregations of the material which gives the PAS positive in the inner layer of the neural lamella. The whole lamella is, clearly, a characteristic part of the connective tissue of insects, as shown by Ashhurst[1] in *Locusta*. Within the smooth capsular layer of the neural lamella there is a well-defined layer of large connective tissue cells (CTS) and fibrous elements associated with them, which surrounds and encloses the neuronal tissue. The neuronal tissue within consists of an outer cortex (NG), which contains numerous large nerve cell bodies, and an inner medulla (MT), composed mainly of nerve cell processes and tracts.

Sites of positive colouration with the PAS reaction

1. In addition to the faint positive reaction on the inner layer of the neural lamella, the connective tissue cell layer often shows a distinct positive colouration, mainly present on small granular elements in the cytoplasm of most of the cells, so that the entire cell layer is brightly coloured. A positive colouration is also often obtained on fibrous elements in between the connective tissue cells (CTS Fig. 16.3: Fig. 16.4).

2. Brightly coloured positives are obtained in clearly circumscribed areas in the cytoplasm of some of the nerve cells. Such areas are illustrated in Fig. 16.4. In semi-thin plastic sections, the colour may be seen to be disposed into tiny granular elements, which are close to the limit of resolution in the light microscope, and fill the circumscribed coloured area.

3. A distinct, but faint, positive reaction may be observed, as stated above, in the inner layer of the neural lamella, and in occasional single fibrous strands scattered throughout it.

4. Within the tracheoles, a distinct positive is obtained on the surface of the minutely folded cuticular layer between the taenidia of the tracheoles.

5. In some, but not all, nuclei, there is a positive colouration detectable in nucleoli and chromatin masses within the nuclear membrane. This is a particularly interesting localization which merits further analysis.

6. In some pupae, a central zone of unusual character occurs within the ganglion, associated with the medullary tracts. It gives a clear-cut PAS reaction. This would seem to consist of a region of cell destruction and reabsorption, where the breakdown of cytoplasmic membranes has liberated groups within the

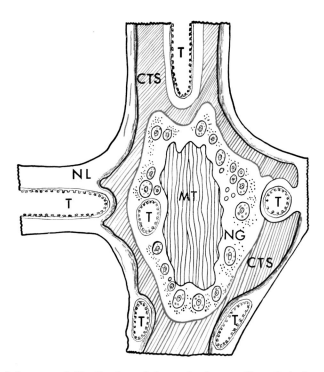

Fig. 16.3 Diagram of the general distribution of tissues in the ganglion of *Antheraea pernyi*. NL, neural lamella; CTS, connective tissue sheath cell layer; T, tracheole; NG, layer of large neurons and 'glial' elements forming a cortex; MT, medullary tract of ganglion, containing numerous cell processes. The sites of PAS positive material in this organ are shown in red, (except for the folded cuticular layer on the inner surfaces of the tracheoles, T, which also gives a positive reaction).

membranes which are capable of responding to the PAS reaction. The positive is given by 'bubbly masses' of cell cytoplasm which, when examined in the electron microscope, prove to be composed of numerous 'whorls' of membraneous elements, resembling the myelin figures of cell breakdown.

The PAS positives in *Antheraea* nerve cells

The illustrative micrographs of *Antheraea* material in this account (in Figs. 16.2 and 16.5) are restricted to the sites of positive PAS reactions in the cytoplasm of the nerve cells, in order to demonstrate the successful application of the technique of combined observations to this histochemical reaction, at easily recognized sites. Fig. 16.2 presents the critical evidence, of both light and electron micrographs taken from exactly the same section. Colour photomicrographs were also taken, successfully, from the histochemically coloured ultra-thin section (of bronze interference colour, thickness about 150 mμ), but could not be reproduced in this volume. Orientation within the plate has been maintained carefully so that it is possible to make an exact point for point correspondence between the light micrographs, and the electron micrographs. Prominent areas within the nerve cell cytoplasm give the clear-cut optical PAS colouration recorded at the areas marked pg in Fig. 16.2, and discussed above (p. 189). These areas

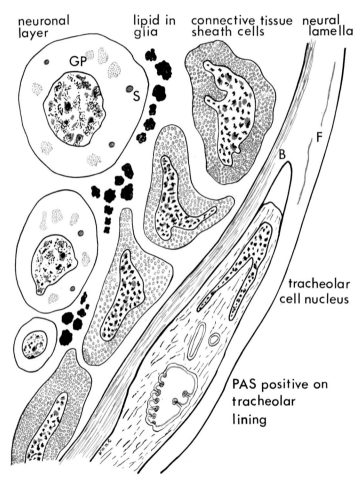

Fig. 16.4   Diagram of the same elements as Fig. 16.3, to show the distribution of the PAS positive material in greater detail. In the neurons note: GP, pools of granular material (glycogen); S, spherical bodies which are PAS positive (compare with electron micrograph of Fig. 16.5, H) and probably lipoprotein; B, basal layer of neural lamella, which is uniformly positive; F, scattered fibrous elements in neural lamella, which are PAS positive.

are unstained by toluidine blue in the thicker sections used for routine light microscopy. It may be noticed that *more* patches of PAS positive material may be detected *more easily* in part B than in part A of Fig. 16.2. Therefore, the resolution, the contrast, and the information-content are all improved, when an Ilford 303 filter is used with anoptral-contrast on this coloured image. In Fig. 16.2, C and D, the regions which appear dark by optical microscopy appear light by electron microscopy. Figure 16.2, C is a low-power electron micrograph of exactly the same section as Fig. 16.2, A and B, and it has been given no further treatment whatever, after the PAS reaction, apart from being viewed consecutively in the light microscope and the electron microscope. In the electron microscope the granular character of these areas may be detected. Figure 16.2, D shows the regions which lie below the nucleus in A, B, and C at higher magnification. The crenated lipid droplets present in the glia may be used as helpful landmarks in comparing these images. The conventional electron microscopical image of this type of area is shown in Fig. 16.5. Figure 16.5, E is a low-power electron micrograph at about the same magnification as A, B, and C, Fig. 16.2. Patchy areas are filled with small granular objects, which at higher magnification (in Fig. 16.5, F, G, H) may be identified, immediately, as characteristic 'rosettes' of glycogen particles. The morphological ultrastructural criteria are therefore in perfect accord with the histochemical positives, and glycogen is convincingly identified by the fully matching criteria. By inference, rather than by the direct proof which is possible by combined observations, Foster[14] has shown that the ultrastructural criteria for glycogen correlate well with the known light-microscopical histochemistry of glycogen. He also demonstrated that positive PAS reactions for glycogen may be obtained on tissues fixed in osmium tetroxide and embedded in methacrylate. Combined observations allow the final step of full correlation to be taken.

The well-marked PAS positives obtained in the connective tissue cell layer are also due to glycogen, as would be expected from the work of Ashhurst,[1] who demonstrated glycogen in these cells in *Locusta*, using standard histochemical methods. Extensive storage and utilization of glycogen is clearly an important feature of the metabolic pattern of the pupa of the Lepidoptera. The most prominent glycogen deposits observed in my study of *Antheraea* were in the connective tissue sheath cells of an early pupa, of a known, timed age of only eight days, after pupation. Other fixations, of pupae many days older, showed much less prominent glycogen deposits in the sheath cells.

It is interesting to note in the nerve cells the frequent close juxtaposition between a 'pool' of glycogen, several of the remarkable mitochondria of these cells, with their ladder-like cristae, and an active Golgi field of stacked membranes and vesicles (Fig. 16.5, F and G). This juxtaposition may indicate an active role for the Golgi field and the mitochondria in the metabolic processing of glycogen in these cells. The occurrence of pools of glycogen in the cytoplasm of invertebrate and vertebrate nerve cells is already well documented (Bullock and Horridge;[11] Berthold[9]), but their significance is not known. Dynamic cytological, ultrastructural and histochemical studies, perhaps coupled with autoradiography, could profitably be carried out on the closed system of the insect pupa, to discover their functional significance.

PART II: HISTOCHEMICAL METHODS FOR CALCIUM APPLIED TO PLASTIC SECTIONS

MATERIALS AND METHODS

In the second part of this article I wish to consider another, rather different, example of substantive histochemistry applied to plastic sections. This is the application of histochemical methods for calcium to suitable tissues embedded in plastic, so that light microscopical histochemistry and electron microscopical morphology may be conducted upon identical material. The work is at an early stage and it is only possible to consider here two aspects; firstly, the value of only one out of a battery of tests for calcium, which were discussed in a previous study (McGee-Russell[29]), and secondly the electron-microscopical features of a suitable 'test tissue', which is of considerable interest in its own right. Of the group of tests, the alizarin red S method for the detection of calcium in tissues is both the most specific and the most sensitive. My own unpublished observations, and the brilliant work of Reynolds[49,50,51] have both shown that this technique may achieve a highly accurate localization of calcium in plastic sections. Reynolds has found that the technique may be most helpful in the dynamic analysis of pathological processes resulting from cell damage, which can involve the accumulation of calcium at high concentrations within the mitochondria. Calcification is a significant aspect of pathological change in organisms, and it is likely therefore to be an important subject for future investigations in dynamic

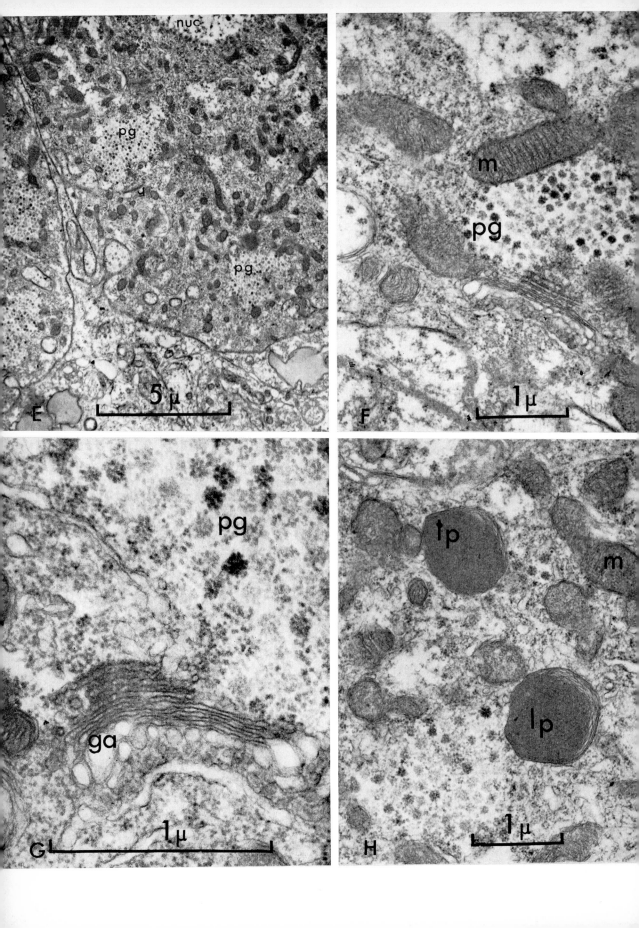

cytology (see Hirsch, pp. 395–404 this volume), through substantive histochemistry and combined observations on plastic sections.

Oocytes of the snail *Helix aspersa* as a test tissue

In the previous study a group of suitable invertebrate tissues was assembled which was appropriate for 'testing the tests' (McGee-Russell[28]). One tissue was a molluscan tissue, the digestive gland of the snail. The secretion and segregation of calcium salts is a significant activity in molluscan tissues. The continuous growth of the external skeleton, the shell, demands a continuous intake of calcium salts by the tissues, and continuous processes of storage and secretion. This situation corresponds more nearly to the metabolic activity of the foetal and infant mammal, in relation to its calcareous internal skeleton, than it does to the more stable situation in the adult, where skeletal growth has ended. Phenomena in the snail may, however, have points of resemblance with pathological processes of calcification. This would be of considerable interest.

The blocks of one of the tissues previously studied, the digestive gland of the snail, frequently included the ovotestis of the snail, and, within the oocytes in the follicles, a positive histochemical localization of calcium, in the form of small granules in the cytoplasm, was obtained (McGee-Russell,[28] Fig. 16.1, D). These oocytes are good cytological and electron microscopical objects, since they may be obtained at most times of the year, by hand dissection of the ovotestis under a stereoscopic binocular, and they are large enough to be brought individually through the processes of fixation, dehydration, and embedding for electron microscopy. They make, individually, excellent small blocks for ultra-micro-tomy, and subsequent electron microscopy and light microscopy. Therefore, the oocytes of the snails *Helix aspersa* and *H. pomatia*, are an appropriate 'test tissue' for electron microscopical investigations of intra-cellular calcium secretion, and the substantive histochemistry of calcium in plastic sections. The principal technical difficulty of the material is the occurrence of calcareous regions in the cytoplasm, which are extremely hard, and difficult to section. Considerable 'tearing' occurs during sectioning with glass knives and, for satisfactory results, it is essential to use a good diamond knife. Material was prepared by fixation in veronal acetate buffered osmium tetroxide with calcium chloride (pH 7·2), or double fixed in glutaraldehyde followed by osmium tetroxide, and embedded in Epon, according to the procedures listed above, in Part I. For study with the light-microscopical histochemical method for calcium, 1–2 μ sections were mounted on glass slides, and handled in accordance with the simple method previously published (McGee-Russell[29]).

The alizarin red S method applied to plastic sections for light microscopy

The glass-mounted Epon plastic section is flooded directly, without removal of the plastic, with the test reagent (2 per cent aqueous sodium alizarin sulphonate solution, pH 4·1–4·3, adjusted with dilute ammonia), and immediately observed by transmitted light, under appropriate light microscopical magni-fication. The sites of calcium are shown by the rapid development of an orange-red precipitate of dye-lake.

For cryostat sections, Reynolds states that a reaction time of only 5 sec gives more satisfactory results than the 30 sec or longer I originally specified for paraffin sections. With plastic sections, the longer times may be required, but it is always essential to observe the reaction *immediately* under the light microscope, in order to obtain the most information about the interaction of the reagent and the tissue. To some extent the reaction time is dependent upon such factors as section thickness and nature of

Fig. 16.5 Electron micrographs of *Antheraea* ganglion. E, low power of ganglion cell cytoplasm to same approximate scale as A, B, and C of Fig. 16.2. Ultra-thin silver section stained conventionally with uranyl acetate; pg, pools of glycogen in characteristic ultra-structural 'rosettes'. F and G, higher power micro-graphs to show: m, mitochondria with 'ladder-like' cristae; ga, Golgi apparatus in close juxtaposition to glycogen pool. H, detail of cytoplasm to show: lp, lipoprotein particles with characteristic ultra-structure, tentatively identified as the organelles responsible for the PAS positives denoted by S in Fig. 16.4.

14—C.S.I.

calcium deposit for optimum results, and hence direct empirical assessment is the best criterion. As found by Reynolds,[49,50,51] once a standard reaction time is chosen the alizarin red S method is sufficiently specific, sensitive, and self-consistent to give histochemical results which correlate well with the results of direct chemical analysis.

Some comment on fixation in relation to the preservation of calcium deposits must be made. Reynolds[51] noted that the preservation of calcium deposits, as judged by the intensity of the alizarin red S positive result, is better in plastic sections of frozen-dried, or frozen-substituted tissue, than in tissue fixed in aqueous osmium tetroxide. He found that calcium chloride in the fixative improves the preservation of calcareous deposits. This is in good accord with earlier observations (McGee-Russell[28]). Because of the findings of Palay and colleagues,[43] it is routine practice in this laboratory to add calcium chloride to fixatives, which may account for our routine success with the alizarin red S method, on plastic sections of suitable tissues. Reynolds[51] stated that phosphate-buffered osmium tetroxide does not preserve intra-mitochondrial electron-opaque deposits (of calcium). Care should therefore be exercised in the choice of fixative and buffer. Nevertheless, where cells secrete well-formed spherulites, or granules, of a calcareous nature, it is likely that the routine fixation and preparative techniques of electron microscopy will preserve the greater part of such deposits, when the fixatives are simple non-precipitant fixatives, maintained at a stable pH of about 7·2 during fixation.

The alizarin red S method is not suitable for combined observations of the *same* section by both light and electron microscopy since, as a dye-lake method producing a precipitate of low stability, it does not have the technical features of precision and persistence which are required. It is, however, an excellent example of a method suitable for the substantive histochemistry and light microscopy of plastic sections, as well as cryostat sections and normal histological sections. Other methods, currently being investigated, may be more appropriate to the technique of combined observations.

RESULTS

Some features of the electron microscopical morphology of snail oocytes

The cytoplasm of the snail oocyte usually contains an abundance of organelles. The mitochondria are numerous and filamentous, as shown in Fig. 16.6, I, which is a high-magnification light micrograph of an Epon plastic section stained with basic fuchsin, and observed by positive phase-contrast microscopy. There are always many spherical and sub-spherical granules in the cytoplasm of the oocyte, of several types, which cannot be discussed at length here. The most highly refractile of these granules give positive results with the alizarin red S method, and appear in electron micrographs as typical calcospherites, consisting of a dense, very high contrast core of material, presumably corresponding to the calcified region, surrounded by a region of low electron density, bounded by a layer of well oriented smooth membranes in closely packed array (Fig. 16.8, Q). Smooth textured, high contrast globules of presumed lipid are also frequent (Fig. 16.7, K, N; Fig. 16.8, O) and would appear to correspond to the 'homogeneous globules' which have been described in the neurons of the snail (McGee-Russell[30]). Spherical globules of mixed character, resembling somewhat the 'compound globules', also described previously in the neurons of the snail, are often seen (Fig. 16.7, K, M). They are usually of the same order of size as the Golgi apparatus (see below) and the lipid globules, that is, about 0·5–1 μ in diameter.

A prominent feature is the presence, in many oocytes, of vacuoles which appear to be empty when viewed by light microscopy, and show no capacity to take up dyestuffs. Two such vacuoles are shown in Fig. 16.6, I (arrows). These vacuoles are interesting, as they possess a characteristic electron-microscopical morphology and may be early stages in the process of intracellular calcospherite secretion.

Calcogenesis in the snail oocyte

The typical stages of what I believe to be the sequence of events in calcium secretion in the cytoplasm of snail oocytes are illustrated in Figs. 16.7 and 16.8, K to Q. The empty-seeming vacuoles are usually between 1–2·5 μ in diameter. In electron micrographs they are immediately seen not to be empty, but

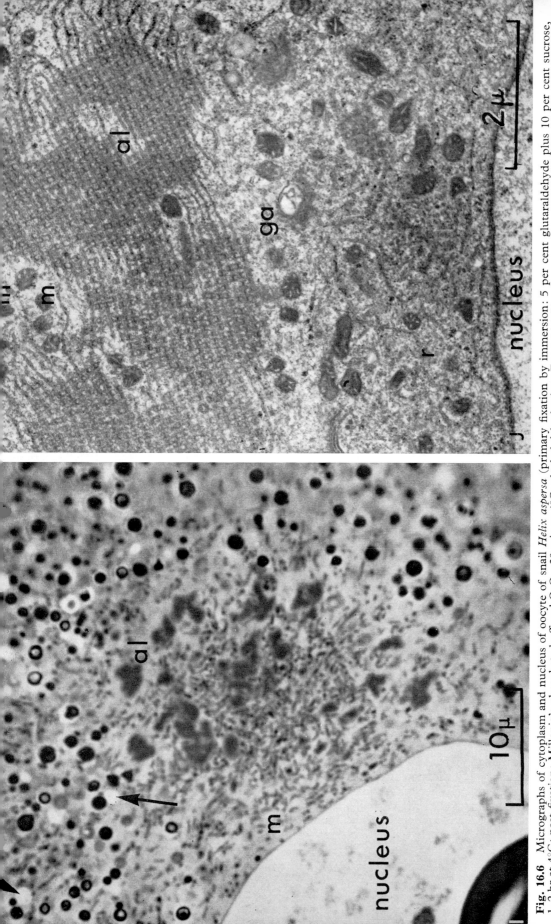

Fig. 16.6 Micrographs of cytoplasm and nucleus of oocyte of snail *Helix aspersa* (primary fixation by immersion: 5 per cent glutaraldehyde plus 10 per cent sucrose, 1–2 hr at 4°C; post-fixation: Millonig's phosphate buffered OsO_4, 20 min at 4°C; alcohol dehydration; Epon 5A/5B).

I, light micrograph on Polaroid P/N 55. 1 μ section stained with basic fuchsin, viewed by positive phase-contrast: m, mitochondria in juxta-nuclear zone; al, cloud-like masses of annulate lamellae stained with basic fuchsin. Arrows mark 'empty-seeming' vacuoles, here named *calcamphorae* (see text).

J, low-power electron micrograph showing ultra-structural detail of similar area. Note: ga, Golgi apparatus of snail oocyte at this stage is distinctly spherical; al, amorphously shaped masses of annulate lamellae, with extensive membrane connections along the edges; in the nuclear envelope, note the remarkable abundance of nuclear pores.

filled with small membrane-bounded spherical vesicles, or membraneous whorls, or complex membrane configurations (Figs. 16.7 and 16.8, K, L, M, N, O, P) which appear to be derived from the coalescence of the small vesicles. The membrane-bounded spherical vesicles are usually about $0 \cdot 1$–$0 \cdot 2$ μ or less in diameter, and do not appear to contain any material of electron contrast greater than 'background'. The margin of the empty-seeming vacuoles is of some interest, since the vacuoles would appear to be surrounded by a single unit membrane, comparable to those of the cytoplasmic membrane system. It has, closely associated with it, a well-ordered double or treble array of smooth membranes, which partly or almost completely surrounds the vacuole (Fig. 11.7, K, L, M, N). These layers of membrane often have connections leading from the vacuolar array into the general cytoplasm of the cell (Fig. 16.7, K. M, N). The innermost membrane of the array would seem to be involved in an active process of 'blebbing', giving rise to the small spherical vesicles, within this specially isolated, probably watery, zone of the cytoplasm. The close packed array of smooth membranes, and the process of blebbing of vesicles is highly reminiscent of the well-known appearance of the Golgi apparatus in many cells. However, it seems clear that these vacuoles and vesicles are a product of the specialized activity of the smooth endoplasmic reticulum of the cell, but not of the Golgi apparatus of these cells, which is itself a highly distinctive entity, seen in Fig. 16.7, M and N. No direct association between the Golgi apparatus and the vacuoles or the vacuolar array has been observed. Spherical Golgi figures are often seen near the vacuoles, as shown in Fig. 16.7, M and N, but no direct connection between them has been proved. The spherical Golgi bodies are usually somewhat smaller than the vacuoles, about $0 \cdot 7$–$0 \cdot 9$ μ in diameter, with a typical cortex of packed smooth membranes and an internum which, in this cell, contains fine fibrous elements, and small vesicles of high electron contrast throughout. The whole configuration is reminiscent of the Golgi zones seen in spermatocytes, but the characters of the components of the 'internum' seem to be somewhat different.

Because of the demonstration, by electron microscopy, of small vesicular components within vacuoles which appear empty by light microscopy, and the suggested significance of the vacuoles and the vesicles in the process of intracellular calcification, it is desirable to have some special terminology to describe them and set them apart from other organelles. For an empty-seeming vacuole I suggest the term *calcamphora* (plural *calcamphorae*) to designate the general character of a container and the probable association with the process of calcium segregation. An amphora was a two-handled Roman vessel. The connections between the boundary array of packed membranes and the general cytoplasmic membrane system may be thought of as 'handles'. From the multiple membraneous wall of the calcamphora small spherical vesicles are budded off into the interior lumen. As shown in Figs. 16.7, L, and 16.8, O, a good descriptive term for these vesicles is *calcampullae* (singular *calcampulla*), since it suggests their shape and mode of development as well as their association with the secretion of intra-cellular calcospherites.

It is now possible to give a brief description of the suggested hypothetical stages in the development of an intra-cellular calcospherite: (1) from a membraneous component of the cell membrane system, or perhaps from the cell surface, a calcamphora arises and segregates within the cytoplasm a watery vacuole; (2) the smooth endoplasmic reticulum (a component of the membrane system of the cell) contributes active membrane elements to the wall of the calcamphora, which carry out the same sort of membrane function as the lamellar components of the Golgi apparatus, resulting in an active budding and blebbing of small vesicles, the calcampullae, into the lumen of the calcamphora (Figs. 16.7 and 16.8, K, L, M, O); (3) the calcampullae accumulate in the lumen, fuse, and interact to produce webs and skeins of membrane which are somewhat concentrically arranged (Fig. 16.7, L and N); (4) within the skeins of membrane, perhaps from the outset of fusion of calcampullae (Fig. 16.8, O), small foci of calcium accumulation develop as precipitation 'seeds', in a physico-chemical sense (Fig. 16.8, P). The high local concentration of calcium ions, by specialized membranes, could be brought about by biochemical mechanisms similar to those invoked by Reynolds[50, 51] to account for the sequestration of calcium by mitochondria, producing a metabolic concentration of inorganic phosphate and a passive sequestration of calcium. The membrane system of the calcampullae may itself constitute the organic calcium-binding component of the developing calcospherite, as well as the physiologically active cell organelle, with specialized enzyme systems, probably phosphatases, which must be necessary to produce directional regulation of growth.

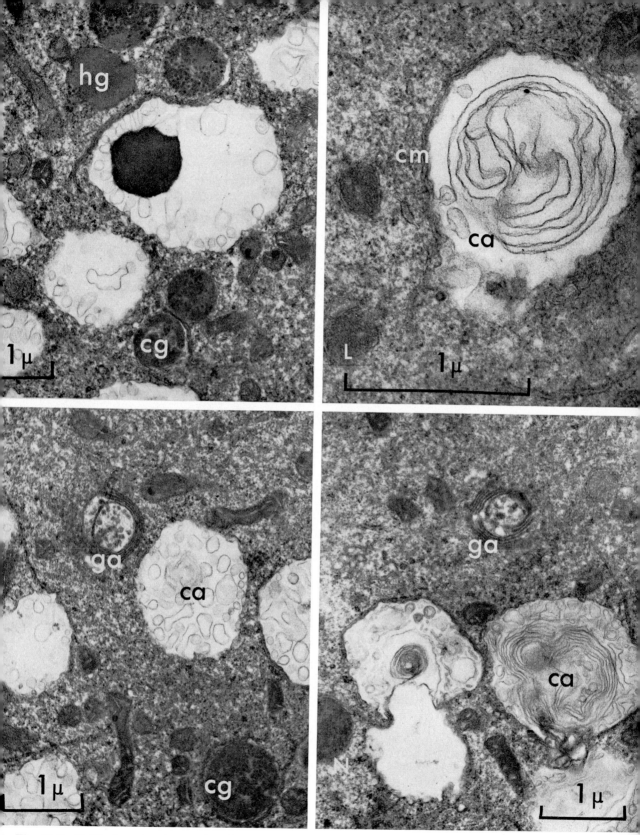

Fig. 16.7 K–N, electron micrographs of snail oocytes showing: hg, homogenous globules; ga, spherical Golgi apparatus; cm, lamellar array of membranes of calcamphora; ca, various aspects of the development and fusion of calcampullae (see text).

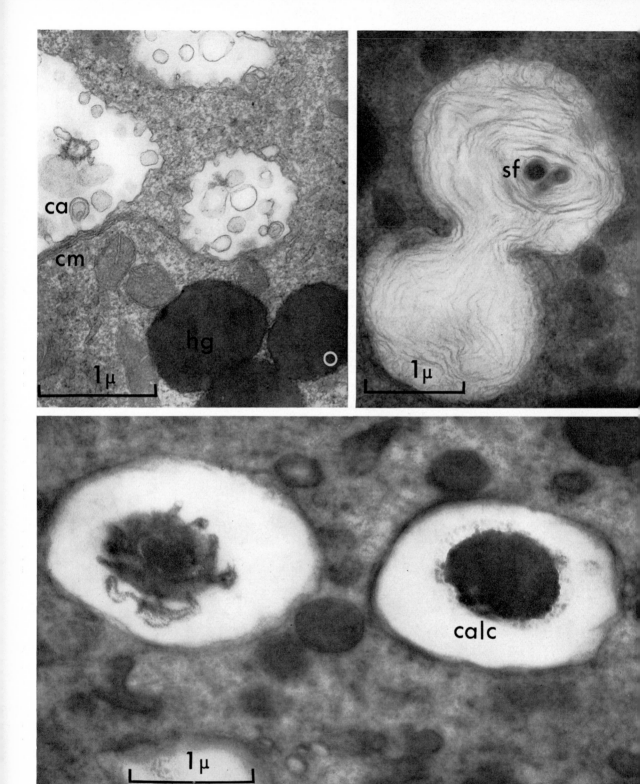

In this particular cell the process of intracytoplasmic calcium accumulation is a normal, controlled, cell activity, and not a response of the cell to pathological damage. Directly comparable mechanisms need not, therefore, necessarily be expected, although it seems likely that similar metabolic activities might be involved; (5) upon the small seed foci, further precipitation of calcium occurs, incorporating more of the calcampullar membranes into the matrix of the developing structure, and building up an amorphously crystalline calcospherite. Some interaction between the smooth membrane system surrounding the calcamphora and the activity of the environmentally isolated membranes of the calcampullae is likely, within the limiting context of the continued accumulation of calcium, the possible depletion of available phosphate or phosphoprotein, and the probable end-point depletion of the membranes of the calcampullar system. Such a balanced, dynamic system, trending towards a terminal end-point, would seem to be necessary to produce the population of intracellular calcospherites of similar size, which is observed in the oocyte. Comparable size implies comparable self-limiting factors.

The precipitation and incorporation of material into a calcospherite would appear to proceed centripetally, towards a seed focus at the centre of the calcamphora, in the middle of the concentric membranes derived from the calcampullae (Fig. 16.8, P). As the calcospherite grows, the central focus achieves an impenetrably high electron contrast because of the accumulation of calcium salts, but the zone immediately inside the bounding membrane of the calcamphora is characteristically of low electron contrast. It is not yet entirely clear whether this is a true space surrounding the calcospherite, inside the bounding membrane, or a material of low contrast. This zone is shown in Fig. 16.8, Q, and in this micrograph there are obvious indications of the incorporation of calcampullar membranes into the central high density, and of the gradual increase in electron density, through the precipitation of calcium salts. If the calcampullar membranes are 'swept up' into the developing calcospherite, the low-contrast zone may well be a true vacuolar space.

Further work will be required to confirm or deny the stages I have outlined above and fill in the earliest and latest stages in the genesis and fate of the calcospherites of the oocyte. It will be interesting, but technically difficult, to establish whether this pattern of development is true for the calcospherites of other cells in this animal, and in other invertebrates. It will be valuable to compare the sequence of events with the events during the earliest stages of calcification in foetal vertebrates. Relationships remain to be established between the new organelles, calcamphorae and calcampullae, and the other better-known organelles of the cell.

The development of intra-cytoplasmic calcospherites is only one aspect of the very active cell secretion and metabolism which characterizes oocytes; and the calcospherite is only one of the several granular organelles involved. The nucleus of an oocyte is characteristically hypertrophied, in association with its control of the great enlargement of the cytoplasm, and its control of the accumulation within the cytoplasm of the metabolically active substances and food stores, which are required for the subsequent development of the daughter offspring. One such substance is, presumably, the calcospherite population, but there is no information whatever on the way in which the nucleus controls this aspect of cell secretory activity. Nevertheless, there is some information demonstrating hypertrophy of the nuclear membrane and its pores, which is the limiting surface for communication between the nucleus and the cytoplasm. This hypertrophy appears to be a characteristic feature of oocytes in animals, whether vertebrate or

Fig. 16.8 Electron micrographs of snail oocytes. O, detail of calcamphora with calcampullae (labels as for Fig. 16.7). P, later stage in ontogenesis of calcampullae, showing concentrically arranged skeins of membrane derived from calcampullae, with high contrast 'seed focus' (sf) for the precipitation of calcium salts at the centre. Note that, for electron microscopy, this section is very thick (yellow-green interference colour), but it gives significant information about a very loosely packed cell organelle. There are advantages in studying sections of *all* thicknesses for the understanding of cell structure and activity.

Q, two reasonably mature calcospherites. This is the only micrograph in the set which is taken from material embedded in methacrylate. In the right-hand spherule note the dense calcified core (calc), the surrounding low contrast zone and the well-defined surface layer of membranes. In the left-hand spherule note the remnant tags of membrane associated with the condensing core of the calcospherite.

invertebrate. To conclude this article I will discuss briefly, and illustrate, some aspects of nuclear envelope hypertrophy in the snail oocyte.

Nuclear envelope hypertrophy in the snail oocyte

Sections of snail oocytes, studied by either light or electron microscopy, frequently show the elements which are illustrated in Fig. 16.6, I and J. In I it may be seen, by high-resolution light microscopy, that a well-defined juxta-nuclear zone of organelles is often developed. Within this zone there is a marked accumulation of small filamentous and rod-shaped mitochondria. Dispersed between the mitochondria are numerous cloud-like masses of polymorphic form, which stain with basic fuchsin, in osmium-fixed Epon sections, and show up with dark contrast when such sections are viewed by positive phase-contrast (Fig. 16.6, I). These elements may be recognized with ease in light-microscopical sections only when they are gathered together into this juxta-nuclear region. When smaller, and dispersed more widely amongst the other organelles of the cell, identification is dubious by light microscopy alone. When such a juxta-nuclear region is studied by electron microscopy, the polymorphic cloud-like masses may rapidly be identified as clouds of annulate lamellae (Fig. 16.6, J), comparable to those described in other oocytes (Rebhun[46,47,48]) and other cell types (Gross;[20] Harrison;[21] see Harrison for additional literature). The accumulations of lamellae do not have well-defined margins, since they tend to send trailing cisternae off into the cytoplasm, either to maintain communication with other 'clouds' further away and, perhaps, the parent nuclear envelope, or to give communication to the general endoplasmic reticulum of the cell.

The mode of origin of these clouds of annulate lamellae from the nuclear envelope has not been observed. It may correspond to the origin described for the annulate lamellae observed in the Salamander, *Necturus*, and an echinoderm, *Thyone briareus*, by Kessel,[24,25] but it may not. The close coincidence in size between the pores of the nuclear membrane and the pores of the annulate lamellae suggest that there may be a more direct 'peeling off' of nuclear envelope, or direct hypertrophy and extension of the nuclear envelope into the cytoplasm. Harrison[21] has found two different modes of origin of annulate lamellae in two different vertebrates, alligator and seagull; it is not yet clear which pattern applies in the snail, or whether a third prevails.

The formation of clouds of annulate lamellar membrane may be regarded as a process of either aggregation or dis-aggregation. It would not be unlikely that sheets of nuclear envelope material with pores (Callan and Tomlin,[12] and subsequent literature) should tend to aggregate together. They would be expected to have the same charge distributions, cohesiveness, viscosity, configurations, and other properties, which could tend to bring them together within the cytoplasmic environment. Equally well, if the process of formation of the annulate lamellae results directly in the cloud-like masses, the images observed may represent a stage in the dis-aggregation and dispersion of the nuclear envelope material throughout the oocyte.

Two hypotheses may be put forward about the hypertrophy of the nuclear envelope which results in such noticeable quantities of annulate lamellae. Firstly, if continuity is maintained between all the groups of annulate lamellae and the nuclear envelope, then the surface area of the nuclear envelope is vastly increased. This vastly increased surface area may be necessary for the increased flood of nuclear messenger material, which is presumably required for the control of the cytoplasmic differentiation of the oocyte. It is well established for vertebrate oocytes that the nuclear chromosomes are highly specialized (Callan, p. 357 this book), and these specialized nuclear elements are, presumably, the source of quantitatively greatly increased outgoing control substances, especially messenger RNA. An unusual form of communication between nucleus and cytoplasm has been observed by Kessel and Beams[26] in oocytes of *Thyone briareus*. If the output of control substances from the nuclear source is hypertrophied, and the site of action of those substances is hypertrophied, it is not unlikely that the transmission channel between the two (which is the nuclear envelope) must be expanded also. However, a second role for the clouds of annulate lamellae may also be suggested. There is reason to suppose that the lipoproteins of the nuclear envelope and the components of the annulate lamellae derived from the nuclear envelope are not identical in composition with the membranes of the endoplasmic reticulum. This may, perhaps, be demonstrated by substantive histochemistry applied to plastic sections, as well as by other methods. If it is correct, then the material of the annulate lamellae constitutes a special additive to the complexly interacting substances present in the cytoplasm, and may have its own direct action and its own special significance, apart from the nucleic acid messenger material which may be associated with it. Such a possibility is somewhat suggested by the close association between the clouds of annulate lamellae and the mitochondria in the juxta-nuclear zone, as well as by the sheer quantity of annulate lamellar material present. Further studies of these cells, during earlier and subsequent differentiation, should reveal interesting patterns of cell activity, and consequently, an interesting biochemistry, histochemistry, and morphology.

SUMMARY

The technique of combined observations involves studying the *same* section in both the light microscope and the electron microscope, and permits an exact cross-correlation of the data obtained by each instrument. Plastic sections may be studied by light microscopy either coloured or uncoloured, by direct microscopy or phase-contrast microscopy. Tissue in plastic sections is accessible to histochemical methods which reveal the identity and localization of substances—'Substantive Histochemistry'—by producing a visible colour. Two such methods are discussed, the PAS (periodic acid/Schiff) technique, and the alizarin red S method for calcium. There are considerable advantages in studying positive histochemical results by phase-contrast and, especially, anoptral-contrast light microscopy, in thin plastic sections, as the level of sensitivity of detection of the positive result is greatly enhanced. The increase in sensitivity is sufficient to allow combined observations of an ultra-thin section coloured by the PAS technique, first with the light microscope, and secondly with the electron microscope. The value of combined observations is illustrated with respect to two tissues: the ganglia of *Antheraea pernyi* studied by the PAS technique in conjunction with the technique of combined observations; and the oocytes of the snail *Helix*, studied as an example of the substantive histochemistry of calcium in plastic sections. Features of the ultra-structure of these two tissues are illustrated and correlated with the findings of light microscopy. An ontogenetic sequence of events, suggested by electron microscopy, is advanced for the intracytoplasmic calcogenesis of oocyte calcospherites, which involves the activity of two new organelles of the cytoplasm, the *calcamphorae* and the *calcampullae*, membrane-bounded entities occurring in the cytoplasm as the first stage of production of a calcospherite.

ACKNOWLEDGEMENTS

I would like to thank Dr. F. K. Sanders, Director, Virus Research Unit, for supporting my work in electron microscopy and for his permission to undertake the joint responsibility for editing this book, which I share with Dr. K. F. A. Ross. I also wish to thank the Medical Research Council for the excellent facilities of their laboratories at Carshalton.

In this book dedicated to a great microscopist and cytologist, Dr. John R. Baker, F.R.S., I take the opportunity to acknowledge my large debt to him and also to another of like stature, Professor Sanford L. Palay, who, many years ago, taught me electron microscopy at the National Institutes of Health, under the Visiting Scientist Programme of the United States Public Health Service. I count myself lucky to have been able to learn something of the interpretation of cell structure from two such scientists.

REFERENCES

1. ASHHURST, DOREEN E. 1959. The connective tissue sheath of the Locust nervous system: a histochemical study. *Quart. J. micr. Sci.* **100**, 401–12
2. —— and RICHARDS, A. GLENN. 1964. A study of the changes occurring in the connective tissue associated with the central nervous system during the pupal stage of the Wax Moth, *Galleria mellonella* L. *J. Morph.* **114**, 225–36

3. —— and ——. 1964. The histochemistry of the connective tissue associated with the central nervous system of the pupa of the Wax Moth, *Galleria mellonella* L. *J. Morph.* **114**, 237–46

4. BAKER, JOHN R. 1945. *Cytological Technique. The Principles and Practice of Methods used to Determine the Structure of the Metazoan Cell*, 2nd edition, Chapter I, pp. 1–21. London: Methuen

5. ——. 1951. Cytological techniques. In *Cytology and Cell Physiology* (edited by Geoffrey H. Bourne). 2nd Edition, Chapter I, p. 7. Oxford University Press

6. BARTL, P. 1966. High water content gels for embedding of electron microscope Specimens. *International Symposium on Electron Microscopy and Cytochemistry*, May 31st–June 4th, Leiden. Programme Abstract, pp. 53–5

7. —— and Bernhard, W. 1966. *J. Microscopie* (in press)

8. BERNHARD, W. 1966. Progress and limitations of ultrastructural cytochemistry carried out on ultrathin sections. *International Symposium on Electron Microscopy and Cytochemistry*, May 31st–June 4th, Leiden. Programme Abstract, pp. 79–81

9. BERTHOLD, CLAES-HENRIC. 1966. Ultrastructural appearance of glycogen in the β-neurons of the lumbar spinal ganglion of the frog. *J. Ultrastruct. Res.* **14**, 254–67

10. DE BRUIJN, W. C. and MCGEE-RUSSELL, S. M. 1966. Bridging a gap in pathology and histology. *J. Roy. micr. Soc.* **85**, 77–90

11. BULLOCK, T. H. and HORRIDGE, G. ADRIAN. 1965. *Structure and Function in the Nervous Systems of Invertebrates*, Vol. I, especially pp. 83 and pp. 81–110. San Francisco: W. H. Freeman and Company

12. CALLAN, H. G. and TOMLIN, S. G. 1950. Experimental studies on amphibian oocyte nuclei. I. Investigation of the structure of the nuclear membrane by means of the electron microscope. *Proc. Roy. Soc. Lond.* Ser. B. **137**, 367–78

13. DODGE, J. D. 1964. Cytochemical staining of sections from plastic embedded flagellates. *Stain Tech.* **39**, 381–6

14. FOSTER, C. L. 1960. The demonstration of glycogen in liver cells fixed in osmium tetroxide. *Quart J. micr. Sci.* **101**, 273–7

15. FRANÇON, M. 1961. Phase contrast microscopy. *Progress in Microscopy*, Chapter II, p. 79. Oxford: Pergamon Press

16. GATENBY, J. BRONTE and LUTFY, R. G. 1956. The Golgi apparatus and the electron microscope. *Nature, Lond.* **177**, 1027–9

17. GRIMLEY, PHILIP M. 1964. A tribasic stain for thin sections of plastic-embedded, OsO_4-fixed tissues. *Stain Tech.* **39**, 229–33

18. ——. 1965. Selection for electron microscopy of specific areas in large epoxy tissue sections. *Stain Tech.* **40**, 259–64

19. ——, ALBRECHT, JOSEPH M., and MICHELITCH, HERMAN J. 1965. Preparation of large epoxy sections for light microscopy as an adjunct to fine-structure studies, *Stain Tech.* **40**, 357–66

20. GROSS, B. G. 1966. Annulate lamellae in the axillary apocrine glands of adult man. *J. Ultrastruct. Res.* **14**, 64–73

21. HARRISON, GLADYS A. 1966. Some observations on the presence of annulate lamellae in alligator and seagull adrenal cortical cells. *J. Ultrastruct. Res.* **14**, 158–66

22. HOUCK, CHARLES E. and DEMPSEY, E. W. 1954. Cytological staining procedures applicable to methacrylate-embedded tissues. *Stain Tech.* **29**, 207–11

23. HUXLEY, H. E. and ZUBAY, G. 1961. Preferential staining of nucleic acid containing structures for electron microscopy. *J. Biophys. Biochem. Cytol.* **11**, 273–96

24. KESSEL, R. G. 1963. Electron microscope studies on the origin of annulate lamellae in oocytes of *Necturus*. *J. Cell Biol.* **19**, 391–413

25. ——. 1964. Electron microscope studies on oocytes of an echinoderm, *Thyone briareus*, with special reference to the origin and structure of the annulate lamellae. *J. Ultrastruct. Res.* **10**, 498–514

26. —— and BEAMS, H. W. 1963. Nucleolar extrusion in oocytes of *Thyone briareus*. *Exp. Cell Res.* **32**, 612–15

27. LUFT, J. H. 1961. Improvements in epoxy resin embedding methods. *J. Biophys. Biochem. Cytol.* **9**, 409–14

28. MCGEE-RUSSELL, S. M. 1957. Tissues for assessing histochemical methods for calcium. *Quart. J. micr. Sci.* **98**, 1–8

29. ——. 1964. Histochemical methods for calcium. *J. Histochem. Cytochem.* **6**, 22–42

30. ——. 1966. On bridging the gap in cytology between the light and electron microscopes by some combined observations on snail neurones. *Quart. J. micr. Sci.* **105**, 139–62

31. ——. 1966. A ring and cup technique for handling thin sections for electron microscopy. *J. Roy. micr. Soc.* **85**, 91–6

32. —— and SMALE, N. B. 1963. On colouring epon-embedded tissue sections with Sudan black B or Nile blue A for light microscopy. *Quart. J. micr. Sci.* **104**, 109–15

33. ——, SMALE, C. J. G., and BANBURY, W. 1964. Two more accessories for the Huxley ultramicrotome. *Quart. J. micr. Sci.* **105**, 375–8

34. MCMANUS, J. F. A. 1946. Histological demonstration of mucin after periodic acid. *Nature, Lond.* **158**, 202

35. ——. 1948. Histological and histochemical uses of periodic acid. *Stain Tech.* **23**, 99–108

36. MARINOZZI, V. J. 1964. Cytochemie ultrastructural du nucléole-RNA et protéines intranucléolaires. *J. Ultrastruct. Res.* **10**, 433–56

37. MEEK, G. A. 1963. Comment during discussion following paper by E. H. Mercer, A scheme for section staining in electron microscopy, pp. 179–86. *J. Roy. micr. Soc.* **81**, 184–5

38. MOE, H. 1962. Removal of a cobalt sulfide stain from thick sections and subsequent staining with specific stains for identification of structures found by electron microscopy of adjacent thin sections. *Acta Anat.* (Basel) **49**, 189–98

39. ——, BEHNKE, O., and ROSTGAARD, J. 1962. Staining of osmium fixed Vestopal embedded tissue sections for light microscopy. *Acta Anat.* (Basel) **48**, 142–8

40. ——, ——, and ——. 1963. New methods for staining of thick sections of osmium-fixed tissue embedded in Vestopal. *J. Ultrastruct. Res.* **8**, 189–96. (Abstract from *Proceedings of the Scandinavian Electron Microscope Society*; also see publications in *Acta Anatomica*)

41. MONNERON, A. 1966. Action of pronase on ultra-thin sections of tissues embedded in epon and glycol methacrylate. *International Symposium on Electron Microscopy and Cyto-Chemistry*, May 31st–June 4th, Leiden. Programme Abstract, pp. 57–8

42. —— and BERNHARD, W. 1966. Action de certaines enzymes sur des tissus inclus en Epon. Personal communication of article in press

43. PALAY, S. L., MCGEE-RUSSELL, S. M., GORDON, S., and GRILLO, M. A. 1962. Fixation of neural tissues for electron microscopy by perfusion with solutions of osmium tetroxide. *J. Cell Biol.* **12**, 385–410

44. PEASE, D. C. 1964. *Histological Techniques for Electron Microscopy*, 2nd edition, particularly p. 261. London: Academic Press

45. POTTS, M. 1966. A method for locating specific histological features for electron microscopy. *J. Roy. micr. Soc.* **85**, 97–102

46. REBHUN, L. I. 1956. Electron microscopy of basophilic structures of some invertebrate oocytes. I. Periodic lamellae and the nuclear envelope. *J. Biophys. Biochem. Cytol.* **2**, 93–104

47. ——. 1956. Electron microscopy of basophilic structures of some invertebrate oocytes. II. Fine structure of the Yolk Nuclei. *J. Biophys. Biochem. Cytol.* **2**, 159–70

48. ——. 1959. Morphological interrelations of basophilic bodies in oocytes of the surf clam *Spisula solidissima*. *Anat. Rec.* **133**, 326–7 (Abstract of paper, American Association of Anatomists 72nd Annual Session)

49. REYNOLDS, EDWARD S. 1963. Liver parenchymal cell injury. I. Initial alterations of the cell following poisoning with carbon tetrachloride. *J. Cell Biol.* **17**, 208–12

50. ——. 1964. Liver parenchymal cell injury. II. Cytochemical events concerned with mitochondrial disfunction following poisoning with carbon tetrachloride. *Lab. Invest.* **13**, 1457–70

51. ——. 1965. Liver parenchymal cell injury. III. The nature of calcium-associated electron-opaque masses in rat liver mitochondria following poisoning with carbon tetrachloride. *J. Cell Biol.* **25**, 53–75

52. RICHARDSON, K. C., JARETT, L., and FINKE, E. H. 1960. Embedding in epoxy resins for ultra-thin sectioning in electron microscopy. *Stain Technol.* **35**, 313–23

53. ROSENBERG, M., BARTL, P., and LEŠKO, J. 1960. Water-soluble methacrylate as an embedding medium for the preparation of ultrathin sections. *J. Ultrastruct. Res.* **4**, 298–303

54. SHALLA, THOMAS A., CARROLL, T. W., and DE ZOETEN, G. A. 1964. Penetration of stain in ultra-thin sections of tobacco mosaic virus. *Stain Tech.* **39**, 257–65

55. SHANKLIN, WILLIAM M. and AZZAM, NABIL A. 1964. Phase-contrast microscopy applied to PAS-stained ground substance. *Stain Tech.* **39**, 110–11

56. WILSKA, A. 1953. A new method of light microscopy. *Nature, Lond.* **171**, 353

57. ——. 1954. Observations with the anoptral microscope. *Mikroskopie Journal of Microscopical Research and Methods* **9**, 1–80

58. WOOD, RICHARD L. and LUFT, JOHN H. 1965. The influence of buffer systems on fixation with osmium tetroxide. *J. Ultrastruct. Res.* **12**, 22–45

Chapter 17

Changing Concepts of the Connective Tissues

J. F. A. McMANUS

Dans quels recoins de tissu cellulaire
Sont les talens de Vergile ou d'Homer ?—VOLTAIRE, *Jeanne d'Arc*

Ever since bodies of animals were first dissected, it has been appreciated that one can distinguish specific and isolated organs and other tissues which hold the organs together to give the animal its form and consistency. These latter comprise the connective tissues which are of great chemical and physiological interest at the present time, as they have been in the past. This presentation will discuss some of the ways in which the connective tissues have been considered in the past and relate the current considerations on the connective tissues to these earlier viewpoints.

When one attempts to assess the level of information in a certain area of study in some period of the past, one is confronted immediately by the problem of how to make the judgement—what sources are to be sought, which statements of information are significant, and where are they to be found. The scientific journals and periodicals would appear, at first glance, to furnish indications of the current knowledge and viewpoints of the particular time. However, one finds, over and over again in the history of science, that there may be a very lengthy 'lag' period between a discovery and its acceptance. Sometimes a significant observation is overlooked, and buried in some obscure journal, or it is of such a sort that the time is not right for its appearance and introduction into the scientific thinking of the time.

After momentary reflection, one can agree with the proposition that what appears in the textbooks of a period can be considered to be the beliefs and viewpoints of the time. In the textbooks can be found the current teachings and beliefs, the viewpoints consistent with those of the teachers or professors, and the information which it is believed the proselyte or student needs to know, in order to begin his education in the area.

This review of the changing concepts of the connective tissues will differ from the numerous and usual reviews, in that it is derived from a series of textbooks, rather than from the original articles as they appeared in the literature. There is much to be said for a bibliography which consists of citations of the original articles in which new information or new viewpoints are first reported. This, however, seems to me to be incomplete, for the purpose of the present article. New knowledge becomes used, and useful, when it is assimilated into, and integrated with, the existing information in the area of study. This happens when knowledge is incorporated into a textbook. For these reasons, the present review quotes liberally from the sometimes disparaged secondary sources, in an attempt to obtain an understanding of the 'state of the art' at several different periods in the past.

The earliest descriptions of the connective tissues emphasized the fibrous nature of many of the parts and also described, spaces, or 'cells', in the meaning of the time. This latter is the origin of the use of the term 'tissu cellulaire', to describe the connective tissues as in the quotation from Voltaire, at the head of this chapter. This usage continued into the time of Bichat[2] in the early nineteenth century, as the following quotation demonstrates:

Cellular System

This system, which many know still, under the name of the cribriform body, the mucous texture, &c. is an assemblage of filaments, and of white soft layers, intermixed and interwoven in different ways, leaving between them spaces communicating together, more or less irregular, and which serve as a reservoir for the fat and serum. Placed around the organs, the different parts of this system act at the same time as a bond to connect, and as an intermediate body to separate them. Carried into the interior of these same organs, they essentially contribute to their structure.

The great extent of this system, which, though everywhere spread, is everywhere continuous, the number of organs it surrounds and the multiplied relations it presents, do not allow me to describe it, as has been done, in one point of view; in order to give a complete view, it is necessary to separate the different points in which it may be examined.

I shall then at first consider abstractedly the general system, as represented by the continuity of all its parts, in order to consider it in relation to the organs that it surrounds, or to whose composition it concurs. I shall examine it afterwards independently of these organs, as it is spread every where in the spaces between them. In fine, its organization, its properties, its relations with other systems, and its development will be the object of my researches.

Further considerations of Bichat[2] on the organization of the connective tissues appear as follows (for 'texture' read 'tissue'):

ARTICLE FOURTH.
Organization of the Cellular System.

The cellular system, like almost all the others, is composed of a peculiar texture and of common parts.

1. *Of the texture peculiar to the organization of the cellular system.*

Much has been written upon the nature of this texture; Bordeu has given some vague ideas upon it, but no experiments. Fontana has made researches which lead but to few results, upon its intimate structure and upon the tortuous cylinders of which, according to him, it is an assemblage. Let us throw aside all hypotheses that examination does not support; let us follow nature in the phenomena of structure that she shows us, and not in those she wishes to conceal. In thus considering the cellular texture, we see that it is very different from a species of glue, with which some have wished to compare it. It is an assemblage of many whitish filaments, crossing very often certain kinds of delicate layers, which form cells with these filaments. To see this organization well, a piece of the cellular portion of the scrotum should be taken, which has no fat, and whose texture is consequently not concealed by this fluid; this portion being stretched into a kind of membrane, is seen very distinctly. Then there may be plainly distinguished, 1st. a transparent network, arranged in layers, which makes the foundation, if we may so say, and the tenuity of which is such, that it has been aptly compared by a physiologist, to the soap bubbles that are thrown into the air with a pipe. It is impossible to distinguish, by the naked eye, any fibre in the texture of these layers; every thing is there uniform. 2nd. They are very evidently crossed by numerous filaments, which running in all directions, are interwoven in every way, all of which touch, when the cellular texture is pressed together, but when stretched out, there can be seen between them the layers of which I have just spoken. The more it is extended, the larger consequently the membrane becomes, the interstices between the filaments are greater, and the intermediate layers are also more apparent.

What is the nature of these filaments ? I presume that some are absorbents, others exhalants, and that many are formed in the places where the layers unite together for the formation of the cells. As the thickness arising from this union is greater, they are distinguished by more evident lines upon the cellular texture stretched into a membrane. What induces me to believe this, is, that when, instead of examining the cellular texture upon a portion taken from the scrotum, and stretched as I have described, it is observed in an artificial emphysema, as in that of slaughter-houses for example, then there is seen upon the covering of each cell, only the non-filamentous layers of which I have spoken, without any of those filaments that were seen crossing it in the other method.

These layers have not the same thickness in all cases; quite dense when the cellular texture is contracted, they become, when it is distended with air or any other means, so fine and attenuated, that the mind cannot conceive that there is any thing organized in them. Their organization is real, however, though some have doubted it. What in fact is a texture that is nourished, inflames and suppurates, which is the seat of very distinct vital functions, and which evidently lives, if it is not an organic texture ? All these vague ideas of concrete juices, of inorganic glue, of hardened juice, that have been applied to the cellular texture, have no solid foundation, and rest neither upon experiment or observation, and ought to be banished from a science in which imagination is nothing, and facts every thing.

The cellular texture has essential differences of organization; everywhere where fat or serum is accumulated, there are real cells which have little sacs communicating with each other, which form reservoirs, the sides of which are composed of the transparent and non-filamentous layers of which we have spoken; it is in these sacs that the serous and fatty depositions take place. On the other hand, in the sub-mucous texture, in that which forms the external membrane of arteries, veins, and excretories, there are none of these sacs, no cells, properly speaking, and no layers to form them. When we carefully raise this texture, and lift it from the surface upon which it is applied, and draw it a little so as to show its structure, we shall see very distinctly numerous filaments interwoven every way, forming a true net-work, meshes, if I may so express myself, but not sacs and cavities. The air distends this net-work when it is driven forcibly into the neighbouring texture; but as soon as an opening is made near it, it escapes and the texture sinks down; when accumulated in the ordinary texture, the sub-cutaneous, intermuscular, &c. it remains in the cells, notwithstanding they have been in part opened, without doubt because the communicating openings are very small. This fact is evident in markets, where we see the cellular texture blown up, around the meats that are stripped of their skin.

It appears that the filaments that are interwoven in every way, and which form about the vessels and under the mucous surfaces, a cellular net-work, are really of the same nature as those spread in different directions in the membranous layers which make the cells, only they are nearer together, and are by themselves.

After what I have said, it is evident that there are two things in the common cellular texture; 1st. a number of fine, transparent layers, found everywhere where the texture is loose, capable of yielding suddenly to different distensions, and of retaining the fluids its cells contain, &c.: 2nd. filaments intermixed with these layers wherever they are, and existing alone in certain places. These layers and cellular filaments have a remarkable tendency to absorb atmospheric moisture. We observe it in dissecting rooms, where a subject dry and easy to dissect in the morning, is often much infiltrated by evening, if the weather has been damp; now this infiltration takes place in the cellular system, which is a real hygrometer.

In 1847, Schwann's elaboration of the cell theory[22] placed a new emphasis on the connective tissue substance. A statement of the cell theory follows:

The elementary parts of all tissues are formed of cells in an analogous, though very diversified manner, so that it may be asserted, *that there is one universal principle of development for the elementary parts of organisms, however different, and that this principle is the formation of cells.* This is the chief result of the foregoing observations.

The same process of development and transformation of cells within a structureless substance is repeated in the formation of all the organs of an organism, as well as in the formation of new organisms; and the fundamental phenomenon attending the exertion of productive power in organic nature is accordingly as follows: *a structureless substance is present in the first instance, which lies either around or in the interior of cells already existing; and cells are formed in it in accordance with certain laws, which cells become developed in various ways into the elementary parts of organisms.*

The intercellular substance and the ground substance of connective tissues became for Schwann the origin of the cells.

Cytoblastema.—The cytoblastema, or the amorphous substance in which new cells are to be formed, is found either contained within cells already existing, or else between them in the form of intercellular substance. The cytoblastema, which lies on the outside of existing cells, is the only form of which we have to treat at present, as the cell-contents form matter for subsequent consideration. Its quantity varies exceedingly, sometimes there is so little that it cannot be recognized with certainty between the fully-developed cells, and can only be observed between those most recently formed; for instance, in the second class of tissues; at other times there is so large a quantity present, that the cells contained in it do not come into contact, as is the case in most cartilages. The chemical and physical properties of the cytoblastema are not the same in all parts. In cartilages it is very consistent, and ranks among the most solid parts of the body; in areolar tissue it is gelatinous; in blood quite fluid. These physical distinctions imply also a chemical difference. The cytoblastema of cartilage becomes converted by boiling into gelatine, which is not the case with the blood; and the mucus in which the mucus-cells are formed differs from the cytoblastema of the cells of blood and cartilage. The cytoblastema, external to the existing cells, appears to be subject to the same changes as the cell-contents; in general it is a homogeneous substance; yet it may become minutely granulous as the result of a chemical transformation, for instance, in areolar tissue and the cells of the shaft of the feather, &c. As a general rule, it diminishes in quantity, relatively with the development of the cells, though it seems that in cartilages there may be even a relative increase of the cytoblastema proportionate to the growth of the tissue. The physiological relation which the cytoblastema holds to the cells may be two-fold: first, it must contain the material for the nutrition of the cells; secondly, it must contain at least a part of what remains of this nutritive material after the cells have withdrawn from it what they required for their growth. In animals, the cytoblastema receives the fresh nutritive material from the blood-vessels; in plants it passes chiefly through the elongated cells and vascular fasciculi; there are, however, many plants which consist of simple cells, so that there must also be a transmission of nutrient fluid through the simple cells; blood-vessels and vascular fasciculi are, however, merely modifications of cells.

★   ★   ★

Cellular (areolar) Tissue. This tissue is known to be composed of extremely minute, tough, smooth fibres, having a pale outline, and usually a serpentine course; they may be seen in their natural state in the mesentery without any dissection. Most areolar tissue may be distended by forcing air into it, and then innumerable cellular spaces are seen communicating with each other in it; it is not known whether these are produced artificially, or whether they existed previously. Areolar tissue also frequently contains fat-vesicles, which, according to Gurlt, are surrounded by a thin and transparent, but not fibrous, pellicle, often have an hexagonal form, and in that respect resemble vegetable tissue. (Gurlt's *Physiologie der Haussaugethiere*, p. 19). In order to become acquainted with the relation which these constituent parts of areolar tissue bear to the elementary cells, we must refer to the formation of the tissue in the foetus.

If we examine some areolar tissue from the neck, or from the bottom of the orbit of a foetal pig measuring three inches and a half in length, we shall find it to be a gelatinous substance, somewhat more consistent than the vitreous humour of the eye, and, in its earliest state, quite as transparent; as development proceeds, however, it becomes more of a whitish colour, and loses its gelatinous quality. When examined with the microscope, small corpuscles of various kinds are seen in greater or less numbers; they are not, however, sufficiently numerous in a foetus of the size specified to form the entire gelatinous substance, but must necessarily be situated in a transparent structureless, primordial substance of a gelatinous nature, which we will for the present call cytoblastema. The whiter this substance appears to the unaided eye, the greater is the number of corpuscles contained in it; their quantity, therefore, is continually increasing during development, while that of the cytoblastema constantly diminishes. As in consequence of its transparency, the cytoblastema cannot be seen, but is only inferred to exist from the circumstance that the corpuscles, which are visible under the microscope, could not, at the period when they are but few, form the entire jelly, and that when moved, it is plainly seen that they are held together by some invisible medium, so it is no longer possible to convince ourselves of its existence, when the corpuscles are very numerous. It is probable, however, that it remains between the fibres of the areolar tissue throughout life.

An advance over the views of Schwann are seen in those of Kolliker.[9] Speaking in 1853 of the connective tissue he says:

The elementary parts which are found in connective tissue may be divided into the essential, never-failing components, and those which are met with only in certain localities. To the former belongs the fasciculated as well as the more homogeneous connective tissue; to the latter, elastic fibres in their different forms and conditions of development, fat cells, cartilage cells, and pigment cells of different kinds. Besides these, connective tissue contains also no inconsiderable quantity of a gelatinous intermediate substance. The bundles of the connective tissue are, among the essential elements, those which occur most frequently; each of them consists of a certain number of very fine fibrils, the connective fibrils, which are distinguished from their nearest allies, the finest elastic fibres and muscular fibrils, by their smaller diameter (0·0003'''–0·0005'''), their pale colour, their homogeneous appearance, and the complete absence of striation. They are united by means of a small quantity of a clear connecting substance, and thus form the bundles in question, which in many respects resemble those of the transversely striated muscles, but differ from them in the absence of any special investment comparable to the sarcolemma, and in their smaller mean diameter (0·004'''–0·005'''). They are either long, slightly wavy cords, of uniform thickness throughout, which are not directly connected together, but arranged in different ways near and above one another, forming great lamellae and bundles; or they coalesce like the elastic networks into meshes, and thus form what I have called the *reticulated connective tissue.* In rare cases the bundles appear not to be composed of fibrils, but are more homogeneous, as in the neurilemma, where they are known as Remak's fibres. Besides this form of connective tissue, there exists a second, rarer kind, in which neither bundles nor fibrils can be clearly distinguished, but only a membranous or more or less solid, finely granulated, or slightly striated, even perfectly homogeneous, clear tissue; *homogeneous* (or Reichert's) *connective tissue.* The other elements which occur in connective tissue present nothing remarkable, and will be more particularly treated of in their proper places in the special part.

The chemical relations of connective tissue are well known: proper connective substances when boiled yields common gelatine, and contains besides a fluid, whose nature, on account of its generally minute quantity, cannot be investigated. Only where it exists in considerable proportion, as in the gelatinous connective tissue of embryos, can the presence of much albumen and mucus be easily demonstrated in it. The chemical qualities of the other constituents of the connective tissue will be spoken of in their place.

Connective tissue is of utility to the organism according to its composition,—sometimes as a solid unyielding substance; sometimes as a soft support for vessels, nerves, and glands; sometimes, finally, as a yielding tissue, filling up spaces and facilitating changes of position. Where elastic elements are present in it in great quantities, its nature alters; and a great abundance of fat or cartilage cells gives it an unusual softness or resistance. The connective tissue is invariably developed from cells, and, in fact, from fusiform or stellate vesicles, which become united into long fibres or networks, and often break up into fibrils before their union. The mode in which this takes place is not yet quite made out, but it is most probable that the cells, as they elongate, change, with their membrane and contents, into a homogeneous softish mass, which subsequently breaks up into a bundle of fine fibrils and some intermediate substance. The development of the homogeneous connective tissue has as yet been little investigated, but it would seem, like the other, to proceed from a fusion of rounded or elongated cells, which are perhaps united by an intermediate substance, in which the metamorphic process has only gone so far as the development of a homogeneous mass, but has not attained the stage of fibrillation.

Further, on Kolliker's concept of the connective tissues, the following paragraphs are important:

The connective tissue is found in all the four classes of the vertebrata, in about the same condition as in man; while, on the other hand, in the invertebrata it is very rare, and when present is more homogeneous, or consists of isolated cells and intermediate substance, rarely more fibrous, as in Cephalopoda, in the mantle of bivalves, in the peduncle of the Lingulae, and of the Cirripeds. Fat-cells also do not occur among the lower animals to the same extent as among the higher. The firm connective tissue is here replaced by a chitinous substance, or by one consisting of cellulose, and by calcareous or horny tissues.

Opinions are still divided as to the structure and development of the connective tissue. Whilst the majority ascribe a distinctly fibrous structure to it, and suppose it to consist of bundles, and these again of fibrils, Reichert considers this tissue to be more homogeneous, and regards the fibrillation partly as artificial, partly as the expression of a folding, a view to which Bidder and Virchow are also inclined. For my own part, I find a certain amount of truth in Reichert's conception, insomuch as it is not to be denied that there also exists a non-fibrillated, more homogeneous connective tissue, which had previously been little investigated; but I am nevertheless of opinion, that, as applied to the great mass of the organs composed of connective tissue, it is incorrect. The possibility of making out fibrils in delicate membranes, even without preparation, the ease with which these may be isolated in tendons and ligaments, and lastly, the circumstance that the fibrils may be demonstrated upon transverse sections of the tendons, and of the more solid connective tissue in general, are for me sufficient reasons for retaining the old view.

With respect to the *development* of the connective tissue, I distinguish two types which correspond with its two principal forms, the solid and the areolated. The former is developed out of masses of cells without any demonstrable matrix, by the elongation of the cells, their breaking up into fibrils, and their coalescence. This is most obvious in the tendons and ligaments, which, as observations upon Batrachian larvae and upon mammalian embryos show, at first consist entirely of common, rounded, formative cells, which about the same time as the transversely striated muscles are formed, (in mammalia in the second month) become fusiform. The further development demonstrates (what had escaped Schwann) that only one portion of these fusiform cells, and in fact cells which are remarkable for their size and paler contours, become bundles of connective tissue, while the others, which Schwann in part depicts rightly (Tab. III, fig. 11; the smallest cell, fig. 6, from connective tissue, the cell b, and the lowest cell upon the right side), remain for a time as fusiform elements, and only subsequently become fused into elastic fibres. There arises, at last, out of cells alone, with no distinguishable matrix, a compact tissue composed of two chemically quite distinct fibres. The *areolated connective tissue* differs from the former in the circumstance that, if not from the beginning yet from the time at which the cells become elongated, an abundant gelatinous intermediate substance is developed between them, which does not yield gelatine, and never becomes converted into it, but contains albumen and a substance similar to mucus; Schwann, indeed, found a substance resembling *pyin*, in this tissue. Although all embryologists know that the areolated connective tissue is at first of a gelatinous consistence, as, for example, under the skin, in the neck, in the omentum, behind the peritoneum, in the orbit, and in the bones, no one has yet drawn attention to the general occurrence of that intermediate substance which was observed by Schwann in a single locality. I originally became acquainted with this tissue between the chorion and amnion, and at first paid more attention to its reticulated anastomosing cells. Subsequently, when I examined it more closely in the enamel organ of the embryonic tooth sac, I paid attention to the peculiar intermediate substance, and at the same time Virchow described this tissue from the umbilical cord, where the gelatinous tissue of Wharton entirely consists of it. Virchow believed that it ought to be distinguished from connective tissue, and proposed the denomination of mucous tissue (*tissu muqueux*) for it. I considered it from the first to be connective tissue, and I now feel the more inclined to remain of this opinion, because I find that every description of the areolated connective tissue of embryos originally commences under this form, and therefore the circumstance that the tissue in the umbilical cord never arrives at perfection, cannot determine its nature.

15—C.S.I.

The mode in which the gelatiniform connective tissue is developed is this: one portion of the cells contained in the gelatinous basis changes into connective tissue by becoming fusiform, and breaking up into common or reticulated, anastomosing connective tissue, which however, as Schwann has already stated, at first yields no gelatine. In this manner a closer or denser network arises, in the interspaces of which the intermediate substance or matrix, and a remainder of the previous formative cells, are contained. In the further course of development, new cells proceed from the matrix, which hereby diminishes by degrees in quantity, and at the same time the original network consolidates, fresh cells being added to it, a part of which also become elastic fibres and vessels. If subsequently the areolated connective tissue includes no adipose cells, the gelatinous tissue ends by completely disappearing, and nothing remains but a loose fibrous tissue, containing at most somewhat less fluid, and loose cells in its meshes; if, on the other hand, it becomes converted into an adipose tissue, the spaces remain, and a great part of the cells which have arisen at the expense of the gelatinous substance, subsequently pass, by the development of fat in their interior, into fat-cells.

In the gelatinous tissue of Wharton, between the chorion and amnion, and in part in the enamel organ, the areolated connective tissue remains more in its foetal condition of a gelatinous tissue, yet there exists no natural line of demarcation from ordinary connective tissue, so much the less, since in the gelatinous substance of Wharton, in older embryos even fibrils are quite evident, and in the enamel organ the passage of a part of the gelatinous tissue into common connective tissue is demonstrable.

So much for the two types of development of the connective tissue. We have yet to state how the bundles become chemically and morphologically what they are. In the first place, I may observe that the formative cells of the connective tissue are not originally distinguishable from the other formative cells of the embryo, do not dissolve by boiling in water, and therefore contain no gelatine. Even when the cells have evidently become fusiform, and have already coalesced into bundles and networks, they still, as Schwann has already stated, yield no gelatine. Therefore, in this case, the change of the cells into a collagenous substance, goes on as slowly as in the matrix of the cartilages, which, according to Schwann, also, at first, yields no gelatine, and therefore it is no objection to the above view of the nature of Wharton's gelatinous tissue, that it yields no gelatin on boiling, as Scherer has found. How the collagenous matter is formed out of cells, whether the contents only, or the membrane also, takes part therein, it is very difficult to say; in any case, from what we know of the contents of embryonic cells, it can hardly be any but a protein substance which yields the gelatine, and, from what takes place in the ossification of the cartilage cells, it seems very probable that the cell-membranes and contents together become metamorphosed into a collagenous substance.

The morphological change, which the formative cells of the connective tissue undergo, in the course of their passage into bundles of fibrils, is very probably this, that after their membranes and contents are fused into a homogeneous semi-solid mass, they then secondarily break up into fibrils; the latter process taking place in the same manner as we see it occur in the contents of the animal muscular fibres. Herewith, as a rule, the nuclei of the cells eventually disappear, or if they remain, as we see occasionally in connective tissue, still they never become changed into the so-called nucleus fibres.

Though in physiological connective tissue, development from cells must be most decidedly affirmed, it does not therefore follow that a substance which chemically and morphologically closely resembles connective tissue, may not arise in a different manner. We know, in fact, that the collagenous basis of cartilage, when it breaks up into fibres, becomes deceptively similar to connective tissue, and furthermore, that fibrous exudations may become changed into a fibrous substance which is scarcely, perhaps not at all, to be distinguished from genuine connective tissue. There also exists, however, *a pathological true connective tissue in cicatrices of all kinds, and perhaps elsewhere, which is developed from cells*; and for my own part, therefore, I am opposed to the classing together of all connective tissues. We must in our classifications not only distinguish similarity or identity in structure and chemical composition, but embrace all the conditions, and especially the genesis; and thence we must distinguish both the collagenous fibrous cartilage and the collagenous organised fibrine, from true connective tissue,—just as we separate the true elastic fibre, from the chemically and morphologically, very similar fibres of the reticulated cartilages and from certain forms of metamorphosed fibrine. On the other hand, the connective tissue which has not been developed from cells may justly and properly be arranged with cartilage.

Virchow in his lectures of 1858, which began Cellular Pathology,[23] gave a new emphasis to the connective tissues. To quote:

The second histological group is formed by the connective tissues (Gewebe der Bindesubstanz). This is the subject in which I take the most interest, because it was here that my own observations, which have led to the result to which I directed your attention at the beginning of these lectures, originated. The alterations which I have succeeded in introducing in the views of histologists with regard to the whole group have, at the same time, enabled me to give a certain degree of roundness and completeness to the cellular theory.

Previously, connective tissue had nearly universally been regarded as essentially composed of fibres. On examining loose connective tissue in different regions, as, for example, beneath the corium, in the pia mater, subserous and submucous cellular tissue, we find wavy bundles of fibres, the so-called wavy connective tissue. This wavy character, which is interrupted at certain intervals, so as to give rise to a kind of fasciculation, could, it was thought, with the less hesitation be attributed to the presence of separate fibres, because at the end of each bundle isolated filaments could in reality be seen to protrude. In spite of this, however, an attack was made upon this very hypothesis, somewhat more than ten years ago, and has proved of very great importance, though in a manner different to that which was intended. Reichert endeavored, namely, to show that the fibres were only an optical illusion produced by folds, and that connective tissue in all parts formed a homogeneous mass, endowed with a great tendency to the formation of folds.

Schwann had, in reference to the formation of connective tissue, assumed that there originally existed spindle-shaped cells, the *caudate corpuscles* (geschwanzte Korperchen) (fibro-plastic corpuscles of Lebert), which afterwards became so famous; and that out of these cells fasciculi of connective tissue were directly developed by the splitting up of the body of the cell into distinct fibrils, whilst the nucleus remained as such. Henle, on the other hand, thought the only conclusion his observations would warrant was, that there were originally no cells at all, but that nuclei only were formed in the blastema at certain intervals; whilst the fibres, which afterwards appeared, were produced by a direct fibrillation of the blastema; and that, whilst the intermediate substance was thus being differentiated into fibres, the nuclei gradually became elongated, so as at length to run into one another, and thus give rise to peculiar longitudinal fibres, *nucleus-fibres* (Kernfasern). Reichert took an extremely important step in opposition to these views. He showed, namely, that originally there were only cells, and those in great abundance, between which intercellular substance was deposited. But the membrane of the cells became, he thought, at a certain period, blended with the intercellular tissue, and then a stage was reached analogous to that described by Henle, in which there no longer existed any boundary between the original cells and the intermediate substance. And, finally, he imagined that the nuclei, too, entirely disappeared in some instances, whilst they were preserved in others. On the other hand, he positively denied the occurrence of the spindle-shaped cells of Schwann, and declared all such, as well as the caudate and jagged cells, to be just as much artificial products as the fibres, which were said to be seen in the intervening substance, were a false interpretation of an optical image.

Now, my own investigations have shown, that both Schwann's and Reichert's observations, up to a certain point, have some foundation in truth. That, in the first place, in opposition to Reichert, spindle-shaped and stellate cells indisputably do exist; and secondly, in opposition to Schwann, and with Reichert, that a direct splitting up of the cells into fibres does not take place, but that, on the contrary, what is afterwards presented to our sight as connective tissue has really taken the place of the previously homogeneous intercellular substance. I have found, moreover, that Reichert, Henle, and Schwann, were wrong in maintaining that ultimately at best only nuclei or nucleus-fibres remained; and that, on the contrary, in most cases the cells themselves preserve their integrity. The connective tissue of a later period is therefore not distinguished in its general structure and disposition in any respect from that of an earlier date. There is not an embryonic connective tissue with spindle-shaped cells and a perfectly developed one without them, but the cells remain the same, although they are often not easy to see.

Essentially, therefore, this whole series of lower tissues may be reduced to one simple plan. Usually, the greater part of the tissue is composed of intercellular substance, in which at certain intervals, cells lie imbedded, which in their turn present the most manifold forms. But these tissues cannot be distinguished by one's containing only round, and another's, on the contrary, only caudate or stellate, cells; but in all connective tissues round, long and angular cells may occur. The simplest case is where round cells lie at certain intervals, and intercellular substance appears between them. This is the form which we see most beautifully shown in hyaline cartilage, as in that lining the joints, for example, in which the intercellular matter is perfectly homogeneous, and we see nothing but a substance which, though, perhaps, slightly granulated here and there, is on the whole quite as clear as water, so that, as long as we do not see the edges of the preparation, doubt may arise as to whether anything at all exists between the cells.

This substance is characteristic of *hyaline cartilage*. Now we find that, under certain circumstances, the round cells became even in cartilage transformed into oblong spindle-shaped ones, as, for example, with great regularity in the immediate neighbourhood of the articular surfaces. The nearer, in the examination of articular cartilage, we approach to the free surface the smaller do the cells become; and, at last, nothing more is seen but small, flatly lenticular bodies, the substance intervening between which sometimes presents a slightly striated appearance. Here, therefore, without the tissue's having ceased to be cartilage, a new type displays itself, which we much more regularly meet with in pure connective tissue, and hence the idea might easily arise that articular cartilage is invested with a special membrane. This is, however, not the case, for there is no synovial membrane spread over the cartilage, but its boundary towards the cavity of the joint is everywhere formed of cartilage itself. The synovial membrane only begins where the cartilage ceases— at the edge of the bone. On the other hand, we see that at certain points the cartilage passes directly into forms in which the cells become stellate, and the way is paved for their final anastomosis; ultimately, spots are met with at which it is no longer possible to say where the one cell ends and the other begins, inasmuch as they communicate so directly with one another that it is impossible to detect a line of separation between their membranes. When such a case occurs, the cartilage, which up to that time had remained hyaline and homogeneous, becomes heterogeneous and striated, and has long since been called *fibro-cartilage*.

From these forms a third has been distinguished, the so-called *reticular* [yellow or spongy] *cartilage*, as seen in the ear and nose, in which the cells are round, but encircled by a peculiar kind of thick, stiff fibres, whose mode of production has not yet been thoroughly made out, but they are, perhaps, derived from the metamorphosis of the intercellular substance.

Under these different types, presented by cartilage in its different localities, all the different aspects which the other connective tissues offer are included. There is also true connective tissue with round, long, and stellate cells. Just in the same manner we find, for example, in the peculiar tissue which I have named *mucous tissue* (Schleimgewebe), round cells in a hyaline, or spindle-shaped ones in a striated, or reticular ones in a meshy, basis-substance. The only criterion we possess for distinguishing them consists in the determination of the chemical constitution of the intercellular substance. Every tissue is called connective tissue whose basis-substance yields gelatine when boiled; the intercellular substance of cartilage produces chondrine; mucous tissue, on expression, a substance, mucin, precipitable by acetic acid, and insoluble in an excess of it, though dissolving in muriatic acid when added in considerable quantity.

Besides these, a few solitary points of difference in regard to peculiarity of form and contents may be presented by individual cells at some later period of their existence. What we concisely designate *fat* is a tissue which it intimately connected with those of which we have been treating, and is distinguished from the rest by the fact that some of the cells enlarge and become stuffed full of fat, the nucleus being thereby thrust to one side. In itself, however, the structure of adipose tissue is precisely the same as that of connective tissue, and, under certain circumstances, the fat may so completely disappear that the adipose tissue is once more reduced to the state of simple, gelatinous connective, or mucous tissue.

Amongst these different species of connective tissue the most important for our present pathological views, are, generally speaking, those in which a reticular arrangement of the cells exists, or, in other words, in which they anastomose with one another. Wherever, namely, such anastomoses take place, wherever one cell is connected with another, it may with some degree of certainty be demonstrated that these anastomoses constitute a peculiar system of tubes or canals which must be classed with the great canalicular systems of the body, and which particularly, forming as they do a supplement to the blood- and lymphatic vessels, must be regarded as a new acquisition to our knowledge, and as in some sort filling up the vacancy left by the old vasa serosa which do not exist. This reticular arrangement is possible in cartilage, connective tissue, bone and mucous tissue in the most different parts; but in all cases those tissues which possess anastomoses of this description may be distinguished from those, whose elements are isolated, by the greater energy with which they are capable of conducting different morbid processes.

In a further significant statement Virchow goes on to say:

I have up to the present time, gentlemen, brought to your notice a series of tissues all of which agree in containing either very few capillary vessels, or none at all. In all these cases the conclusion to be drawn seems to be very simple—that, namely, the peculiar cellular, canalicular arrangement which they possess serves for the circulation of juices. It might, however, be supposed that this was an exceptional property, appertaining only to the non—or scantily—vascular and, generally speaking, hard, parts, and I must therefore add a few words concerning the soft textures which possess a similar structure. All the tissues which we have hitherto considered, belong, in accordance with the classification which I have already given you, to the series of connective tissues; fibro-cartilage, fibrous or tendinous tissue, mucous tissue, bone and the teeth, must one and all be considered as belonging to the same class. But to the same category belongs also the whole mass of what has usually been included under the name of *cellular tissue* (Zellgewebe), and for which the name proposed by Johannes Muller, *connective tissue* (Bindegewebe), is the most appropriate; that substance, which fills up the interstices in the most different organs, sometimes in greater, sometimes in less, quality—which renders the gliding of parts one upon the other possible, and formerly was imagined to enclose considerable spaces (cells in an inexact sense of the word), filled with a gaseous vapour or with moisture.

Of this kind is the peculiar interstitial, or connective, tissue, such as we meet with in the interior of the larger muscles between the several primitive fasciculi and in a still larger quantity between the several parcels, or bundles, or primitive fasciculi. Numerous arteries, veins, and capillaries lie in it; and the arrangements for its nutrition are the most favourable that can be imagined. Notwithstanding this, however, there exists in it also, in addition to its blood-vessels, a more delicate system of nutrient channels, precisely similar to that with which we have just become acquainted; only that, wherever it is specially required, in particular parts a peculiar change takes place in the cells, the place of the simple cell-networks and fibres being gradually occupied by a more compact structure, which originates in a direct transformation of them, namely, the so-called *elastic tissue*.

One of the more popular histology textbooks of the last quarter of the nineteenth century was that edited by Stricker and written by many authors. Rollet[19] contributed the chapter (number 2) on the connective tissues. It was translated into English by a number of distinguished scientists among whom one recognizes Delafield the pathologist. The beginning of the chapter on the connective tissues follows:

It has become customary in histology to associate together a series of tissues under the term connective tissues. From these tissues all those portions of the animal body are formed, which can be regarded in the most general significance of the terms as the basement membrane, supporting layer or investment for epithelial structures, blood, lymph, mucles, and nerves. In the Vertebrata the group of connective substances includes connective tissue, cartilage, bone, the tissue of the cornea and dentine.

The connective tissues are developed from the middle germinal layer, in which blood and muscle also originate. The typical connective substances are recognized histologically by the circumstance that they contain extensive and continuous layers of material (intercellular substance), which, when compared with the cellular structures distributed through its substance (protoplasma), or the morphological elements in other tissues, always appears as a more passive substance, and one which participates but slightly in the processes characteristic of life. These masses consist for the most part of gelatine-forming substances, such as collagen, chondrogen, and ossein. The connective tissues frequently pass by substitution or genetic succession into one another; they appear therefore to be morphologically equivalent; so that in many instances certain organs, or parts of organs, belonging to animals nearly allied to one another, are formed sometimes of one, sometimes of another of these tissues.

Even if our knowledge of such facts disposed us to collect the tissues into a single category, this is still not the immediate and primary reason that has led to the formation of a group of connective substances. This last has become customary since the experimental investigation of these tissues has shown that they present similar modes of development, and possess consequently an homologous significance in regard to their microscopic constituents.*

The fate of the connective tissue theories thus originating has been very variable. Reichert first appeared with his doctrine of continuity of substance. According to this the connective tissues contain a matrix, originating in the fusion of cells, or of certain portions of cells, with an amorphous intercellular substance. Reichert associated with this mode of development the peculiar connective tissue formerly regarded as fibrous, but considered by him to be destitute of structure, and pointed out that in both there was an absence of any apparent boundary line between the allied tissues where they were in contact with one another, or, as he expressed it, there was a 'continuity' of their matrix.

This theory was, even from the first, strongly opposed by Henle, and did not in the first instance meet with general acceptance. If the views on the absence of structure in connective tissue taught by Reichert, and now disproved, found certain adherents, amongst them Virchow himself may be included, it can scarcely be held, as however is frequently done, that the connective tissue theory promulgated by Virchow in 1850, was only a modification of that of Reichert. We are indebted to Virchow and Donders for directing attention to the persistence of cells in mature connective tissue. Virchow, whilst he regarded the cells of connective tissue (connective tissue corpuscles) as the analogues of the cells of cartilage and bone, constructed a simple scheme for the structure of connective tissues; and, upon the other hand, sought to attribute to the excitation, growth, and proliferation of these tissue cells a series of the most important pathological processes, and was thus led to the profound views contained in his cellular pathology.

According to Virchow's idea the greater part of the tissues belonging to the group of connective tissues consists of intercellular substances, the latter indeed varying in regard to their chemical nature in the several members of the series, and containing variously formed but similar cells imbedded in their substance. The views of Virchow obtained general acceptance. The special methods which he employed in his researches caused him, however, to described forms which had nothing to do with connective tissue cells, and induced him in the case of connective tissue, as had been done by earlier inquirers in regard to osseous tissue, to admit the existence of cell processes frequently anastomosing with one another, which he regarded as forming a plasmatic canal system traversing the tissue in all directions. Henle in both instances expressed determined and persistent opposition to the existence of connective tissue corpuscles in the sense understood by Virchow. The point in question required an exact appreciation of appearances exhibited under the microscope, and the final result was that inquirers for the most part convinced themselves of the existence of persistent cells in mature connective tissue.

In the meanwhile, however, through the investigations of Max Schultze, Brucke, and others, the doctrine of cells founded by Schwann, and up to that time generally received, experienced some important modifications. It was no longer possible to describe animal cells as uniform elementary parts of a vegetative character, constructed according to a certain scheme. The new opinions held in regard to the structure of connective tissue substances could not remain without influence upon the general conception of a cell. Still more directly was the connective tissue question affected by the views which were coincidentally expressed by Max Schultze upon the solid intercellular substances of the animal tissues. Up to that time the majority

*[*Editorial note:* The reader is referred to the article by Pantin (pp. 5–6) for a discussion of the significance of homology in present-day considerations of the problems of cell structure. The modes of thought in cytology today, and almost a century ago, are often strikingly parallel.]

of observers regarded the matrix of hyaline cartilage as the prototype of an amorphous intercellular substance, and indeed very generally as the starting-point for its consideration. Max Schultze, on the other hand, opposed to this the hitherto little recorded views of Remak and Furstenburg, on the matrix of cartilage, and sought to show that we have not here to deal with an intercellular substance in the sense of a hardened secretion between the cells, but rather that the so-called intercellular substance, from its very commencement, proceeds from the protoplasm of the cells. This, in its turn, immediately led to renewed investigation respecting the generic significance of the matrix of bone, and of the fibrillar substance of connective tissue.

Max Schultze forthwith stated his opinion that the fibrillar substance of connective tissue originates from 'embryonal cells composed of protoplasm, and destitute of any investing membrane, which have amalgamated with one another.' A thin layer only of the protoplasm remains lying around the nucleus of the primary cell, representing with this nucleus a connective tissue cell, destitute of cell wall (connective tissue corpuscles). It should also be mentioned that, quite independently of the discussion maintained on these points in Germany, similar views respecting the development of connective tissue were expressed in England by Beale. According to Beale's peculiar terminology, connective tissue is originally composed of elementary parts (cells), which consist of germinal matter (Keimstoffe, protoplasm); but subsequently a part of the germinal matter is converted into formed material (in connective tissue the fibrillar substance), which was itself in the first instance germinal matter, and was developed at the cost of that matter. Beale, whose statements were of a somewhat general nature, admitted a similar genetic relation between the matrix of bone and cartilage, and the cells of those tissues.

Waldeyer especially endeavored to confirm these views, in the case of bone, by his beautiful researches on the process of ossification. It is obvious that, in the event of the above-described mode of development being demonstrated in the several cases of bone, cartilage, and connective tissue, a similar genetic agreement for all these tissues, though undoubtedly in a different sense from that advanced by Virchow, would also be obtained. But to what extent satisfactory replies have been given to these questions will hereafter receive consideration when these tissues are severally described.

As observers gradually acquired these views respecting the histogenesis of the connective tissue substances, a new starting-point for important general considerations respecting the living processes taking place in connective tissue was obtained, in quite another mode, by the investigation of living connective tissue. Von Recklinghausen demonstrated that, in living connective tissue, cells are present which agree in their characters with the white blood corpuscles (lymph or pus corpuscles), and, in consequence of the amoeboid movements they are capable of performing, constantly change their situation in the tissue. Von Recklinghausen further proved that when suppuration occurred in connective tissue, in opposition to the doctrine propounded by Virchow of the formation of pus by multiplication of the tissue cells, a migration of these movable cells from without into the substance of the tissue must be admitted to take place. These facts have attracted a proportionately greater interest since Stricker established the permeability of the walls of the vessels for red blood corpuscles. Cohnheim, indeed, has recently referred to the older observations of Waller on the relation of the white blood corpuscles in inflammation, which have hitherto remained unnoticed; and, supported by these older and his own more recent observations, has propounded the view that purulent infiltration really consists only in the migration of colorless blood cells through the vascular walls into the tissues. The relations thus shown to exist between the blood and the tissues must, as we shall see, still be held in view in discussing other questions bearing upon the connective tissue substances in the following pages. For this reason, the three typical connecting substances—connective tissue, cartilage, and bone—will now be separately described. The consideration of the peculiar tissue of the cornea, on the other hand, with dentine, and some others, will, on account of their more limited and special distribution in certain organs, be postponed to a later period.

Of Connective Tissues

A series of various forms of tissue must be included under the term connective tissue. This name was originally given in 1830, by Johann Muller, to the tela cellulosa of the older anatomists; but as at that time observers had already convinced themselves that this tissue is essentially composed of very fine fibres, which may be proved to be the chief constituent of tendons, ligaments, membranes, and other formed portions of the organism, all these tissues, together with the tela cellulosa, were included amongst those portions of the organism which are composed of connective tissue. Formerly, however, the description of this tissue was limited to a fibrous form of the tissue, possessing very definite histological and chemical characters.

This limitation has, however, been greatly extended by custom, and just as, in consequence of their functional agreement and continuity of substance, a series of microscopically different structures are combined under a common term—as muscle, nerve, etc.—we are on similar grounds led to a general application of the term connective tissue, and to distinguish its several forms. Amongst the microscopic morphological constituents thus distinguishable in connective tissue may be enumerated cells; networks and trabeculae, developed from cells consisting of peculiar delicate unbranched fibres (connective tissue fibrils), for the most united into fasciculi; and, lastly, fibres which are differentiated from those above named by the resistance they offer to the action of acetic acid and alkalies, by their repeated division, by their forming networks, and by their fusing into lamellae (elastic fibres).

Rollet's viewpoints on the connective tissues can be said to represent the teaching continuing into the first quarter of the present century until the recent past. One may quote from Maximow and Bloom (1930)[16] as evidence for this:

Common Connective Tissues

The Loose, Irregularly Arranged, Connective Tissue

This tissue is the direct product of a transformation of the mesenchyme which remains after all the other types of the connective tissue have been formed. Its study is especially important because it contains almost all the cellular and intercellular elements which occur in all the other kinds of connective tissue. It serves, therefore, as a prototype of the connective tissue in general. It fills out the spaces between the organs; it occurs in varying amounts in different places and penetrates, together with the blood vessels, into the interior of the organs. On macroscopical examination it is seen as a whitish, sticky, soft mass which is easily torn during dissection. When the organs are separated from one another it is stretched between them in thin membranes and threads. Like a collapsed sponge it contains innumerable small cavities or, better, potential cavities, which can be easily filled artificially with liquids or air and are the 'cells' of the old anatomists; they are responsible for the name 'areolar' tissue, which even now is sometimes used.

To study the loose, irregular connective tissue in the fresh, supravital condition, an artificial edema is produced by the injection into it of some drops of an isotonic salt solution faintly stained with neutral red. A small particle of the edematous mass is then placed under the microscope in the same solution.

The Intercellular Substance.—The intercellular substance forms the main mass of the tissue; three parts can be distinguished in it: (1) The collagenous or white fibers; (2) the elastic or yellow fibers, and (3) the amorphous ground substance.

The Collagenous or White Fibers.—These elements are the most characteristic constituent of all types of connective tissue. In the loose tissue they are long, straight or wavy threads or ribbons of varying thickness (1 to 12 μ), which run in all directions. Their ends cannot be found. In cross section they are but rarely round, most of them being rod or spindle shaped. Thus, their form, as a rule, is not cylindrical, but prismatic or flattened. The collagenous fibers are colorless and have a low refractive index; they always show a more or less distinct, longitudinal striation, while in cross sections they seem granular or dotted. The reason for this is that they consist of extremely fine, uniform, parallel, collagenous fibrils, 0·3 to 0·5 μ in thickness, which are believed to be held together by a cementing substance. The more fibrils in a fiber, the thicker it is. On the surface of the fiber the cement substance forms a thin membrane. While the fibrils are supposed not to branch, the fibers do branch in many places and separate the fibrils into smaller bundles. The fibers may even split up into single, isolated fibrils. After the cementing substance between the fibrils and the bundles is dissolved, through maceration in lime water or picric acid, the bundles disintegrate into separate fibrils.

Several observers have recently insisted that the idea of the absence of branchings in the fibrils is quite erroneous. They claim that in intensely stained slides the fibrils are seen to separate in many places from the fibers and to form an extremely fine network between nearby fibers by branching and anastomosing. The finest fibrils are at the limit of microscopical visibility; it is quite possible that these fibrils are really bundles of still finer fibrils.

The collagenous fibers are very soft and flexible; they offer a great resistance to a pulling force; they are not elastic in the common sense of the word. In polarized light both they and their fibrils show uniaxial, positive bi-refringence. In artificially stretched fibers this optical phenomenon becomes more prominent.

The collagenous substance is an albuminoid which, although giving some of the reactions of protein substances, is lacking in several chemical groups typical of these compounds. When boiled with water it dissolves and yields a colloidal solution of animal glue or gelatin, which on cooling becomes jelly-like. This property gave it the name of *collagen*. Weak acids and alkalis cause the collagenous fibers to swell. A very typical microscopical reaction is obtained with dilute acetic acid. It causes the bundles to swell considerably, lose their longitudinal striation, and become transparent; in many places these swollen bundles are constricted transversely or obliquely. A satisfactory explanation of this phenomenon has not been given as yet. The swelling may be due to the rupture of the membrane ensheathing the bundle, while those parts of the sheath which remain unruptured cut deeply into the soft, swelling, collagenous mass, and form the constrictions. Another explanation is based on the claim of the existence of the elastic fibers spirally surrounding the collagenous bundles.

Among other reactions, collagenous fibers are destroyed by the action of concentrated acids and alkalis. Strong acids are therefore used, in the process called maceration, to soften the tissues of different organs so that the cells may be separated from the interstitial connective tissue. The collagenous bundles are easily digested by pepsin in acid solution, but resist alkaline trypsin solutions. Collagen gives an insoluble product with salts of heavy metals and especially with tannic acid; the tanning of leather is based on the treatment of the collagenous feltwork of the skin with tannic acid. Collagenous fibers have no specific staining reactions; however, acid aniline dyes, as for instance the acid fuchsin of the van Gieson stain or the aniline blue in Mallory's mixture, stain the bundles sharply, especially after mordants. Collagen may present physical and chemical differences in various parts of the body and in different species of animals.

The Elastic or Yellow Fibers.—In the loose, irregular connective tissue, elastic fibers are scarce. They run in various directions; their beginnings and endings cannot be traced; they appear as brilliant, highly refractive, cylindrical threads or flat ribbons of varying thickness. In the loose connective tissue they are always much thinner than the collagenous fibers. In contrast to the latter, the elastic fibers are not fibrillar, but are usually homogeneous, although the larger fibers may stain more deeply at their periphery; they branch and anastomose freely and form a very loose network. If the tissue is fixed in its natural position the elastic fibers are straight, while in teased preparations they often appear wavy or spiral. They are highly elastic, that is, they yield easily to stretching, but when released at once reassume their former length. When assembled in large numbers they have a yellowish color on macroscopical examination.

The characteristic constituent of the elastic fiber, *elastin*, is also an albuminoid and may vary slightly in its qualities according to its origin. It is highly resistant toward boiling water, acids, and alkalis, and through the action of the latter it can be isolated from the other constituents of the tissue. Elastin is slowly digested by both pepsin and trypsin. Unlike the collagenous fibers, elastic fibers can be stained in a fairly selective way by certain dyes, as orcein or resorcin fuchsin.

The Amorphous Ground Substance.—Most investigators believe that the collagenous and elastic fibers are embedded in a jelly-like, amorphous substance which is related to the cement substance keeping the fibrils together in the fibers. In some places, as in the pulp of the tooth, it may form a continuous mass. Usually, however, it is believed to be arranged in thin lamellae or membranes, between which cleftlike spaces with a minute amount of tissue liquid remain. These appear as the 'cells' which can be filled with water or air. In cross section through the subcutaneous tissue, such lamellae are seen in many places as thin lines whose thickenings are the cross sections of the fibers embedded in the lamellae. The tissue cells are supposed to be adherent to the surface of the lamellae.

It is very difficult to demonstrate the amorphous substance. In a fresh preparation usually nothing is seen between the fiber and cells. A thin, pale film between the fibers can be noticed only after drying a sheet of loose connective tissue on a slide and staining it heavily.

Some recent investigators have questioned the existence of an amorphous substance in the adult. They claim that the fibers are not embedded in a special, amorphous membrane, but that the spaces between the fibers are filled with a continuous, usually liquid, colloidal mass. According to them, although the tissue, after the injection of liquid or the penetration of air into its substance, appears 'cellular' or 'areolar' with lamellae between the cavities, this is merely due to an artificial compression of the diffuse networks of fibers into more or less fibrillar sheets.

In many places, and especially at those surfaces where the connective tissue meets another kind of tissue, the consistency of the interfibrillar sol may increase and the tissue liquid become a gel of varying consistency. Through a similar condensation the membranes of the lamellated tissue arise. The nature of the amorphous ground substance requires further elucidation.

Basement membranes between epithelium and connective tissue always contain networks of collagenous, elastic, or reticular fibers and cannot be cleanly isolated from the connective tissue.

Modern study of the connective tissues began in 1934 with Sylvia Bensley's description[1] of a metachromatic ground substance, relating it to the mucus material described by the earlier authors. Bensley's observation of the localization or non-escape of paramecia injected into the sub-cutaneous tissue and McMasters[15] experiments with vital dyes and india ink, showing a continuous ground substance impeding free flow of injected material, have combined to demonstrate the functional integrity of the inter-cellular ground substance.

Histochemistry, as it developed, was applied to the connective tissues. The findings have been amply reviewed (McManus,[12] Fulmer[6]) and need not concern us here in detail. Suffice it to say that a carbohydrate component in the ground substance was amply confirmed by a variety of oxidation Schiff techniques (e.g. McManus[10,11]) and that acidic groups in the carbohydrates could be demonstrated by the Alcian Blue and Colloidal Iron methods (Mowry[18]).

During this time a very great deal of chemical investigation was undertaken to identify the particular configuration of the carbohydrates of connective tissue. Studies particularly by Meyer[17] and his associates, by Hale,[4] by Dorfman,[5] and by Schubert[20,21] have made it clear that the connective tissue carbohydrates include hyaluronic acid, at least three chondroitin sulphates, keratin sulphate, and perhaps other carbohydrate macromolecules somewhat resembling pectin, the polygalacturonide of plants.

As summarized by Schubert,[21] a very great deal of difficulty arises, when classical chemical separation techniques are applied to the carbohydrates of connective tissue. It is difficult to isolate a completely protein-free carbohydrate. It would seem that molecular weight determinations are full of error under these circumstances, but such have been made and reported.

At least part of the difficulty in the chemical study of connective tissue ground substance appears to be conceptual. For the classical chemist, the purposes of analyses are the isolation, identification, and characterization of the molecules in the material being studied. When he examines living tissues of the complexity of the connective tissues, the chemist is dealing with molecules with bindings and degrees of hydration yet to be determined, and quite possibly possessing characteristics requiring new sets of pre-suppositions for their understanding. Living material has water for the continuous phase of all metabolic activities. Degrees of hydration of ions in the biological systems may differ, and the same is true of the hydration of the molecules. The chemist's methods are essentially dehydrating and disruptive, and destroy the very materials about which information is sought. Histochemistry has certain advantages in this regard, in that the histology and cytology are preserved, to a degree, in the tissue in which the reactions are demonstrated, and on which the reactions are performed. The relative insensitivity of histochemical methods, requiring fairly large amounts of material for a positive reaction, have limited their usefulness, as has the fact that histochemical reactions are often group reactions: e.g. the PAS reaction for 1, 2 glycols, whether in glycolipids, glycoproteins or carbohydrate (in McManus and Mowry[14]).

If one were to attempt to summarize the findings in the chemistry of the connective tissues to date it would be difficult to escape the conclusion that pure substances exist in the mind of the chemist, and, perhaps after the destructive handling of tissues in the test tube, but not in nature. The consistent presence of water in each portion of living material, and the pervasive occurrence of physical and chemical bindings between the chemical constituents of living matter, result in an organized complexity with which the techniques and the concepts of the chemist are at present unable to cope. Indications of fragments of the complexity are beginning to appear, in the studies of the ions and polyelectrolytes in the connective tissues (Catchpole and colleagues,[4] see Bland[3]) where the complex carbohydrates serve as polyanions. It is probable that the dynamic state of the physiology of the connective tissues will become more important in our consideration of water and electrolyte balance in the organism, and, in time, in medicine and pathology.

Electron microscopy has been applied to the study of the connective tissues. Most success has been obtained with the fibrillar elements, and with the cells, where consistent patterns of appearance have been obtained. The relationship of collagen to the reticulin fibrils seems clear, with an identical periodicity of banding being possessed by both. A characteristic ultrastructure has not as yet been obtained for the elastic fibres of classical histology, although recent studies of the microfibrils (cf. Haust[8]) suggest that variety of ultrastructure is the rule.

The ground substance of the connective tissues has been much studied by electron microscopy, but little thought has been given to the effects of the conditions required for the study. The deformation of a connective tissue spread, when immersed in the usual fixatives (often mixtures of osmium), can be observed directly. A change in the consistency of the tissue is immediate, and this signifies the chemical and physical alterations which the tissues undergo. What happens to the 'bound' water, to the ions and electrolytes, under these circumstances, has not been studied to my knowledge, and is quite unknown. This distorted and disturbed tissue is then embedded in plastic, cut on an ultramicrotome onto acetone water, and then examined in vacuo in the electron microscope. The interpretation of the ultrastructure of the ground substance, in terms of 'canals', 'connections', the relationship of ground substance to fibrils, and so on, under these conditions seems hazardous if not frankly hallucinogenic.*

* [*Editorial note*: We hope that this book may help to show that without hazardous, even hallucinogenic interpretations, and open-mindedness to new approaches (despite the often disturbing facts they produce), the progress of cytology would be much slower.]

It is only fair to say that the methods of classical histology are quite as inadequate as those of electron microscopy. The present emphasis is on the important ions, electrolytes, and water of the connective tissues, while the classical histological methods of fixation, dehydration, embedding, sectioning and staining, etc. could not be planned better to disturb the relationships of these materials. The freeze-dry methods originating with Altmann and recently revived in various forms, have an initial promise of obviating very many of these difficulties, but freezing and drying certainly disturb the consistency (i.e the physical chemistry) of the tissue. I will be prepared to believe that freeze-drying does not disturb the chemistry of a tissue when it is proven to me that the dry paint on the wall has the same chemical constitution as that same paint, fluid in the can.

It is clear that a very rapid freezing process may be used to arrest living tissues in a form resembling the active state. What has not been developed as yet, and what is needed badly, is a method of microscopic study of the still frozen tissue with the crystals still in place.* Sections could be cut by cryostat, and the tissue examined on the cold stage of the microscope, in the cold, with a search being made for particular special conditions of microscopy such as wave-length variation from ultraviolet to infra-red, etc. which would allow the visualization of the tissue and cellular structure with the water, ions and electrolytes in place, and cellular relationship undisturbed.

In this review of the changing concepts of the connective tissues, we have seen the initial descriptive phases using the available methods of the times change the naked-eye studies of Bichat[2] being replaced by the microscopy of Schwann,[22] Kolliker[9] and Virchow.[23] Comparative studies, and embryological investigations, resulted, in time, in the viewpoints of Rollet,[19] and culminated, in the recent past, in the descriptions of Maximow and Bloom.[16] The more recent dynamic concept of the connective tissues has again utilized the contemporarily available methods: histochemistry, biochemistry, and electron microscopy. The informational limitations of histochemistry, and the technical and conceptual difficulties of biochemistry, and electron microscopy, have delayed the development of an adequate contemporary appreciation of the connective tissues.

It now seems clear that the connective tissues are a reservoir for the body water, ions, and electrolytes; that they serve as an intermediate area for cellular building blocks and metabolites; and that they are in continual dynamic interchange with the closed circulation, on the one hand, and with the body cells on the other (cf. McManus[13]). We know a little, perhaps a very little, about the structure, and ultrastructure, and about the chemistry of the connective tissues; but very much more needs to be known before a whole and adequate concept of the connective tissues can be obtained.

In investigations of the connective tissues, it seems clear that the conceptual aspects of the problem have been determined in very large degree, by the technical capacities and capabilities of the time. With advancing technology, new methods of study will be developed. This will in time allow an understanding of the constituent and interacting molecules of the living connective tissues at the levels where structure and function are one.

REFERENCES

1. BENSLEY, S. H. 1934. On the presence, properties and distribution of the intercellular ground substance of loose connective tissue. *Anat. Rec.* **60**, 93
2. BICHAT, XAVIER. 1822. *General Anatomy, applied to Physiology and Medicine* (trans. by George Hayward), Vol. I. Boston: Richardson and Lord
3. BLAND, JOHN (Editor). 1966. *Biophysics and Biochem. of Connective Tissue.* Seminar, Fed. Proc. **25**, 937
4. CATCHPOLE, H. R. and colleagues. In Bland (1966) above
5. DORFMAN, A. and MATHEWS, M. B. 1956. The physiology of connective tissue. *Ann. Rev. Physiol.* **18**, 69
6. FULLMER, H. M. 1965. Histochemistry of connective tissue. In *Internat. Rev. Conn. Tissue Res.* (edited by D. A. Hall), Vol. **3**. New York, London: Academic Press

* [*Editorial note:* The remarkable present results, and the undoubted future development of the freeze-etch procedure in electron microscopy, form a close approximation to these requirements; but, undoubtedly, *direct* observation of frozen tissue (never thawed), with both light and electron microscopical techniques, is an exciting future technical problem, and necessity.]

7. HALE, C. W. 1946. Histochemical demonstration of acid polysaccharides in animal tissues. *Nature, Lond.* **157**, 802

8. HAUST, M. D. 1965. Fine fibrils of extracellular space (microfibrils). *Am. J. Path.* **XLVII**, 1113

9. KOLLIKER, A. 1853. *Manual of Human Histology* (trans. and edited by George Busk and Thomas Huxley), Vol. 1. London: The Sydenham Soc.

10. MCMANUS, J. F. A. 1946. Histological demonstration of mucin after periodic acid. *Nature, Lond.* **158**, 202

11. ——. 1948. Histological and histochemical uses of periodic acid. *Stain Technol.* **23**, 99

12. ——. 1954. Histochemistry of connective tissue. In *Connective Tissue in Health and Disease* (edited by G. Asboe-Hansen), 31. Copenhagen: Munksgaard

13. ——. 1966. Chapter 15, The connective tissues and the internal environment. In *General Pathology: The Biological Aspects of Disease*. Chicago: Year Book Medical Publishers

14. —— and MOWRY, R. W. 1960. *Staining Methods. Histologic and Histochemical*. New York: Paul B. Hoeber, Inc.

15. MCMASTER, PH. D. 1941. An inquiry into the structural conditions affecting fluid transport in the interstitial tissue of the skin. *J. Exper. Med.* **74**, 9

16. MAXIMOW, A. A. and BLOOM, W. 1930. *A Textbook of Histology*. Phila: W. B. Saunders

17. MEYER, K. 1958. Chemical structure of hyaluronic acid. *Fed. Proc.* **17**, 1075

18. MOWRY, R. W. 1963. The special value of methods that color both acidic and vicinal hydroxyl groups in the histochemical study of mucins. *Annals of New York Acad. of Sciences* **106**, 402

19. ROLLET, A. 1872. The connective tissues. Chapter in *Manual of Histology* (edited by F. Stricker). By various authors. Translated by A. H. Buck *et al.* New York: William Wood and Co.

20. SCHUBERT, M. 1964. Intercellular macromolecules containing polysaccharide. In *Connective Tissue: Intercellular Macromolecules*. Proc. of a Symposium by the New York Heart Assoc. Boston, 1964; also publ. as *Suppl. Biophys. J.* **4**, No. 1, part 2; 119

21. —— in Bland (1966) above

22. SCHWANN, TH. 1847. *Microscopical Researches into the Accordance in the Structure and Growth of Animals and Plants* (translated by Henry Smith). London: Sydenham Soc.

23. VIRCHOW, RUDOLF. 1860. *Cellular Pathology* (translated by Frank Chance). London: John Churchill

Chapter 18

Apparent Intracellular Collagen Synthesis

G. A. MEEK

INTRODUCTION

Collagen, the 'glue-former', is the commonest protein in the vertebrate organism. The study of collagen secretion is currently receiving a great deal of attention, for it may turn out that disorders of this process give rise to the family of crippling diseases known as 'collagen diseases', the best known of which are rheumatism and arthritis. The problem is peculiarly suited to the modern techniques of ultrastructural and biochemical research. These techniques, over the past ten years, have yielded a vast amount of information concerning the secretion of vertebrate collagen. Many problems, however, remain to be solved. One of these problems is whether collagen is ever laid down within the cytoplasm of cells, or whether, as the current hypotheses have it, it is invariably laid down extracellularly from precursor materials synthesized and secreted by cells. This paper reports some observations which have been made on the connective tissues of an invertebrate, the common garden snail, *Helix aspersa*—which appear to demonstrate that this animal is capable of synthesizing fully formed collagen fibres within the cytoplasm of some of its connective tissue cells.

Connective tissue is basically a matrix of 'non-living' substance or 'ground substance', containing living cells which manufacture the constituents of the ground substance. The 'grounds substance' can be classified into fibrous and non-fibrous components. The non-fibrous components are: tissue fluid together with mucopolysaccharide-protein complexes, such as hyaluronic acid and chondroitin sulphate; the fibrous components are: collagen ('reticulin' has the same structure as collagen as revealed by the electron microscope[10]), and elastin, which does not show the characteristic periodic banding of collagen under the electron microscope.[18] Only collagen will be considered in this paper. The cells which secrete the ground substance are generally known as 'fibroblasts'; specialized fibroblasts are called 'chondroblasts' in dense connective tissue, and 'osteoblasts' in bone. All these cell types have a very similar ultrastructural morphology.[6, 9, 23, 27, 32] All cell types believed to be responsible for the formation of collagen fibres will be referred to collectively in the following text as fibroblasts.

One of the more puzzling features of the present hypothetical mechanism for the synthesis of collagen by fibroblasts is the absence of any evidence for the formation of mature collagen fibres within the cytoplasm of these cells. There appears to be no *a priori* reason why this should be so. The pioneering work of F. O. Schmitt[31] and his co-worker J. Gross,[10, 11] who were the first workers to apply the electron microscope to the examination of the structure of collagen, showed that it has a unique banded structure. In vertebrates it has a periodicity of approximately 640 Å, which enables it to be recognized amongst all the other sub-optical microscopical fibres which the electron microscope has revealed, in all the fibrous connective tissues so far studied. Since it bears this 'fingerprint', collagen can be identified with reasonable certainty both in teased preparations and in ultrathin sections. The fact that this banded structure has never been clearly demonstrated within the cytoplasm of fibroblasts is widely known, and biological electron microscopists tend to be on the lookout for it. The vast majority of investigators, however, examine vertebrate material, so it therefore appears to be an established fact that the fibroblasts of higher animals do not lay down mature collagen fibres within their cytoplasm. Mature, banded collagen is frequently seen in very close association with the plasma membrane of fibroblasts, but, so far as the author is aware, this collagen always appears to be extracellular. In order to explain this observation, the following mechanism has been postulated and is now generally accepted.

Collagen is believed to be built up from a long-chain protein monomer called 'tropocollagen', which has been shown to be composed of three amino acid chains twisted into a triple helix approximately 2800 Å long by 14 Å wide. The amino acid composition of tropocollagen is unusual, since it contains a very high proportion of hydroxyproline, an amino acid seldom found in other proteins, together with a high proportion of glycine. For a short review of the biochemical and morphological evidence for the existence of tropocollagen, the reader is referred to Gross.[11] It is believed to be the precursor of collagen, and is synthesized by the ribosomes of the ergastoplasm of the fibroblasts. It is secreted as such into the extracellular milieu, and is then built up in an overlapping, staggered fashion like the fibrils of a rope, to give rise to the typically banded mature collagen fibres. That such polymerization can take place extra-cellularly has been demonstrated by Cox and others,[7] who has been able to show *in vitro* successive stages in the assembly of striated collagen fibres from tropocollagen molecules by the technique of negative staining.

Evidence supporting the hypothesis of intracellular tropocollagen synthesis and secretion, followed by extracellular collagen formation, has been put forward by Revel and Hay[26] and Ross and Benditt,[28] who have used the technique of autoradiography at the electron microscope level, to follow the path of synthesis of collagen precursor within the cytoplasm of fibroblasts. This technique is used to follow the passage of an amino-acid through the cell by labelling it with the hydrogen isotope tritium, which emits β-particles of very low energy, suited to localization by electron microscope autoradiography. The parti-cular amino-acid chosen for labelling has, so far, been proline. It must be stressed, however, that proline is by no means a specific precursor of collagen; most other proteins have been demonstrated to contain small proportions of it.[17] Hydroxyproline is far more specific, but does not appear to be incorporated into body protein.[35] Collagen also contains a greater proportion of glycine than other proteins, but a labelled form of glycine has not been used in the experiments, presumably also because of its non-specificity. Revel and Hay[26] injected tritiated proline into developing salamander larvae and examined the connec-tive tissue cells of the limb buds, by fixing them 15 min after injection, and preparing electron microscope autoradiographs. These showed silver grains localized over the cytoplasm of the chondrocytes, particu-larly over the granular endoplasmic reticulum. Owing to the difficulties of resolution associated with this method, it is hard to be certain of the exact localization of the label, but the authors claim to have found it mainly over the cisternae of the endoplasmic reticulum. If the tissue was fixed 30 min after injection of the labelled precursor, grains were now seen clearly over the Golgi region of the cells, particularly over the large vacuoles. If the tissues were fixed 2 hr after injection, very little label was seen over the chondro-cytes, but many grains could be seen over the extracellular matrix, which contained striated collagen fibres. Ross and Benditt[28] have made a similar study, using guinea pig fibroblasts. They wounded animals by making skin incisions, and then allowed them to heal. Seven days after the initial wounding, tritiated proline was injected intraperitoneally, and parts of the healing wounds were removed at inter-vals ranging from 15 min to 24 hr. Electron microscope autoradiographs were then made of the wound specimens, and micrographs were made, at random, at relatively low powers, of fields containing fibro-blasts and extracellular collagenous material. The relationship of the grains to the cell organelles was determined by superimposing a grid over the micrographs, and making grain counts. The quantitative accuracy of this method is open to criticism, because the grain yield of a tritiated ultrathin section is directly proportional to section thickness, which can vary over wide limits in any individual section, due to the differential hardness of the embedded tissue.[36] However, with these reservations, the results of Ross and Benditt are reasonably consistent, and depict the sequence of events after the administration of the tritiated proline. The concentration in the ergastoplasm was maximal after 15 min, remained steady for about 2 hr, and then dropped steadily throughout the remainder of the 24 hr. The concentration in the Golgi complex rose to a maximum after 2 hr, and then dropped steadily. The concentration in the large cisternae of the ergastoplasm, close to the periphery of the cell, rose to a maximum after 4 hr, and thereafter dropped steadily. The concentration in the collagen-containing extra-cellular matrix was very small in the first 30 min, but rose rapidly to a maximum at 4 hr, after which it remained steady or rose very slowly.

The interpretation placed on the results of these investigations is that the collagen precursor, proline, is taken up by the fibroblast, either by pinocytosis, or by diffusion through the plasma membrane. It then enters the protein-synthesizing ribosome-ergastoplasm complex, where it becomes incorporated into a collagen precursor, presumably tropocollagen. It is then transferred to the Golgi complex, where the collagen precursor is concentrated in vacuoles bounded by non-granular membranes; these secretory vacuoles then approach the periphery of the cell, fuse with the plasma membrane, and the contents are discharged into the extracellular milieu. Finally, the soluble collagen precursor polymerizes, in close proximity to the surface of the secretory cell, into banded fibrils, recognizable as mature collagen. Some minor points of disagreement remain; Ross and Benditt[28] maintain that the soluble precursor can be secreted directly from the cisternae of the endoplasmic reticulum, by fusion of the granular membranes with the plasma membrane, a result not in accordance with the findings of Porter,[23] Revel and Hay[26] and Goldberg and Green.[9] However, these autoradiographic results show that the process is apparently one of merocrine secretion, i.e. the cell secretes a substance which it has manufactured, but loses none of its actual cytoplasm in the process. This, again, is at variance with the findings of other workers in this field,[24, 6] who maintain that the evidence from ultrastructural morphology points to a process of apocrine secretion, in which cytoplasmic processes containing the secretion product are shed by a pinching-off process, or by deposition of the peripheral part of the cell. The classical observations of Stearns,[33, 34] who followed the deposition of collagen by fibroblasts in a rabbit ear chamber with the light microscope, point to apocrine secretion. Porter[23] has published micrographs of fibroblasts in a hepatoma, in which small vesicles are seen in linear formation across cytoplasmic blebs, and infers that the fusion of these vesicles would separate the bleb from the body of the parent cell as a form of apocrine secretion.

To summarize then, the majority of published evidence supports the hypothesis that the vertebrate fibroblast synthesizes a soluble precursor of collagen in the ergastoplasm; that this product is amassed within cytoplasmic vesicles; and that the product is secreted into the extracellular milieu by a merocrine or apocrine process. Mature collagen fibres are believed not to exist within the cytoplasm of vertebrate fibroblasts.

In contrast to the vast amount of work which has been done on the structure and synthesis of collagen by vertebrates, very little similar work has been done on invertebrate collagen. A review of this field by Dr. Doreen Ashhurst is published elsewhere in this book, to which the reader is directed for further information. One significant fact emerges: the periodicity of the repeated banding is less than is found in vertebrate collagen, and in some cases the banding has been reported to be lacking completely. Reed and Rudall[25] originally reported that banding was absent in teased, shadowed preparations of collagen from earthworm cuticle; but a later report by Ruska and Ruska,[29] using sectioned material from the same source shows clearly that banding is present, at a periodicity of 560 Å. Pedersen[21] reports the complete absence of striations in sections of planarian connective tissue, but a much earlier report by Schmitt, Hall, and Jakus[31] shows the collagen from cephalopod molluscs to be striated in shadowed preparations, with a periodicity of 640 Å. A report by Bradbury and Meek[5] on the structure of connective tissue fibres in the leech, *Hirudo medicinalis*, shows that banding could not be distinguished in sections of the material, but an ill-defined banding of about 300 Å periodicity was demonstrated in shadowed sections from which the embedding medium had been removed by a solvent. It is possible that the use of Araldite, rather than methacrylate, as an embedding medium, would presrve the banded structure, which could then be visualized by modern staining techniques. X-ray diffraction data, however, as distinct from electron microscope data, have indicated spacing of around 640 Å (normal vertebrate spacing) in certain species of Echinodermata, Coelenterata, and Porifera.[15]

There is even less published literature on the process of fibrogenesis in the invertebrates than there is on the structure of invertebrate collagen. The paper by Bradbury and Meek,[5] already referred to, was mainly concerned with fibrogenesis, and showed that the fibrils of the ground substance appear to 'shred off' the surface of the fibroblasts; evidence pointing to merocrine rather than apocrine secretion.

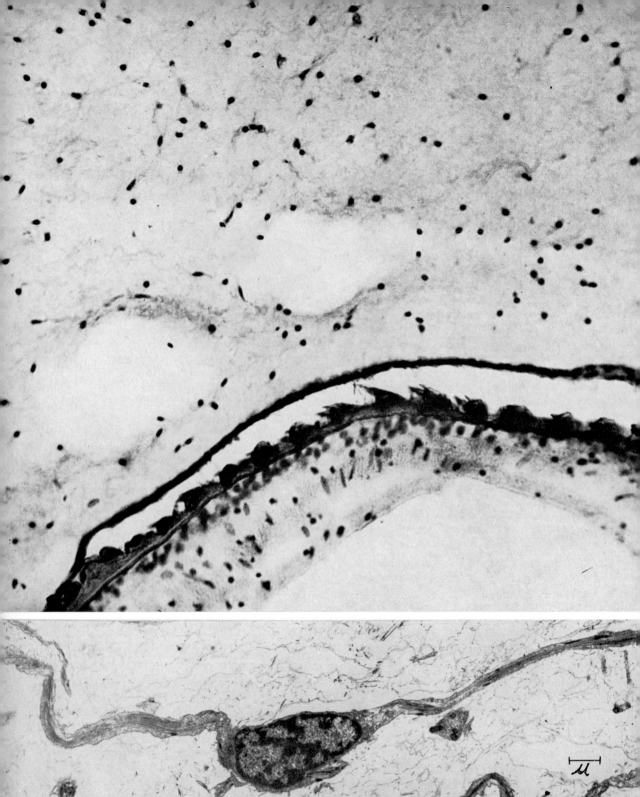

Fig. 18.1 (*above*) Light micrograph of a paraffin section of a hibernating snail radula sac, stained with H. & E. The upper part is the cartilage matrix, supporting the serrated radula. Long cell processes can be seen ramifying throughout the matrix substance. ×65

Fig. 18.2 (*below*) Electron micrograph of a bipolar cell from the same tissue as Fig. 18.1, embedded in Araldite and stained with phosphotungstic acid plus lead. Fibrils can be seen in both processes. ×7500.

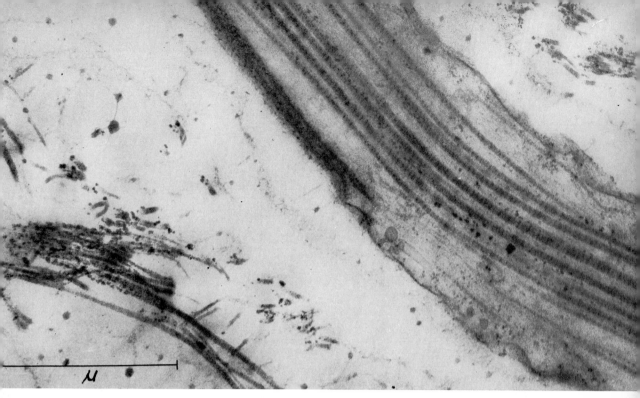

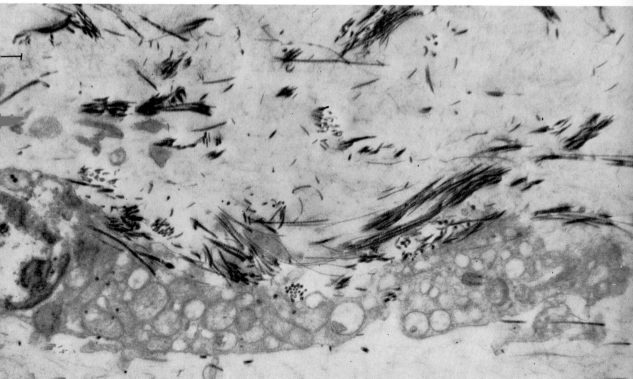

Fig. 18.3 (*above*) The distal end of a cell process showing intracytoplasmic, asymmetrically striated fibrils with a periodicity of 400 to 500 Å. Extracellular fibrils running more or less parallel to the process and which appear to be morphologically identical, can be seen in the matrix material surrounding the process. × 58,000.

Fig. 18.4 (*below*) Part of a fibroblast process close to the nucleus. Collagen fibrils are very closely associated with the cytoplasm on the upper edge of the cell; no cell membrane is visible in this region. On the lower side, the cell membrane is clearly visible; no association of collagen fibrils with the cell surface can be seen. Some fibrils are clearly extracellular, others are clearly intracellular. × 15,000.

Ashhurst has studied fibrogenesis in locust embryos[2] and pupal wax moths.[1] Her observations support the hypothesis of merocrine secretion. She reports specifically that she did not see any intracellular banded fibrils, but describes banded intracellular cytotubules.[14]

It appears, therefore, that there are a number of questions regarding the structure and biosynthesis of collagen in invertebrates which require answers. Most important of these are: (a) is invertebrate collagen morphologically the same as vertebrate collagen? and (b) is the process of synthesis and secretion the same as it is in vertebrates?

RESULTS

During the course of studies on the cerebral ganglia of snails, the author noticed that sections of the connective tissue sheath surrounding the ganglia appeared to show fibroblasts containing intracellular periodically striated fibres; and also, that the collagen, especially in material which had been block stained in phosphotungstic acid/absolute alcohol, showed clearly marked striations, of a periodicity varying between 400 and 500 Å. Negatively stained preparations of teased collagen fibres showed a periodicity of about 470 Å. This suggested, in agreement with previous findings, that vertebrate and invertebrate collagens may differ morphologically, and indicated that an investigation into the connective tissue of the snail might be worth while. Accordingly, other tissues containing collagen, e.g. foot muscle, oviduct, and so on were examined. The radula sac cartilage received particular attention, since this is the only part of the animal referred to specifically as 'cartilage' in the literature.[4] It was found to contain an abundance of bipolar and stellate cells (Fig. 18.1) embedded in a hyaline matrix, which stained heavily after the periodic acid-Schiff test (PAS). The cells have long processes running through the hyaline ground substance, some of which are seen to be birefringent under the polarizing microscope. These processes stain bright green after the Masson technique, blue after Mallory, red after Lison's Alcian blue/chlorantine fast red, and reddish brown after van Gieson's method, which indicates that collagen may be present in the cell processes. When the material is fixed in buffered osmium tetroxide solution, or 3 per cent glutaraldehyde followed by post-osmication, and stained in a 1 per cent solution of phosphotungstic acid in absolute alcohol, as the final stage of dehydration (a method which stains collagen fibres heavily[13]), and then is examined in the electron microscope, some of the cell processes are clearly seen to contain fibrils (Fig. 18.2). When these intracellular fibrils are examined at higher magnification, they are clearly seen to be cross-striated (Fig. 18.3), with a major banding periodicity of approximately 500 Å. On comparing the intracellular fibrils with the extracellular fibrils which are found free in the matrix of the ground substance (Figs. 18.3, 18.4 and 18.5), the morphological correspondence is most striking. The periodicity of the extracellular fibrils, when measured in sections, varies from 400 to 500 Å; when measured in teased material after negative staining it averages 470 Å. The precise number of inter-bands is difficult to determine in the case of the intracellular fibrils, since these can only be studied in sectioned material, and even in the thinnest sections, the banding is not as sharp or as even as is the case with the extracellular collagen fibrils. The intracellular fibrils give the impression of being less mature, and at an earlier stage of formation, than the fully-formed matrix fibrils. The extracellular collagen fibrils are frequently seen in very close association with the surface of the cell processes containing the intracellular fibrils. In many places, the cell membrane appears to be incomplete, and the extracellular fibrils appear to be in continuity with the cytoplasmic process (Figs. 18.4, 18.5, and 18.6). It is always possible of course, that this appearance is simply an artifact due to oblique sectioning of the plasma membrane of the cell. However, the appearance is so frequently observed, that one is left with the impression that this cannot always be the explanation. Cell processes, especially those distal from the nucleus, often present an appearance which can only be described as 'disintegrating' (Fig. 18.4), in which intra- and extracellular banded fibrils appear parallel with one another. So many breaks are apparent in the cell membrane, that they can hardly be explained by invoking artifacts of sectioning.

It must be emphasized, that of all the material so far examined, banded intracellular fibrils are *best* seen in the radula sac cartilage tissue of hibernating snails. The cells do not contain the abundant vesicular cytoplasm associated with active fibroblasts, but resemble rather the dormant 'fibrocytes' described by

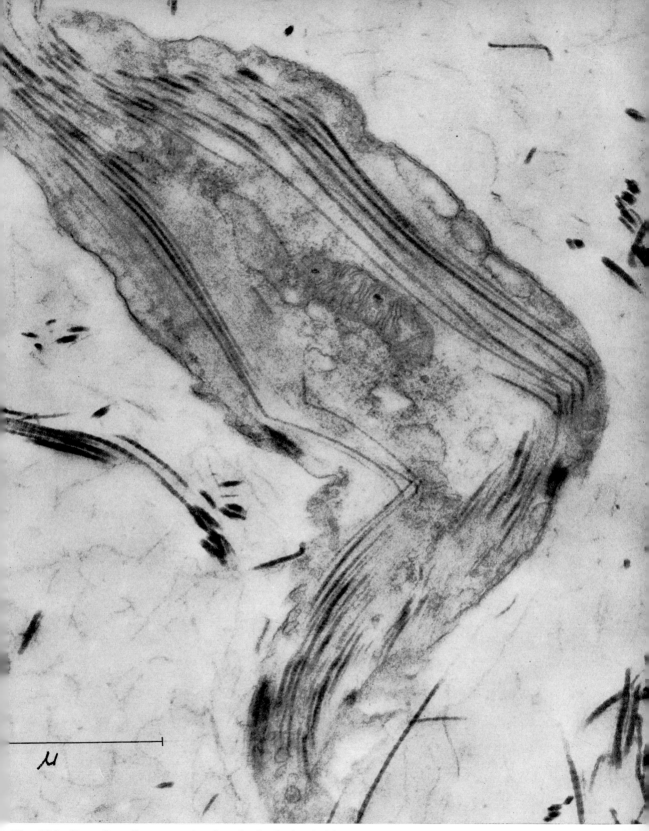

Fig. 18.5 Part of a cell process, showing clearly the banded intracellular fibrils and their close relationship with other cell organelles, such as mitochondria, endoplasmic reticulum, and ribosomes. × 48,000.

Porter.[23] These have long, tenuous processes, in which the volume of cytoplasm is strikingly reduced when compared to active fibroblasts, the ribosomal particles are fewer, and the Golgi membranes insignificant to the point of vanishing.

In contrast to the appearance of fibrocytes in the hibernating animal, the same cells in the active animal present an appearance much more like that seen in vertebrate fibroblasts. (It must be assumed that these are the same cells, since the tissue appears to contain only one cell type.) The membranes of the endoplasmic reticulum are studded with numerous ribosomes; the cisternae form vacuoles, especially close to the nucleus; there is an abundance of free ribosomes, together with large, electron-dense bodies bounded by a single membrane (Fig. 18.6). These bodies frequently display an internal structure very similar to the internal structure of the acid phosphatase-containing inclusions described as lysosomes by Meek and Lane[16] in the neurons of the same animal. The processes of these cells frequently show cell membrane discontinuities, at which it is impossible to distinguish whether the collagen fibril matrix (in which the cells lie) is continuous with the cytoplasm or not. The appearance shown in Fig. 18.6 is typical. It is possibly an artifact resulting from the close apposition of two fibroblast processes. The cell membrane at the point of contact may be sectioned so obliquely that it is not visible. The problem of interpretation here can be stated thus: are the collagen fibrils seen in the centre of the micrograph part of the cytoplasm of a single fibroblast process bounded by a single cell membrane, which can clearly be distinguished on the outer sides; or do we have here two separate cell processes, one containing dense bodies and ribosomal aggregates, and the other containing granular endoplasmic reticulum membranes, plus mitochondria and many free ribosomes? If the latter interpretation is accepted, where are the two cell membranes in between the processes? If the arrangement of the collagen fibrils in the centre of the micrograph is closely studied, it does not seem possible for a cell membrane to be present, however obliquely sectioned it may have been. If it is accepted that no cell membrane is present, how can the conclusion be avoided that the fibrils are intracellular, and are almost certainly part of a single cell process?

Further evidence in favour of the possibility of intracellular fibrils is the presence of cytotubules in the cytoplasmic processes of the active fibroblasts. These cytoplasmic inclusions were first described in plant cells by Ledbetter and Porter,[14] and are known to be a component of many vertebrate cells after glutaraldehyde fixation and post-osmication. Tubules around 200 Å in diameter are a prominent feature in fibroblasts of the summer snail; they run parallel with the direction of the processes, and preliminary results indicate that they may become banded and merge into the banded fibrils. This finding is in accordance with a report by Ashhurst[1] of the finding of banded cytotubules in the cytoplasm of fibroblasts of the wax moth, which she describes as 'fibrils with hollow cores', 125 to 200 Å in diameter, and which sometimes show a banding, spaced at 150 to 200 Å. These dimensions are, as nearly as can be measured, identical with those found in *Helix*.

In order to prove this hypothesis of intracellular collagen, it has to be demonstrated beyond reasonable doubt that (a) the cells are in fact fibroblasts or fibrocytes; and (b) that the banded fibrils are in fact collagen. The whole weight of evidence supports the thesis that the cells must be fibroblasts. The use of the various optical staining methods already referred to, indicates that the cells lie in a matrix of mucopolysaccharide, bound together by collagen fibres. The tissue itself appears to have no function other than that of supporting the developing radula, although observations on transparent, newly hatched animals, while feeding, indicate that it may be used to plug the oesophagus while the animal is masticating its food. The only apparent function of the cells is to secrete the mucoprotein-collagen tissue in which they are embedded. A very remote possibility exists that they may be related to developing muscle cells, since it has been reported[30] and confirmed by the author that a periodicity of around 250 Å can, with some difficulty, be demonstrated in the fibres of some invertebrate smooth muscle cells. It is well known that intracellular muscle fibrils exhibit periodicity (for a review see Hodge[12]) and that invertebrate smooth muscle gives X-ray diffraction patterns corresponding to an axial spacing of about 400 Å.[3] Myosin and related muscle proteins, from both vertebrate and invertebrate sources, can be prepared in such a way as to give rise to periodic striations, which are, on superficial inspection not dissimilar to collagen. However, the banding is of a shorter periodicity; it is symmetrical, and has far less complex

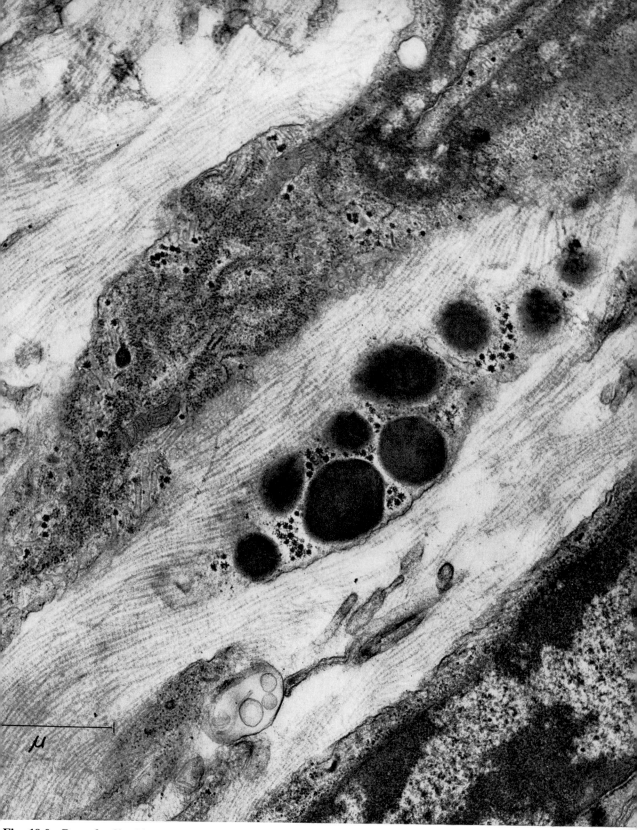

Fig. 18.6 Part of a fibroblast process from an active, summer snail. Masses of collagen fibrils appear to originate from the dense aggregation of ribosomes in the upper right-hand area of the micrograph, and follow a course parallel to the process until they cannot be distinguished from the extracellular fibrils. See text for further discussion of this micrograph. × 34,000.

interbanding. This raises a very interesting point, because Pease[19,20] has suggested that the smooth muscle cells of vertebrate arteries are capable of producing collagen, and also Pineda[22] has suggested that the collagen in peripheral nerve tumours must be derived from Schwann cells. It is therefore a possibility that both fibroblasts, and myoblasts in the snail, have a common precursor cell which produce banded intracellular fibrils. There seems, however, to be no very obvious reason why muscle cells should be present in a supporting tissue such as the snail radula sac cartilage.

The mode of secretion by these cells of the intracellular collagen fibrils appears to be apocrine. The processes of the active cells shed their cytoplasm, blebs of which are frequently seen amongst the collagen fibres (see Figs. 18.4 and 18.6), leaving behind bundles of collagen fibrils. The cell membrane seems to disappear first; if the dense bodies in the processes are in fact lysosomes, they may be causing the apocrine secretion process by release of their hydrolytic enzymes. Whether or not the whole cell disintegrates (holocrine secretion) cannot as yet be stated, but pycnotic nuclei, stripped of cytoplasm, are occasionally seen in the matrix material. The evidence as a whole suggests that apocrine secretion is more likely than holocrine.

To establish beyond doubt that the cells are fibroblasts, and the intracellular striated fibrils are collagen, is a difficult problem. Biochemical evidence would be almost impossible to obtain, since the volume of material in a radula sac cartilage is less than 1 mm.[3] Experiments are therefore in progress in this laboratory, designed to follow the incorporation of radioactive collagen precursors into the cells and to study the effects of enzymes, e.g. collagenase, upon them. Preliminary results indicate that tritiated proline is taken up by snail connective tissue cells, but that the product appears to be concentrated in the mucopolysaccharide slime-forming tissues rather than in the structural proteins of the animal.

One conclusion, however, is certain. The processes by which invertebrates synthesize their collagen are not the same as the corresponding process in vertebrates—a not surprising conclusion. Whether the difference is fundamental or only one of detail remains to be discovered.

REFERENCES

1. ASHHURST, D. 1964. Fibrillogenesis in the wax-moth, Galleria mellonella. *Quart. J. Micr. Sci.* **105**, 39
2. ———. 1965. The connective tissue sheath of the locust nervous system: its development in the embryo. *Quart. J. Micr. Sci.* **106**, 61
3. BEAR, R. S. 1945. Small-angle diffraction studies on muscle. *J. Amer. Chem. Soc.* **65**, 1625
4. BORRADAILE, L. A., EASTHAM, L. E. S., POTTS, F. A., and SAUNDERS, J. T. 1959. *The Invertebrata*, 3rd edition, p. 607. Cambridge University Press
5. BRADBURY, S. and MEEK, G. A. 1958. A study of fibrogenesis in the Leech, *Hirudo medicinalis. Quart. J. Micr. Sci.* **99**, 143
6. CHAPMAN, J. A. 1962. Fibroblasts and collagen. *Brit. Med. Bull.* **18**, 223
7. COX, R. W., GRANT, R. A., and HORNE, R. W. Native collagen fibrils and their formation from tropo collagen. *J. Roy. Micr. Soc.*, **87**, 123
8. FITTON JACKSON, SYLVIA. 1964. Connective tissue cells. In *The Cell* (edited J. Brachet and A. E. Mirsky) Vol. VI, p. 387. Academic Press
9. GOLDBERG, B. and GREEN, H. 1964. An analysis of collagen secretion by established mouse fibroblast lines. *J. Cell Biol.* **22**, 227
10. GROSS, J. 1950. A study of certain connective tissue constituents with the electron microscope. *Ann. New York Acad. Sci.* **52**, 964
11. ———. 1961. Collagen. *Scientific American* **204**, 120
12. HODGE, A. J. 1958. Principles of ordering in fibrous systems. *Proc. IV Int. Cong. For Electron Microscopy, Berlin*, p. 119
13. JAKUS, MARIE. 1956. Studies on the cornea: II. The fine structure of Descemet's membrane. *J. Biophys. Biochem. Cytol.* **2** supp., 243
14. LEDBETTER, M. C. and PORTER, K. R. 1963. A 'microtubule' in plant cell fine structure. *J. Cell Biol.* **19**, 23
15. MARKS, M. H., BEAR, R. S., and BLAKE, C. H. 1949. X-ray diffraction evidence of collagen-type protein fibres in the Echinodermata, Coelenterata, and Porifera. *J. Exper. Zool.* **111**, 55
16. MEEK, G. A. and LANE, NANCY. 1964. The ultrastructural localization of phosphatases in the neurons of the snail, *Helix aspersa. J. Roy. Micr. Soc.* **82**, 193

17. PARTRIDGE, S. M. and ELSDEN, D. F. 1961. Dissociation of the chondroitin sulphate-protein complex of cartilage with alkali. *Biochem. J.* **79**, 26

18. ——. 1962. Elastin. In *Advances in Protein Chemistry* **17**, 227

19. PEASE, D. C. and MOLINARI, S. 1960. Electron microscopy of muscular arteries: pial vessels of cat and monkey. *J. Ultrastruct. Res.* **3**, 447

20. —— and PAULE, W. J. 1960. Electron microscopy of elastic arteries: the thoracic aorta of the rat. *J. Ultrastruct. Res.* **3**, 469

21. PEDERSEN, K. J. 1961. Studies on the nature of planarian connective tissue. *Z. Zellforsch.* **53**, 569

22. PINEDA, A. 1965. Collagen formation by principal cells of acoustic tumours. *Neurology* **15**, 536

23. PORTER, K. R. 1964. Cell fine structure and biosynthesis of intercellular macromolecules. In *Connective Tissue: Intercellular Macromolecules*, p. 167. New York Heart Ass. Symp. London: Churchill

24. —— and PAPPAS, G. D. 1959. Collagen formation by fibroblasts of the chick embryo dermis. *J. Biophys. Biochem. Cytol.* **5**, 153

25. REED, R. and RUDALL, K. M. 1948. Electron microscope studies on the structure of earthworm cuticle. *Biochem. Biophys. Acta* **2**, 7

26. REVEL, J. P. and HAY, E. D. 1963. An autoradiographic and electron microscopic study of collagen synthesis by differentiating cartilage. *Z. Zellforsch. und mikroskop. Anat.* **61**, 110

27. ROBINSON, R. A. and CAMERON, D. A. 1964. Bone. In *Electron Microscopic Anatomy* (edited by S. M. Kurtz), p. 315. New York: Academic Press

28. ROSS, R. and BENDITT, E. P. 1965. Wound healing and collagen formation. V. Quantitative electron microscope radioautographic observations of proline H³ utilisation by fibroblasts. *J. Cell Biol.* **27**, 83

29. RUSKA, CARLA and RUSKA, H. 1961. Die Cuticula der Epidermis des Regenwurms. *Z. Zellforsch.* **53**, 759

30. SCHOLTE, F. W. 1962. Die Muskulatur von *Helix pomatia* und ihre Innervation. *Proc. First Europ. Malac. Cong.*, London, p. 113

31. SCHMITT, F. O., HALL, G. E., and JAKUS, MARIE. 1942. Electron microscope investigations of the structure of collagen. *J. Cellular. Comp. Physiol.* **20**, 11

32. SHELDON, H. 1964. Cartilage. In *Electron Microscopic Anatomy* (edited by S. M. Kurtz), p. 295. New York: Academic Press

33. STEARNS, M. L. 1940. Studies in the development of connective tissue in transparent chambers in the rabbit's ear. Pt. I. *Am. J. Anat.* **66**, 133

34. ——. 1940. *Ibid.*, Pt. II. *Am. J. Anat.* **67**, 55

35. STETTEN, MARJORIE R. 1949. Some aspects of the metabolism of hydroxyproline, studied with the aid of isotopic nitrogen. *J. Biol. Chem.* **181**, 31

36. WILLIAMS, M. A. and MEEK, G. A. 1966. Studies in thickness variation in ultrathin sections for electron microscopy. *J. Roy. Micr. Soc.* **85**, 337.

NOTE ADDED IN PROOF

The author of this chapter has requested the editors to point out that it was written and submitted for publication in February 1966, and that not all the views expressed therein are currently held by the author.

Chapter 19

Fibroblasts—Vertebrate and Invertebrate

DOREEN E. ASHHURST

The supporting tissues of the majority of animals contain the fibrous protein, collagen. This protein is an important constituent of vertebrate bone and tendon, and of the layers of loose connective tissue which surround the internal organs of vertebrates. In the invertebrates, collagen is not found in such large quantities, but it has now been found in the cuticles of both annelid and nematode worms (Bird;[12] Maser and Rice;[34] Ruska and Ruska;[52] Watson;[58] Watson and Silvester[59]), in the body walls of coelenterates, molluscs, and sponges (Chapman;[16] Gross and Piez;[24] Melnick[36]), in the byssus threads of molluscs (Jackson and others[27]), in the Cuverian tubules of *Holothuria* (Gross and Piez;[24] Watson and Silvester[59]) and in the internal connective tissues of arachnids, insects, leeches, molluscs, planarians and *Peripatopsis* (Amoroso and others;[1] Ashhurst;[2,3] Ashhurst and Chapman;[6] Ashhurst and Richards;[8] Baccetti;[9,10] Bradbury and Meek;[13] Dumont, Anderson, and Chomyn;[18] Meek;[35] Pedersen;[43] Robson;[47] Smith and Wigglesworth[55]) and in the silk of some Hymenoptera (Rudall[51]). With the increasing number of investigations, collagen will undoubtedly be found to have a considerably wider distribution among the invertebrates.

The collagen molecule is peculiar in that it contains the amino-acid, hydroxyproline. This amino-acid is readily identified by chromatographic techniques and, since it has not been found in any other animal protein, its presence is usually taken to be synonymous with that of collagen. The collagen molecule, tropocollagen, is, in vertebrates, approximately 3000 Å long and 15 Å in diameter. It is thought that the tropocollagen molecules are staggered in respect to each other by one-quarter of their length, and that this ordered staggering is responsible for the 640 Å periodicity of banding, which is typical of many native collagen fibrils. There are some collagens with larger molecules (Maser and Rice[34]), and such variations in the size of collagen molecules may account for the variations in the periodicity of the banding (or, in some instances, the complete lack of banding) which are observed in collagens from such tissues as earthworm cuticle, neural lamella of moths, cornea, and some immature vertebrate fibrils (Ashhurst;[4] Jackson and Smith;[28] Jakus;[29] Watson and Silvester[59]).

Collagen fibrils are found in association with acid and neutral mucopolysaccharides; invertebrate collagens usually have a higher carbohydrate content than those of vertebrates (Gross and Piez[24]). The various chondroitin sulphates and hyaluronic acid are the best characterized, and probably the most important of the mucopolysaccharides of connective tissue. Chondroitin sulphate affects the rate of polymerization of tropocollagen molecules to form fibrils (Wood[62]).

Collagen and its associated mucopolysaccharides are produced by the fibroblasts. These are present in all connective tissues, but, as will be discussed later, the production of collagen and mucopolysaccharides may be only one of their functions. While much is now known about the collagen-producing activity of the fibroblasts, there is little information about the production of mucopolysaccharides.

The fibroblast may be described as a cell engaged in protein synthesis, and it will be seen that it shares many features in common with other cells engaged in protein synthesis, such as the exocrine pancreas cell. Fibroblasts from many different vertebrate tissues have been studied during collagen formation, but only five invertebrate fibroblasts have been examined. The purpose of this essay is to bring together information about vertebrate and invertebrate fibroblasts, and to see to what extent they are comparable cells.

VERTEBRATE FIBROBLASTS

The fibroblasts of vertebrates can be divided into two groups: one group which produces the collagen of the connective tissue layers around the various tissues, and the other group comprised of the chondroblasts and osteoblasts, which produce the collagen of bone and cartilage. The main difference between these two types of cell is that the chondroblasts and osteoblasts produce a much greater amount of acid mucopolysaccharide. The shape of fibroblasts is irregular. At least one part of the cell membrane is adjacent to the extracellular matrix, and hence their shape is only partially determined by the presence of other cells.

As is consistent with a cell engaged in protein synthesis, the endoplasmic reticulum of fibroblasts is well developed. In a typical vertebrate fibroblast, the endoplasmic reticulum fills a large portion of the cytoplasm, and the cisternae are dilated to form vesicles. The endoplasmic reticulum of an active fibroblast from a guinea-pig skin wound is seen in Fig. 19.1, A; it presents an essentially similar pattern in fibroblasts from such widely divergent tissues as amphibian skin, carrageenin granuloma, regenerating tendon and skin wounds of guinea-pigs, developing aorta and dermis of chick embryos, and the skin of mouse embryos (Chapman;[17] Gieseking;[20] Kajikawa;[30] Karrer;[31] Peach and others;[41] Ross and Benditt;[48, 50] Weiss and Ferris[60]). In most of these tissues, the dilatations of the endoplasmic reticulum contain material more electron dense than the surrounding cytoplasm. This suggests that the dilated cisternae contain the products of the synthetic activity of the endoplasmic reticulum; Chapman[17] and Kajikawa[30] describe filaments, about 50 Å in diameter, in the dilatations.

The organization of the ribosomes on the cisternae has received little attention, but Ross and Benditt[49] showed that, in guinea-pig skin wounds, the ribosomes of active fibroblasts were arranged in precise patterns in relation to the endoplasmic reticulum (Fig. 19.1, B).

The site of protein synthesis in cells engaged in the secretion of proteins is on the ribosomes. This has been established by such studies as those of Palade and co-workers (Caro and Palade;[15] Palade[38]) on the exocrine pancreas cell, which secretes zymogen. A first attempt to establish the ribosomes as the site of collagen production was made by Lowther and others.[33] They found, after the differential centrifugation of fibroblasts from carrageenin granuloma, that a high proportion of the collagen in the microsome fraction, is soluble in dilute neutral salt solutions. Previous investigations had shown that only newly formed collagen is soluble in dilute neutral salt solutions (Jackson and Bentley[26]). Thus it appeared that the site of collagen synthesis is on the ribosomes of the endoplasmic reticulum.

Since that time, the techniques of electron autoradiography have been applied to this problem by Hay and Revel,[25] Revel and Hay,[45] and Ross and Benditt.[48, 50] Tritiated proline was used as the radioactive label. In fibroblasts from regenerating limbs of *Amblystoma*, and in guinea-pig skin wounds, the radio-

Fig. 19.1 A. Radioautograph of a portion of a fibroblast from a 7-day guinea-pig skin wound 30 min after intraperitoneal administration of proline-H^3. In this cell, the silver grains can be seen to lie over both endoplasmic reticulum (er) and Golgi complex (g). The collagen fibrils (c) in this area are not labelled.

B. Micrograph of part of a fibroblast from a 7-day skin wound in a guinea-pig. Several profiles of rough endoplasmic reticulum (er) can be seen. Where the cisternae are sectioned normally, a single layer of ribosomes may be seen attached to the membranes, but where the membranes are sectioned tangentially (arrows), the ribosomes are seen to be arranged in curved and spiral patterns.

C. Micrograph of part of a fibroblast in a 7-day carrageenin induced guinea-pig granuloma. The profiles of the endoplasmic reticulum are irregular and distended. Many fine filaments (f) occur in the cytoplasm, particularly in the outer regions of the cell. The cell membrane (cm) adjacent to these filaments is ill-defined and disrupted.

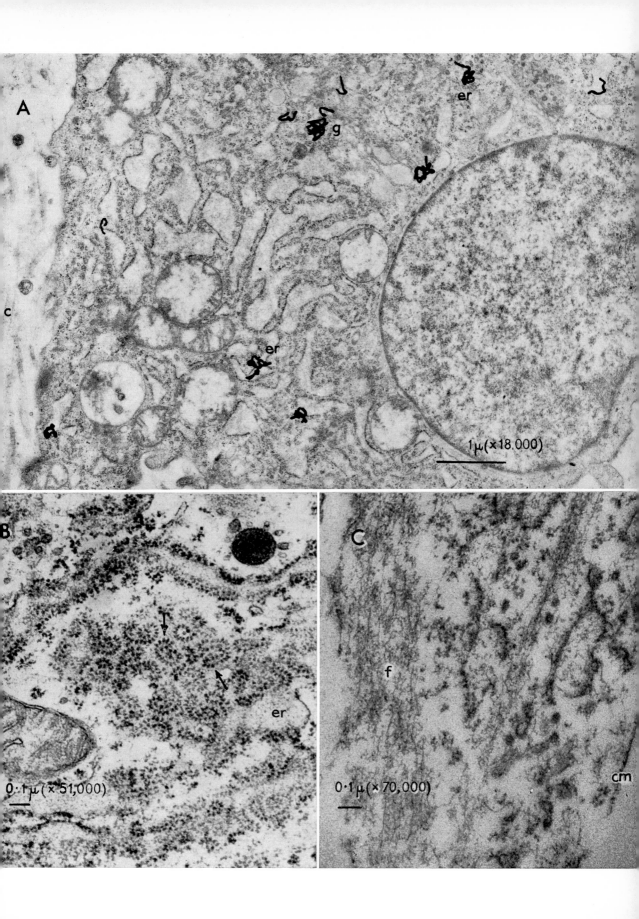

active label appears over the endoplasmic reticulum within 10 to 15 min after the injection; this result confirms that the endoplasmic reticulum is the site of collagen production.

The passage of secreted proteins from the endoplasmic reticulum to the Golgi complex for 'packaging' has become accepted as the usual sequence of events in a protein-synthesizing cell.* This pattern has been confirmed for both the regenerating limb and basement lamella of *Amblystoma*, since, 20 to 30 min after the injection of H^3-proline, radioactivity is present in the vacuoles of the Golgi region (Hay and Revel;[25] Revel and Hay[45]). In chondrocytes, the Golgi complex is invariably well developed, but this is not necessarily so in other fibroblasts. Quantitative estimations of the number of silver grains over the different inclusions of the guinea-pig skin wound fibroblasts, during fibrillogenesis, suggested to Ross and Benditt[50] that while some of the synthesized protein passes into the Golgi vacuoles, another portion bypasses the Golgi complex and passes from the endoplasmic reticulum directly into the extracellular space (Fig. 19.1, A).

Before pursuing further the evidence on the release of collagen from the cell, it is pertinent to mention here another function of the Golgi complex in fibroblasts. Dziewiatkowski[19] demonstrated that chondroitin sulphate is produced intracellularly by chondrocytes; the synthesis of chondroitin sulphate is simultaneous with, and at a comparable rate to the synthesis of collagen (Campo and Dziewiatkowski[14]). While the site of the initial production of the polysaccharide molecule is not known, it has been shown, by following the incorporation of $S^{35}O_4$ in chondrocytes by electron autoradiography, that the sulphation of chondroitin sulphate occurs in the vacuoles of the juxtanuclear Golgi complexes; the label appears over these vacuoles three minutes after the injection of the tracer (Godman and Lane[21]).

The existence of two routes by which the protein molecules synthesized by the fibroblast may reach the extracellular space, has already been mentioned. This problem is complicated still further by the fact that in some fibroblasts, for example those of carrageenin granuloma in the guinea-pig, in the hind limbs of rat embryos, in heart cells of chick embryos in tissue culture, and skin of mouse and chick embryos (Chapman;[17] Gieseking;[20] Godman and Porter;[22] Kajikawa;[30] Yardley and others[64]), polymerization of the tropocollagen molecules may start within the cytoplasm, since aggregations of small fibrils are clearly visible in regions near the cell membrane (Fig. 19.1, C). These aggregations usually occur in the cortical region of the cytoplasm adjacent to the cell membrane, and in such areas it often appears that the cell membrane is no longer present. The intracellular fibrils seem to merge with similar extracellular fibrils, in such a manner as to suggest that the intracellular fibrils are passing through a break in the cell membrane. Such discontinuities in the cell membrane have been questioned by Goldberg and Green,[23] who consider that the apparent disappearance of the cell membrane could be due to the oblique plane of sectioning in these regions. The collagenous nature of the intracellular fibrils has not been established; autoradiographic studies have not so far provided any information about their nature, but this technique would probably provide the most unequivocal evidence. Sheldon and Kimball[54] found banded fibrils, with a periodicity similar to that of fibrous 'long-spacing' collagen, in the Golgi vacuoles of the fibroblasts of rabbit ear cartilage. The nature of the intracellular fibrils is a problem which requires further investigation, especially since much of the work on intracellular microtubules has been done since the majority of these studies on fibroblasts.

The next question is concerned with the passage of the synthesized collagen from the fibroblast to the extracellular matrix. This again seems to be very variable. In discussing intracellular fibrils, it was mentioned that, in many instances, they appear to pass through gaps in the cell membrane, a mechanism which is a form of apocrine secretion. In some fibroblasts, it appears that the portion of cytoplasm which is pinched off includes vesicles of the endoplasmic reticulum, in addition to the fibrils (Gieseking[20]). Another method by which collagen passes from the cell is a form of merocrine secretion. Several workers have evidence that Golgi vacuoles and vesicles of the endoplasmic reticulum may come to the cell membrane, join with it, and discharge their contents into the intercellular space (Goldberg and Green;[23] Revel and Hay;[45] Ross and Benditt[50]). A particularly good series of micrographs illustrating these events, in fibroblasts of the hind limb cartilage of *Amblystoma* larvae, has been published by Revel and Hay.[25]

* Editorial note: see, for example, chapter 31 in this volume.

It would appear that the method used for the extrusion of the secretory products from the cell is determined by the location of the 'finished' products within the cell. The reason for the variation in the site and the form of the collagen molecules which are ready to be passed from the cell is not known. Further investigations using new techniques may show that these differences are not so great as they appear at present.

The polymerization of collagen molecules to form collagen fibrils occurs in two stages: first, the tropocollagen molecules come together to form the 'nucleus' of the fibrils, and then, in the second phase, this nucleus grows by the accretion of further tropocollagen molecules on to its surface (Wood[61]). The intracellular fibrils might represent the first stage in the tissues in which they occur, although, as has been mentioned, there is as yet no conclusive evidence that they are collagenous. Porter and Pappas[44] suggested that the outer surface of the cell membrane might act as a 'template' for the polymerization of tropocollagen molecules. The second stage in the formation of the fibrils takes place in the extra-cellular space. Wood[62] showed that this stage is affected by the presence of chondroitin sulphate, and recently, the presence of this acid mucopolysaccharide has been demonstrated, by means of an electron histochemical technique, around the outer surface of the cell membrane of actively synthesizing fibro-blasts (Yardley and Brown[63]).

INVERTEBRATE FIBROBLASTS

Invertebrate fibroblasts have so far received very little attention, though the presence of collagen in widely divergent phyla would suggest that much interesting information on methods of collagen produc-tion could be obtained from studies of fibroblasts from representative species. So far, the fibroblasts from only one annelid, two insects, one mollusc, and a sponge, have been examined. This discussion will be concerned mainly with the insect fibroblasts. The activity of the molluscan fibroblasts from the radula sac cartilage of *Helix aspersa* is described by Dr. G. A. Meek elsewhere in this volume, (Chapter 18).

The presence of collagen in insects was not established until the late 1950s, when it was shown, by various techniques, to be present in the layer of connective tissue, the neural lamella, which surrounds the ventral nerve cord of insects (Ashhurst;[2] Ashhurst and Chapman;[6] Baccetti;[9] Richards and Schnei-der;[46] Smith and Wigglesworth[55]). Nothing was known of its formation, and the small amount of connective tissue present in most insects made this a rather difficult problem. In the Lepidoptera, how-ever, a large mass of connective tissue is formed on the dorsal side of the abdominal region of the central nervous system, during the pupal stage (Ashhurst and Richards[7,8]), and the fibroblasts of this connective tissue in *Galleria mellonella* were chosen for a first study (Ashhurst[4]). As Lepidopteran collagen fibrils do not have clear banding, and are very thin, the fibroblasts which produce the banded collagen fibrils of the neural lamella in the embryo locust, *Schistocerca gregaria*, were also studied (Ashhurst[5]).

The fibroblasts in *Galleria*, which were studied in detail, are packed tightly together in a pile, on the dorsal side of the abdominal nerve cord. There is virtually no extracellular space between adjacent cells, so the development of large fibrous areas is at the expense of the cytoplasm of the cells. In *Schisto-cerca*, the fibroblasts form a layer around the whole nervous system, and they secrete the neural lamella on their outer surface, which is exposed to the haemolymph. In contrast, the fibroblasts of the snail radula sac cartilage are dispersed in a hyaline matrix.

The endoplasmic reticulum is well developed in both the moth and locust fibroblasts. At a stage when collagen production is at its peak, the cytoplasm of the moth cells is almost filled with a series of large vesicles, containing a substance more electron dense than the surrounding cytoplasm. Ribosomes are not present on the membranes surrounding these vesicles, but their identification as dilatations of the cisternae of the endoplasmic reticulum was established, because, in favourable sections, their membrane is seen to be continuous with the outer nuclear membrane, and also with the ribosome covered cisternae of the endoplasmic reticulum (Figs. 19.2, B, 19.3, B). The dilatations of the endoplasmic reticulum of the locust fibroblasts were easily identified, as their membrane is covered with ribosomes, and the confluence between the cisternae and dilatations is very often observed (Fig. 19.2, A). Actively synthesizing snail fibroblasts also have the typical fibroblast endoplasmic reticulum, with the cisternae dilated to form

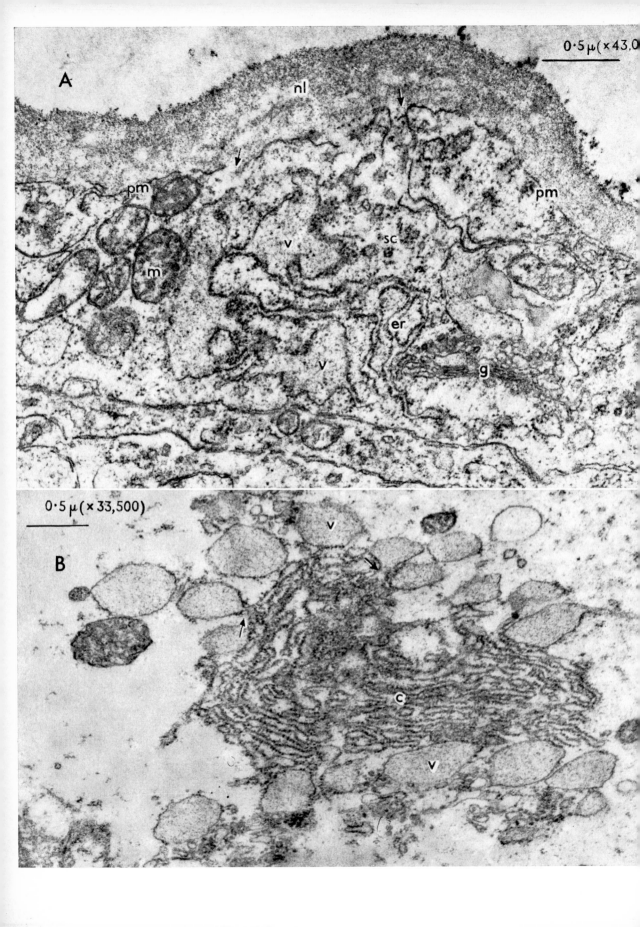

vesicles, and many ribosomes. When collagen production ceases, the endoplasmic reticulum is much reduced, in all these cells (Figs. 19.3, A, 19.5, A).

No structures which could be designated as Golgi complexes were observed in *Galleria*, although several hundred cells at varying stages of fibrillogenesis were observed. Typical Golgi regions were, however, observed in the locust fibroblasts (Fig. 19.2, A), but their lack of development during the secretion of the neural lamella suggests that they play little or no part in the collagen secretory activity of the cells.

Intracellular fibrils are clearly seen in the cytoplasm of moth fibroblasts. These may be several microns in length; their diameter is between 120 and 200 Å, and occasionally indications of banding, with a periodicity of about 120 to 200 Å are visible (Fig. 19.4, A). These fibrils are hollow (Fig. 19.4, B). While they bear some resemblance to the microtubules described by several authors (Behnke;[11] Ledbetter and Porter;[32] Sandborn and others[53]), their appearance in large numbers, at the time of great synthetic activity, suggests that they may be the intracellular precursors of the collagen fibrils. It has already been mentioned that there is no intercellular space, in which to shed the products of these fibroblasts. Areas of fibrils develop at the expense of the cytoplasm of the cells, and it is invariably found that the cell membrane is either missing or extremely ill-defined in the regions where the cytoplasm is adjacent to an area of fibrils. Where the cytoplasm of two cells come together, the two plasma membranes are very clearly defined. The lack of cell membrane next to the fibrous areas, together with the presence of intracellular fibrils, suggests that these fibrils may come to the periphery of the cells, and, with the break-down of the cell membrane, be incorporated into the fibrous areas. The structure of the fibrils in the areas of connective tissue is obscured by deposits of acid and neutral mucopolysaccharides, though, occasionally, a small length of fibril devoid of mucopolysaccharides may be seen (Fig. 19.4, C). These appear similar to the intracellular fibrils. A group of fibrils no longer within a cell, but still not surrounded by mucopolysaccharides, is seen in Fig. 19.4, D. Further elucidation of the mechanism of secretion of the fibrous tissue in *Galleria* must await the application of new techniques to the problem.

The situation in the locust is entirely different. No intracellular fibrils are present. In places, however, the cell membrane appears to be absent. The plane of the sections and the nature of the breaks suggests that they may be real, and not artifacts, though there was no evidence for material passing through them. The material in the endoplasmic reticular vesicles and that in the developing neural lamella is very similar, and occasionally, indentations of the neural lamella into the cells can be seen (Fig. 19.3, C). This suggests that vesicles might come to the surface, join with the plasma membrane and discharge their contents. At first the neural lamella appears homogeneous, but gradually small banded collagen fibrils are formed (Fig. 19.5, B) and after hatching, the neural lamella contains a mass of fibrils with a periodicity between 550 to 600 Å; their diameter is between 150 and 400 Å (Fig. 19.5, C).

The fibroblasts of *Helix aspersa* seem to be unique among the fibroblasts so far studied, since fibrils, with banding identical to that of the extracellular collagen fibrils, are found within the cytoplasm of fibroblasts of hibernating snails (Meek[35]). In active fibroblasts from summer snails, the intracellular fibrils appear to be continuous with the extracellular collagen fibrils; the cell membrane is completely absent in these regions. In addition, the cytoplasm of the summer snail fibroblasts contains microtubules about 150 Å in diameter, which run parallel to the long axis of the cell. Preliminary results suggest that these may be the precursors of the banded fibrils (Meek, this volume). This bears a striking resemblance to the situation in *Galleria*, where it has been suggested that the intracellular fibrils, which are similar to microtubules, are the precursors of the extracellular fibrils (Ashhurst[4]). The mature collagen fibrils of

Fig. 19.2 A. An electron micrograph of the sheath in an 11-day locust embryo. The neural lamella (nl) has a distinct form. The sheath cells or fibroblasts (sc) have a well-developed endoplasmic reticulum (er) with vesicles (v) containing a somewhat electron-dense material. Mitochondria (m), a Golgi area (g), and a lipid droplet are also visible. The plasma membrane (pm) has many discontinuities (arrows).

B. A micrograph of part of a fibroblast from a 90 hr moth pupa, with a typical aggregation of cisternae (c) with attached ribosomes, and vesicles (v) of the endoplasmic reticulum. The membranes of the cisternae and vesicles are continuous (arrows).

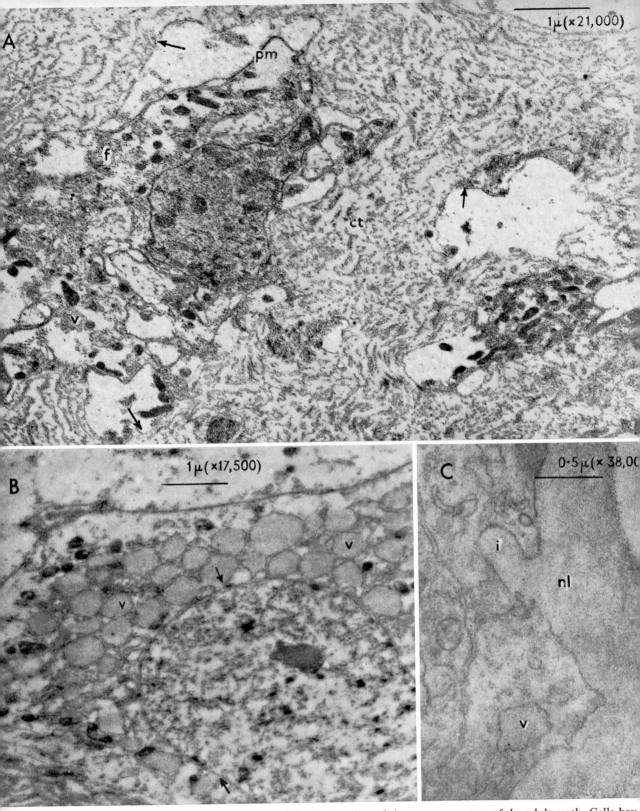

Fig. 19.3 A. Part of the dorsal mass of the abdominal region of the nervous system of the adult moth. Cells have decreased in size with formation of fibrous tissue. Only the remnants of the vesicles (v) of the endoplasmic reticulum can be seen. Many transverse sections of the intracellular fibrils (f) occur in the cytoplasm. The plasma membranes (pm) are well defined between the cells, but are discontinuous next to the fibrous connective tissue (ct) which is now extensive (arrows). B. A fibroblast from a 108-hr moth pupa, showing the great development of the vesicles (v) of the endoplasmic reticulum. The continuity between the membranes of the vesicles and the outer nuclear membrane can be seen (arrows). C. Part of the sheath of an 11-day locust embryo to show an indentation (i) of the neural lamella (nl) with contents similar to those of a neighbouring vesicle (v) of the endoplasmic reticulum.

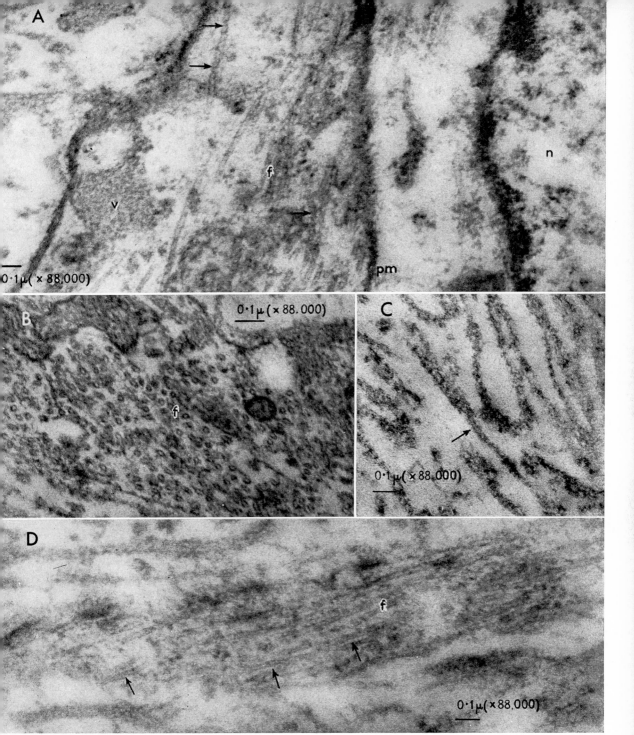

Fig. 19.4 A.   A micrograph of a cell from a 144 hr pupa, with a series of intracellular fibrils (f) in longitudinal section. Indications of banding can be seen in certain areas (arrows). The plasma membrane (pm), the nucleus (n), and the vesicles (v) confirm the intracellular situation of the fibrils.

B. Another section from a pupa 144 hr old, with intracellular fibrils (f) in transverse section. Their hollow cores are clearly seen.

C. Part of an area of fibrous connective tissue from an adult moth. A portion of one of the fibrils is incompletely surrounded by the ground substances (arrow).

D. A bundle of intracellular fibrils (f) from a pupa 144 hr old which are no longer surrounded by a plasma membrane, but are not yet covered by the ground substances. Indications of banding can be seen (arrows).

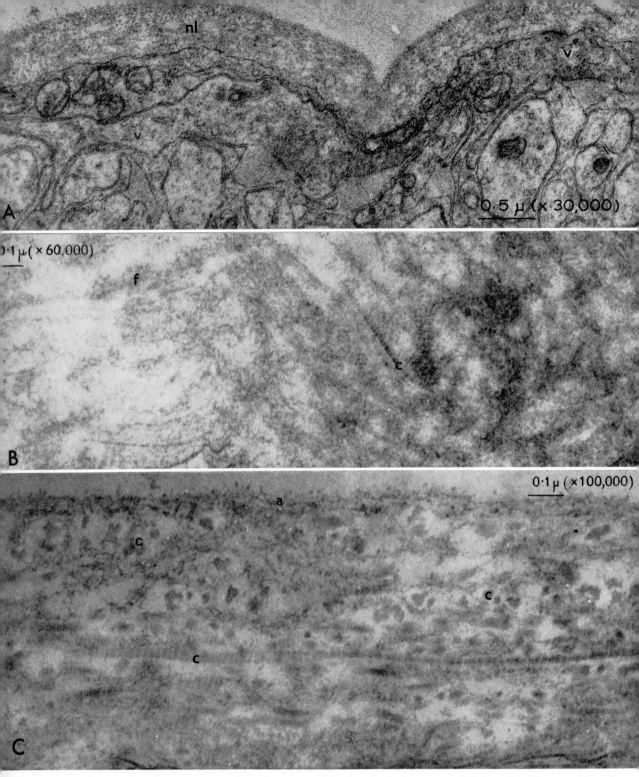

Fig. 19.5 A. A low-magnification micrograph of the sheath of a locust nymph 2 days old. The neural lamella (nl) contains many fibrils. The cytoplasm of the sheath cells is now more compact and only a few small vesicles (v) of the endoplasmic reticulum are present.

B. A micrograph of part of the neural lamella of an 11-day locust embryo. A distinctly banded collagen fibril (c) is seen amongst the amorphous material. Many other fibrils (f) with less distinct banding are present.

C. A micrograph of part of the neural lamella of a 2-day locust nymph at higher magnification. Collagen fibrils (c) are present in both longitudinal and transverse section. A thin layer of amorphous material (a) is present on the outer surface.

the moth and snail are dissimilar, as the former are not clearly banded, while the latter display banding with a periodicity of 540 Å.

The variation among the invertebrate fibroblasts appears to be as great as the variation among vertebrate fibroblasts. The two kinds of insect fibroblasts differ, since, in the moth, intracellular fibrils are present, and these are probably discharged from the cell by a form of apocrine secretion, whereas in the locust, the secretory products appear to be passed straight from the vesicles of the endoplasmic reticulum to the exterior, a form of merocrine secretion. The secretory mechanism in *Helix aspersa* bears a close resemblance to that in the moth. Collagen fibrils, with banding periodicity of 460 Å, have recently been reported in vacuoles in the cytoplasm of amoeboid cells in the gill of the bivalve mollusc, *Lampsilis ventricosa* (Sun and Wikman[56]). Thus, intracellular fibrils which can be identified as collagen, may be a common phenomenon among the invertebrates. In the leech, *Hirudo medicinalis*, and the sponge, *Tethya lyncurium*, however, the cell membrane appears to have an active role in fibril formation (Bradbury and Meek;[13] Pavans de Ceccatty and Thiney[39, 40]). The newly formed fibrils of the sponge are hollow, but as they move away from the cell membrane, the hollow core disappears and they appear as typical banded collagen fibrils.

CONCLUSIONS

The ability, common to cells from a wide variety of tissues in diverse animals, to produce collagenous proteins does not appear to impose stringent limitations on the organization of the synthesizing cell. A well-developed endoplasmic reticulum, with cisternae dilated to form vesicles, is of universal occurrence among fibroblasts, but from the foregoing account, it is obvious that there is a wide range of variation in the mechanisms for the expulsion of the collagenous protein from the cell, and also, in the form in which it passes to the extracellular matrix. The diversity of mechanism and form is as great among the vertebrate as among the invertebrate fibroblasts, but, as yet, clearly banded fibrils have not been found inside a vertebrate cell.

Another interesting fact is that the synthesis of collagen may be only one of the functions of a particular cell. Schwann cells can produce collagen (Nathaniel and Pease[37]), and it appears that certain smooth muscle cells secrete the collagen found in parts of the pial vessels of the brain of cats and monkeys (Pease and Molinari[42]). In the developing aorta, Karrer[31] observed that some fibroblasts lose their inclusions in the later stages of fibrillogenesis, and transform into smooth muscle cells. Similarly, the fibroblasts which form the connective tissue around the nervous system in insects may have a role in the maintenance of the ionic environment inside the nervous system (Treherne[57]). Whether further studies on fibrillogenesis will show an even greater diversity among fibroblasts from different animals, or whether they will indicate a greater uniformity, remains to be seen. Nevertheless, the characteristics shared in common by the insect, mammalian, and snail fibroblasts, are quite striking.

ACKNOWLEDGEMENTS

I should like to thank Drs. E. P. Benditt, J. A. Chapman, and R. Ross and the Rockefeller University Press for permission to reproduce the micrographs in Fig. 19.1, and the *Quarterly Journal of Microscopical Science* for permission to reproduce Figs. 19.2 to 19.5. The micrographs are from the following papers:

Fig. 19.1, A.   Ross, R. and Benditt, E. P. 1965. *J. Cell Biol.* **27**, 83
Fig. 19.1, B.   Ross, R. and Benditt, E. P. 1964. *J. Cell Biol.* **22**, 365
Fig. 19.1, C.   Chapman, J. A. 1961. *J. Biophys. Biochem. Cytol.* **9**, 639
Figs. 19.2, B, 19.3, A and B, 19.4, A–D.   Ashhurst, D. E. 1964. *Quart. J. micr. Sci.* **105**, 391
Figs. 19.2, A, 19.3, C, 19.5, A–C.   Ashhurst, D. E. 1965. *Quart. J. micr. Sci.* **106**, 61

REFERENCES

1. AMOROSO, E. C., BAXTER, M. I., CHIQUOINE, A. D., and NISBET, R. H. 1964. The fine structure of neurons and other elements in the nervous system of the giant African land snail, *Archachatina marginata. Proc. Roy. Soc.* B, **160**, 167–80
2. ASHHURST, D. E. 1959. The connective tissue sheath of the locust nervous system: a histochemical study. *Quart. J. micr. Sci.* **100**, 401–12

3. ——. 1961. A histochemical study of the connective-tissue sheath of the nervous system of *Periplaneta americana. Quart. J. micr. Sci.* **102**, 455–61

4. ——. 1964. Fibrillogenesis in the wax-moth, *Galleria mellonella. Quart. J. micr. Sci.* **105**, 391–403

5. ——. 1965. The connective tissue sheath of the locust nervous system: its development in the embryo. *Quart. J. micr. Sci.* **106**, 61–73

6. —— and CHAPMAN, J. A. 1961. The connective-tissue sheath of the nervous system of *Locusta migratoria:* an electron microscope study. *Quart. J. micr. Sci.* **102**, 463–67

7. —— and RICHARDS, A. G. 1964. A study of the changes occurring in the connective tissue associated with the central nervous system during the pupal stage of the wax-moth, *Galleria mellonella. J. Morph.* **114**, 225–36

8. —— and ——. 1964. The histochemistry of the connective tissue associated with the central nervous system of the pupa of the wax-moth, *Galleria mellonella*, L. *J. Morph.* **114**, 237–46

9. BACCETTI, B. 1956. Lo stroma di sostegna di organi degli insetti esaminate a luce polarizzata. *Redia* **41**, 259–76

10. ——. 1961. Indagini comparative sulla ultrastruttura della fibrilla collagene nei deversi ordini degli insetti. *Redia* **46**, 1–7

11. BEHNKE, O. 1964. A preliminary report on 'microtubules' in undifferentiated and differentiated vertebrate cells. *J. Ultrastruct. Res.* **11**, 139–46

12. BIRD, A. F. 1957. Chemical composition of the nematode cuticle. Observations on individual layers and extracts from these layers in *Ascaris lumbricoides* cuticle. *Exptl. Parasitol.* **6**, 383–403

13. BRADBURY, S. and MEEK, G. A. 1955. A study of fibrogenesis in the leech, *Hirudo medicinalis. Quart. J. micr. Sci.* **99**, 143–8

14. CAMPO, R. D. and DZIEWIATKOWSKI, D. D. 1962. Intracellular synthesis of protein-polysaccharides by slices of bovine costal cartilage. *J. Biol. Chem.* **237**, 2729–35

15. CARO, L. G. and PALADE, G. E. 1964. Protein synthesis, storage, and discharge in the pancreatic exocrine cell. An autoradiographic study. *J. Cell Biol.* **20**, 473–95

16. CHAPMAN, G. 1959. The mesoglea of *Pelagia noctiluca. Quart. J. micr. Sci.* **100**, 599–610

17. CHAPMAN, J. A. 1961. Morphological and chemical studies of collagen formation. I. The fine structure of guinea-pig granulomata. *J. Biophys. Biochem. Cytol.* **9**, 639–51

18. DUMONT, J. N., ANDERSON, E., and CHOMYN, E. 1965. The anatomy of the peripheral nerve and its ensheathing artery in the horseshoe crab, *Xiphosura (Limulus) polyphemus. J. Ultrastruct. Res.* **13**, 38–64

19. DZIEWIATKOWSKI, D. D. 1962. Intracellular synthesis of chondroitin sulphate. *J. Cell Biol.* **13**, 359–64

20. GIESEKING, R. 1960. Electronenoptische Beobachtungen an Fibroblasten. *Structur und Stoffwechsel des Bindegewebes* (edited by W. H. Hauss and H. Losse), p. 131. Stuttgart: G. Thieme

21. GODMAN, G. C. and LANE, N. 1964. On the site of sulphation in the chondrocyte. *J. Cell Biol.* **21**, 353–66

22. —— and PORTER, K. R. 1960. Chondrogenesis, studied with the electron microscope. *J. Biophys. Biochem. Cytol.* **8**, 719–60

23. GOLDBERG, B. and GREEN, H. 1964. An analysis of collagen secretion by established mouse fibroblast lines. *J. Cell Biol.* **22**, 227–58

24. GROSS, J. and PIEZ, K. A. 1960. The nature of collagen. I. Invertebrate collagens. *Calcification in Biological systems* (edited by R. F. Sognnaes), pp. 395–409. Washington: American Association for the Advancement of Science

25. HAY, E. D. and REVEL, J. P. 1963. Autoradiographic studies of the origin of the basement lamella in *Amblystoma. Develop. Biol.* **7**, 152–68

26. JACKSON, D. S. and BENTLEY, J. P. 1960. On the significance of the extractable collagens. *J. Biophys. Biochem. Cytol.* **7**, 37–42

27. JACKSON, S. F., KELLY, F. C., NORTH, A. C. T., RANDALL, J. T., SEEDS, W. E., WATSON, M., and WILKINSON, G. R. 1953. The byssus threads of *Mytilus edulis* and *Pinna nobilis. Nature and structure of collagen* (edited by J. T. Randall), pp. 106–16. London: Butterworth

28. —— and SMITH, R. H. 1957. Studies on the biosynthesis of collagen. I. The growth of fowl osteoblasts and the formation of collagen in tissue culture. *J. Biophys. Biochem. Cytol.* **3**, 897–912

29. JAKUS, M. A. 1962. Further observations on the fine structure of the cornea. *Invest. Ophthalmol.* **1**, 201–25

30. KAJIKAWA, K. 1961. The fine structure of fibroblasts of mouse embryo skin. *J. Electronmicr.* **10**, 131–44

31. KARRER, H. E. 1960. Electron microscope study of developing chick embryo aorta. *J. Ultrastruct. Res.* **4**, 420–54

32. LEDBETTER, M. C. and PORTER, K. R. 1963. A "microtubule" in plant cell fine structure. *J. Cell Biol.* **19**, 239–50

33. LOWTHER, D. A., GREEN, N. M., and CHAPMAN, J. A. 1961. Morphological and chemical studies of collagen formation. II. Metabolic activity of collagen associated with subcellular fractions of guinea-pig granulomata. *J. Biophys. Biochem. Cytol.* **10**, 373–88

34. MASER, M. D. and RICE, R. V. 1962. Biophysical and biochemical properties of earthworm-cuticle collagen. *Biochim. Biophys. Acta* **63**, 255–65
35. MEEK, G. A. 1966. Intracellular collagen fibrils. *J. Physiol.* **182**, 3P
36. MELNICK, S. C. 1958. Occurrence of collagen in the Phylum Mollusca. *Nature, Lond.* **181**, 1483
37. NATHANIEL, E. J. H. and PEASE, D. C. 1963. Collagen and basement membrane formation by Schwann cells during nerve regeneration. *J. Ultrastruct. Res.* **9**, 550–60
38. PALADE, G. E. 1961. The secretory process of the pancreatic exocrine cell. *Electron Microscopy in Anatomy* (edited by J. D. Boyd, F. R. Johnson and J. D. Lever), pp. 176–206. London: Arnold
39. PAVANS DE CECCATTY, M. and THINEY, Y. 1963. Microscopie électronique de la fibrogénèse cellulaire du collagène, chez l'éponge siliceuse *Tethya lyncurium*. *C.R. Acad. Sci.* **256**, 5406–8
40. —— and ——. 1964. Essai d'interprétation de la fibrogénèse cellulaire du collagène, chez l'éponge *Tethya lyncurium*. *Vie et Milieu* suppl. **17**, 129–46
41. PEACH, R., WILLIAMS, G., and CHAPMAN, J. A. 1961. A light and electron optical study of regenerating tendon. *Amer. J. Path.* **38**, 495–514
42. PEASE, D. C. 1960. Electron microscopy of muscular arteries; pial vessels of the cat and monkey. *J. Ultrastruct. Res.* **3**, 447–68
43. PEDERSEN, K. J. 1961. Studies on the nature of planarian connective tissue. *Z. Zellforsch.* **53**, 569–608
44. PORTER, K. R. and PAPPAS, G. D. 1959. Collagen formation by fibroblasts of the chick embryo dermis. *J. Biophys. Biochem. Cytol.* **5**, 153–66
45. REVEL, J. P. and HAY, E. D. 1963. An autoradiographic and electron microscopic study of collagen synthesis in differentiating cartilage. *Z. Zellforsch.* **61**, 110–44
46. RICHARDS, A. G. and SCHNEIDER, D. 1958. Über den komplexen Bau der Membranen des Bindegewebes von Insekten. *Z. Naturforschung.* **13**b, 680–7
47. ROBSON, E. A. 1964. The cuticle of *Peripatopsis moseleyi*. *Quart. J. micr. Sci.* **105**, 281–99
48. ROSS, R. and BENDITT, E. P. 1961. Wound healing and collagen formation. I. Sequential changes in components of guinea-pig skin wounds observed in the electron microscope. *J. Biophys. Biochem. Cytol.* **11**, 677–700
49. —— and BENDITT, E. P. 1964. Wound healing and collagen formation. IV. Distortion of ribosomal patterns of fibroblasts in scurvy. *J. Cell Biol.* **22**, 365–89
50. —— and ——. 1965. Wound healing and collagen formation. V. Quantitative electron microscope radioautographic observations of proline-H^3 utilization by fibroblasts. *J. Cell Biol.* **27**, 83–106
51. RUDALL, K. M. 1962. Silk and other cocoon proteins. *Comparative Biochemistry* (edited by M. Florkin and H. S. Mason), pp. 397–433, Vol. 4. New York: Academic Press
52. RUSKA, C. and RUSKA, H. 1961. Die Cuticula der Epidermis des Regenwurms (*Lumbricus terrestris*, L.). *Z. Zellforsch.* **53**, 759–64
53. SANDBORN, E., KOEN, P. F., MCNABB, J. D., and MOORE, G. 1964. Cytoplasmic microtubules in mammalian cells. *J. Ultrastruct. Res.* **11**, 123–38
54. SHELDON, H. and KIMBALL, F. B. 1962. Studies on cartilage. III. The occurrence of collagen within vacuoles of the Golgi apparatus. *J. Cell Biol.* **12**, 599–613
55. SMITH, D. S. and WIGGLESWORTH, V. B. 1959. Collagen in the perilemma of insect nerve. *Nature, Lond.* **183**, 127–8
56. SUN, C. N. and WIKMAN, J. 1965. The connective tissue in *Lampsilis ventricosa* as revealed by electron microscopy. *J. Cell Biol.* **27**, 145A
57. TREHERNE, J. E. 1962. Distribution of water and inorganic ions in the central nervous system of an insect (*Periplaneta americana*, L.). *Nature, Lond.* **193**, 750–2
58. WATSON, M. R. 1958. The chemical composition of earthworm cuticle. *Biochem. J.* **68**, 416–20
59. —— and SILVESTER, N. R. 1959. Studies of invertebrate collagen preparations. *Biochem. J.* **71**, 578–9
60. WEISS, P. and FERRIS, W. 1956. The basement lamella of Amphibian skin. Its reconstruction after wounding. *J. Biophys. Biochem. Cytol.* **2** suppl., 275–82
61. WOOD, G. C. 1960a. The formation of fibrils from collagen solutions. 2. A mechanism of collagen fibril formation. *Biochem. J.* **75**, 598–605
62. ——. 1960b. The formation of fibrils from collagen solutions. 3. Effect of chondroitin sulphate and some other naturally occurring polyanions on the rate of formation. *Biochem. J.* **75**, 605–12
63. YARDLEY, J. H. and BROWN, G. D. 1965. Fibroblasts in tissue culture. Use of colloidal iron for ultrastructural localisation of acid mucopolysaccharides. *Lab. Invest.* **14**, 501–13
64. ——, HEATON, M. W., GAINES, L. M., and SHULMAN, L. E. 1960. Collagen formation by fibroblasts. Preliminary electron microscopic observations using thin sections of tissue cultures. *Bull. Johns Hopkins Hosp.* **106**, 381–93

NOTE ADDED IN PROOF

The author of this chapter has asked the editors to point out that the literature survey that it contains was completed in February 1966.

Chapter 20

The Secretion, Structure, and Strength of Echinoderm Calcite

D. NICHOLS AND J. D. CURREY

INTRODUCTION

The skeleton of echinoderms arouses special interest because it is laid down in the form of an open meshwork. The cells which secrete it are a syncytium, and permeate the spaces between the struts of skeletal material. Thus, the whole element, in the living state, superficially resembles a two-phased skeletal system. Such a system is thought to be particularly strong (Currey[5]). But the scale of the two components is very different from that of the two-phased skeletons possessed by other animal phyla. Further, the protoplasmic strands are hardly to be considered a skeletal phase in the true meaning of the term: the protoplasmic component of each element is sufficiently loosely packed to allow the free passage of coelomocytes among the channels in which it lies. One must, therefore, regard the echinoderm skeleton as unusual, because it is single-phased (Currey and Nichols[6]). It derives its inherent strength from certain features of its construction, such as its meshlike pattern, assisted by the intimate ramification of living tissue within almost all elements.

GENERAL NATURE OF THE ECHINODERM SKELETON

In all echinoderms the skeleton is internal. In some classes it forms a rigid theca or cuirass, as in the Echinoidea; in others it may form vertebra-like supporting ossicles, as in the Ophiuroidea and Crinoidea; or it may constitute a loosely-connected covering or support for the body, as in the Asteroidea. It is least developed in the Holothuroidea, where it is reduced to mere isolated ossicles; in at least one pelagic holothuroid, *Pelagothuria*, skeletal ossicles are apparently totally absent (Chun[2]).

In all groups the meshlike microscopic structure is seen (Fig. 20.1, A and B). In most skeletal elements, the inorganic component accounts for about half the volume, and the channels permeating it the other half (West[28]). Where connection between ossicles occurs, it is through connective tissue fibres 'sewn' into the channels of the ossicles (Fig. 20.1, D).

In echinoderm larvae the skeleton is generally in the form of needle-like solid spicules which help to maintain the shape of, for instance, the arms of plutei. Similar very thin spicules may also occur in some adult echinoderms, particularly holothuroids and in some parts of the teeth of echinoids; but such instances are rare. The great majority of echinoderm skeletal elements have a fenestrated structure.

CHEMICAL COMPOSITION

The echinoderm skeleton is made of calcium carbonate, in the physical form of calcite rather than aragonite. In common with other organic calcites, there may be a high proportion of magnesium carbonate in the crystal, varying between 4 per cent and 16 per cent (Clarke and Wheeler[3]; Chave[1]). Such high proportions of magnesium are rarely found in inorganically produced crystals, and, interestingly, the magnesium content of fossil echinoderm calcite is usually low, suggesting a metastable condition which cannot be maintained outside the living animal (Chave[1]). The proportion of magnesium in the skeleton shows a variation according to the temperature of the water in which the animals are found (Clarke and Wheeler[3]; Chave[1]; Vinogradov[25]; Raup and Pilkey, quoted by Raup[21]). We shall refer to the skeleton as calcite, although it must be understood that there is always a certain amount of magnesium present.

CRYSTALLOGRAPHY

While most authorities consider each morphological element of the skeleton to be composed of a single crystal, it has recently been suggested (Garrido and Blanco[8]; Nissen[17]) that each crystal is, in fact, a collection of many submicroscopic crystallites, oriented the same way. There seems to be no compelling evidence that this is the case.

The orientation of the crystallographic axes is interesting. In general, there is no constant relationship between the morphological orientation of skeletal elements and their crystallographic axes (Raup[19]), though in echinoid spines the c-axis (optic axis) does appear to be always nearly parallel to the long axis (West[28]). The situation in the plates of the echinoid apical disk is interesting, in that the crystallographic axes of the individual plates define an axis of bilateral symmetry corresponding to that of the larva (Raup[20]).

The crystallographic homogeneity of the skeleton is shown strikingly under some conditions of fossilization, where new (secondary) calcite is deposited in the spaces within the ossicle, in continuity with the original (primary) calcite. However, the original calcite has a slightly different appearance from that of the infilling—it is less transparent—so that the form of the original skeletal material is preserved. Sometimes the secondary calcite also crystallizes beyond the limits of a plate or other ossicle (Fig. 20.1, C), and may even completely fill the interior of a fossil echinoderm test, or theca. When a fossil ossicle is broken, it usually splits along cleavage planes; when a non-fossil ossicle is broken, it cracks irregularly, the lines of cleavage being interrupted by the spaces. This manner of cracking may be of importance in imparting strength to the skeletal elements, as is argued below.

SECRETION AND ABSORPTION OF THE SKELETON

Echinoderm skeletal material is laid down by primary mesenchyme cells. The primordium is thought by Woodland,[32] Chun,[2] Theel,[23] and others to be secreted intracellularly, while Okazaki[18] has suggested that, in larvae at least, the cells secrete an organic envelope within which calcification occurs. Okazaki[18] has raised a further interesting point with regard to the organic envelope which has implications in the present work: she questions whether it is a mere envelope or whether it intrudes into the skeletal element to act as a ground substance on which mineral components are deposited. In an attempt to answer this she dissolved away the larval spicules with acid and showed that there remained sheaths of the same shape as the skeletal elements. In some larvae the interior of these sheaths was frequently tinged with orange-red, suggesting that some organic substance had remained behind after the inorganic material of the spicule had been removed. Further, Okazaki isolated larval spicules by centrifugation, removed any organic matter adhering to the external surface with caustic soda, then dissolved away the calcite. Here too there remained a 'cord of the same shape' as the spicule, a cord which could be seen by phase-contrast, or after staining with Nile blue sulphate. These observations strongly suggested to Okazaki that 'the skeletal envelope of sea-urchin larvae functions as an organic matrix for the spicule; that is, the growth of the spicule is not a purely physical crystallization of calcium carbonate within the envelope but a successive arraying of calcareous elements onto an organic substratum' (her p. 316).

Most of the change in shape during growth seems to occur by the differential addition of calcite where required, rather than by its resorption (Durham[7]); nonetheless, the available evidence does suggest that resorption can and does occur under some circumstances. For instance, Lovèn[15] described the resorption of some plates adjacent to the enlarging peristome in *Toxopneustes* (= *Strongylocentrotus*)

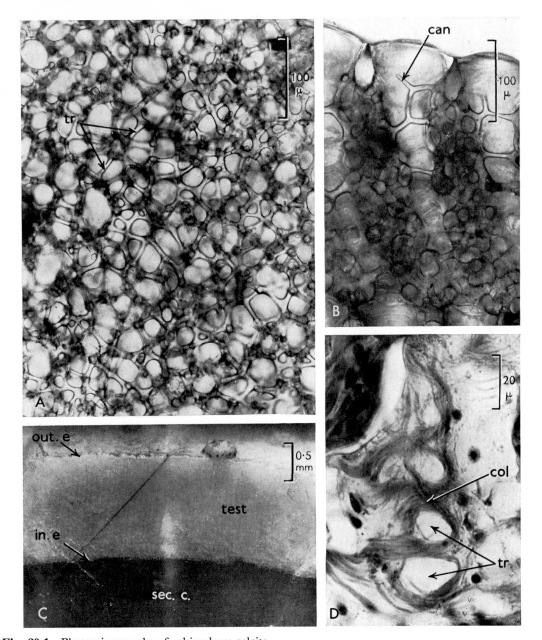

Fig. 20.1 Photomicrographs of echinoderm calcite.

A. T.S. centre of spine of the echinoid *Heterocentrotus mammillatus*, showing the typical open meshwork.

B. T.S. same spine as in A, but at its periphery, showing dense calcite permeated in places by canals. In A and B the organic material has been removed from spaces and canals.

C. Fossil calcite. Section across test of the Cretaceous spatangoid echinoid *Micraster cortestudinarium*, in which the primary calcite (top centre, with crack across it and spine tubercle on its outer surface) is in crystallographic continuity with the secondary calcite, filling the spaces and continuing beyond the limits of the original plate towards the interior of the animal (dark region, bottom).

D. The open calcite meshwork used as an anchorage for collagen. L.S. distal end of tube-foot of *Echinus esculentus*. Collagen fibres of the stem retractor system 'sewn' round trabeculae of the skeletal rosette.

KEY. *can*: canals permeating calcite; *col*: collagen fibres; *in.e*: inner edge of test; *out.e*: outer edge of test; *sec.c*: secondary calcite extending beyond inner edge of test; *test*: fossil test with trabeculae visible at inner edge; *tr*: trabeculae of calcite.

droebachiensis; Théel[24] described the reduction in size of spicules of the echinoid *pluteus* larva over a two-hour period under the influence of amoebocytes; and Gordon[9] described the resorption that occurs in various plates of the developing post-larval echinoid, in this case *Echinus* (= *Psammechinus*) *miliaris*.

THE PROBLEM AND METHODS OF STUDY

The evidence given so far concerning the crystalline state of the skeletal elements would appear to be conflicting. The evidence of mode of fossilization, manner of cleavage, and much of the crystallographic

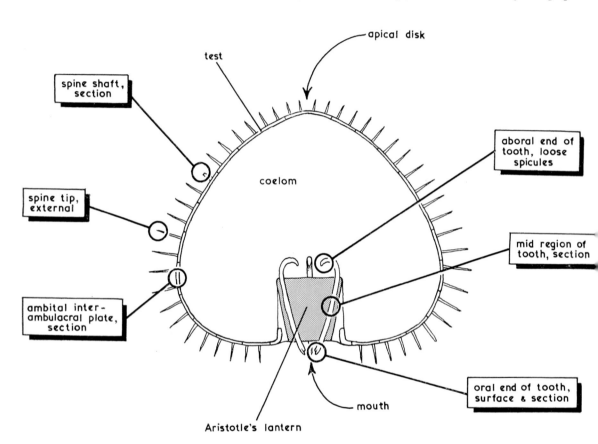

Fig. 20.2   Diagrammatic section through the sea-urchin *Echinus esculentus*, the animal used most widely in this study, showing the regions of the skeleton which have been examined by the scanning electron microscope. The diameter at the ambitus (the widest part) of the animal used was 10 cm.

evidence, seems to support the more conventional view, that each skeletal element consists of a single crystal. But the X-ray work of Nissen[17] and of Garrido and Blanco,[8] together with the embryological work of Okazaki,[18] would tend to support a 'polycrystal' hypothesis.

Normally, this problem could be resolved easily by replication of a broken surface of a skeletal element, and examination in the electron microscope. This has been done for many other skeletal materials, for instance, that of molluscs (Grégoire[10, 11]), but so porous is the skeleton of echinoderms, that a surface replica cannot easily be made, because of the difficulty of peeling away the replicating material. The recent invention of the scanning electron microscope, which requires no surface replication

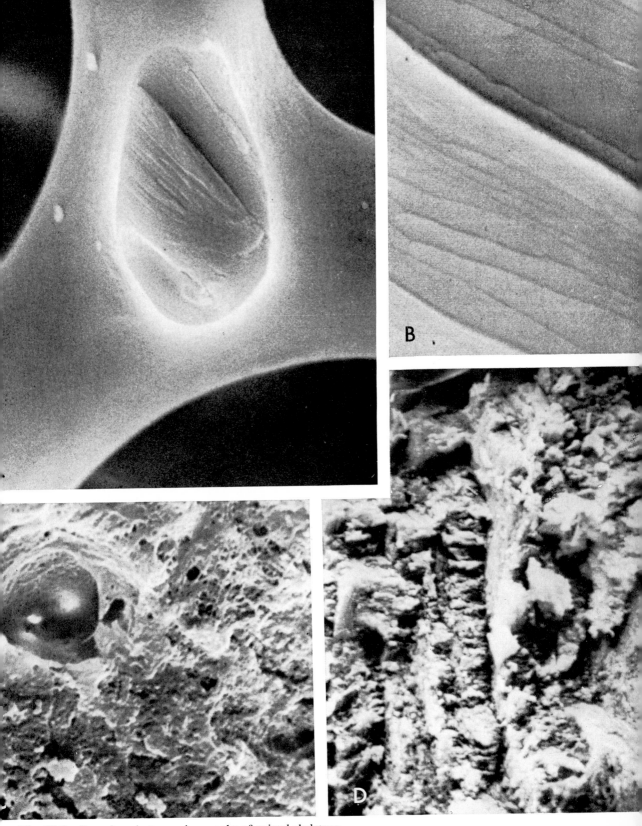

Fig. 20.3 Scanning electron micrographs of animal skeletons.

A. Echinoderm calcite. T.S. broken trabecula of interambulacral plate of *Echinus esculentus*, showing the comparatively clean break interrupted only by cleavage steps. × 5,000.

B. Natural calcite. Broken crystal, showing cleavage steps. × 5,000.

C. Vertebrate bone. T.S. broken femur of rabbit, showing irregular surface of break. × 2,500.

D. Mollusc skeleton. Edge of broken shell of the gastropod *Cepaea nemoralis*, showing highly accidented broken surface. × 2,500.

for the examination of surface structure, has meant that this problem can now be resolved, and an easily reproducible method be used for comparative work. The purpose of this contribution is to present results obtained with such an instrument, which show, first, that within the present limits of instrumental resolution, the elements of the skeleton do, indeed, appear to be single crystals; and, secondly, that a suggestion we made elsewhere (Currey and Nichols[6]), that the skeletal tissue is maintained by the cells permeating it, is strongly supported.

Pieces of skeleton from the sea-urchins *Heterocentrotus*, *Cidaris*, and *Echinus*, pieces of the shell of a gastropod mollusc *Cepaea*, a piece of rabbit bone, and a natural crystal of calcite, were all treated and examined in the following way. The pieces were first cleaned of all adhering organic matter with sodium hypochlorite solution; then washed in running tap water and several changes of distilled water; and finally dried. After mounting in a convenient attitude on a chuck, they were coated with about 500 Å of aluminium or palladium. In some cases this was evaporated on while the specimens were rotated quickly. The pieces were then examined using the Cambridge 'Stereoscan'. All the electron micrographs of echinoderm material reproduced here are of skeletal elements of *Echinus esculentus* (L.) (Fig. 20.2).

FINE STRUCTURE

The fracture surfaces of the echinoderm ossicles and those of the natural calcite are very similar in being smooth, whereas the fracture surfaces of the mollusc shell and the rabbit bone have a highly accidented (irregularly broken) appearance.

Fig. 20.3, A shows the broken surface of a spine of *Echinus*. Where the material has cracked, it is evident that the surface is smooth, except for the occasional cleavage steps that one would expect from a single crystal, which, in fact, are also shown by the broken crystal of natural calcite (Fig. 20.3, B). Fig. 20.3, D shows the broken surface of the shell of *Cepaea*, and demonstrates the appearance, at the same magnification, of a skeletal material which is known to be two-phased; the resolution of the instrument is insufficient to demonstrate the protein sheath of the individual crystals that can be shown by viewing a replica with a conventional electron microscope (Grégoire[11]). Similarly, bone (Fig. 20.3, C) viewed with the 'Stereoscan' shows a complex surface even though it is beyond the capabilities of any electron microscope to show, using replicative techniques, the individual apatite crystals at a surface.

The smooth appearance of the broken surface was characteristic of every echinoderm spine and thecal plate we looked at, and of every region of these elements.

These findings strongly suggest that each skeletal element is composed of calcite uninterrupted by a second phase of organic material binding small crystallites together. They further suggest, though less strongly, that each element is a single crystal, since, if it were composed of many entities, with markedly different orientations, it is unlikely that so smooth a broken surface would be produced.

In the echinoids there is one notable exception to the single-crystalline nature of single skeletal elements: the teeth of Aristotle's lantern. Each tooth grows continually from the aboral end, where it is laid down as a series of separate small crystals, which later become united. The nature of this union is unclear, but Salter[22] has suggested that small particles of calcite bridge the gap between separate crystals, and become fused to both. Whatever the mechanism by which the crystals are joined, they do not fuse completely into a single large crystal, but may be firmly united by trabeculae, which, according to Salter, have arisen as the tooth grows. This comes to resemble the familiar open meshwork system of other skeletal

Fig. 20.4 Scanning electron micrographs of echinoderm calcite.
    A. General view of a broken ambital interambulacral plate of *Echinus esculentus*, showing the smooth appearance of the trabecular surfaces when they have remained in contact with living material throughout life. × 2,800. [Note: Fig. 20.3A is an enlargement of part of this picture.]
    B. General view of the broken distal end of a tooth of *Echinus esculentus* at the same magnification as A, showing the rough appearance of the trabecular surfaces after they have lost contact with living material for some time. × 2,800.

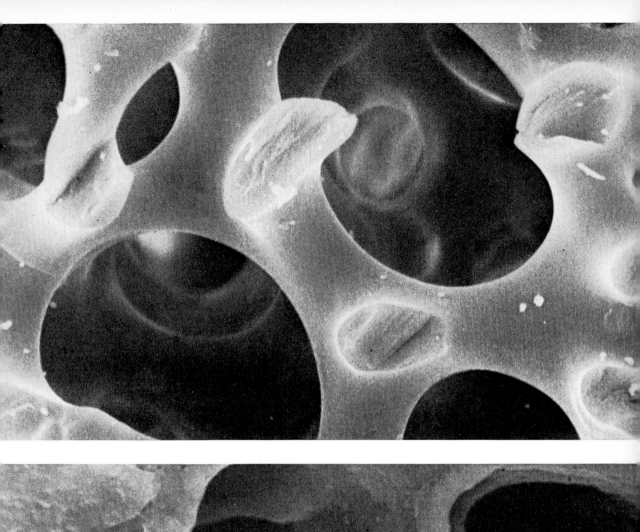
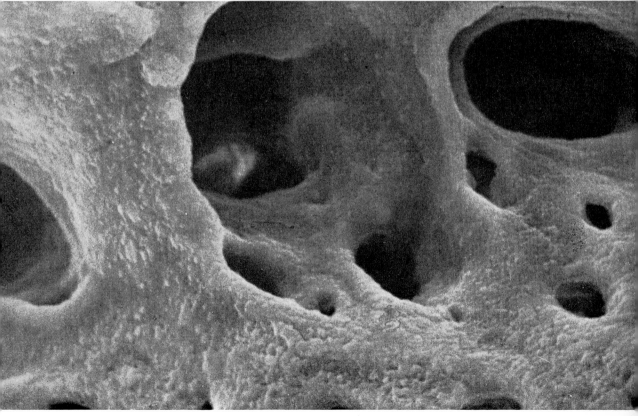

elements, though it is important to note that it appears to have arisen by a rather different mechanism. In other places in the tooth the separate crystals are simply pressed closely together (Fig. 20.5, A and B).

ECHINODERM CALCITE AS A STRUCTURAL SYSTEM

If echinoderm skeletal elements are single crystals, the question arises: what gives them their inherent strength? Virtually all other hard skeletons are two-phased materials, with one a stiff strong element and the other much less stiff, which may or may not be strong. For instance, in bone, these two elements are apatite and collagen (McLean and Urist[16]); in molluscs, calcite or aragonite, and conchiolin (Wilbur[29]; Grégoire[10,11]); in brachiopod shell, calcite and a protein (Jope[13]; Williams[30,31]); and in coral, aragonite and chitin (Wainwright[26,27]).

These systems presumably function like typical two-phased engineering materials. In the skeletons there are either elongated needles (or 'fibres' as they are generally called) or, less commonly, sheets of a strong stiff material, each embedded in a matrix of flabby material. The fibrous phase provides rigidity and is inherently strong, while the matrix may be comparatively weak. The matrix imparts its own share of strength to the system in the following way: inherently strong materials almost inevitably possess flaws, particularly on the surface. These cause local concentrations of stress at their ends and lead to very high values of stress locally, while the stress in the element as a whole may be quite small. The high local stresses may cause cracks to start running, under the influence of stresses on the element as a whole, which are much less than the material may otherwise theoretically bear. Once such a crack starts to run, it could easily spread right through the material, causing it to break; but in two-phase materials the cracks run out of the stiff fibres into the surrounding matrix. This, being a flabby material, will then distort, using up the energy of the crack as strain energy, but it will not break. In this way failures are kept local (Currey[4]; Kelly[14]). The matrix binds all the stiff fibres together, and also transmits stresses from one fibre to its neighbours, by shearing forces.

Since the echinoderms do not have the normal two-phase system, we must consider the mechanical properties when a single phase forms a meshwork. We say 'single phase', because the stresses in the skeleton will be transmitted almost entirely through the calcite—there will be almost no transmission of stress by the protoplasm, because the protoplasm has a very low stiffness, and, more importantly, because the calcite acts as a single block. We may conclude that the echinoderm system is not comparable with the fibre-matrix system. Any crack forming in the meshwork will very rapidly run out of the solid phase, into the interskeletal tissue. The force which can no longer be borne by the broken trabecula will now be transmitted to neighbouring trabeculae, through the calcite rather than through the organic tissue, unlike the case in conventional two-phase systems. The only effect on the skeletal element as a whole will be to raise somewhat the stress in the surrounding trabeculae. But as each trabecula is small compared with the overall size of the element, and there are many per small volume, then the extra stress in the surrounding trabeculae will not be large. This situation is quite different from what would be the case if the element were solid, for then the crack, once spreading, would travel through the entire member and destroy it.

The second consequence of the open meshwork system is that the surface area of the skeleton is much larger than that of a solid skeleton with the same overall dimensions. Since dangerous flaws usually start

Fig. 20.5 Scanning electron micrographs of the tooth of *Echinus*.
   A. Tooth broken approximately midway down its length, to show closely apposed stacked plates in the broken surface. × 330.
   B. Similar region, × 3,300.
   C and D. Needle-like ossicles from the tooth at different stages of development, both × 2,000. C. From the aboral (newest) end of the tooth, showing that they are unattached and that while surrounded by living tissue they have smooth surfaces. D. From the oral (oldest) end of the tooth, the part used for feeding. Here, the 'needles' have become attached to an underlying plate by cross connections, and the surfaces, now no longer surrounded by living tissue, are rough.

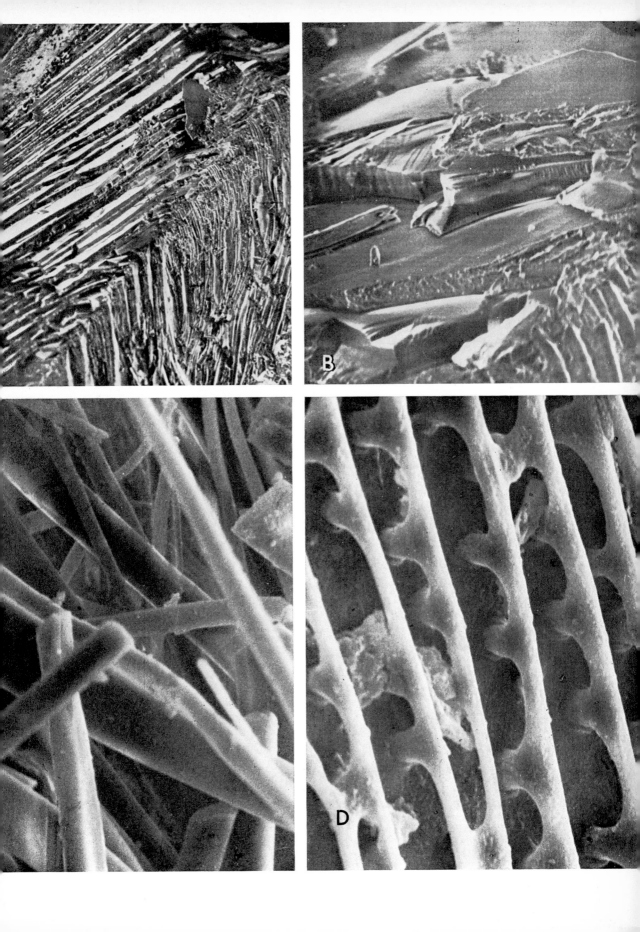

at the surface, this would seem to be a disadvantage. But all the surfaces are here in contact with living tissue. We have mentioned above (on p. 252) that calcite resorption takes place during post-larval development. There seems no reason to suppose that this ability would be lost in adulthood. The living tissue should, therefore, be able to cure dangerous flaws when they arise, by erosion and redeposition of calcite. Extending this idea somewhat, possibly even a broken trabecula could be made good.

This suggestion is supported by the following evidence: one would expect that the ability to repair a surface would be lost in any part of the echinoderm skeleton that is not permeated by living tissue. Such a situation arises at the oral end of the echinoid tooth. The tooth is used mainly for scraping food from the substratum, and is constantly replaced as it is worn away by abrasion. So far as can be seen with the optical microscope, no living tissue is present at the oral end. Figs. 20.4, B and 20.5, D show the surface of different parts of the oral end of an echinoid tooth. The trabecular structure can be seen; though, as mentioned above, such a pattern was formed in a different way from that of other skeletal elements. The surface is very rough when compared with the unbroken surfaces of trabeculae which are in contact with living material (compare Fig. 20.4, A with B). Furthermore, many of the trabeculae are broken, although this is not visible in the micrographs shown here. Of course, the end of the tooth, by virtue of its function, is likely to become broken and abraded, but it is noteworthy that the roughness continues to the deeper levels of the tooth, suggesting that even away from the outer surface there is erosion of some sort that cannot be made good in the absence of living material. Similarly, the distal ends of echinoderm spines often lose their covering of epithelium (Hyman[12]) though living tissue usually remains in the deeper spaces of the meshwork. Here too, at the outer surfaces of the spines we examined, surface flaws were very evident.

One further function of the open meshwork system is seen at the edges of the ossicles, and concerns the use of the cavities in the skeleton as 'lace holes' for joining together adjacent skeletal elements: collagen fibres intertwine between the two elements to hold them firmly together, while still allowing a certain amount of hingeing, where this is desirable. This is shown particularly clearly in the photomicrograph, Fig. 20.1, D.

CONCLUSIONS

In the Phylum Echinodermata the skeletal elements show a wide morphological diversity. Nonetheless, the microscopic structure is remarkably constant, and has been since the beginning of phanerozoic time. This constancy clearly reflects a selective advantage. In this article we have suggested what this advantage might be. The echinoderms, unique in so many features, have apparently also solved the problem of skeletal strength in an original way. Why they have apparently never adopted the more usual two-phased skeletal system is at present impossible to say.

This study is in its early stages; that it has been at all successful is largely due to the recent development of the scanning electron microscope.

SUMMARY

The recent invention of the scanning electron microscope has provided a most convenient technique for the comparative examination of the broken and natural surfaces of skeletal elements of echinoderms. This has shown, first, that within the limits of resolution of the machine, each element most likely consists of a single crystal of calcite, and, secondly, that the natural surfaces of the open meshwork may be actively maintained in a flaw-free condition by the living tissue permeating its cavities.

ACKNOWLEDGEMENTS

We have had to rely heavily on help from colleagues working with the scanning electron microscope. Particularly, we have been assisted in a most generous way by Dr. J. Sikorski and Mr. T. Buckley of the Textile Physics Department, University of Leeds; by Dr. G. R. Booker, Department of Metallurgy, University of Oxford; and by Miss P. J. Killworthy and Miss J. Farrow, Cambridge Instrument Company. In addition, we have had help from Mr. J. McCrea and Mr. J. Haywood of the Department of Zoology, Oxford. Lastly, and most importantly, we have both benefited over many years, in a manner, and to a degree which cannot adequately be acknowledged, by the unfailing help, friendship and inspiration of Dr. John R. Baker, F.R.S.

REFERENCES

1. CHAVE, K. E. 1954. Aspects of the biochemistry of magnesium. 1. Calcareous marine organisms. *J. Geol.* **62**, 266–83
2. CHUN, C. 1900. *Aus den Tiefen des Weltmeeres.* Berlin
3. CLARKE, F. W. and WHEELER, W. C. 1922. On the inorganic constituents of marine invertebrates. *Prof. Pap. U.S. geol. Surv.* **124**, 1–62
4. CURREY, J. D. 1962. Stress concentrations in bone. *Quart. J. micr. Sci.* **103**, 111–33
5. ——. 1964. Three analogies to explain the mechanical properties of bone. *Biorheology* **2**, 1–10
6. —— and NICHOLS, D. 1967. Absence of organic phase in echinoderm calcite. *Nature, Lond.* **214**, 51–83
7. DURHAM, J. W. 1955. Classification of clypeasteroid echinoids. *Bull. Dept. Geol. Univ. Calif.* **31**, 73–198
8. GARRIDO, J. and BLANCO, J. 1947. Structure cristalline des piquants d'oursin. *C.R. Acad. Sci. Paris,* **224**, 485
9. GORDON, I. 1926. The development of the calcareous test of *Echinus miliaris. Phil. Trans. Roy. Soc.* B **214**, 259–312
10. GRÉGOIRE, C. 1957. Topography of the organic components in mother-of-pearl. *J. Biophys. Biochem. Cytol.* **3**, 797–808
11. ——. 1961. Sur la structure submicroscopique de la conchioline associée aux prismes des coquilles de mollusques. *Inst. roy. Sci. nat. Belge* **37**, 1–34
12. HYMAN, L. H. 1955. *The Invertebrates.* Vol. IV. Echinodermata. New York: McGraw-Hill
13. JOPE, H. M. 1965. Pt. H, Brachiopoda. In *Treatise on Invertebrate Paleontology.* University of Kansas Press.
14. KELLY, A. 1966. *Strong solids.* Oxford University Press
15. LOVÈN, S. 1874. Études sur les Echinoidées. *K. svenska VetenskAkad. Handl.* new series, **11** (7), 91
16. MCLEAN, F. C. and URIST, M. R. 1961. *Bone.* University of Chicago Press
17. NISSEN, H.-U. 1963. Röntgengefügeanalyse am Kalzit von Echinodermenskeletten. *N. Jb. Geol. Paleont. Abh.* **117**, 230–4
18. OKAZAKI, K. 1960. Skeleton formation of sea-urchin larvae. II. Organic matrix of the spicule. *Embryologia* **5**, 283–320
19. RAUP, D. 1959. Crystallography of echinoid calcite. *J. Geol.* **67**, 661–74
20. ——. 1965. Crystal orientations in the echinoid apical system. *J. Pal.* **39**, 934–51
21. ——. 1966. The endoskeleton. In *Physiology of Echinodermata* (edited by R. A. Boolootian), pp. 379–95. New York: Wiley
22. SALTER, S. J. A. 1861. On the structure and growth of the tooth of *Echinus. Phil. Trans. Roy. Soc.* **151**, 387–407
23. THÉEL, H. 1892. On the development of *Echinocyamus pusillus. Nova Acta Reg. Soc. sci. upsal.* ser. III, **15**, 1–57
24. ——. 1894. Notes on the formation and absorption of the skeleton in the echinoderms. *Öfvers K. VetenskAkad. Förh. Stockh.* **8**, 345–54
25. VINOGRADOV, A. P. 1953. The elementary composition of marine organisms. *Sears Foundation Marine Research, New Haven Mem.* No. 2, 647 pp.
26. WAINWRIGHT, S. A. 1963. Skeletal organization in the coral, *Pocillopora damicornis. Quart. J. micr. Sci.* **104**, 169–83
27. ——. 1964. Studies of the mineral phase of coral skeleton. *Exp. Cell Res.* **34**, 213–30
28. WEST, C. D. 1937. Note on the crystallography of the echinoderm skeleton. *J. Pal.* **11**, 458–9
29. WILBUR, K. M. 1960. Shell structure and mineralization in molluscs. *Amer. Assoc. Adv. Sci. Pub.* **64**, 15–40
30. WILLIAMS, A. 1956. The calcareous shell of the Brachiopoda and its importance to their classification. *Biol. Rev.* **31**, 243–87
31. ——. 1966. Growth and structure of the shell of living articulate brachiopods. *Nature, Lond.* **211**, 1146–8
32. WOODLAND, W. 1906–1908. Studies in spicule formation, Pts. I, II, III, and IV. *Quart. J. micr. Sci.* **49**, 305–25; 533–59; **51**, 45–54; 483–509

Part 2

The Quantitative Interpretation of Cell Structure

Chapter 21

Quantitative Histochemistry: Cell Structure revealed through Cell Function

J. CHAYEN

INTRODUCTION

Basically, the aim of all histochemistry is quantitative: no sooner has a certain substance been demonstrated in a tissue than the worker wants to know something of its abundance in that tissue. Even the simplest studies compare the intensity of histochemical colour reactions obtained in various tissues, or after different treatments. There is an inherent assumption that a definite stoichiometry exists between the amount of a substance in a section or smear, and the amount of colour produced. It may well be that this assumption is not always valid, even when the reacting substance is fully available for histochemical demonstration. However, even in such cases where a definite number of dye-molecules will be produced in unit time, by each reacting group or enzyme present in the tissue section, the amount of dye produced is still only a measure of the reacting moieties which were available, or free to react, at the time the section was tested. One of the clear rules that seems to be emerging, is that remarkably few potentially reactive moieties are free to react with exogenously added material, until some sort of damage has liberated them from their linkage with the rest of the tissue. This damage is often caused by the methods used in preparing the tissue sample; it may also be induced by the procedure used for demonstrating the reactive molecule. Examples of each type will be discussed. The point to be stressed at this juncture, is that histochemistry, if performed with its fullest subtlety, can disclose not only the total amount of reactive moieties, be they reactive groups or enzymes, but also how much (or how little) is freely available in life until 'unmasked' by physiological, damaging or pathological changes. Consequently a quantitative assessment of the ratio of free (or 'manifest') to bound (or 'latent') activity, may form the most delicate indicator of the nature and integrity of cellular structure. For example, electron microscopy can show lysosomal membranes but, up to the present, only histochemistry can assess the permeability, and hence the integrity, of those membranes. Similarly, electron microscopy has shown the detailed structure of mitochondria, but recourse has to be made to quantitative enzyme histochemistry if the permeability of the membrane to cofactors and substrates, or if the functioning of the electron-transport system, is to be evaluated. Yet the permeability is a measure of the integral *structure* of the outer membrane, and the electron-transport function may depend on the precise *structure* of the electron-transport particle (for a biochemical discussion of this problem see [1]).

DENSITOMETRY

The histochemist observes a weak, moderate, strong, or intense colour as a consequence of his reaction, and may report them as $+$, $++$, $+++$, and $++++$ respectively. In such studies, it is rare to find evidence that the colour observed bears a linear relationship to the amount of reaction produced. Moreover, it deserves to be emphasized that the eye tends to be attracted by the amount of stain per unit area of the cell, and very often of a selected unit area at that. It seems incapable of integrating the total amount of dye in a cell.

A step in converting histochemistry from an exercise in the instructive colouring of tissue sections, or smears, to an exact science, came with densitometric measurement of the images seen by microscopy. Among the earliest, and most successful, were those of Caspersson and his associates (e.g. see [2]). These depended on the facts that nucleic acids absorb light of the shorter range of the ultraviolet, and that the amount of the light absorbed, per unit weight of nucleic acid, is considerably more than for most naturally occurring molecules. In more precise terms, at 260 mμ the molecular extinction coefficient of the nucleic acids is far greater than that for most substances likely to occur in cells, so that the amount of absorption shown by cells, at around this wavelength, could be related directly to the actual mass of nucleic acid present (but see [3,4]). These measurements could be made in two ways. In the first, the cells were photographed at 265 mμ, and the amount of blackening shown in the negative was measured by the use of a microdensitometer. The conversion of blackening on the negative into the equivalent absorption by the cells, required comparison with some standard reference system; the amount of apparent absorption at 265 mμ had also to be corrected for non-specific absorption. This was usually done by estimating the scattered light, by photographing at a wavelength, such as 310 mμ, at which there was no true absorption (see [2]).

The second way of making these measurements was by the use of suitable photocells to detect the ultraviolet light, with photo-multipliers to magnify the effect to a level that could be measured accurately (see [2,4]). These procedures became less simple when it was found that nucleic acids in living cells, being associated with protein, had very different absorption characteristics than when fixed, or isolated and purified.[4,6] On the other hand, although the precise stoichiometry of staining methods for DNA, such as the Feulgen reaction[7] or the methyl green method of Kurnick[8] may also be variable, they were found to give reproducible results which had considerable meaning in cell biology. For use with these techniques, photometric methods were devised; also a most ingenious photographic procedure has been described in which the nuclei are photographed and the enlarged photographic image is weighed, to give a measure of the Feulgen stain.[9] The earlier studies used a simple photocell, or photomultiplier, and the image was projected on to the light-sensitive surface. Unfortunately, this type of simple measurement could produce considerable error, owing to the physical nature of the protoplasm which was stained. A very clear exposition of this was given by Gomori[10] (see also [4,9,11]). To overcome the optical heterogeneity of the stained tissue, Deeley[12] produced a scanning micro-densitometer which is available commercially, and American workers developed rather complicated two-wavelength procedures (see [9]).

Consequently, it is now possible to stain DNA by a fairly stoichiometric method, and measure the amount of the dye (and thus, arbitrarily, the relative amount of DNA), present in a selected nucleus. Since the methods are integrating methods, that is amount of dye at each point in the nucleus is integrated, the total amount of DNA per nucleus can be measured, in arbitrary terms. The terms are arbitrary, because it is not usually possible to relate the number of dye molecules to the actual weight of DNA. This factor will vary with the fixation, the conditions under which the reaction is carried out and, possibly, with the nature of the binding between the nucleic acid and the rest of the chromatin. However, for most purposes, the important question is the degree of ploidy of the nuclei, and work of tremendous value has come from comparison of the amount of Feulgen stain per nucleus with that of the spermatid (e.g. see [13,14]).

Direct measurement of the amount of stain observed in a nucleus or in an intra-cellular structure, has been developed considerably. Sandritter and his associates[15] have studied the stoichiometry of a gallocyanin reaction for DNA. Since DNA frequently occurs in association with histone, an arginine-rich protein, quantitative methods for basic protein,[16,17] and for arginine,[18] have been developed, and studied by microdensitometry. This type of study has been useful in measuring the proportion of DNA to basic protein.[19] It becomes less easy to evaluate when the amount of arginine, or of some other amino-acid, is measured in a non-chromatin structure. For example, Danielli[20] developed a number of precise reactions for amino-acids, and these could be measured by scanning microdensitometry. The question is, what do such measurements mean? Suppose, for simplicity, you are confronted with a cell which contains a single region which stains intensely for arginine. The amount of arginine present can be measured very accurately, to the nearest picogramme (10^{-12} g) with a reproducibility of ± 3 per cent (see [18]). Yet you cannot determine from such a measurement whether this means that the cell contains an abnormal protein which is especially rich in this amino acid, or whether it only has an exceptional amount of a normal protein which contains the normal proportion of arginine. In more exact terms, the measurement of 0·5 unit of arginine in the cell could be the result of 1 unit of a protein containing 50 per cent of its weight as arginine or it could be produced by 10 units of a more normal protein of which only 5 per cent was arginine.* This could be resolved if the amount of stain were measured and calculated as against the total dry mass of the cell, or of the relevant part. Thus it may be that this type of measurement will become of importance only when it is done in conjunction with interference microscopy. However, even then, it will not disclose the true composition of the protein, but only how many of its reactive amino-acids have been rendered available for the particular histochemical reaction. For all that, the method could be of considerable value, in that it would permit an assessment to be made, of the extent to which altered environmental conditions, either physiological or pathological, had affected the structural binding of these amino-acids.

ENZYME HISTOCHEMISTRY: 'MANIFEST' ACTIVITY AS COMPARED WITH 'LATENT' ACTIVITY

Assessment of the availability of active groups has found powerful significance in the study of those changes in sub-cellular structure which can be demonstrated by enzyme histochemistry. The simplest example is found in the study of the lysosome. The history of these sub-cellular particles has been much reviewed (e.g. [21,22]). It was found by de Duve and his co-workers, that the mitochondrial fraction, isolated by differential centrifugation of homogenates of rat liver, could be split into two. One contained the mitochondrial respiratory enzymes, while the other lacked these, but contained the hydrolytic enzymes which worked optimally at an acidic pH value. Since the particles in the second fraction contained the hydrolytic, and indeed the potentially cell-autolytic enzymes, they were named lysosomes. A special property of these bodies was that their enzymes were not able to act on the intact cytoplasm, because they were isolated from the cell itself by the lysosomal membrane; yet this membrane was so altered by acidic conditions (as well as by substances which could disturb lipid-protein associations), that the enzymes could be liberated free into the suspending media used by biochemists. At first they were regarded as 'suicide capsules': little sacs inside the cytoplasm, filled with destructive and active enzymes which were 'latent' only because the sac membrane separated them from the cytoplasm. Once they were liberated into the cytoplasm, as in the biochemical procedures used for estimating their presence, then, it was believed, they might digest the cell. It was later found that the amount of 'free' activity, under identical conditions of biochemical testing, varied in different tissues, and even under different conditions. The use of unfixed sections allowed Bitensky[23] to emphasize that the fragility of the membrane might be not only an indication of the potential likelihood that these destructive enzymes might leak into the cytoplasm, but rather, that it might show how readily substrate could enter these particles, and be 'detoxicated' inside the membrane. Thus, it now seems likely that these enzymes work inside the lysosomal membrane, on substances which enter the particle; and that the structural nature of the membrane controls how readily substrate can pass to the intra-lysosomal enzymes. Consequently, it is the structure of the membrane of this organelle that must vary to permit different manifest activities

---

* [*Editorial note:* Refractive index measurements made by immersion refractometry with phase-contrast microscopy, or by interference microscopy, have shown that the cytoplasm in a very wide variety of different kinds of living cells contains between 10 per cent and 15 per cent weight/volume of solid material. Much of this material is protein, which indicates that in most cells the amount of cytoplasmic protein probably does not vary by even as much as $\times 2$. Consequently, in the hypothetical case cited, the editors consider that it is much more likely that an intense staining reaction for arginine would indicate the presence of an abnormally high proportion of this substance, and agree with the author that the dilemma may be resolved by a combination of methods of study.]

of the enclosed enzymes. Changes in the structure of this membrane have been sought, without avail, by electron microscopy; this is not surprising since the preparative methods used for examinations by this instrument are of such a kind as to disturb the lipid-protein association more than would be necessary for delicate changes in permeability to be effected. Similarly, the use of fixed sections for enzyme histochemistry has shown only that intracytoplasmic bodies exist which contain lysosomal enzymes. Whether such bodies are real, or represent coagulated cytoplasm, cannot be established: the same effect would be observed if lysosomes were present, with the enzymes fixed at their site (Fig. 21.1, a), or if the organelles were destroyed and the enzymes evenly distributed (Fig. 21.1, b) throughout a very unevenly coagulated cytoplasm (Fig. 21.1, c). A situation of this sort has been considered in relation to the lysosomal enzyme β-glucuronidase.[24]

Yet changes in the properties of the lysosomal membrane, under various conditions, have been shown by histochemical examination of unfixed cryostat sections. Such changes have indicated that lysosomes are remarkably and reversibly activated by a wide range of physiologically and pathologically

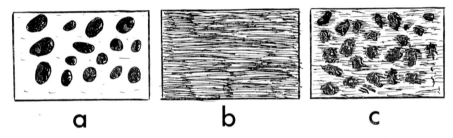

a   b   c

Fig. 21.1   Diagrammatic representation of possible effects of fixation on lysosomal enzymes.
1a. Lysosomal enzymes, unfixed, stained black, inside the lysosomes.
1b and c. The possible effect of fixation. The fixative could first destroy the organelles and disperse the enzymes uniformly throughout the cytoplasm (1b). The secondary effect might be to coagulate the cytoplasm, each 'lump' of coagulum having a high concentration of the lysosomal enzymes, so simulating a true lysosome (1c).

abnormal conditions (e.g. see [23, 25, 26, 27, 28, 29]). Indeed, the altered permeability of the lysosomal membrane is a very delicate indicator of cellular disturbance, and so has a significant place in cell biology, and in interpreting the functional structure of cells. This depends on demonstrating the latency of intralysosomal enzymes. However, the concept requires some elaboration:—

The sections of 'unfixed tissue' must be prepared in such a way that the lipid-protein associations of the cells are relatively intact (e.g. by the methods described by [30, 31, 32]). Suppose such sections are stained for lysosomal acid phosphatase by the Gomori technique (as modified by Holt[33] and Bitensky[32]), and stained particles are not produced in a cell type until the reaction has proceeded for 20min. Suppose that in contrast, a strong reaction can be obtained in macrophages after only 5 min. Suppose further that there is a progressive increase in the number of particles seen, and in the intensity of the staining. This does not prove that the enzyme is latent within the particles. Two interpretations are possible (see Fig. 21.2).

All that is assumed to be known is that, after 20 min incubation, the intensity of the staining of the particles has reached a point X which is of sufficient intensity for us to see the particles. This same level (M) is reached in the macrophages with only 5 min incubation. We require to know how this intensity is achieved. In macrophages it may be reasonable to assume that OM is a straight line, so that staining continued linearly (or nearly so) until sufficient lead was deposited to reach the threshold of detection, when converted into the sulphide. With the other cells this could also have been true: OX could be a straight line, so that it was only when this slowly acting enzyme had worked for 20 min that its reaction product was sufficient for us to detect. In this case the enzyme is not latent; it is just weak. However, an alternative possibility is that the enzyme is quite active, but that it is 'latent', so that an initial interval, say 10 min, was taken up in allowing the incubation medium to modify a lysosomal membrane so as to render the 'enclosed enzyme' manifest.

To test this second possibility in practice, sections are treated with incubation medium which lacks either the lead, or the glycerophosphate, or even both. If the enzyme is 'latent' in this way, then it will be found that a short pre-treatment in buffer alone (for example for 10 min), will allow the same intensity of staining to be reached after only 10 min in the incubation medium as would otherwise be reached in 20 min incubation. We could see, then, that the enzyme was quite active, but that its 'true' activity could be asserted, or made manifest, only by altering some physical barrier in the section (by treatment with buffer). (In this connection, comparison of the angles XLP and XOP in Fig. 21.2, could be significant.) In our theoretical example, we could say, therefore, that the enzyme in the macrophage was fully manifest (or nearly so) whereas the 20 min required to show activity in the other cells in the rest of the section was due to its being in a latent form, which became manifest only with prolonged incubation. We can say this because we have shown in the example that, once the permeability factor has been overcome by exposure for 10 min to the acidic buffer alone, it takes only 10 min to achieve the same degree

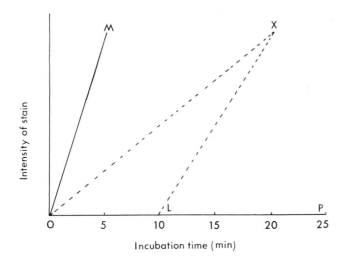

Fig. 21.2 Latency: interpretation of the intensity:time sequence in the staining of acid phosphatase activity of lysosomes.

OM: The intensity of stain in macrophages goes from zero to intensity M in a very short time; i.e. there appears to be no latency.

X: This level of staining in other tissues could be achieved either by slow, continuous increase of lead phosphate (OX) in which case there is no evidence of latency. Alternatively it could have been achieved by a rapid reaction (LX) which occurred only after a period (OL) during which the latency of the enzyme (i.e. the impermeability of the lysosomal membrane) had first to be overcome.

of staining (i.e. of activity) in other cells as is seen in macrophages, or as is seen after 20 min in the full medium. We would then know that the activity: time curve for a 20 min incubation does not follow a line like OX but a line, or a curve, which approximates to OLX.

This type of reasoning was the basis of Diengdoh's[25] evidence that lysosomes could be demonstrated in mouse skin. This tissue is very refractory to homogenization, so that biochemical proof of the presence of such delicate organelles is very difficult. However, Diengdoh demonstrated that particles could be shown histochemically that had acid phosphatase in a latent state, which could be rendered manifest by the same procedures of damage, by osmotic or surface-active agents, as had been used by de Duve for lysosomes isolated from liver. More recently, Chayen and others (in preparation) have used this method to demonstrate the lysosomal nature of β-glucuronidase.

The particular value of this type of study is that it allows some evaluation of the 'strength' of the permeability barrier that must be modified before the full activity of an intra-particulate enzyme can be made manifest. It may give us knowledge of what may be termed 'functional structure' rather than the 'morphological structure', shown by electron microscopy.

This concept of 'functional structure' is easier to understand if the membranes of mitochondria are studied. The membrane structure, as seen by electron microscopy, is remarkably constant however the amount of activity detected is changed within the limits of the physiological range of a given normal cell. It seems likely that some of the control of cellular metabolism may lie in changes in the mitochondrial membrane (e.g. see [34,35]). To study these changes in functional structure, some more quantitative histochemistry had to be devised.

QUANTITATIVE ENZYME HISTOCHEMISTRY

One particular technique has been developed, derived from an earlier procedure learned while working in Professor Brachet's laboratory in Brussels. In that laboratory, it was shown that the amount of nucleic acid in a tissue section could be found, if the section was stained with toluidine blue,[36] and the bound dye was then extracted, and measured in a spectrophotometer. A similar method was later used by Jarditsky and Glick[37] for enzyme histochemistry.

Thus, it seemed reasonable to ask whether, after tissues had been 'stained' for dehydrogenases, the amount of formazan which produced the stain, could be eluted and estimated, in the same way as had been done with toluidine blue. This question seemed to us to be of practical importance when it became apparent that malignant growths show greater pentose-shunt dehydrogenase activity than normal tissues.[38]

In the procedure for dehydrogenase activity, the unfixed tissue section is incubated at 37°C in the presence of the following: a buffer which is selected to give optimal conditions for the enzyme; substrate for the enzyme; co-factors or activators that may be found to be necessary, and a hydrogen-acceptor. The principle of the technique is that the dehydrogenase removes hydrogen from the substrate, and passes it, through oxidation-reduction systems, to a system which can transfer the hydrogen, or the electrons, either to another oxidation-reduction system or to oxygen, or to the hydrogen acceptor which is used in the histochemical method. The hydrogen-acceptor is selected to compete, more or less effectively, for the hydrogen, and, in accepting it, becomes deposited on the section as a coloured dye. Tribute must be paid predominantly to Professor A. M. Seligman for developing the various tetrazolium salts for this purpose. The reduced, insoluble dyes (the formazans) from some tetrazolium salts such as nitroblue tetrazolium precipitate so rapidly, and so finely on to the section, and they have such staining properties, that they give remarkably accurate localization of activity within cells. However, they stain so well that they cannot be extracted readily. Consequently, neotetrazolium which precipitates more coarsely was selected for the procedure. The formazan of neotetrazolium does not yield precise subcellular localization, but it may be extracted, quantitatively, into a mixture of 10 per cent heptan-1-ol in tetrachloroethane.

STOICHIOMETRY

The first problem was to see if the activity of an enzyme bore a direct relationship to the amount of formazan which it produced. To test this it was necessary to show that the amount of formazan deposited was directly proportional to the activity of the enzyme. Enzymes are more or less active depending on whether they are tested at optimal, or less optimal pH values. Hence, the fact that a true pH: activity curve could be made for histochemical succinate dehydrogenase activity[39] was one way of demonstrating stoichiometry. Another was to show a linear increase of activity, as measured by the amount of formazan produced, with time. Lastly it was found possible to derive a Michaelis constant for this enzyme activity. The particular advantage of succinate dehydrogenase for this demonstration is that it is a tightly bound enzyme which is entirely 'manifest' owing to the trauma induced in the sections by the incubating procedures.

THE USE OF UNFIXED SECTIONS

Enzyme histochemists face a dilemma, whether to use fixed or unfixed sections. The advantage of fixed sections is that the soluble enzymes, or the lyo-form[40] of less soluble enzymes, will not escape into the incubation medium with the alacrity shown by enzymes from unfixed sections. However, fixation causes various degrees of inactivation, and may change not only the physical properties of the enzyme itself, but also its co-ordination with the metabolic pattern of the rest of the cell. Moreover, it may make membranes totally permeable, so that changes in permeability, (which may be critical in the control of living metabolism) become impossible to study.

These difficulties were overcome when unfixed cryostat sections were used, because of two facts. First, enzyme loss is in some measure reduced by the use of a protective agent in the medium during incubation*. A colloid stabilizer, polyvinyl alcohol (PVA) was therefore used in our work. This had previously been shown to act as a narcotic in 'mummifying' cells[41]; and had been used to protect isolated cells, and for sub-cellular matter during homogenization[5, 42]. Second, and more importantly, it was found that soluble dehydrogenases in the sections were lost from them if these enzymes were not fully active. This is true even when they could act on a substrate in the presence of coenzyme, provided there was no hydrogen acceptor. It is necessary for the substrate, coenzyme, hydrogen acceptor, and a sufficient concentration of the stabilizer (PVA) all to be present[43, 44]. When they are all present however, it was found that none of the soluble dehydrogenases tested, nor any of the measurable nitrogenous matter in the section, was lost, to any detectable degree, into the incubation medium. When these conditions occur, it is as if a 'molecular cohesion' holds the reacting groups in the protoplasm of the cut sections together, and allows active biochemical work to be done. There is also some evidence that, if the tissue blocks are treated with a fairly dilute solution of PVA before they are chilled, this 'molecular cohesion' may be further protected.

THE EFFECT OF DAMAGE ON THE PERMEABILITY OF MITOCHONDRIA

The need for refined techniques, in which membranes are protected against the damage inflicted by the incubation medium, was shown by studies on glutamic dehydrogenase.[45, 35] In these investigations, blocks of rat liver were prepared, and some were pressed, more or less severely, with a flat instrument. The object of this was to test the effect of mechanical damage, of the type associated with homogenization, but far gentler, in that the pressure was insufficient to distort the histology of the tissue. The effect of such pressure on the 'manifest activity' of a mitochondrial, NAD-dependent enzyme, glutamic dehydrogenase, was studied. Consequently all blocks of tissue were processed similarly, and tested for this enzyme activity, by quantitative histochemistry. To make rigorous comparison between the results obtained from each treatment, five sections were tested from each block, and the results were examined for significance by Student's statistical test. By this means it was shown that mechanical damage can increase the 'manifest activity' of this dehydrogenase, at least three-fold. This increased activity would seem to be due to increased permeability of the mitochondrial membrane to the co-factor, and possibly to the substrate (see [34, 46, 35]).

CONCLUSION

It would seem, then, that a non-disruptive biochemistry has been developed, that can be applied to intact tissue slices. It is quantitative, and its results can be subjected to statistical tests for significance. It is, perhaps, more valuable than conventional biochemistry, in that it can detect changes in, for example, the permeability of the mitochondria smaller than those which can be studied with isolated mitochondria. If gentle pressure induces a three-fold increase in activity of a mitochondrial enzyme, a full homogenization procedure is likely to exceed this (see [46]). Yet the early stages of cell

* See pp. 38–39 of David and Brown's chapter in this book.

damage[47] seem to involve changes in mitochondrial permeability only up to that produced by the gentle pressure used by Wells and others.[45] Thus it is not surprising that homogenate biochemistry has not been able to detect the biochemistry of cell damage, but only that of cell death. It seems probable that the responses of cells to damaging agents or conditions, or their responses to physiological and pharmacological agents, begin with changes in the functional structure of membranes, or of linked enzyme systems. Such changes have been reported for the human endometrium prior to menstruation,[27] and for human myocardium during prolonged open-heart surgery.[48] It seems likely that 'functional structure' can best be studied by this new form of 'non-disruptive biochemistry' which we term quantitative histochemistry.

SUMMARY

The development of histochemistry, from the art of colouring tissues into the science of tissue chemistry, has necessitated the measurement of the chemically active groups of intact tissue sections. Methods for doing this are described. As a consequence of such techniques it is now possible to meld micro-biochemistry with histology, and so obtain a measure of the chemical and physiological functioning of cells. It is claimed that this assessment of cellular function gives information about the molecular architecture of cells, which cannot be obtained, as yet, even by electron microscopy.

ACKNOWLEDGEMENTS

Mr. H. J. Cotes played a decisive part in the development of the kind of quantitative histochemistry described here: from his personal money, he presented us with a spectrophotometer. His understanding of the need for this development and his very great and immediate generosity, has earned our deep gratitude. In a short discussion apparently he understood what various erudite committees had not; the fact that this form of quantitative histochemistry is now beginning to be established is due very largely to his understanding. I am also grateful to be allowed to acknowledge the debt I owe to my immediate colleagues who joined me in this adventure in the quest for a basis for quantitative histochemistry; it is their work, skill and enthusiasm which has developed this subject. In particular I acknowledge my great indebtedness to Dr. Lucille Bitensky, Roger Butcher, Peter Altmann, and Len Poulter.

REFERENCES

1. LEHNINGER, A. L. 1965. *The Mitochondrion.* New York: Benjamin
2. CASPERSSON, T. 1950. *Cell Growth and Cell Function.* New York: Norton
3. CHAYEN, J. 1953. Ascorbic acid and its intracellular localisation, with special reference to plants. *Int. Rev. Cytol.* **2,** 77–131
4. WALKER, P. M. B. 1958. Ultraviolet absorption techniques. In *Physical Techniques in Biological Research* (edited by G. Oster and A. W. Pollister), Vol. **3,** pp. 402–87. New York: Academic Press
5. CHAYEN, J. 1960. The localisation of desoxyribose nucleic acid in cells of the root meristem of *Vicia Faba. Exp. Cell Res.* **20,** 150–71
6. DAVIES, H. G. 1954. The action of fixatives on the ultra-violet-absorbing components of chick fibroblasts. *Quart. J. micr. Sci.* **95,** 433–57
7. LESSLER, M. A. 1951. The nature and specificity of the Feulgen reaction. *Arch. Biochem. Biophys.* **32,** 42–54
8. KURNICK, N. B. 1950. The quantitative estimation of desoxyribose nucleic acid based on methyl green staining. *Exp. Cell Res.* **1,** 151–8
9. POLLISTER, A. W. and ORNSTEIN, L. 1959. The photometric chemical analysis of cells. In *Analytical Cytology* (edited by R. C. Mellors), 2nd edition, pp. 431–518
10. GOMORI, G. 1952. *Microscopic Histochemistry.* University of Chicago Press
11. CHAYEN, J. and DENBY, E. F. 1968. *Biophysical Techniques.* London: Methuen
12. DEELEY, E. M. 1955. An integrating microdensitometer for biological cells. *J. Sci. Instr.* **32,** 263–7
13. SWIFT, H. H. 1953. Quantitative aspects of nuclear nucleoproteins. *Int. Rev. Cytol.* **2,** 1–76
14. LEUCHTENBERGER, C. 1957. Quantitative cytochemistry (microspectrophotometry). A fruitful approach to the study of disease. *J. Mount Sinai Hospital* **14,** 971–82

15. SANDRITTER, W., KIEFER, G., and RICK, W. 1963. Über die Stöchiometrie von Gallocyaninchromalaun mit Desoxyribonukleinsäure. *Histochemie* **3**, 315–40
16. ALFERT, M. and GESCHWIND, I. I. 1953. A selective staining method for the basic proteins of cell nuclei. *Proc. Nat. Acad. Sci. U.S.* **39**, 991–9
17. DEITCH, A. D. 1955. Microspectrophotometric study of the binding of the anionic dye, naphthol yellow S, by tissue sections and purified proteins. *Lab. Invest.* **4**, 324–51
18. MCLEISH, J., BELL, L. G. E., LA COUR, L. F., and CHAYEN, J. 1957. The quantitative cytochemical estimation of arginine. *Exp. Cell Res.* **12**, 120–5
19. ——. 1959. Comparative microphotometric studies of DNA and arginine in plant nuclei. *Chromosoma* **10**, 686–710
20. DANIELLI, J. F. 1952. *Cytochemistry: A Critical Approach.* John Wiley: New York
21. DE DUVE, C. 1959. Lysosomes, a new group of cytoplasmic particles. In *Subcellular Particles* (edited by T. Hayashi), pp. 128–59. New York: Ronald Press
22. ——. 1963. The lysosome concept. In *Ciba Found. Symp. on Lysosomes* (edited by A. V. S. de Reuck and M. P. Cameron), pp. 1–31. London: Churchill
23. BITENSKY, L. 1963a. The reversible activation of lysosomes in normal cells and the effects of pathological conditions. In *Ciba Found. Symp. on Lysosomes* (edited by A. V. S. de Reuck and M. P. Cameron), pp. 362–75. London: Churchill
24. CHAYEN, J., BINES, S., and BITENSKY, L. (in preparation)
25. DIENGDOH, J. V. 1964. The demonstration of lysosomes in mouse skin. *Quart. J. micr. Sci.* **105**, 73–8
26. BITENSKY, L. and COHEN, S. 1965. The variation of endometrial acid phosphatase activity with the menstrual cycle. *J. Obstet. Gynaec. Brit. Comm.* **LXXII**, 769–74
27. COHEN, S., BITENSKY, L., CHAYEN, J., CUNNINGHAM, G. J., and RUSSELL, J. K. 1964. Histochemical studies on the human endometrium. *Lancet*, July 11, 56–9
28. BITENSKY, L., CHAYEN, J., CUNNINGHAM, G. J., and FINE, J. 1963. Behaviour of lysosomes in haemorrhagic shock. *Nature, Lond.* **199**, 493–4
29. GAHAN, P. B. 1965. Reversible activation of lysosomes in rat liver. *J. Histochem. Cytochem.* **13**, 334–8
30. SILCOX, A. A., POULTER, L. W., BITENSKY, L., and CHAYEN, J. 1965. An examination of some factors affecting histological preservation in frozen sections of unfixed tissue. *J. Roy. Micr. Soc.* **84**, 559–64
31. BITENSKY, L. 1962. The demonstration of lysosomes by the controlled temperature freezing-sectioning method. *Quart. J. micr. Sci.* **103**, 205–9
32. ——. 1963b. Modifications to the Gomori acid phosphatase technique for controlled-temperature frozen sections. *Quart. J. micr. Sci.* **104**, 193–6
33. HOLT, S. J. 1959. Factors governing the validity of staining methods for enzymes and their bearing upon the Gomori acid phosphatase technique. *Exp. Cell Res.* Suppl. **7**, 1–27
34. LEHNINGER, A. L. 1951. Phosphorylation coupled to oxidation of dihydrodiphosphopyridine nucleotide. *J. Biol. Chem.* **190**, 345–59
35. CHAYEN, J., BITENSKY, L., and WELLS, P. J. 1966. Mitochondrial enzyme latency and its significance in histochemistry and biochemistry. *J. Roy. Micr. Soc.* **86**, 69–74
36. DAVIDSON, J. N. and WAYMOUTH, C. 1944. The histochemical demonstration of ribonucleic acid in mammalian liver. *Proc. Roy. Soc. Edinburgh* **62**, 96–8
37. JARDETSKY, C. D. and GLICK, D. 1956. Studies in histochemistry. XXXVIII. Determination of succinic dehydrogenase in microgram amounts of tissue and its distribution in rat adrenal. *J. Biol. Chem.* **218**, 283–92
38. CHAYEN, J., BITENSKY, L., AVES, E. K., JONES, G. R. N., SILCOX, A. A., and CUNNINGHAM, G. J. 1962. Histochemical demonstration of 6-phosphogluconate dehydrogenase in proliferating and malignant cells. *Nature, Lond.* **195**, 714–15
39. JONES, G. R. N., MAPLE, A. J., AVES, E. K., CHAYEN, J., and CUNNINGHAM, G. J. 1963. Quantitative histochemistry of succinate dehydrogenase in tissue sections. *Nature, Lond.* **197**, 568–70
40. NACHLAS, M. M., PRINN, W., and SELIGMAN, A. M. 1956. Quantitative estimation of lyo- and desmoenzymes in tissue sections with and without fixation. *J. Biophys. Biochem. Cytol.* **2**, 487
41. CHAYEN, J. and MILES, U. J. 1954. The preservation and investigation of plant mitochondria. *Quart. J. micr. Sci.* **94**, 29–35
42. —— and DENBY, E. F. 1960. The distribution of deoxyribonucleic acid in homogenates of plant roots. *Exp. Cell Res.* **20**, 182–97
43. ALTMANN, F. P. and CHAYEN, J. 1965. Retention of nitrogenous material in unfixed sections during incubation for histochemical demonstration of enzymes. *Nature, Lond.* **207**, 1205–6
44. —— and ——. 1966. The significance of a functioning hydrogen-transport system for the retention of 'soluble' dehydrogenases in unfixed sections. *J. Roy. Micr. Soc.* **85**, 175–180
45. WELLS, P. J., BITENSKY, L., ALTMANN, F. P., and CHAYEN, J. 1965. The effect of mechanical damage on the latency of glutamate dehydrogenase. *Biochem. J.* **96**, 36P

46. BENDALL, D. S. and DE DUVE, C. 1960. Tissue fractionation studies. 14. The activation of latent dehydro-
    genases in mitochondria from rat liver. *Biochem. J.* **74**, 444–50
47. KIRKBY, W. W. 1965. The use of histochemical examination of maintenance-cultured tissue in assessing
    potential cytotoxic substances. *Biochem. J.* **94**, 24–25P
48. NILES, N. R., BITENSKY, L., CHAYEN, J., CUNNINGHAM, G. J., and BRAIMBRIDGE, M. V. 1964. The value of
    histochemistry in the analysis of myocardial dysfunction. *Lancet*, May 2, 963–5

Chapter 22

The Study of Cell Differentiation by Quantitative Microscopic Methods

K. F. A. ROSS AND D. E. JANS

PROBLEMS CONNECTED WITH QUANTITATIVE CYTOLOGICAL STUDIES

The difficulties inherent in almost all quantitative investigations on cellular material are formidable, and nearly all the methods that have been evolved are subject to appreciable errors. For this reason many people have been discouraged from using them at all, and others have spent a disproportionate amount of time and energy in improving quantitative techniques which may inherently be unperfectible. Two of the quantitative cytological techniques especially familiar to the present writers, interference microscopy and microdensitometry of living and fixed cells, provide numerous instances of experimental approaches that are not wholly satisfactory, if a high level of absolute accuracy in the measurement of cell data is the ultimate aim.

In interference microscopy, phase-changes of cytological material cannot be measured by ordinary visual means with a repetitive accuracy greater than the nearest 1/60 of a wavelength,* and these measurements themselves may be subject to appreciable systematic errors which can be extremely difficult to quantify with any great accuracy (see Davies,[17] Richards,[36] Ross[41]). In many investigations with the interference microscope, the limiting factors determining the accuracy of the data finally arrived at, depend not on the accuracy with which phase-changes can be measured, but on the much lesser accuracy with which the linear dimensions of cytological objects can be measured under a light microscope. Measurements are seldom obtained with an accuracy greater than $\pm 0.2 \mu$ (see Ross and Galavazi,[43] Ross[42]). If measurements of integrated phase-change are made by interference microscopy over the entire area of a cell, to find its dry mass, the difficulties are even greater, because the relationship between the phase-change given by an object and the intensity of the interference microscope image is normally linear only over a range of a little more than a quarter of a wavelength. Many cell regions and objects give phase-changes considerably greater than this. Nevertheless it should be noted that, very recently, Smith[46,47] has developed an electronic device, by which phase-changes of up to nearly a whole wavelength may be represented in linear terms by electric signals derived from a photo-metric scan. This can enable accurate integrated phase-change measurements to be made over the entire area of many of the less refractile kinds of cells.

In the field of microdensitometry two main difficulties arise. Firstly, no microdensitometer so far made has a slit narrow enough to record faithfully the changes in optical density of the finest details that are optically resolvable in a microscope preparation of cellular material. This means that, if this fine detail

---

* Densitometric methods of measuring phase-changes are, however, marginally more accurate than visual methods. Methods involving phase-modulated light with polarizing interference microscopes, recently developed independently by Allen and his associates 1963,[1] 1966,[2] and by Dyson, 1966,[19] enable phase-changes to be measured with enormously greater accuracy. This is not always of much practical value in cytological investigations, in view of the limitations in the accuracy with which values for thicknesses can be obtained (see above). The equipment required is, moreover, elaborate, delicate, and costly.

is to be recorded, the preparation must be photographed, and a densitometer trace must be made from a photographic transparency of considerably greater magnification than the size of the object. The density values in the photograph must faithfully represent all the density values in the original preparation. The kind of film, the exposure time, and the development time must be selected so that the range of density values in the original preparation are linearly represented in the photograph. This, fortunately, can usually be achieved by comparing the densitometric traces of photomicrographs of a series of objects of known absorption (such as a step-wise arrangement of superimposed neutral density filters), with traces made on photomicrographs of the cytological material to be studied.

Secondly, and more seriously, the optical densities in each region of a cell may not truly represent the amounts of material of interest that are, in fact, present in these regions. In estimating the total mass of material in each region, it has already been mentioned that when preparations of living cells are examined by interference microscopy, the relationship between the phase-change given by a cell region, and the intensity of its image, is linear only over a range of about $100°$ (just over a quarter of a wavelength). Yet many specialized regions in living cells can give phase-changes of up to $220°$ or more (e.g. the lipid droplets in the mouse ascites tumour cells measured by Ross.[38] It therefore follows that densitometry of a single interference photomicrograph of a living cell cannot be made to provide a fully accurate indication of the dry mass of every region within its area easily or without ambiguity.*

Densitometry is, however, more frequently employed on photomicrographs of fixed cells, stained by recognized histochemical procedures which reveal the presence of specific chemical substances by interaction to produce a coloured product. In these cases, it is important to know whether the reaction of the dye-forming substance with the substance in the tissue is truly stoichiometric; that is to say, whether the depth of staining in a given reaction truly indicates the amount of the cell material which is present. There are two separate aspects of this last question to be considered; firstly, do the molecules of the dye react with the molecules of the substance that they are intended to display, on a one-to-one basis; and, secondly, can the deposition of dye molecules on the surface of an organelle or inclusion, containing a substance with an affinity for it, effectively prevent the penetration of further molecules of the dye, and thus prevent the reactive substance situated more deeply within it from reacting at all?

Strict stoichiometric reaction of a dye with a substrate, on a molecule for molecule basis, has only been established for a limited number of histochemical reactions (e.g. for the Feulgen reaction for DNA from the work of Walker and Richards,[54] and of others; and for the reaction of Azure A with polysaccharides from the work of Szirmai and Balazs,[53] and Balazs and Szirmai[5]). Although it is now believed by some workers that, under optimum conditions, many standard histochemical reactions are in fact stoichiometric (Chayen[16]) this cannot be regarded as proven. Some of the objections to this contention are argued lucidly, if a little vehemently, by van Duijn.[18] The question of blocking a reaction by the deposition of dye molecules on the surfaces of dense organelles and cell inclusions is a serious problem, and one which would seem to be difficult to begin to investigate without creating some disruption and dispersal of the contents of such bodies within the cell. This is undesirable if accurate localization and quantification is the aim.† (This experimental approach would be justified if the aim was simply to see whether the absorption of the dispersed material was the same as that of the dense body from which it was derived.)

Thus the really important question, which may legitimately be asked in connection with almost all quantitative cytological techniques, is: just how rigorously quantitative do they in fact have to be, in order to provide significant and interesting biological information about cells? Can experiments be designed to yield such information, in spite of the fact that the quantitative data they give is known to be

---

* Although it is true that a curve might be constructed to indicate the (non-linear) relationship between image-intensity and object phase-change, and the intensities of different cell regions could be related to this curve, in most instances a single intensity value could indicate two quite different possible phase-changes.

† Swift[52] has demonstrated that this possible difficulty does not affect the densitometric evaluation of the staining of most nucleoli, because very thin sections of these bodies, in which their central regions are exposed to allow direct access to the dye, show the same staining patterns as uncut nucleoli lying entirely between the surfaces of thicker sections.

subject to certain unquantifiable errors? In this article it will be argued that they can be so designed. That even if all the formidable objections to the absolute accuracy of the measurements, discussed above, are applicable, valuable information may nevertheless be forthcoming, provided one is content to devise experiments that demonstrate significant quantitative *differences*, rather than absolute quantitative values. In other words, in the present stage of refinement of many quantitative cytological techniques, *comparative* measurements are both more reliable, and more informative, than attempts to obtain absolute quantitative measurements.* Examples of the information gained by such comparative measurements, which throw light on some of the processes involved in cell differentiation, will be discussed below.

THE PROCESSES OF CELL DIFFERENTIATION, ILLUSTRATED PARTICULARLY BY DEVELOPING MUSCLE TISSUE

The exact processes by which the cells of a developing organism become different from each other, to form the different elements of that organism's tissues (which in their turn define its ultimate phenotypic characteristics) has been, and remains, one of the most challenging and fascinating fields of study in the whole of biological science. It is only in the last twelve years that the beginnings of a real understanding of these processes have been arrived at, and the importance of RNA in these processes (clearly foreshadowed by the work of Caspersson and of Brachet in the 1930's) fully appreciated.

The initial impetus for the work which has led to this new understanding, was provided by the famous interpretation of the purine-pyrimidine linked double helical structure of the DNA molecule from the data of Wilkins and Randall,[58] by Watson and Crick in 1953[56]. The self-replicating potentialities of such molecules of genetic material were almost at once perceived after their structure had been determined, since it was appreciated that if the two helical components became separated through the purine/pyrimidine linkages being broken, two other DNA molecules exhibiting the same purine/pyrimidine sequences could be made, with each of the separated helical elements forming a template for their assembly. It was postulated in exactly the same way that the separated molecules of DNA could act as templates for the synthesis of long molecules of RNA with very similar purine/pyrimidine configurations. These long molecules of RNA, known as 'messenger RNA', in turn form templates for the assembly of amino-acids in predetermined sequences, in the synthesis of specific proteins. This is affected through the mediation of shorter molecules of RNA (also synthesized on the DNA templates) known as 'soluble or transfer RNA' which exists in 64 possible configurations each capable of attaching itself specifically to one of the 20 amino-acids, and each possessing also within its molecule a specific sequence of three purines and pyrimidines capable of reacting only with a complementary triple group on the 'messenger RNA' template. Thus through the interaction of 'transfer RNA' and 'messenger RNA' the amino-acids can only be assembled in specific sequences predetermined by the sequences of purines and pyrimidines on the 'messenger RNA' coded by the genetic material.

The individual contributions to our present knowledge of the genetic code embodied in the DNA, and of the role of RNA in implementing it in the activity of protein synthesis, are far too numerous to be discussed individually, and only a few of the major discoveries in the field can be mentioned here. Jacob and Monod[27] appreciated that RNA is the vehicle conveying the information coded in the DNA to the seat of protein synthesis, and gave the name 'messenger RNA' to this form of RNA. Weiss and Nakamato[57] established that this RNA was synthesised in the nucleus and had sequences of purines and pyrimidines complementary to those in the DNA. Spencer and others[48] established the structure of 'transfer or soluble RNA' as being much smaller helical molecules containing about 70 purine/pyrimidine sub-units. Nirenberg and Mathaei[33] were principally instrumental in 'breaking the genetic code' when, in a series of highly ingenious experiments involving the addition of synthetic RNA to media containing single amino-acids and simple mixtures of amino-acids, and analysis of the proportion of the amino-acids in the polypeptide end product, they established that certain specific triple sequences of purines and pyrimidines in the RNA were capable of building polypeptides from specific amino-acids. Then, in a very elegant experiment, (involving changing an amino-acid molecule attached to a particular transfer RNA molecule into another),

---

* It should be pointed out, however, that this is not necessarily true of all kinds of quantitative cytological investigations. Caspersson, for example, has so refined the techniques of ultra-violet microspectroscopy, over the years, that most of the sources of error in these techniques have been appreciated and corrected (see Caspersson[13, 14]), so that in this case more reliance can be placed on the absolute values measured than with most quantitative cytological techniques.

Chapeville and others[15] established that it was through the medium of triple sequences of purine and pyrimidines on the molecules of 'soluble' or 'transfer' RNA that the amino-acids could be assembled upon the template provided by the 'messenger' RNA. Finally, Goodman and Rich[20] conducted a series of experiments to establish the function of the ribosomes in this process, and concluded that each ribosome makes a separate protein molecule by travelling along the length of a molecule of 'messenger RNA' and incorporating the amino-acids provided by the 'transfer' RNA in the pre-indicated places, and that several ribosomes could thus operate simultaneously along the length of one molecule of 'messenger' RNA.

It is now universally accepted that the several different kinds of RNA, by mechanisms of the kind briefly summarized above, form the all-important metabolic link by which the coding contained in the DNA that comprises the hereditary material can initiate the synthesis of all the complex variety of specific substances that are required at a particular stage in the normal differentiation of a tissue. Furthermore, from studies of the 'giant' or polytene chromosomes of *Diptera*, which are particularly suitable material, it has been demonstrated that RNA is synthesized in the 'Balbiani rings' (Balbiani[6]) or 'puffs' at particular recognizable positions on the chromosomes, representing specific gene loci, and that in cells of different tissues at definite stages of development, characteristic 'puffs' develop at definite loci (Pavan and Bruer,[34] Beerman[10]). It thus seems probable that these 'puffs' are in fact a visible demonstration of the production of the specific RNA required to initiate a particular process of cellular differentiation.

In ordinary cells, however, the interphase DNA occurs in a dispersed condition and not as 'giant' chromosomes. It is not normally visible as a distinct entity. The only visible nuclear organelles which contain RNA are the nucleoli, which take the form of irregularly shaped aggregations of rather dense material, consisting mostly of protein. During cell division this nucleolar material usually becomes dispersed.

It is, consequently, often very difficult to trace the fate of nucleolar material during the processes of cell differentiation, since these processes are almost always accompanied by cell division, which normally forms part of the process of tissue-building. It would thus obviously be desirable if a cytological material could be studied, in which cell differentiation occurred, without the additional complexities introduced by cell division.

Such a material is, fortunately, provided by developing striated muscle fibres. When fully differentiated, these fibres are multi-nucleate syncytia. This multi-nucleate condition is derived not, as was once widely believed, by 'amitotic' divisions of nuclei within the developing muscle fibres nor, (except in the earliest stages of the differentiation of this tissue, not here considered), by ordinary mitoses of nuclei, but by the union, and fusion, of many uninucleate cells, known as myoblasts, to form multi-nucleate 'muscle straps' which later develop the striations and peripherally disposed nuclei and sarcoplasm, characteristic of fully differentiated muscle fibres. Thus a muscle fibre is a *true syncytium* in the sense defined by Baker.[3] Almost all the nuclei retain their identity throughout all the later stages of its differentiation, and their contents can be studied through many stages of the tissue's development, without being involved in cell division.

These stages of development may be reproduced in tissue cultures of myoblasts derived from explants of embryonic muscle tissue, or from regenerating adult muscle tissue. The changes shown in the earliest stages of this process, namely the actual fusions of individual myoblasts to form muscle straps, have proved especially convenient for study. These fusions occur quite rapidly and frequently during the first two days of the life of a culture, and form a very striking stage in the tissue's development. Since all the subsequent discussion of the processes of cell differentiation is based on studies of myoblast material at these early stages of differentiation, it is just as well to bear in mind that it is possible that this particular differentiation process may be so atypical, and that conclusions drawn from its study are not applicable generally to the differentiation of other tissues. Further experiments with other tissues alone can determine this. Although myoblast material is uniquely suitable for studying changes in the nucleoli, the technique which the present authors have developed for measuring the changes in the cytoplasmic RNA (described below) could undoubtedly be applied to other developing tissues. There is, however, no reason yet to believe that the differentiation process need be different, when, as in this case, cell division is absent. It is safe to presume that there are three main processes which constitute 'differentiation': (i) the development of characteristic internal cellular structures, (ii) the association of cells in a characteristic way to form a tissue, and (iii) the development of characteristic extracellular products. Each process must sometimes involve bursts of metabolic activity. This metabolic activity must in its turn be controlled by the syntheses of specific proteins, controlled by the genetic material through the mediation of RNA. Muscle tissue is unlikely to be exceptional in this respect.

CHANGES IN THE DRY MASS OF LIVING MYOBLAST NUCLEOLI DURING DIFFERENTIATION AS MEASURED BY INTERFERENCE MICROSCOPY

Six years ago one of the present writers (K.F.A.R.) started the series of studies of differentiation reported here by measuring the dry mass of the nucleoli of myoblasts, growing in fresh tissue cultures, before, during, and after their fusion. Full details of the experimental techniques employed have been reported elsewhere (Ross[39, 41, 42]). They will only be discussed here in so far as they are relevant to the problems of accuracy and interpretation considered in the first section of this article (pp. 275–277). Briefly, the techniques involved obtaining values for the refractive index and the volume of each nucleolus, and deriving a value for its dry mass from the product of these two parameters. Values for the refractive index of each nucleolus were obtained by measuring the phase-change in the light passing through their central regions, and their two measurable dimensions in the plane of the microscope slide. To measure the phase-changes through the nucleoli unaffected by the overlying or underlying cytoplasm and nuclear sap, it was necessary to mount the living cells in an isotonic saline/protein medium, with a refractive index equal to that of both cytoplasm and nuclear sap. (It was fortunate that the refractive indices of both were nearly always found to be identical.) The constitution of this medium has been described by Ross.[39]

A value for the refractive index of each nucleolus, n_0, was then obtained from the formula

$$n_0 = \phi \times \frac{\lambda}{t} + n_m \qquad (22.1)$$

where ϕ was the phase-retardation, in fractions of a wavelength, measured in the direction of the optical axis of the microscope, of the light passing through the centre of each nucleolus, λ the mean wavelength of the light by which these measurements were made, and t the estimated thickness of the nucleolus in the direction of the microscope's optical axis. This latter value could not be measured directly, and instead it was assumed to be equal to the mean of the two measurable dimensions in the plane of the slide.* In the majority of cases such an assumption was probably justified, since, although irregular in shape, these nucleoli were seldom grossly irregular, for only 10 per cent of them had measurable dimensions that differed from each other by more than a factor of 2, and 50 per cent of them appeared to be nearly equidimensional (see Ross[39]). Nevertheless, it is undoubtedly true that, in some cases, this assumption must have introduced appreciable errors into some of the individual refractive index values of the nucleoli, and the values of their (weight/volume) percentage solid content derived from these refractive indices. The assumed values for nucleolar thickness were also used, together with the two measurable nucleolar dimensions, to obtain values for nucleolar volumes: so that, individually, all the parameters used for calculating the values for nucleolar dry mass were subject to an unknown error of finite size or order.

Values for the weight/volume percent concentration of solid material in each nucleolus, c_s, were obtained from the formula:

$$c_s = \frac{n_0 - n_w}{\alpha} \qquad (22.2)$$

where n_w is the refractive index of water (1·334) and α the mean specific refraction increment of the

---

* This assumption is used once (as a denominator) to calculate values for refraction index, and again as one of the terms (as a numerator) to derive the values for volume. The assumption must lie within finite limits suggested by the two measured dimensions for a given nucleolus, since it has never been observed that nucleoli of this type are shaped like plates.

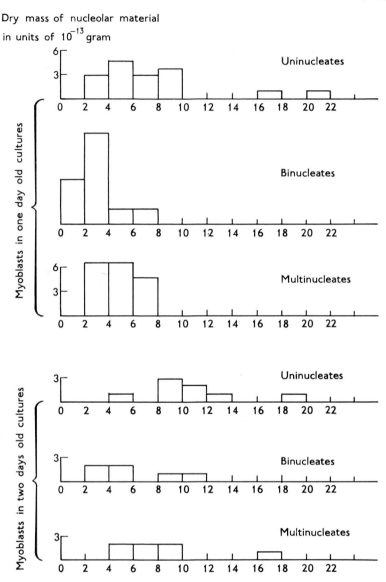

Fig. 22.1   Histograms showing the distribution of the values for the total dry mass of nucleolar material per nucleus, measured by interference microscopy, in uni-nucleate, bi-nucleate and multi-nucleate living myoblasts in one-day-old and two-day-old tissue culture preparations, grown from muscle explants obtained from 10–12 day old chick embryos. Ordinate: numbers of nuclei. Abscissa: dry mass of nucleolar material per nucleus, in tenths of a picogram.

material of which the nucleoli were composed (taken as being equal to 0·0018). The individual values for the volumes of the nucleoli, v, were obtained from the formula:

$$v = a \times b \times t \times 0\cdot524 \tag{22.3}$$

where a and b were the measured nucleolar dimensions and t the assumed one. The products of the volumes and the weight/volume percentage solid content of the nucleoli ($v \times c_s$) give values for their dry mass (and not, as erroneously reported by Hale[21] and by Walter,[55] for their 'wet mass').

The accuracy of individual dry mass values obtained in this way was questionable, since they depend not only on the accuracy of the estimated dimension, t, but on the accuracy with which it is possible to measure a microscopic object under a high resolution 2 mm objective with a wavelength of around 0.55 μ (see Ross and Galavazi[43]). In the rather less than optimum conditions under which these nucleoli were measured, the error was probably more nearly ± 0.25 μ. However, because the values for the dry mass of the nucleoli were obtained from the product of the values for their percentage solid content and their volume, the dry mass values were in fact rather *more* accurate than the data from which they were derived. For example, if, through an error in linear measurement, the dimensions of a nucleolus were underestimated, this would of course have given rise to a proportionally greater underestimation of its volume: but it would at the same time have given an overestimation of its refractive index, and hence of its percentage solid content. Similarly, an overestimation of its linear dimensions would have given an overestimation of its volume and an underestimation of its percentage solid content. Thus the errors of opposite sign in these parameters, when multiplied together to give dry mass values, tended to cancel each other out (but did not do so completely; see Ross[42]).

The level of accuracy of the individual dry mass values obtained with the technique was, however, of little consequence. The important thing was that, when large numbers of nucleolar dry mass values were obtained in this way, their distribution was found to provide interesting and important information when the nucleoli in myoblasts at one stage of development, were compared with the nucleoli of myoblasts at a different stage. In such comparisons, distributional differences of even as little as 0.2 picograms were found to have a measure of significance.

When this technique was used to study myoblasts from explants of 9–12 days old chick embryo muscle tissue in cultures 1–2 days old, it was immediately apparent that myoblasts which were binucleate, and which were, in the main, still actively fusing with each other, mostly contained strikingly less nucleolar material per nucleolus than uninucleate myoblasts that had not yet begun to fuse; and the values for nucleolar dry mass in multinucleate myoblasts exhibited a distribution which was intermediate between that of uni-nucleate and that of bi-nucleate, myoblasts (see Fig. 22.1). The most likely explanation of these results, was that, during the process of the active fusion of two myoblasts, material was removed from a nucleolus at a greater rate than it was being synthesized and replaced; and that later, when cell fusion was completed, the dry mass of the nucleolus tended to return again to approximately its former level (Ross[39]). (This explained the intermediate distribution of dry mass values exhibited by the multinucleate myoblasts, since the chain of cells would be expected to contain nuclei (and nucleoli) derived, both from myoblasts which were still actively fusing, and from those in which the fusions had been completed.)

The opposite trend was observed, however, in the distributions of the values for the percentages of solid matter in the nucleoli. These were significantly higher in the binucleate myoblasts than in the uninucleate myoblasts (and intermediate values were again exhibited in the multinucleate myoblasts), see Fig. 22.2. This indicated that the nucleolar material being removed was, for the most part, rather less dense than the material left behind. Since nucleoli are known to contain both RNA and a considerable amount of protein, it rather suggested that it might be the RNA, which in all its forms has a rather lower molecular weight than protein, that was being removed from the nucleoli during the fusion process, and made available elsewhere, probably in the cytoplasm.*

---

* Swift[52] has expressed the opinion that much of the material being removed from the nucleoli and simultaneously appearing in the cytoplasm of active fusing cells (see below p. 298) must be ribosomal RNA. If this is so, it is possible that the material in the nucleoli of diminished mass found in myoblasts from dystrophic individuals (see p. 287) could be deficient in the 150–300 Å granules which are now believed to be associated with ribosome production (Hay and Gurdon[22]). It would be extremely interesting if electron microscopy of these nucleoli were to reveal a preponderance of the 'dense fibrous masses' reported by Jones[28] in the nucleoli of the homozygous mutant larvae of *Xenopus laevis* which are incapable of synthesizing ribosomes.

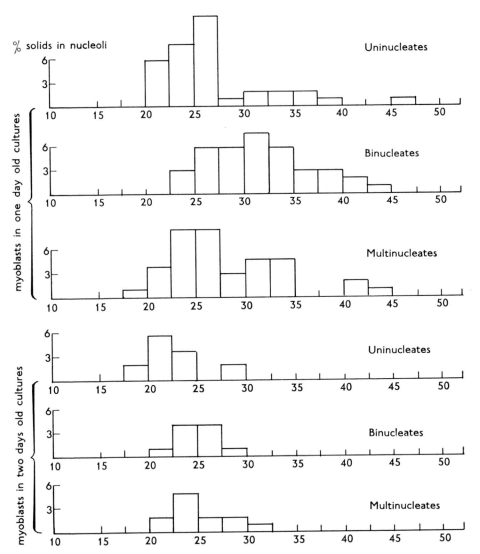

Fig. 22.2 Histograms showing the distribution of the values for the percentages of solid material found by interference microscopy in the nucleoli of uni-nucleate, bi-nucleate, and multi-nucleate living myoblasts in one-day-old and two-day-old tissue culture preparations grown from muscle explants obtained from 10–12 day old chick embryos. Ordinate: numbers of nucleoli. Abscissa: percentages of solid material (weight/volume) in the nucleoli.

The distributions of the values for nucleolar volume (the other parameter from which the values for nucleolar dry mass were derived), exhibited the same trends in uni-nucleate, bi-nucleate, and multi-nucleate myoblasts as the values for nucleolar dry mass, see Fig. 22.3. The volumes of the nucleoli in the bi-nucleate, usually actively fusing, myoblasts were very strikingly lower than in the uni-nucleate myoblasts. So striking were these differences in volume, that it seemed likely that the nucleoli might be expected to exhibit a measurable linear shrinkage when two cells fused, and a measurable enlargement when fusion was completed. Some preliminary experiments reported by Ross,[39] in which two pairs of fusing myoblasts were photographed at successive intervals, appeared to demonstrate such a linear

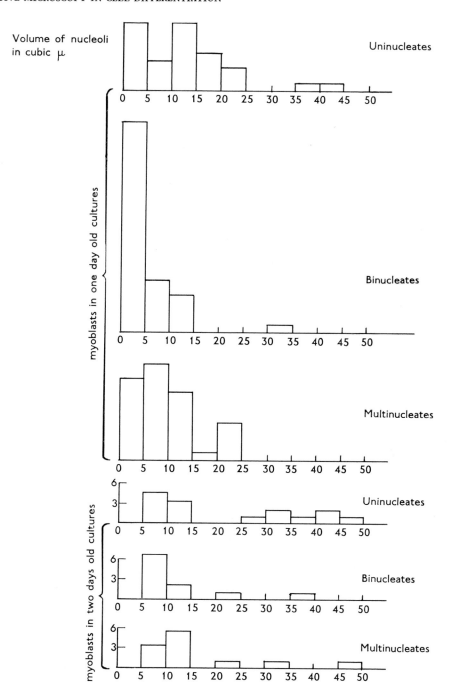

Fig. 22.3 Histograms showing the distribution of the values for the total volume of nucleolar material per nucleus, obtained from linear measurements, in uni-nucleate, bi-nucleate, and multi-nucleate living myoblasts in one-day-old and two-day-old tissue culture preparations grown from muscle explants obtained from 10–12 day old chick embryos. Ordinate: numbers of nuclei. Abscissa: volume of nucleolar material per nucleus, in cubic μ.

shrinkage of nucleoli during the fusion process. The conditions under which the observations were made were far from ideal. It was necessary to remove the growing culture from the incubator, where it was kept at 37°C, and expose it to room temperature for several minutes for photography, at approximately 1 hour intervals. It could therefore be argued that temperature fluctuations might have affected adversely the normal process of cell fusion. An additional disadvantage was that a continuous record of the changes could not be obtained.

These difficulties were later overcome by constructing (with the generous help of Mr. A. Taylor, of Taylor Industries, Rowlands Gill, County Durham, England) a constant temperature chamber around an interference microscope (illustrated and described by Ross,[41] and using time-lapse cine-micrography to make a continuous record of the changes in the linear dimensions of the nucleoli during and after

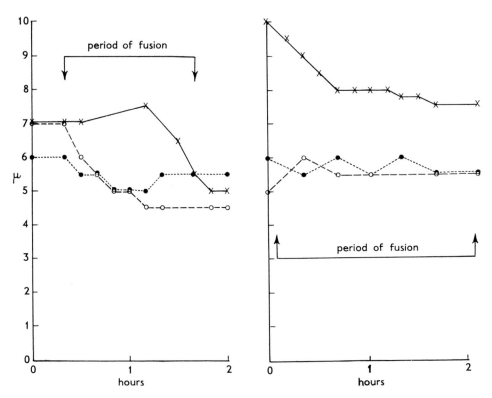

Fig. 22.4 Graphs showing the changes in the mean diameter of the nucleolus over periods of two hours in chick myoblasts in two cases of their fusion to form a binucleate cell (arrived at by averaging the larger and smaller dimensions of the nucleoli measured on successive cine-photo-micrographs). In A all the nucleoli from both myoblasts exhibit some degree of shrinkage, and in B the shrinkage appears to be confined to a single nucleolus in one myoblast. (In each case, one cell had only one nucleolus, represented by a full line, and the other had two, represented by broken lines.) Ordinate: mean nucleolar diameter (μ). Abscissa: time (hr).

fusion. To follow the full process, it was necessary to select two uninucleate myoblasts that looked as if they were about to fuse with each other, and film them continuously for several hours. Such attempts were frequently frustrated. The cells selected often did not undergo fusion at all, but simply slid past each other, or made only temporary contact with one another. However, eventually six films were made of chick and mouse myoblasts undergoing fusion in developing tissue cultures and one film was obtained of the subsequent nucleolar changes in a freshly-formed binucleate cell. The standard dimensions (see above) of sharply-focused nucleoli were measured on the projected images of cine-micrograph frames taken at selected time intervals. The mean values of the two measurable dimensions were plotted graphically against time, as shown in Figs. 22.4 and 22.5. Fig. 22.4, A and B, shows the changes in the linear dimensions of nucleoli in two of a different pair of fusing cells, (the nucleoli from each nucleus of the pair of cells are distinguished from each other by linking their dimensional measurements respectively by continuous or broken lines).

The pattern exhibited in Fig. 22.4, B was the one most frequently found. The nucleolar shrinkage seems to be confined to the nucleoli in only one of the pair of fusing cells. Fig. 22.5, conversely, shows the progressive increase in size of eight nucleoli in a newly formed mouse binucleate cell where all the nucleoli appear to be involved to some extent. (Minor fluctuations in the apparent diameters of the nucleoli could, of course, be accounted for by rotations in the event of their being irregular in shape, but

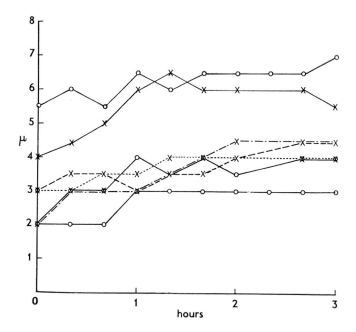

Fig. 22.5 Graph showing the changes in the mean diameter of the nucleoli over a period of 3 hr in a newly formed binucleate myoblast from a normal mouse (arrived at by averaging the larger and smaller dimensions of the nucleoli measured on successive cine-photomicrographs). Every nucleolus exhibits some degree of expansion during the period. (There were 3 nucleoli in each nucleus, those from one nucleus are represented by circles and those from the other by crosses, and individual nucleoli are distinguished by different kinds of lines connecting these points.) Ordinate: mean nucleolar diameter (μ). Abscissa: time (hr).

such fluctuations could hardly be expected to produce the consistent trends shown here.) Thus, two separate sets of evidence (change of dry mass and linear shrinkage) seem to suggest that the fusion of two myoblasts is accompanied by a marked depletion of nucleolar material in at least one of the cells participating in the fusion, and that this is followed by a slower subsequent build-up of material again.

If, however, the fluctuations of nucleolar dry mass in the first day and second day cultures of myoblasts are compared, as here in Figs. 22.1–22.3 (which was not done with this data when it was first obtained and presented by Ross[39]), a second trend becomes apparent, namely, a slow and steady overall increase in the mass of the nucleoli of both the uni-nucleate, bi-nucleate, and the multi-nucleate myoblasts. This process of steady nucleolar growth (which may be the one principally involved in the evidence presented in Fig. 22.5) could indicate a more or less steady accumulation of RNA or RNP at the site of the nucleoli, which is partially obscured by the more dramatic fluctuations in nucleolar mass, size, and density exhibited during cell fusion. It will be shown that this observation assumes some importance when one is interpreting the quantitative changes in other kinds of nucleolar material reported

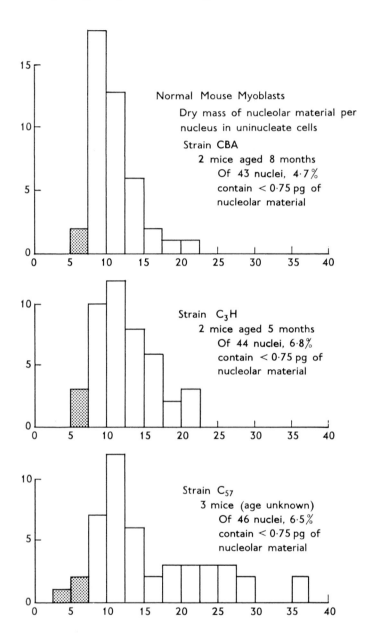

Fig. 22.6 Histograms showing the distributions of the total dry mass of nucleolar material per nucleus found in nuclei of uninucleate myoblasts cultured from lesions in the adductor muscles of three different strains of normal mice. Ordinates: numbers of nuclei. Abscissa: dry mass of nucleolar material in tenths of a picogram (units of 10^{-13}g). Shaded areas: proportion of nuclei containing less than 0·75 pg of nuclear material.

by other workers (see p. 298), or considering the present authors' findings about the changes in the cytoplasmic RNA in the myoblast material, described later.

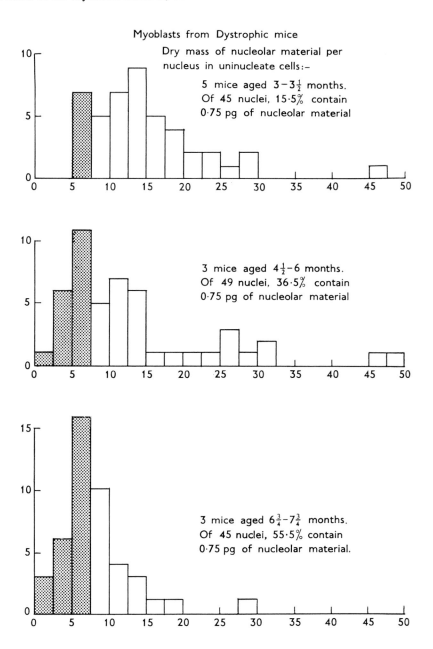

Fig. 22.7 Histograms showing the distribution of the total dry mass of nucleolar material per nucleus found in the nuclei of uninucleate myoblasts cultured from the muscles of Bar Harbor dystrophic mice of different ages. Ordinates: numbers of nuclei. Abscissa: dry mass of nucleolar material in tenths of a picogram (units of 10^{-13} g). Shaded areas: proportion of nuclei containing less than 0·75 pg of nuclear material.

A COMPARISON OF THE DRY MASS OF NUCLEOLI IN MYOBLASTS FROM NORMAL AND DYSTROPHIC MICE

Further evidence of the probable importance of nucleolar material in the growth and development of muscle fibres was obtained by one of the present writers,[40] by applying the technique described to measuring the nucleolar dry mass in myoblasts from animals suffering from muscular dystrophy. This provides an example of how the study of pathological conditions can often give some insight into the functioning of normal life processes. It also illustrates the contention, strongly held by both the present authors, that it is often so difficult to distinguish between so-called 'pure' and so-called 'applied' science that this classification of scientific activities is not helpful.

Like most other diseases, muscular dystrophy is characterized by a considerable number of symptomological criteria. The most important of these criteria from the point of view of the present investigation are that it is *hereditary*; that it is associated with a progressive tendency for the muscle fibres to fail to develop and differentiate normally during the growth of the individual; and that those fibres that have at first formed normally, undergo a recognizable morphological breakdown, and become defective in function.

A hereditary condition, in many respects similar to one of the kinds of muscular dystrophy found in Man (dystrophia myotonica), is found in a strain of mice (Bar Harbor dy dy). Explants from their muscle tissue can, without too much difficulty, be made to yield myoblasts in tissue culture. There is some evidence to suggest that these myoblasts may originate as attempts at regeneration in the fibres undergoing breakdown as a result of the disease (Hudgson[24]). They are very rarely observed to undergo the processes of cell fusion and muscle-strap development found in explants of normal embryonic muscle, and, when this occurs, it is only in young dystrophic mice.

From the hereditary character of the disease it is reasonable to assume that an abnormality in the coding (contained in the sequences of purines and pyrimidines in the DNA comprising the genetic material) gives rise to abnormalities in protein synthesis. These cause the muscle fibres to fail to become fully or stably differentiated, and this leads to the ultimate breakdown of many fibres and to the failure of others to develop at all. One may expect this condition to be associated with abnormalities in RNA synthesis at certain critical stages in the fibre's development, since RNA is the vehicle through which protein synthesis is implemented in the pattern predetermined by the DNA. Abnormalities in the RNA could, quite possibly, be manifest in the development of abnormal nucleoli.

Consequently, it was very interesting to find that, when the values for nucleolar dry mass in myoblasts derived from explants from traumatized regions of normal mice* were compared with those for myoblasts from Bar Harbor dystrophic mice, the latter myoblasts had an abnormally high proportion of nucleoli with a very low dry mass. This was true even in young dystrophic individuals less than 4 months old, in which the disease had not yet progressed far. Older dystrophic individuals exhibited this abnormality to a more pronounced degree. In dystrophic mice around 7 months old more than 50 per cent of the nucleoli in their myoblasts were found to contain less than 0·75 of a picogram of solid material, compared with only about 5 per cent in normal mice and 15 per cent in young dystrophic mice, see Figs. 22.6 and 22.7.

It was also interesting to find that the solid concentration in the nucleoli was significantly higher in

---

* Pearce[35] found that whereas explants from dystrophic mouse muscle would yield myoblasts in tissue culture preparations, explants from the muscle of normal mice seldom did so. If, however, a region of normal mouse muscle was traumatized by crushing with forceps 2 days before excision, explants from this injured region yielded a plentiful supply of myoblasts: and the material was used as a normal control in the experiments described here. In 1965 Hudgson[24] found that explants from similarly traumatized regions of muscle from dystrophic mice also yielded a plentiful supply of myoblasts, and that morphologically these tissue cultures exhibited characteristics common to both those shown by the traumatized normal and the untraumatized dystrophic explants. He concluded that all or most myoblasts were derived from attempts at regeneration of the tissue, but that the traumatized explants from dystrophic tissue yielded a mixed population of myoblasts from abnormal fibres, and from fibres which were still functioning normally (see also Ross Hudgson[61]).

myoblasts cultured from older dystrophic mice, than in those derived from younger age groups (see Fig. 22.8) which indicated that, as the disease progresses, the nucleoli in myoblasts from dystrophic muscle tend to become denser as well as smaller. This suggests (from the same argument as that used on p. 281) that the nucleoli in dystrophic mice, at advanced stages of the disease, are more likely to be

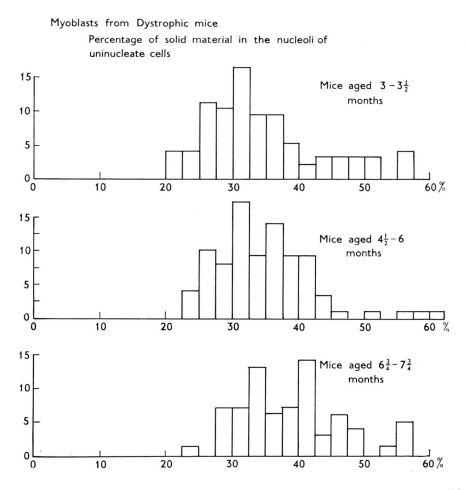

Fig. 22.8 Histograms showing the distributions of the percentage concentrations of total solids in the nucleoli of uninucleate myoblasts cultured from the muscles of Bar Harbor dystrophic mice of different ages. Ordinates: numbers of nucleoli. Abscissa: per cent solid content of the nucleoli.

deficient in RNA than in the (relatively denser) protein component of the material of which they are comprised. Thus it looks as if the inability of these cells to fuse with each other to form normal muscle straps may be correlated with nucleolar RNA deficiency.*

* As this article goes to press, the authors have been able to demonstrate that the cytoplasmic RNA in myoblasts from dystrophic mice (which appear to have lost the ability of fusing with each other) is also significantly lower than in myoblasts from normal mice. (See Ross, Jans, and Susheela,[45] Ross and Jans.[44])

A DENSITOMETRIC TECHNIQUE FOR MEASURING CHANGES IN CYTOPLASMIC RNA IN DIFFERENTIATING MYOBLASTS

Although the work described above strongly suggests that RNA is involved in the processes of growth and development of myoblasts in tissue culture, especially during the active fusion of two myoblasts in muscle strap formation, it provides no direct evidence that it is, in fact, RNA that undergoes the fluctuations in mass and concentration that were found. An interference microscope can only measure the *total* dry mass of all the solids in a particular cell region. It cannot measure the amount of any specific cell substance that may be present. Furthermore it is, normally, only of value to make such measurements of total dry mass on distinct and recognizable cell regions, or organelles. The only discrete bodies of the kind, which were of interest to the investigation were nucleoli, and, as has already been said, nucleoli may contain a high proportion of substances other than RNA.* Consequently in order to try to detect and study changes in the RNA content of the cytoplasm and nucleoli, an entirely different experimental approach had to be devised.

The method adopted involved making two photomicrographic records of fixed cultures of myoblasts, one directly after staining with a basic dye, and one after removing the dye, digesting with RNAse, and then re-staining. Both photographs were subjected to detailed densitometric analysis over their entire area. Although this was rather a lengthy and tedious technique, it had the advantage of yielding a great deal of information of biological interest; and, since, as far as the present authors know, it does not appear to have been employed or reported elsewhere,† it may be helpful to describe it in some detail.

The principal difficulty lay in evolving a staining technique selective enough, yet powerful enough, to enable satisfactory densitometric measurements to be made on the myoblasts. Two factors made this rather difficult: the extreme thinness of the myoblasts, when growing on glass surfaces in tissue culture (see below, p. 294), and the fact that, in order to get good growth, cock's plasma was always added to the culture media (for details see Ross and Hudgson[61]), and was often present as a plasma clot, in the form of a thin film, in the immediate vicinity of the growing cells. The staining technique had to be capable of staining even the thinnest regions of the myoblasts adequately, and yet at the same time stain any surrounding non-cellular material as little as possible. Standard staining procedures for RNA enploying pyronine, or Azure B, were unsatisfactory in this respect: pyronine stained the background plasma clot quite strongly and both reagents stained the thinner regions of the cells too faintly. Toluidine blue, used directly on simply fixed preparations, was also too faint. The staining procedure finally adopted, which was successful, but which is not claimed to be ideal, was a modification of the method developed by Love and Liles.[30] This method was suggested to us by Professor A. G. Everson Pearse. Love and Liles used toluidine blue, with and without ammonium molybdate, to distinguish different types of nucleoproteins in a variety of cells. The nucleoproteins were classified by the varying amounts of bathochromasy (Baker[3]) that they exhibited. The addition of ammonium molybdate, which was found by Love and Liles to give the deepest, and most uniform, staining of nucleolar material and cytoplasmic RNA, was unsatisfactory when applied to myoblast material, because it appeared to leave a fine granular deposit on the associated plasma clots. Toluidine blue alone was employed. The coverslip preparations were stained in a solution of toluidine

---

* Stenram (1957)[49] using Barer's 'double immersion' method, with interference microscopy, obtained values for the refractive indices of the nucleoli of rat liver cells in paraffin wax sections, before, and after, ribonuclease digestion. These values led him to conclude that the nucleoli contained only about 14 per cent by weight of RNA. Although his experimental approach is not entirely above criticism (see Ross[41]) this figure may not be far wrong, and could be typical of the proportion of RNA in nucleoli of other cell types also.

† Hydén,[25] and Brattgard and Hydén[12] have used a very similar experimental approach for determining the amounts of RNA in neurons and glial cells from ultra-violet absorption measurements, but their apparatus included an elaborate and extremely elegant automatic photometric scanning and matrix-printing device, developed by Hydén and Larsson.[26] Facilities for the construction of extremely costly equipment of this kind have not been available to the present authors, who consequently had to develop the lengthier but simpler experimental technique here described.

blue buffered to pH 3, after being fixed for 1 hr in mercuric-formol. According to Love and Liles,[30] the formaldehyde in this mixture inactivates the protein-bound amino groups in the nucleic acids, and frees the maximum amounts of RNA and DNA to react with the molecules of the dye. It is, however, somewhat puzzling that these authors used the dye in a buffer as acid as pH 3, since Hermann and others,[23] whom they quote, obtained evidence to suggest that the maximum amounts of nucleic acid will combine with toluidine blue at pH 4·2; and, on their own admission, they obtained deeper staining reactions at this more alkaline pH. Although, in the experiments described below, the toluidine blue was buffered to pH 3, the staining was controlled by RNAse extraction, and the method was intended not so much to determine the total RNA in the myoblasts, as to detect its fluctuation during growth and development. Hence the fact that a proportion of the RNA present might escape staining was not so important.*

The full procedure used was as follows:†

1 Photograph selected cells in the preparation, alive, under 4 mm phase-contrast objective, so as to record stages of cellular activity.
2 Remove coverslip with attached cells from culture medium, and place successively in the following solutions:
3 0·9 per cent saline solution   3 min
4 Fix, and deaminate, in formol-sublimate at 37°C   1 hr
5 Running tap water   5 min
6 Lugol's iodine   5 min
7. 5 per cent sodium thiosulphate   5 min
8 Running tap water   5 min
9 Van Slyke's nitrous acid at 3°C (a 50:50 mixture of 10 per cent acetic acid and 5 per cent sodium nitrite)   18 hr
10 0·01 per cent toluidine blue in acetate buffer at pH 3 $\frac{1}{2}$ hr
11 Mount in this toluidine blue solution and photograph the same cells under a 4 mm ordinary objective. (This mounting medium was used in order to try to ensure that the dye was retained in the tissue for the few minutes required for finding and photographing the cells.)
12 Remove coverslip and cells from the slide and place, successively, in the following solutions:
13 60 per cent alcohol at 37°C (until all the dye is removed).   c. 20 min
14 Running tap water   5 min
15 0·1 per cent RNAse in dist. water at 37°C   3 hr
16 Running tap water $\frac{1}{2}$ min
17 0·01 per cent toluidine blue in acetate buffer at pH 3 $\frac{1}{2}$ hr
18 Mount in this toluidine blue solution and photograph the same cells, under an ordinary 4 mm objective

The cells were first photographed alive under a phase contrast objective because it was frequently easier then to recognize whether they were fusing or simply lying in contact with each other. It was also necessary to obtain a record of the appearances of the living cells, so that they could be found again, quickly and easily, after they were fixed and stained. (The film of the living cells was developed soon after it had been taken, and was projected on to the wall in front of the microscope, frame by frame. It formed an invaluable aid in the search for the same microscopic field in the stained material.) A

---

* Experiments are now being undertaken by the authors using similar myoblast material stained with toluidine blue at pH 4. The cells exhibit appreciably deeper staining, but so far show exactly the same pattern in the relative amounts of RNA recorded in the cytoplasm of uni-nucleate and fusing cells.

† The method described here accords fairly closely with the optimum criteria for fixation and staining for the quantitative cytochemistry of RNA recommended by Swift.[51] Unfortunately the book containing this excellent comprehensive survey of the field had not been published when the authors evolved their technique: but, had it been, they would have used neutral formalin as a fixative instead of formol-sublimate, an alcohol solution of van Slyke's nitrous acid instead of an aqueous one, and a toluidene blue solution buffered to pH 3·5 or 4, instead of pH 3.

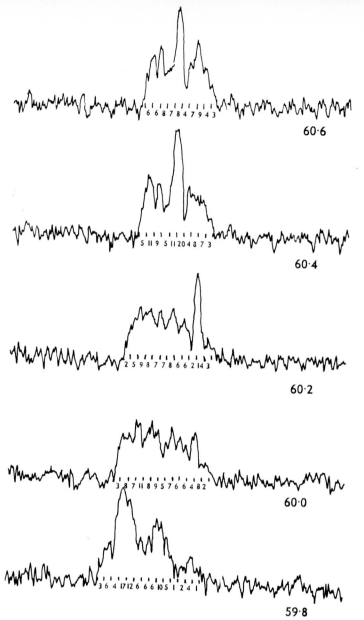

60·6

60·4

60·2

60·0

59·8

Fig. 22.9 Five contiguous traces made with a microdensitometer across the lower part of cell b, indicated by arrows in Fig. 22.10 D_1 (opposite), while scanning the negative of the photomicrograph shown in Fig. 22.10 B. Between each trace, the stage carrying the negative was shifted up by 0·2 mm, which was also the width of the negative scanned by each trace. (These shifts are indicated by the vernier scale readings, 59·8–60·6 mm, recorded at the right-hand side of each trace.) Where each trace crossed the image of the cell, considerable departures from the optical density of the background were recorded. These were evaluated at intervals of 0·2 mm along each scan on the negative, and are represented by the small figures in the middle of each trace. From values such as these, density maps, like those shown in Figs. 22.10 D and 22.11 D, were constructed for each cell stained both before and again after digestion with RNAse.

Fig. 22.10 Photomicrographs of a single myoblast, a, and an adjacent pair of actively fusing myoblasts, b and c; and optical density maps of these cells derived from densitometric measurements.

A, a photomicrograph of these cells alive under a phase contrast microscope.

B, the same cells, photographed under an ordinary light microscope after being fixed and stained with toluidine blue.

C, the same cells, restained with toluidine blue after being incubated for 3 hr in RNAse solution.

The optical density maps, D_1 and D_2, were constructed from successions of contiguous density traces made with a microdensitometer across the photomicrographic negatives of B and C respectively. (A sample of five such traces, made across the negative of B, is shown in Fig. 22.9 (on this page), and from it the five rows of figures indicated by arrows in cell b in D_1 are derived.)

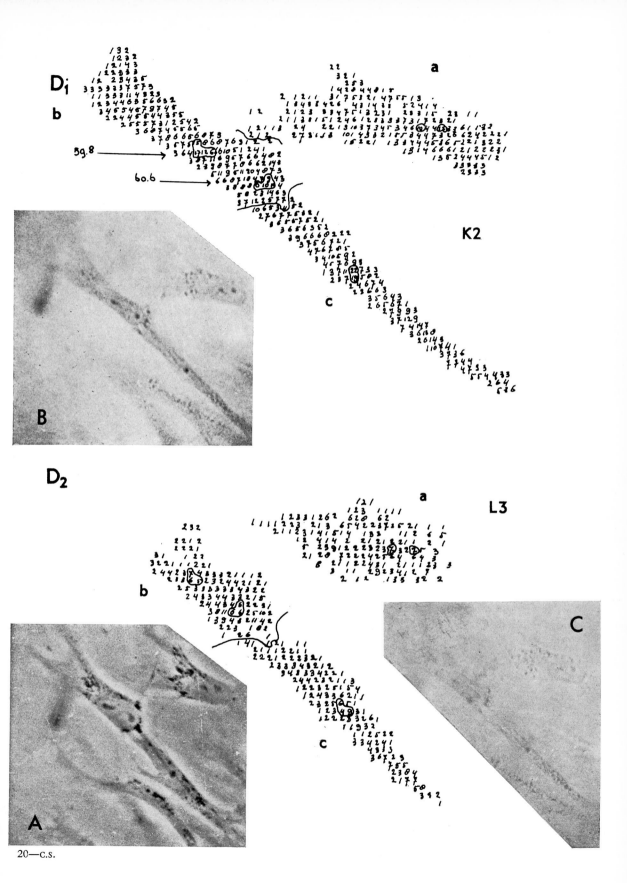

4 mm objective was used, in preference to a 2 mm objective, for all the photographs, because of its greater depth of focus.

It was of great importance that every part of these relatively flat and extended cells should be photographed in sharp focus if a true densitometric assessment of the stained material they contained was to be obtained. No figures are available for the thickness of living myoblasts, but some measurements were made by Barer and Dick,[9] on clear regions of the cytoplasm in living chick fibroblasts (which are similar in size to myoblasts and often difficult to distinguish from them) by interference microscopy with the 'double immersion' method of Barer.[7] They obtained values ranging from 0·5 μ in the thinnest extremities of the pseudopodia to nearly 2·5 μ in the vicinity of some of the nuclei. Authorities differ about the depth of focus obtainable by a 4 mm objective with a numerical aperture of around 0·75. Martin and Johnson[31] give the figure of 8 μ as the maximum depth of focus through which no appreciable image deterioration occurs (based on an assumption by Rayleigh that this will happen if the lengths of the paths of light from the top and bottom of an object differ by more than $\lambda/4$, which they describe as 'rather rigorous'). But Barer,[8] more cautiously, prefers the figure of 2 μ, as the effective depth of focus of such an objective. Thus it seems almost certain that every part of the fixed and stained myoblasts could be sharply focused by the 4 mm objective that was used, because the myoblasts certainly must have shrunken, and become even thinner after fixation, than when they were alive (see Ross[37]). All the same, it was obvious that, in many cases, the margin of error permissible in focusing the cells was narrow. An experiment was performed, in which it was found that densitometric traces made across a cell very slightly out of focus, gave values for its total absorption, which were 12 per cent lower than those made across the same cell sharply in focus. Consequently, great care was taken to ensure that every part of each fixed and stained cell, photographed before and after RNAse digestion, was sharply focused. Any that could not be so focused, or were not in focus, were rejected for subsequent densitometric scanning.

For densitometry, it was necessary to select the level of illumination in the microscope, the exposure time, the film, the developer, and the development time and temperature, so that the optical densities exhibited by the cellular material on the photographic negatives gave a true representation of the range of density exhibited by the stained regions of the cells themselves. This was done by photographing selected materials of known density (a progression of super-imposed thicknesses of a neutral density filter) under the 4 mm objective, and making traces of these negatives to form a standard of comparison. The variables of illumination, exposure time and development time were adjusted, in photographing both this standard, and the cells, until the ranges of density normally exhibited by the cell material were linearly represented as density differences on the photomicrographic negatives.* As every microscope, set up for photography, has its own individual characteristics, there is little point in describing the particular procedure which was used in detail, but it is worth mentioning that with Kodak 'Plus X' film developed in 'Promicrol' (May and Baker), it was possible to obtain a linear representation of densities over a range considerably greater than that exhibited by the cellular material (see also Ross[41]).

Thus with the technique described above, it was possible to make two photomicrographic negatives of each cell, or group of cells, both sharply focused, and both truly representative of the density differences exhibited in the staining of the original; the first stained before the removal of RNA with RNAse, and the second stained again after its removal (see Figs. 22.10 and 22.11). The cells in each of the photomicrographs were then scanned over their entire areas with a double-beam automatic recording microdensitometer (manufactured by Joyce Loebl Ltd, Gateshead, Co. Durham, England), at a trace magnification of $\times 10$. The height of the scanning slit was carefully adjusted to make a linear scan 0·2 mm wide across the photographic image of the cell on the negative, and, as the mechanical stage, on which the negative was mounted, was moved up by exactly this amount between each successive scan, it was

* Had the relationship not been linear, it would have been possible to obtain values for the optical density of the cellular material if the exact non-linear qualities of the negatives had been determined by comparison with a density standard photographed in the same field in the manner described by Woodward, Gorovsky and Swift.[59] A very suitable standard for this purpose is the opaque rotating disc with its circumference cut in steps first used by Caspersson.[14]

possible to obtain a succession of densitometer traces of contiguous areas of the cell image 0·2 mm apart (see Fig. 22.9). Usually about forty such traces, depending partly on the orientation of the cell relative to the direction of trace, were required for each cell, and this took several hours to obtain.

Prior to this, prints were made from the photographic negatives, showing the cells selected for scanning, at exactly the same magnification as that of the densitometer trace. These were used both as an aid to finding the place to begin making the traces on the negatives, and subsequently in the construction of density distribution charts of the cell's area. For the latter purpose, these prints were covered, first with a transparent graticule of squares of 2 mm side on a sheet of film base, and then with a sheet of tracing paper. Figures indicating the differences in density, between a particular point on the cell and the background density, were marked on the tracing paper over each square. The density differences themselves were measured directly from the densitometer traces, each being the mean departure from the background level of intensity exhibited over a length of 2 mm on the trace (see Fig. 22.9). It was necessary to select cells, and groups of cells, that were reasonably separate from each other, so as to be able to include appreciable portions of the background field, on each densitometer trace. Thus, the entire area of the cell was covered with density measurements in squares, representing squares with a side of 1·6 μ or areas of 2·56 sq. μ in the cells themselves. One advantage of this method was that it did not greatly matter if there was a regular gradient in the background density in the immediate vicinity of the cells being measured, because such a gradient showed itself as a regular slope in the background trace, and, by drawing a line connecting the parts of the trace on each side of that given by the cell, the departures in density given by each cell region could be related to this base-line. Background gradients of this kind were not very often found, but sometimes occurred, probably as a result of the presence of a plasma clot in the vicinity of the cell.

Usually 100–300 individual density measurements were required to cover the entire area of a cell. As each of these represented an area of 2·56 sq. μ this meant that the areas of the cells could be measured to a considerable degree of accuracy, along with the other parameters (see column b in Tables 1–5). It was only necessary to subtract the density values obtained over a particular area of a cell after RNAse digestion, from the values obtained over the same area before digestion, to obtain values indicative of the amount of RNA present in that particular cell area, at the time of fixation. Although it is, of course, very unlikely that all the RNA in any given area was stained with the basic dye, the use of this strictly standardized procedure, for all the cells studied, makes it probable that the same proportion of RNA was stained in each; and it was likely that the values obtained from these subtractions provided a reliable indication of the *relative* amount of RNA present in selected regions of cells, at different stages of development.

THE CHANGES IN CYTOPLASMIC RNA FOUND IN DIFFERENTIATING MYOBLASTS AND THEIR POSSIBLE INTERPRETATION

Although the experimental method described above can be made to yield a considerable amount of comparative cell data, the information so far obtained has been almost wholly confined to making comparative measurements of RNA in the cytoplasm of uni-nucleate and fusing myoblasts,* in one or two

---

* In the course of the investigation, figures were obtained for the relative amounts of RNA in the nucleoli as well as in the cytoplasm (column f in Tables 1–5), but these were not regarded as very accurate because of the relatively small area of these nucleoli. In contrast to the cytoplasm, where measurements were made and summated over 100–300 squares, each nucleolus occupied only 3 or 4 of the squares of 1·6 μ side, into which the image of each cell was divided, in making the density maps described above. This means that there is a strong possibility that the recorded areas and densities of nucleoli subject to appreciable errors and these values cannot therefore be expected to reflect the trends exhibited by the much more precise values for the total dry mass and volumes of the nucleoli obtained with interference microscopy.

Another parameter that it is possible to measure with this technique, is the proportion of RNA in the cytoplasm of fusing cells to within a short distance (say 5 μ) of the line of their fusion. Preliminary measurements, however, showed nothing very significant about the levels of RNA in these regions, compared with those in the rest of the cytoplasm, and this investigation has not, at present, been pursued any further.

day-old cultures, from explants of 10–12 day-old chick embryos. Cells were selected at random for measurement, in each culture, and, as far as is known, only myoblasts were measured. Every possible care was taken to try to ensure that the explants consisted of muscle as free from connective tissue as possible, but there is always a small proportion of fibroblasts in such cultures, and these cannot always be distinguished easily from myoblasts. Consequently, it should be borne in mind that some of the uninucleate cells measured could have been fibroblasts, although the proportion accidentally measured in this way is unlikely to have been large enough to have affected the results seriously.† (It is hardly likely that any fusing or bi-nucleate cells could have been mis-identified in this way, since fibroblasts, unlike myoblasts, rarely, if ever, fuse with each other.) It is also possible that a small proportion of 'actively fusing' myoblasts were wrongly classified, when they were, really, only passing each other, or establishing a temporary contact. The pairs of myoblasts undergoing true active fusion, were recognized by making careful studies of photographs of the appearance under phase-contrast while still alive, and by studying the photographs of the same pairs of cells, after fixation and staining. Both record photographs were helpful in determining whether the cells were truly fusing or not. In the phase-contrast pictures in particular, it was nearly always possible to distinguish a 'boundary line' still partially separating each cell from its fusing partner.‡ This was extremely useful in determining which parts of the (fusing) cytoplasm had belonged to each cell. It enabled the levels of the cytoplasmic RNA in each of two fusing cells to be measured, separately, with a reasonable degree of certainty. In the relatively rare cases where isolated bi-nucleate cells were observed in which fusion was apparently complete, no trace of such a boundary could be made out. In these cases, to obtain comparable information about the cytoplasmic

---

† It is, perhaps, unfortunate that the extremely elegant clonal culture techniques of Konigsberg,[29] by which unambiguous cultures of pure myoblasts can be obtained, had not been published when the series of investigations reported here was initiated: although it should also be appreciated that these techniques require considerable skill and experience to carry out successfully, and can be very time-consuming.

From the culture of colonies derived from single cells, Konigsberg and his collaborators have found that it is usually possible to distinguish myoblasts from fibroblasts by their morphological appearance: myoblasts being, for example, normally bipolar and elongated, and fibroblasts more spread out and irregular. Their figures in fact reveal that the criteria they employed were more effective for recognizing myoblasts than fibroblasts: 96·5 per cent of the cells they designated as myoblasts that grew into colonies gave rise to 'muscle straps', but 15 per cent of the irregular cells they designated as fibroblasts also did so. So a fibroblast-like cell had nearly a one in six chance of being a myoblast.

In the present authors' investigations, by far the majority of the uninucleate cells randomly selected for RNA measurements (88 per cent) were of the bipolar 'myoblast' form, but not all, and indeed a few of the actively fusing cells were of the more irregular and spread out form more typical of fibroblasts (see Fig. 22.10). It should be emphasized, however, that Konigsberg's morphological criteria do not necessarily apply with different culture media and substrates to those that he used. For example, the 'ruffled membrane', which he describes and includes in his criteria, is typically present in both types of cell when they are growing on plain glass surfaces or on a substrate of synthetic collagen, but is very nearly completely absent in cells growing on a substrate of avian plasma, as were those in the present investigation. It is therefore the authors' opinion that it would have been premature to have attempted to apply rigid morphological classificatory criteria to their material(see also Ross and Hudgson[61]).

‡ This observation seems to accord with the findings of Murray and his co-workers (Murray[32]), who found that dyes introduced into myoblasts by micro-injection were, in many cases, confined to restricted regions of apparently bi-nucleate and multi-nucleate cells. They concluded that the moment of time when the cytoplasm of fusing cells became 'truly continuous' may quite often occur after all the signs of a visible boundary have disappeared.

Fig. 22.11 Photomicrographs of an incipient 'muscle strap' composed of three myoblasts, a, b, and c, together with optical density maps of these cells derived from densitometric measurements. Cells a and b are still actively fusing, while cells b and c fused with each other earlier.

A, photomicrograph of these cells alive, taken under a phase contrast microscope.

B, the same cells, photographed under an ordinary light microscope, after being fixed and stained with toluidine blue.

C, the same cells, restained with toluidine blue, after being incubated for 3 hr in RNAse solution.

The optical density maps, D_1 and D_2, were constructed from successions of contiguous density traces made with a microdensitometer across the photomicrograph negatives of B and C respectively.

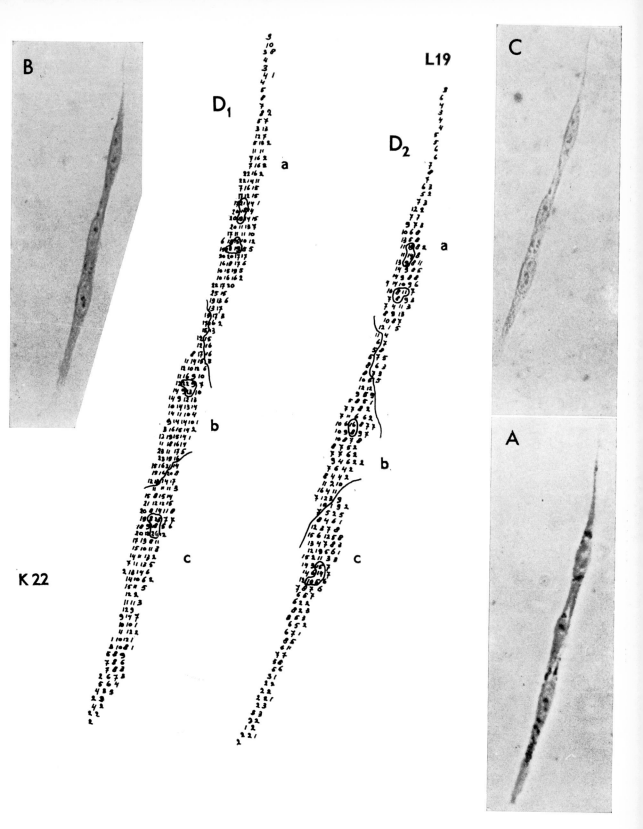

RNA, it was necessary, either to summate this over the whole pair and divide by two, or to draw an arbitrary dividing line, as was done with the data shown in Table 22.5. (Pairs of fully-fused cells were rare, because, once two cells had fused, or even while they were still fusing, another uni-nucleate cell would normally begin to fuse with that pair in the process of ordinary muscle-strap formation (see, for example, Fig. 22.11). The particular configuration of two fused cells thus represented a rather rare and transitory condition.) Cells that had made only very small areas of contact with each other were not classified as actively fusing cells (although some of them could have been in the initial stages of fusion), and they were treated as a separate category.

The results of comparing the amounts of cytoplasmic RNA in uni-nucleate and fusing pairs of cells over the initial two days of their development in tissue culture, are shown in Tables 22.1–22.4 and in Fig. 22.12. It can be seen clearly that the expected elevation in cytoplasmic RNA is indeed found in the actively fusing cells. An unexpected finding was that, in nearly every case, *one* cell of each fusing pair and not the other had a cytoplasmic RNA content which was appreciably higher than that found in the uni-nucleate cells. It was also odd that the cell in which the elevated level of cytoplasmic RNA was found was, almost always, the one which had the larger area. The increase in RNA in one cell and not the other could indicate that one cell rather than the other was taking the 'metabolic initiative' in starting the fusion process, by releasing RNA from its nucleoli into the cytoplasm in order to start off the protein syntheses necessary for the activity of fusion. In this way the observations may be correlated with the previous findings (reported on p. 285) that, in most cases, the nucleoli in only one of the fusing cells in a pair undergo appreciable shrinkage during cell fusion. It may also be significant that, in the few cases observed of binucleate cells where fusion appeared complete, the level of cytoplasmic RNA in each cell was not significantly greater than that of uni-nucleate cells (compare Tables 22.1 and 22.3 with Table 22.5). This suggests that the elevation of cytoplasmic RNA in one cell may be a transitory phenomenon associated with active cell fusion.

When the levels of cytoplasmic RNA in one-day-old cultures of myoblasts were compared with those in two-day-old cultures, a second pattern appeared to be discernible. Namely, a small steady rise in the amount of RNA in the cytoplasm of both uni-nucleate and fusing cells, but particularly in the latter,* during the two-day period. This suggests that RNA may be synthesized in the nucleoli and pass into the cytoplasm (or possibly be synthesized in the cytoplasm) at a fairly steady rate, but that the levels in both regions of the cell fluctuate more markedly during bursts of metabolic activity associated with a particular growth or differentiation process.

In this connection, it is interesting to find that Stowell, in 1961[50] reported that he and his collaborators found that the nucleoli in the liver cells of protein-starved rats showed a marked increase in dry mass, and that they returned to their normal size again in rats which had been re-fed with protein, after a period of starvation. Also that the nucleoli in cells of the blastulae of *Amblystoma* were twice the size in embryos cooled to 9°C than in embryos kept at 18°C. Stowell and his co-workers did not interpret these findings, but in both cases the normal metabolism of the cells was probably retarded under the circumstances that led to nucleolar enlargement. Liver cells are predominantly concerned with metabolizing food products, and early embryonic cells with the metabolism of growth and differentiation. It is therefore not unreasonable to expect the former to be slowed down by starvation, and the latter by cooling. This indicates that, in both cases, the synthesis of RNA in the nucleoli may have proceeded at a more or less steady rate, but that its ultimate utilization, in initiating cellular metabolic processes, may have been slowed, leading to an accumulation of nucleolar RNA. This accords with the picture of fairly steady RNA synthesis, and its utilization in a more irregular manner during spurts of growth, differentiation or development, that appears to emerge from the results which we have reported here.

In 1960 Brachet[11] summarized many of his findings over the preceding twenty years by saying 'RNA is accumulated, and is most actively synthesized in the regions of the embryo which have the greatest importance for morphogenetic processes'; and, as long ago as 1950 Caspersson[14] wrote 'increase in nucleolar masses is a most conspicuous phenomenon during cytoplasmic protein synthesis (and cell

* See note added in proof, p. 304.

growth) . . . in rare cases one finds cells with large nucleoli and heavily ultra-violet absorbing cytoplasm which are not growing, but which will start growing immediately when the proper stimulus reaches them. One example is the plant seed before germination; another is found in the imaginal discs of *Drosophila*. . . .' It is to be hoped that the findings reported in this paper can be regarded as providing further experimental confirmation of the conclusions drawn and the insight exhibited by these two great

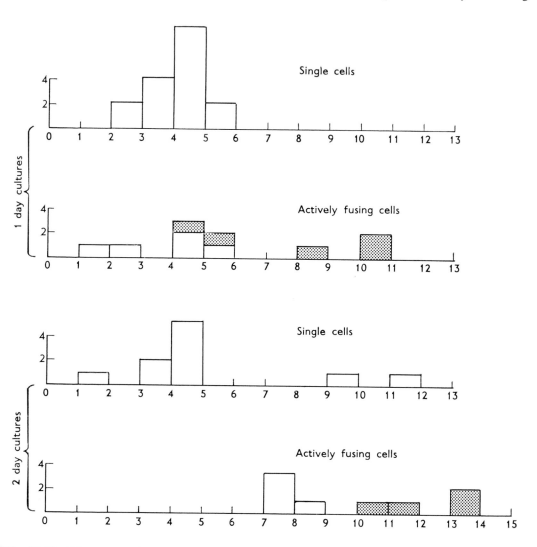

Fig. 22.12 Histograms, showing the distribution of the amounts of RNA found in single and in actively-fusing chick myoblasts in one-day-old and two-day-old tissue culture preparations, derived from the data displayed in Tables 22.1–22.4. Ordinate: number of cells. Abscissa: amounts of RNA, in units of optical density, as shown in Tables 22.1–22.4. In this histograms showing the distribution of RNA in the actively-fusing cells, the levels found in one of each fusing pair are represented by shaded areas, and the levels found in the other of each pair by unshaded areas. From this it can be seen that, in the one-day-old cultures, the cytoplasmic RNA found in one of each pair of fusing cells, but not in the other, tends to be appreciably higher than that found in the single cells. In the two-day-old cultures the RNA in both cells of the fusing pairs tends to be higher than in the single cells, but again more markedly so in one cell of each fusing pair than in the other.

cytologists, long before the molecular structures of DNA and RNA had been determined, or their significance appreciated.*

Table 22.1 Values for cell area and nucleolar and cytoplasmic RNA obtained by summating the measurements obtained from optical density maps of 16 single uninucleate chick myoblasts in one-day-old tissue culture preparations. In this table and the subsequent ones (Tables 22.2–5) the units of optical density shown in columns b–i from which the comparative values for cytoplasmic and nucleolar RNA are derived are the same as those shown on the density maps illustrated in Figs. 22.10 and 22.11 multiplied by a factor of 1/80

a	b	c	d	e	f	g	h	i
				(c − d)			(f − g)	(e − h)
Cell No.	Total cell area (μ^2)	Total basiphil material	Total basiphil material after RNAse	Total RNA in cell	Nucleolar basiphil material	Nucleolar basiphil material after RNAse	Nucleolar RNA	Cytoplasmic RNA
AK21 L18	225	10·8	6·8	4·0	0·9	0·5	0·4	3·6
BK21 L3	280	10·0	5·4	4·6	0·9	0·3	0·6	4·0
AK2 L3	600	8·5	4·9	3·6	0·6	0·3	0·3	3·3
AK7 L8	275	9·0	3·6	5·4	0·7	0·2	0·5	4·9
AK12 L11	290	8·16	3·03	5·13	0·62	0·28	0·34	4·79
AK13 L12	255	5·50	2·99	2·51	0·48	0·17	0·31	2·20
BK13 L12	285	6·46	2·54	3·92	0·46	0·32	0·14	3·78
BK32 L26	215	7·62	3·84	3·78	0·56	0·18	0·38	3·40
AK25 L21	265	10·28	5·32	4·96	0·65	0·50	0·15	4·81
AK15 L14	675	13·42	8·05	5·37	0·82	0·42	0·40	4·97
CK20 L17	395	10·61	6·00	4·61	0·72	0·32	0·40	4·21
CK18 L16	415	11·46	7·07	4·39	0·75	0·46	0·29	4·10
AK26 L27	300	10·01	5·62	4·39	0·55	0·32	0·23	4·16
AK33 L27	402	11·11	5·30	5·81	0·71	0·37	0·34	5·47
AL'1a L'31	295	13·30	6·72	6·58	1·60	0·55	1·05	5·53
CJ'27 L'4a	157·5	7·23	4·96	2·27	0·27	0·17	0·10	2·17

Table 22.2 Values for cell area and nucleolar and cytoplasmic RNA obtained by summating the measurements obtained from optical density maps of 5 pairs of actively fusing chick myoblasts in one-day-old tissue culture preparations

a	b	c	d	e	f	g	h	i
				(c − d)			(f − g)	(e − h)
Cell No.	Total cell area (μ^2)	Total basiphil material	Total basiphil material after RNAse	Total RNA in cell	Nucleolar basiphil material	Nucleolar basiphil material after RNAse	Nucleolar RNA	Cytoplasmic RNA
AK18 L16	370	16·69	5·17	11·52	0·90	0·31	0·59	10·93 ⎫
BK18 L16	275	6·93	5·28	1·65	0·83	0·56	0·27	1·38 ⎬
AK19 L17	265	9·73	5·44	4·29	0·58	0·41	0·17	4·12 ⎫
BK19 L17	300	15·81	5·12	10·69	0·61	0·23	0·38	10·31 ⎬
BK2 L3	510	13·07	4·32	8·75	0·97	0·43	0·54	8·21 ⎫
CK2 L3	385	9·96	4·51	5·45	0·51	0·23	0·28	5·17 ⎬
BK7 L8	275	11·24	5·96	5·28	0·87	0·33	0·54	4·74 ⎫
CK7 L8	340	11·84	3·52	8·32	0·83	0·25	0·58	4·74 ⎬
AJ'24a AL'1a	112·5	6·41	3·48	2·93	0·48	0·30	0·18	2·75 ⎫
BJ'24a BL'1a	215	11·24	5·93	5·31	0·45	0·33	0·12	5·19 ⎬

* The reader is referred to Ross & Hudgson[61] for a review of the literature on myogenesis, a critique of the methods and the merits of distinguishing myoblasts and fibroblasts, and an account of muscle tissue culture related to pathology.

Table 22.3 Values for cell area and nucleolar and cytoplasmic RNA obtained by summating the measurements obtained from optical density maps of 10 single uninucleate chick myoblasts in two-day-old tissue culture preparations

a Cell No.	b Total cell area (μ^2)	c Total basiphil material	d Total basiphil material after RNAse	e (c − d) Total RNA in cell	f Nucleolar basiphil material	g Nucleolar basiphil material after RNAse	h (f − g) Nucleolar RNA	i (e − h) Cytoplasmic RNA
AN11 O13	402	6·22	4·41	1·81	0·36	0·23	0·13	1·68
BL'24a M'20	505	13·76	9·55	4·21	0·39	0·30	0·09	4·12
AL'24a M'20	362·5	13·81	8·96	4·85	0·74	0·41	0·33	4·52
AL'30a M'26	312·5	16·04	12·73	3·31	0·70	0·52	0·18	3·13
BL'14a M'8	780	22·87	13·24	9·63	1·6	1·07	0·53	9·10
AL'14a M'8	922·5	26·54	15·16	11·38	0·95	0·65	0·30	11·08
AL'25a M'21	515	13·09	8·37	4·72	0·60	0·39	0·21	4·51
AL'23a M'19	705	14·68	10·05	4·63	0·55	0·37	0·18	4·45
AL'16a M'10	682·5	17·70	12·82	4·88	0·53	0·36	0·17	4·71
AL'26a M'22	325	11·05	7·1	3·95	0·55	0·38	0·17	3·78

Table 22.4 Values for cell area and nucleolar and cytoplasmic RNA obtained by summating the measurements obtained from optical density maps of 4 pairs of actively fusing chick myoblasts in two-day-old tissue culture preparations

a Cell No.	b Total cell area (μ^2)	c Total basiphil material	d Total basiphil material after RNAse	e (c − d) Total RNA in cell	f Nucleolar basiphil material	g Nucleolar basiphil material after RNAse	h (f − g) Nucleolar RNA	i (e − h) Cytoplasmic RNA
AN27 O29	1170	28·12	14·02	14·10	1·9	1·27	0·63	13·47 ⎞
BN27 O29	697	15·83	8·01	7·82	0·67	0·30	0·37	7·45 ⎠
AN22 O27	840	16·85	8·86	7·99	1·21	0·71	0·50	7·49 ⎞
BN22 O27	1212	28·61	14·82	13·79	0·71	0·55	0·16	13·63 ⎠
AL'22a M'18	965	30·71	18·57	12·14	1·21	0·80	0·41	11·73 ⎞
BL'22a M'18	800	26·91	18·05	8·86	0·68	0·44	0·24	8·62 ⎠
AN12 O15	390	12·85	4·66	8·19	1·05	0·31	0·74	7·45 ⎞
BN12 O15	625	18·85	7·32	11·53	1·39	0·42	0·97	10·56 ⎠

Table 22.5   Values for cell area and nucleolar and cytoplasmic RNA obtained by summating the measurements obtained from optical density maps of 4 binucleate chick myoblasts from 1–2 day-old tissue culture preparations in which the process of cell fusion appeared to have been completed. (As no boundary lines were recognizable between the fused cells, an arbitrary one was drawn between them in each case, and the figures for the areas and RNA content of each of the component cells were derived for this assumption)

a	b	c	d	e (c−d)	f	g	h (f−g)	i (e−h)
Cell No.	Total cell area (μ^2)	Total basiphil material	Total basiphil material after RNAse	Total RNA in cell	Nucleolar basiphil material	Nucleolar basiphil material after RNAse	Nucleolar RNA	Cytoplasmic RNA
AK36 L29	210	8·63	4·45	4·18	0·75	0·40	0·35	3·83
BK36 L29	200	8·13	3·95	4·18	0·71	0·31	0·40	3·78
BN11 O13	500	11·83	4·80	7·03	0·51	0·26	0·25	6·78
CN11 O13	900	12·08	4·68	7·40	0·61	0·28	0·33	7·07
AL′31a M′27	355	13·26	9·49	3·77	0·96	0·62	0·34	3·43
BL′31a M′27	174	9·49	5·92	3·57	0·55	0·41	0·14	3·43
AJ′27a L′4a	87·5	5·53	2·75	2·78	0·42	0·24	0·18	2·60
BJ′27a L′4a	145	6·99	4·31	2·62	0·36	0·22	0·14	2·48

SUMMARY

Measurements made by interference microscopy of the dry mass of the nucleoli of living myoblasts grown in tissue culture preparations from normally differentiating and dystrophic muscle fibres have revealed two main findings:

(1) The process of cell-fusion which characterizes the initial stages of the differentiation of a normal muscle fibre is associated with a characteristic fluctuation in the dry mass of the nucleoli of the myoblasts involved. The total solids in these nucleoli falls while the cells are actually fusing with each other, and rises again after fusion is completed.

(2) A considerable proportion of similar myoblasts derived from animals with muscular dystrophy (a hereditary disease in which the muscle fibres are unable to differentiate normally) have nucleoli with an abnormally low dry mass. This proportion increases as the disease progresses.

It was thought that these changes in nucleolar mass could be due to changes in the amount of nucleolar RNA, since RNA is the vehicle by which the coding contained in the DNA can initiate specific protein syntheses concerned with particular differentiation processes. In order to establish this, a densitometric scanning technique was developed, and optical density measurements were made on photomicrographs of fixed myoblasts stained with a basic dye both before and after digestion with RNAse. This has revealed that, when two myoblasts fuse during normal differentiation, RNA seems to leave the nucleoli of one of the fusing cells and appear in extra large amounts in its cytoplasm. After fusion is complete, the RNA levels tend to return to normal.

It is suggested that nuclear RNA may similarly be involved in initiating the metabolism associated with other active differentiation processes.

ACKNOWLEDGEMENTS

The authors wish to thank Miss A. E. Brown, Miss A. Lackenby, Miss M. Jenkisson and Mr. J. J. Fulthorpe for their very skilled technical assistance. All the investigations reported here were financed by generous grants from the Muscular Dystrophy Group of Great Britain, the Muscular Dystrophy Association of America and the Medical Research Council. They were carried out in the Muscular

Dystrophy Group Research Laboratories at Newcastle upon Tyne General Hospital under the direction of Dr. J. N. Walton, F.R.C.P., to whom we are also indebted for reading and making some very helpful comments on our manuscript. The latter was carefully typed by Miss P. Ruddick.

REFERENCES

1. ALLEN, R. D., BRAULT, J., and MOORE, R. D. 1963. A new method of polarization microscopic analysis. *J. Cell Biol.* **18**, 223
2. ——, ——, and ZEH, R. 1966. Image contrast and phase-modulated light methods in polarization and interference microscopy. In *Recent Advances in Optical and Electron Microscopy* (edited by R. Barer and V. Coslett). New York and London: Academic Press
3. BAKER, J. R. 1952. The cell theory, a restatement, history and critique. Part 3. The cell as a morphological unit. *Quart. J. micr. Sci.* **93**, 2
4. ——. 1958. *Principles of Biological Microtechnique*. London: Methuen; and New York: John Wiley
5. BALAZS, E. A. and SZIRMAI, J. A. 1958. Quantitative determination of cationic dyebinding in connective tissue. *J. Histochem. Cytochem.* **6**, 278
6. BALBIANI, E. G. 1881. Sur la structure du noyau des cellules salivaires chez les larves de *Chironomus. Zool. Anz.* **4**, 637
7. BARER, R. 1953. Determination of dry mass, thickness, solid and water concentration in living cells. *Nature, Lond.* **172**, 1098
8. ——. 1956. *Lecture Notes on the Use of the Microscope*, 2nd edition. Oxford: Blackwell Scientific Publications
9. —— and DICK, D. A. T. 1957. Interferometry and refractometry of cells in tissue culture. *Exp. Cell Res. Suppl.* **4**, 103
10. BEERMANN, D. W. 1962. Riesenchromosomen. In *Protoplasmatologia*. **VI/D**. Berlin: Springer-Verlag
11. BRACHET, J. 1960. *The Biochemistry of Development*. Oxford: Pergamon
12. BRATTGARD, S. O. and HYDÉN, H. 1954. The composition of the nerve cell studied with new methods. *Int. Rev. Cytol.* **3**, 455
13. CASPERSSON, T. O. 1936. Die Absorptionsmessung im mikroskopischen Preparat als mikrochemische Methode. *Skand. Arch. Physiol.* **77**
14. ——. 1950. *Cell Growth and Cell Function*. New York: Norton & Co.
15. CHAPEVILLE, F., LIPMANN, F., VON EHRENSTEIN, G., WEISBLUM, B., RAY, W. J., JNR., and BENZER, S. 1962. On the role of soluble ribonucleic acid in coding for amino-acids. *Proc. Nat. Acad. Sci.* **48**, 1086
16. CHAYEN, J. 1966 (personal communication)
17. DAVIES, H. G. 1957. The determination of mass and concentration by microscope interferometry. In *General Cytochemical Methods* (edited by J. F. Danielli), Vol. I. New York: Academic Press
18. VAN DUIJN, C., JNR. 1962. Differential staining of the nucleic acids ? *Nature, Lond.* **193**, 999
19. DYSON, J. 1966. An interference microscope for precise measurement of thin films. *Proc. Roy. Micr. Soc.* **1**, 139
20. GOODMAN, M. and RICH, A. 1963. Mechanisms of polyribosome actions during protein synthesis. *Nature, Lond.* **199**, 318
21. HALE, A. J. 1958. *The Interference Microscope in Biological Research*. Edinburgh: Livingstone
22. HAY, E. D. and GURDON, J. B. 1967. Fine structure of the nucleolus in normal and mutant *Xenopus* embryos. *J. Cell Sci.* **2**, 151
23. HERMANN, H., NICOLAS, J. S., and BORICIOUS, J. K. 1949. Toluidene blue binding by developing muscle tissue: assay and data on the mechanism involved. *J. Biol. Chem.* **107**, 321
24. HUDGSON, P. 1965. (personal communications)
25. HYDÉN, H. 1955. The chemistry of single neurons: a study with new methods. In *Biochemistry of the Nervous System* (edited by H. Waelsch). New York: Academic Press
26. —— and LARSSON, S. 1956. The applications of a scanning and computing cell analyser to neurocytological problems. *J. Neurochem.* **1**, 134
27. JACOB, F. and MONOD, J. 1961. Genetic regulatory mechanisms in the synthesis of proteins. *J. Molec. Biol.* **3**, 318
28. JONES, K. W. 1965. The role of the nucleolus in the formation of ribosomes. *J. Ultrastruct. Res.* **13**, 257
29. KONIGSBERG, I. R. 1963. Clonal analysis of myogenesis. *Science* **140**, 1273
30. LOVE, R. and LILES, R. H. 1959. Differentiation of nucleoproteins by inactivation of protein-bound amino groups and staining with toluidene blue and ammonium molybdate. *J. Histochem. Cytochem.* **7**, 164
31. MARTIN, L. C. and JOHNSON, B. K. 1949. *Practical Microscopy*, 2nd edition. London: Blackie
32. MURRAY, M. R. 1960. Skeletal muscle tissue in cultures. In *Structure and Function of Muscle* (edited by G. H. Bourne), Vol. 1. New York: Academic Press

33. NIRENBERG, M. W. and MATTHAEI, J. H. 1961. The dependence of cell-free protein synthesis in *E. coli* upon naturally occurring or synthetic polyribonucleotides. *Proc. Nat. Acad. Sci.* **47**, 1588
34. PAVAN, C. and BREUR, M. E. 1956. Differences in the nucleic acids content of the loci in polytene chromosomes of *Rynchosciara angelae* according to the tissues and larval stages. In *Cell Secretions* (edited by G. Schreiber, Belo Horizonte)
35. PEARCE, G. W. 1963. Tissue culture in the study of muscular dystrophy. In *Muscular Dystrophy in Man and Animals* (edited by G. H. Bourne and N. Golarz). Basle: Krager
36. RICHARDS, O. W. 1963. *A. O. Baker Interference Microscope Reference Manual*, 2nd edition. Buffalo, New York: American Optical Co.
37. ROSS, K. F. A. 1953. Cell shrinkage caused by fixatives and paraffin-wax embedding in ordinary cytological preparations. *Quart. J. micr. Sci.* **94**, 125
38. ——. 1961. The chemical and physical nature of the visible regions of mouse ascites tumour cells, and of the cells of solid tumours grown from them. *Quart. J. micr. Sci.* **102**, 59
39. ——. 1964. Nucleolar changes in differentiating myoblasts. *Quart. J. micr. Sci.* **105**, 423
40. ——. 1965. Changes in the nucleolar material of normal and dystrophic myoblasts. In *Research in Muscular Dystrophy. Proceedings of the Third Symposium*. London: Pitman
41. ——. 1967a. *Phase Contrast and Interference Microscopy for Cell Biologists*. London: Edward Arnold
42. ——. 1967b. The dry mass of living cell inclusions, and the value of comparative measurements. *J. Roy. micr. Soc.* (in the press)
43. —— and GALAVAZI, G. 1965. The size of bacteria, as measured by interference microscopy. *J. Roy. micr. Soc.* **84**, 13
44. —— and JANS, D. E. (in the press). The nucleolar and cytoplasmic RNA in normally differentiating and dystrophic myoblasts. In *Research in Muscular Dystrophy. Proceedings of the Fourth Symposium*. London: Pitman Medical
45. ——, ——, and SUSHEELA, A. K. 1968. The nucleolar and cytoplasmic RNA in normally differentiating and dystrophic myoblasts. In *Report of the 2nd International Congress of Neuro-Genetics and Neuro-Ophthalmology of the World Federation of Neurology*
46. SMITH, F. H. 1966. A photo-electronic phase-measuring microscope. *Proc. Roy. micr. Soc.* **1**, 139
47. ——. 1967. *J. Roy. micr. Soc.* (in the press)
48. SPENCER, M., FULLER, W., WILKINS, M. H. F., and BROWN, G. L. 1962. Determination of the helical configuration of ribonucleic acid molecules by x-ray diffraction study of crystalline amino-acid-transfer ribonucleic acid. *Nature, Lond.* **194**, 1014
49. STENRAM, U. 1958. Interferometric determination of ribose nucleic acid concentration in liver nucleoli of protein-fed and protein-deprived rats. *Exp. Cell Res. Suppl.* **15**, 174
50. STOWELL, R. E. 1963. The relationship of nucleolar mass of protein synthesis. *Exp. Cell Res. Suppl.* **9**, 164.
51. SWIFT, H. 1966. The quantitative cytochemistry of RNA. In *Introduction to Quantitative Cytochemistry* (edited by G. L. Wied). New York: Academic Press
52. ——. 1967. (personal communications)
53. SZIRMAI, J. A. and BALAZS, E. A. 1958. *Act histochemica*. Supplement band **1**, 556
54. WALKER, P. B. M. and RICHARDS, B. M. 1957. A method for investigating the stoichiometry of Feulgen stain. *Exp. Cell Res. Suppl.* **4**, 97
55. WALTER, F. 1964. The use of micro-interferometry as a quantitative method in biological research. *Scientific and Technical Information, English edn.*, **1**, 9. London: E. Leitz
56. WATSON, J. D. and CRICK, F. H. C. 1953. A structure for deoxyribose nucleic acid. *Nature, Lond.* **171**, 737
57. WEISS, S. B. and NAKAMOTO, T. 1961. On the participation of DNA in RNA biosynthesis. *Proc. Nat. Acad. Sci.* **47**, 694
58. WILKINS, M. H. F. and RANDALL, J. D. 1953. Crystallinity in sperm heads: molecular structure of nucleoprotein in vivo. *Biochem. et Biophys. Acta* **10**, 192
59. WOODWARD, J., GOROVSKY, M., and SWIFT, H. 1966. DNA content of a chromosome of *Trillium erectum*: effect of cold treatment. *Science* **151**, 215
60. OKAZAKI, K. and HOLTZER, H. 1965. An analysis of myogenesis in vitro using fluorescein-labelled antimyosin. *Journ. Histochem. Cytochem.* **13**, 726.
61. ROSS, K. F. A. and HUDGSON, P. (in the press) Tissue culture in muscle disease. In *Disorders of Voluntary Muscle, 2nd edn.*, edited by J. N. Walton, London, Churchill.

NOTE ADDED IN PROOF

Our finding that the cytoplasmic RNA is elevated more markedly in actively fusing myoblasts in 2-day-old cultures than in 1-day-old cultures, while this elevation is not so apparent in the single cells (p. 298 and Fig. 22.12) seems to fit in with the findings of Okazaki and Holtzer[60], who demonstrated that myoblasts are particularly active in the synthesis of myosin after they begin to fuse with each other. RNA must be involved in initiating all the major patterns of protein synthesis and cellular metabolism; and one can expect to find a high level of cytoplasmic RNA in cells that are both actively fusing and synthesizing myosin. It is also reasonable to expect more active myosin synthesis on the second day.

Chapter 23

A Method for the Assessment of Surface Roughness in Landschütz Ascites Tumour Cells

W. GALBRAITH

INTRODUCTION

The inhibitory action of Gum Tragacanth on the Landschütz ascites tumour has been studied in this laboratory.[5] Some of this work indicates that the primary effect of the gum is on the cell surface.[4] The Landschütz ascites tumour consists of completely separate cells, each of which is spherical and covered with fine processes or microvilli. It was therefore of interest to discover whether Gum Tragacanth produced any changes in the surface roughness of these cells.

The diameter of the microvilli is comparable to the resolution limit of the microscope, and even under dark-ground microscopy, which is the best method for showing them, they are only visible around the perimeter of the cell. It is therefore difficult to make a reliable estimate of their incidence, particularly as the cells rapidly attach to any glass surface that they settle on to, and this may be expected to change their surfaces. If left on a glass surface, they lose their spherical shape, becoming star or spindle shaped.

A technique was required which would give a repeatable number representing the surface roughness of the cells. It was not essential that this number should have an obvious physical meaning such as the number of microvilli per square micron. The technique chosen was based on consideration of the movement of a body through a fluid. Essentially, the technique is the inverse of the falling sphere type of viscometry, in which the viscosity of a fluid is found from the speed with which a body of known shape sinks through it. In the present experiments, the viscosity is known, and the shape of the body, a sphere with unknown surface roughness, is not.

When a body moves through a fluid, it is possible to calculate a dimensionless number called the Reynolds number, which is the ratio of the inertial to the viscous forces involved. If two bodies of the same shape move at the same Reynolds number the fluid flows are dynamically identical, even though the speeds, sizes, densities, and viscosities may be widely different.

Another dimensionless quantity may also be calculated, called the drag coefficient, which is a measure of the force per unit area required to drive the body through the fluid. If these two dimensionless quantities are the two co-ordinates of a graph, the curve relating them depends only on the *shape* of the body in question. Shape is also a dimensionless quantity. Fig. 23.1 shows this curve, on logarithmic axes, for spheres. A logarithmic plot is used to cover the very wide range involved. The line is taken from Bayley,[1] and the four points were found by a calibration technique using smooth-surfaced glass spheres of about 250 μ diameter, as described in this paper under Experimental Procedure. The abrupt change in the curve at about Reynolds number 10^5 is due to the flow becoming turbulent. In the region of interest in this work, the curve approximates to a straight line on logarithmic co-ordinates.

In general, if a body is not a smooth sphere, then, at any particular Reynolds number, its drag coefficient will differ from that of a sphere. If these two dimensionless quantities are found for cells sinking under gravity through a liquid, the difference between their drag coefficient and that for a smooth sphere at the same Reynolds number, is a measure of their departure from sphericity. Landschütz ascites cells when not in contact with a substrate are close approximations to spheres, except for the microvilli distributed over their surfaces. The rougher the surface, the greater will be the drag coefficient at a given Reynolds number.

Therefore in order to quantify the surface roughness of these cells, measurements are taken which allow the Reynolds number and drag coefficient to be calculated. The logarithms of these are used throughout, both for convenience, and because any errors will be of the multiplicative rather than the additive type, making a logarithmic plot more appropriate. The experimental value of the drag coefficient is then compared with that for smooth spheres at the same Reynolds number. The difference between the two, in logarithmic units, is a number which in this paper will be called the *roughness index*. The larger the roughness index, the rougher is the surface of the cell, although it does not define the type of roughness. For instance, it will not differentiate between many short microvilli and a few long ones, which is the reason for the deliberately non-committal name.

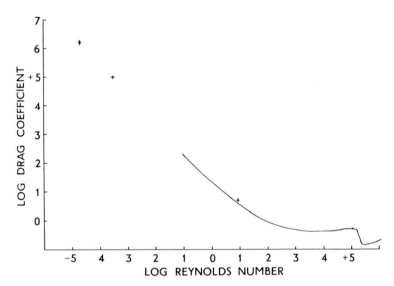

Fig. 23.1   The relationship between the logarithms of Reynolds number and drag coefficient for smooth spheres. The line is taken from Bayley[1] and the four points are the calibration points referred to in this paper.

In the course of the experiments, it was found that the roughness of the untreated tumour cells changed with the age of the tumour. This paper is therefore concerned only with untreated tumours. It is hoped to publish work on tumours treated with Gum Tragacanth at a later date.

APPARATUS

Four main pieces of apparatus were used in the experiments. The balance was a Stanton model MCIH on a brick base, weighing to micrograms (10^{-6} g). This was used to weigh a specially made glass pyknometer in the form of a very fine straight pipette, with a capacity of about 10 ml and a weight of 2·9 g. Its two tubes had internal and external diameters of 0·36/1·33 mm, and it could be filled very accurately to a fiducial mark on the upper limb. It was weighed suspended horizontally on the hooks of the balance.

A Cooke, Troughton and Simms ultra-violet microscope was used because it was horizontal and massive. It was used with visible light and a 95 × phase-contrast objective. The eyepiece was a Watson WISE image-shearing measuring eyepiece, which permitted measurement of the diameter of the cells, as they sank under gravity, between two parallel lines on the eyepiece graticule.

An Ostwald capillary viscometer was used to find the viscosity of the liquid through which the cells were allowed to sink. This viscometer is a comparison instrument, so it was necessary to measure the time of flow both with the liquid and with glass-distilled water whose viscosity and density could be found from tables.

EXPERIMENTAL PROCEDURE

Balb C+ male and C− male and female mice, C.B.R.I. strain, were used in these experiments. Each was injected intraperitoneally with about two million Landschütz ascites tumour cells. Tumours were used ranging in age from 2 to 17 days. Each experiment required only one mouse, and it was usually possible to extract sufficient tumour without killing the mouse, except when the tumour was very young.

The apparatus was assembled at least 2 hr before the start of the experiment in order to allow it to stabilize at room temperature, and for the same reason the balance case was ventilated for the same period. It was found convenient to prepare in this way, in the morning, and start the experiment in the afternoon, so that any possible diurnal changes would not affect the results.[3] Most of the apparatus was assembled inside a perspex-topped handling box to avoid temperature changes caused by the breath. The room heating and ventilation system was left unchanged throughout the day.

At 2.00 p.m. the room temperature was taken, and the pyknometer weighed empty. Five ml of glass-distilled water were placed in the viscometer and the time of flow taken (mean of five readings). Ten ml of Earles fluid[2] were placed in a pointed, graduated centrifuge tube. This completed the preliminary phase of the experiment.

The second phase had to be completed rapidly, to prevent deterioration of the cells. It could usually be completed in 30 min. 0·2 ml of ascites fluid were extracted from the mouse and mixed thoroughly with the Earles fluid in the centrifuge tube. The pyknometer was filled with this cell suspension and weighed. From the contents of the pyknometer two microscopical preparations were made:

(a) a Bürker haemocytometer was filled;
(b) a slide was made using two thicknesses of lens tissue (about 75 μ) as spacers under the coverslip, and waxed. It was inverted for 30 sec to bring the cells away from the slide, and was placed on the stage of the horizontal microscope.

The rest of the contents of the pyknometer were replaced in the centrifuge tube and spun for 10 min on an M.S.E. laboratory centrifuge at No. 3, about 600 g. Meanwhile the cells in two large squares of the Bürker haemocytometer were counted and the mean taken. On the horizontal microscope, the time taken for each of 25 cells to sink between two horizontal lines of the image shearing eyepiece was recorded, at the same time as its diameter was also measured. The distance travelled by the cells, and the units in which the diameter was measured, had previously been calibrated against a micrometer slide.

The third phase of the experiment could be done at leisure, like the first, as cells were not involved. The pyknometer was rinsed out with Earle's fluid, filled with supernatant from the centrifuge tube and weighed. The supernatant was replaced, 5 ml placed in the viscometer, and the mean of five measurements of time of flow taken. Room temperature was again recorded, and the mean found of the two readings. These never varied by more than 1°C.

Four calibration experiments were done, using smooth-surfaced glass spheres of about 250 μ diameter. These could be handled individually by means of a suction tweezer. In this way each could be measured with a travelling microscope, and then transferred to an apparatus in which it sank through a liquid. In one experiment the spheres were allowed to sink through 50 cm of glass-distilled water. This gave a result closely agreeing with the curve from Bayley[1] (Fig. 23.1), but not in the range required by these experiments. In two other experiments the fluid was glycerine, and the spheres were observed with a vertically travelling microscope as they sank through 0·5 cm. Two slightly different experimental methods gave close agreement (Figs. 23.1 and 23.3). A very similar experiment, using a 9:1 mixture of glycerine and distilled water gave another point which, together with the points just described, defined a line for the behaviour of smooth spheres. This could be used to calculate the roughness index. The accuracy of these calibration experiments is greater than that of the cell experiments, as the procedure is simpler, and the weighings less critical. The densities of the glass spheres and of the glycerine, were found by weighing, in a density bottle, and the viscosity of the glycerine, by using the Ostwald viscometer.

CALCULATIONS

In Table 23.1 are shown the different factors which must be taken into account in these experiments, and below is shown the way in which they are combined to calculate the Reynolds number and drag coefficient. For simplicity of calculation, the factors may be separated into three groups:

(1) Physical constants and calibration constants which are the same for all the experiments, depending on the microscope used, and so on.

(2) Factors which are the same for all the cells in any one experiment, such as the viscometer times of the medium and distilled water, and the number of cells per ml of the suspension.

(3) Factors relating only to one cell. There are only two of these—its diameter in arbitrary units and the time taken for it to sink an arbitrary, but fixed distance.

Table 23.1

(1)	Physical and calibration constants	
	Acceleration due to gravity	a
	Distance moved by cells	b
	Bürker factor	c
	Diameter factor	d
(2)	Factors constant during the experiment	
	Room temperature	θ
	Weight of pyknometer	e
	Weight of pyknometer plus cell suspension	f
	Weight of pyknometer plus supernatant	g
	Viscometer time, distilled water	h
	Viscometer time, supernatant	i
	Bürker count	j
	Density of distilled water at room temperature	k
	Viscosity of distilled water at room temperature	l
	Mean of all cell diameters, arbitrary units	m
(3)	Cell measurements	
	Cell diameter, arbitrary units	n
	Time for cell to sink arbitrary distance	p

$$\text{Reynolds number} = \frac{\text{density} \times \text{velocity} \times \text{diameter}}{\text{viscosity}} = bd \times \frac{kh}{li} \times \frac{n}{p}$$

$$\text{Drag coefficient} = \frac{2 \times \text{force per unit frontal area}}{\text{density} \times \text{velocity squared}} = \frac{8a}{\pi b^2 cd^2} \times \frac{(f-g)}{jm^3(g-e)} \times np^2$$

To simplify the calculations, factors in the first two groups are combined. The Reynolds number and drag coefficient for each individual cell may then be found by combining the results with the factors in the third group.

An example of the results at this stage is shown in Fig. 23.2. Each point represents one cell. The best straight line is fitted to these points by the method of least squares. This line is extended in each direction to one standard deviation of the Reynolds number from the mean value. If the mean Reynolds number is X and the drag coefficient is Y, the results of the whole experiment may be summarized in four numbers—the constants A and B in the straight-line equation $Y = BX + A$, the mean Reynolds number X and its standard deviation S. The constant B denotes the slope of the line shown in the figure. It is a physical, not a biological parameter, relating the Reynolds number and drag coefficient of near spheres at the values being considered. The values for different experiments are therefore very similar, as may be seen from Fig. 23.3, variations being due to experimental errors. The value of S is relatively constant, and is due to variation in cell diameter within the experimental population. Neither A, X, nor Y are by themselves a measure of surface roughness. This is shown by the displacement between the line for the

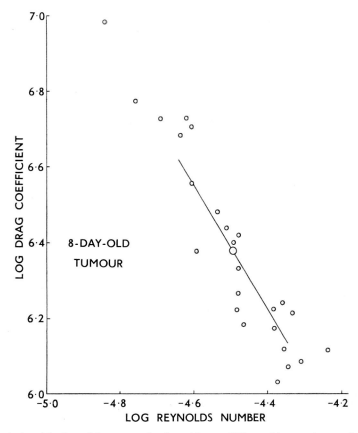

Fig. 23.2 The relationship found between the logarithms of Reynolds number and drag coefficient for the cell population in a single experiment. Each point represents a single cell. The straight line of best fit is shown. In each direction it is extended by one standard deviation of the Reynolds number from the mean value.

cells and that for smooth spheres. The simplest way to calculate a single number to represent this displacement is to take the difference between the drag coefficient at the cell mean value and that of a smooth sphere at the same Reynolds number, a shown in Fig. 23.4. The calibration experiments give good values of the drag coefficient at Reynolds numbers close to the desired one and on either side of it, so the straight line joining these points can be taken as representing the relationship between the two quantities over the limited range which is of interest in these experiments. In this way a single number can be derived for each population of cells, which gives a quantitative measure of their surface roughness, and which will be referred to as the roughness index R.

21—C.S.I.

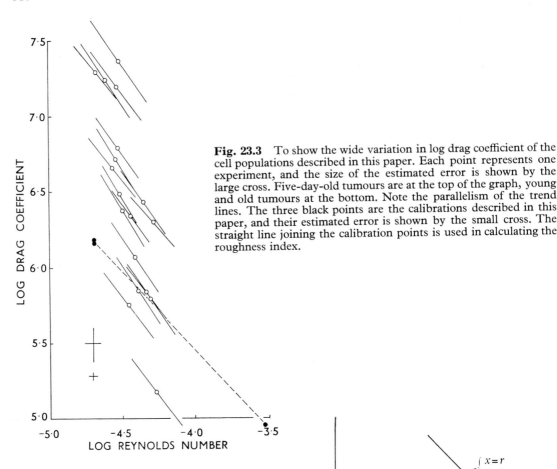

Fig. 23.3 To show the wide variation in log drag coefficient of the cell populations described in this paper. Each point represents one experiment, and the size of the estimated error is shown by the large cross. Five-day-old tumours are at the top of the graph, young and old tumours at the bottom. Note the parallelism of the trend lines. The three black points are the calibrations described in this paper, and their estimated error is shown by the small cross. The straight line joining the calibration points is used in calculating the roughness index.

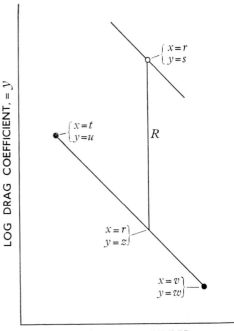

Fig. 23.4 Calculation of roughness index. The white point is the mean value for a single experiment, the black points are calibrations using glass spheres. The line joining them is $y = a + bx$, where $a = u - bt$ and $b = \dfrac{u-w}{t-v}$. Therefore $z = a + br$. Roughness index $R =$ $s - z = (s - u) + \dfrac{(t-r)(u-w)}{(t-v)}$.

CONCLUSIONS

In Fig. 23.6 is shown the way in which the roughness index changes during the life of the tumour. Although it would be preferable if the accuracy of the experiment could be improved further, it is clear that the young tumour has relatively smooth cells, but that these become rapidly rougher, reaching a maximum at 5 days after tumour inoculation. The cells of the population then become slowly smoother again.

This behaviour is interesting, particularly when compared with the results described in reference [3]. There it is shown that this tumour undergoes a marked change at 5 days after tumour inoculation. At

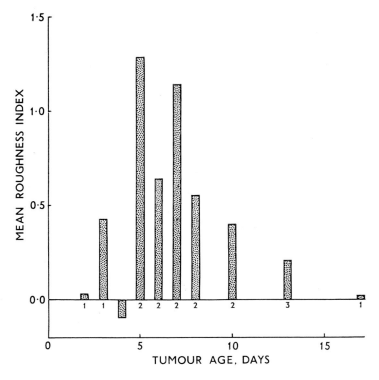

Fig. 23.5 Histogram to show the change in the roughness index with the age of the tumour. Each bar is the mean value for all experiments at that tumour age, and the small number below each bar indicates the number of experiments. Note the sharp increase in cell surface roughness at 5 days of tumour age, and its subsequent decline.

this time there is a minimum of cells entering mitosis, and the increase in total cell number per unit time changes also.

Therefore it is important that the possibility of change with age of the characteristics of a tumour should be borne in mind, and that values should only be compared at comparable ages. To give an idea of the magnitude of possible variation, a 5 day-old tumour cell requires ten times as much force to drive it through a fluid at a given Reynolds number as a young or old tumour cell.

It is hoped to continue this work and to study the effect on the roughness index of the surface-active tumour inhibitor, Gum Tragacanth.

SUMMARY

A description is given of a method for the assessment of the incidence of microvilli on the surface of Landschütz ascites tumour cells. The cells are allowed to sink under gravity through a medium of known viscosity, and their Reynolds number and drag coefficient are calculated. The relationship between these two quantities is compared with that for smooth spheres.

It is found that the surface roughness changes with the age of the tumour, reaching a maximum at 5 days after tumour inoculation.

ACKNOWLEDGEMENTS

The author wishes to thank Professor A. Haddow and Dr. E. M. F. Roe for their encouragement, Dr. R. Lumley Jones for reading this paper, and Dr. A. Knowles for making the pyknometers.

This investigation has been supported by grants to the Chester Beatty Research Institute (Institute of Cancer Research: Royal Cancer Hospital) from the Medical Research Council and the British Empire Cancer Campaign for Research, and by the Public Health Service Research Grant No. CA-03188-08 from the National Cancer Institute, U.S. Public Health Service.

REFERENCES

1. BAYLEY, F. J. 1958. *An Introduction to Fluid Dynamics*, p. 152. London: George Allen and Unwin.
2. EARLE, W. R. 1943. *J. natn. Cancer Inst.* **4**, 165
3. GALBRAITH, W. and MAYHEW, E. 1965. *Br. J. Cancer* **XIX**, 603
4. MAYHEW, E. and ROE, E. M. F. 1964. *Br. J. Cancer* **XVIII**, 537
5. ROE, E. M. F. 1959. *Nature, Lond.* **184**, 1891

Part 3

The Qualitative Interpretation of Specialized Cell Activities

Chapter 24

Some Observations on the Structure and Function of the Reissner's Membrane of the Inner Ear

J. T. Y. CHOU

INTRODUCTION

Reissner's membrane is attached to the upper end of the stria vascularis on one side, and the dorsal surface of the vestibular lip of the lamina osceca spiralis on the other. From the basal whorl along the cochlear duct to the apical whorl, it is a continuous sheet of tissue, two cells thick. The ectodermal layer faces the scala media, and the mesothelial layer faces the scala vestibuli. It separates the endolymph, which contains a high concentration of potassium (140 meq/litre) and a low concentration of sodium (26 meq/litre) (Smith, Lowry, and Wu;[16] Citron, Exley, and Hallpike[7]), from the perilymph.

In order to explain the difference in potassium and sodium contents of the two fluids, it has been suggested that active transport across Reissner's membrane may be necessary (Citron and Exley[8]). Naftalin and Harrison[13] suggested that liquid flow may proceed from the perilymph through Reissner's membrane to the endolymph and that the stria vascularis acts as a selective absorbing site. In this scheme, Reissner's membrane would prevent the flow of potassium from the scala media to the scala vestibuli; potassium would thus accumulate in high concentration in the endolymph. Metabolic studies revealed that Reissner's membrane has a high oxygen utilization (Chou[4]).

All these investigations indicate, that for a full understanding, it is necessary to know the state of Reissner's membrane in life.

In the present investigation, therefore, a study has been made of the living cells of Reissner's membrane, using positive phase-contrast microscopy.

MATERIALS

Reissner's membrane was removed from guinea-pigs anaesthetized by intraperitoneal injection, as described by Chou.[3, 4] A piece of Reissner's membrane was obtained by opening the tip of the cochlea, and removing the membrane with fine watchmaker's forceps. A total of twelve guinea-pigs were used.

The tissues were transferred on to a slide on which there was a small amount of perilymph, previously collected through the round window from the same animal. Endolymph was sometimes used. Physiological saline (0·9 per cent NaCl) was used as an alternative mounting medium, in certain cases. The tissues were immediately examined by phase-contrast microscopy.

Five rabbits (age between 2–4 years) were also used for the investigation. Reissner's membrane was obtained from the animals by opening the wall of the cochlea from the round window. This part of Reissner's membrane was taken, instead of that at the tip of the cochlea, because, in the rabbit, the cochlea does not protrude into the middle ear, and the cochlear walls are much thicker than those in the guinea-pig. The whole procedure, and further separation from other tissues, was carried out under a binocular microscope.

RESULTS

When Reissner's membrane removed from the guinea-pigs was examined by light microscopy, the cells appeared transparent, and contained no pigmented objects.

When examined by positive phase-contrast microscopy, Reissner's membrane appeared as a sheet of polygonal cells, with well-defined cell membranes. Figure 24.1, A shows a low power phase-contrast photomicrograph. Figure 24.1, B shows these cells under higher magnification. Most cells are 35–45 μ long with oval or elongated nuclei, and three or four nucleoli. The nucleus is very dense, and always situated in the centre of the cell.

Three distinct cytoplasmic inclusions were identified:

(1) *Mitochondria,*
(2) *Spherical Vacuoles,*
(3) A few Refractile Granules.

The mitochondria are fine filaments scattered at random in the cytoplasm; they are of uniform thickness, and either straight or curled. They averaged 1–5 μ in length; they did not aggregate in any particular part of the cells, and are easily identified.

Figure 24.1, C is a higher magnification phase-contrast photomicrograph, to show the fine, long, filamentous mitochondria.

Janus black, dissolved in well-oxygenated saline, stained the mitochondria as fine greyish filaments. Neither Janus green nor dahlia violet, stained the mitochondria of Reissner's membrane.

The spherical vacuoles vary in size from 0·5 to 2 μ in diameter. They appeared as homogeneous bright spherical objects, scattered through the cytoplasm (Fig. 24.1, B and Fig. 24.1, C). They could be stained by neutral red. Nile blue, and methylene blue, did not stain these inclusions.

Some small refractile granules appeared, very rarely, as grey spheres, using positive phase-contrast microscopy. They were stained by neutral red, methylene blue, and Nile blue.

The Reissner's membrane of the rabbit is much thicker than that of the guinea-pig. These cells are attached to each other in a very irregular way (Fig. 24.1, D). The distinct cytoplasmic inclusions consist of fine filamentous mitochondria, and spherical refractile granules. The mitochondria sometimes aggregated together as a group of fine thread-like objects. They often had a small granule at one end, or both ends.

Groups of irregularly shaped bodies are seen in most of the cells. They could be coloured by neutral red.

DISCUSSION

The present investigation of Reissner's membrane of the guinea-pig illustrates that the visible inclusions comprise mitochondria and spherical granules. In all the normal mammalian cochlea so far examined by phase-contrast microscopy, Reissner's membrane is a firm and well-organized structure, though in rabbit it appears to be much thicker, and the cells are irregular.

Histochemical localization of enzyme activity in Reissner's membrane has not been examined satisfactorily. Although certain enzymes have been located in various tissues in the inner ear (Vosteen;[18] Spoendlin and Balogh;[17] Baloch and Nomura;[1] Nomura, Gacek, and Balogh;[14] Gacek, Nomura, and Balogh[9]), none has been described as present in Reissner's membrane.

A metabolic investigation of Reissner's membrane, using the Cartesian diver technique (Chou[4]), showed that the average respiration rate was 7·8 μl./μl. tissue per hour, at 24°C. This value is slightly lower than that found for the stria vascularis, which was 10·1 μl./μl. tissue per hour (Chou and Rodgers;[5] Chou and Hughes[6]). It is known that all metabolic reactions receive their energy from the oxygen

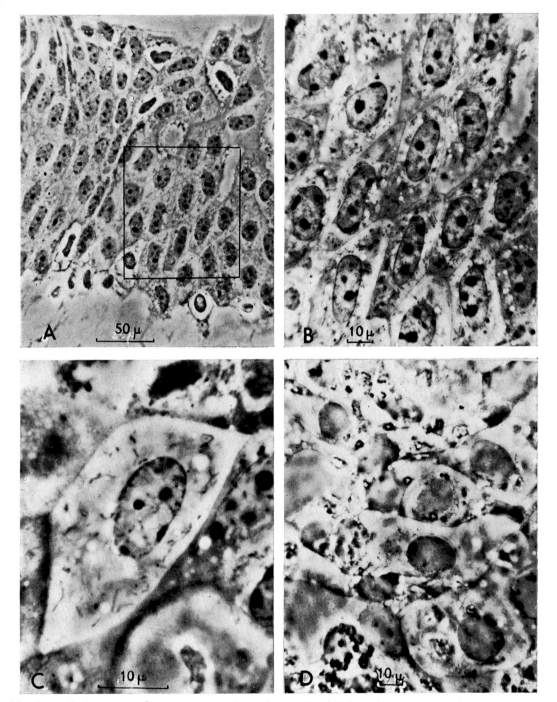

Fig. 24.1 **A** Low-power [phase-contrast photomicrograph of Reissner's membrane of the guinea pig. Mounting medium perilymph.

B Medium-power phase-contrast photomicrograph (area shown in A) of Reissner's membrane of the guinea pig showing the well-defined cell membrane, fine mitochondria and spherical vacuoles.

⊢ **C** High-power phase-contrast photomicrograph of Reissner's membrane of the guinea pig showing curled filamentous mitochondria, large homogenous spherical vacuoles and a few refractile granules.

D Medium-power phase-contrast photomicrograph of Reissner's membrane of the rabbit showing few fine mitochondria, spherical refractile granules.

metabolism, which depends upon the activity of enzyme systems. The high rate of respiration of Reissner's membrane suggests that this tissue is engaged in some kind of activity, which is, most probably, maintaining the equilibrium of the ionic concentrations between endolymph and perilymph in the cochlear duct. Whittam[19] showed that, in kidney and brain, as much as 50 per cent of the energy produced by the tissues, in the form of adenosine triphosphate, may be used for ion transport. Green[10] explained that the adenosine triphosphate so used is produced by oxidative phosphorylation, and Lehninger[12] demonstrated that this process is within the mitochondria. It appears that a high rate of production of adenosine triphosphate depends upon the type, rather than the number of mitochondria present. The oxygen used by Reissner's membrane gives a measure of the amount of work the membrane has accomplished.

It is noteworthy that the cells in Reissner's membrane from the guinea-pig contain few, but long filamentous mitochondria. The cells are not so fully packed with these inclusions as those in the stria vascularis (Chou[3]). It is, however, possible to consider that the total volume of mitochondria in each tissue is one of the important factors governing the metabolic rate.

Recent studies on the ionic concentrations of the cochlear endolymph and utricular endolymph (Rodgers and Chou[15]), showed that the sodium concentration in cochlear endolymph is significantly higher than in the utricular endolymph. It was therefore suggested that the endolymphatic spaces in the utricle and cochlea do not directly intercommunicate. This supports the evidence that the high rate of oxygen consumption showed by Reissner's membrane must, probably, be used for balancing the different ionic concentrations of the two fluids. In these studies it was also shown that, within 50 min, the concentration of potassium and sodium in utricular endolymph becomes approximately equal to the concentrations in perilymph. It is considered that the walls of the utricle, which have a high metabolic rate (Chou and Rodgers[5]) may also take part in maintaining the differences of the ionic concentrations in life.

Grüneberg, Hallpike, and Ledoux[11] studied the inner ear of the shaker-1 mouse. They observed that the degeneration process in the cochlea is initiated in a fully differentiated organ. The changes first appear in Corti's organ, then in the stria vascularis, and lastly in the ganglion cells. No changes were observed in Reissner's membrane, and the tectorial membrane. More recently, Bosher and Hallpike[2] studied inner ear degeneration of the deaf white cat. They recorded that Reissner's membrane may even collapse, when the osmotic relationships between endolymph and perilymph varied.

Deformation of Reissner's membrane may occur very late and slowly in pathological cases, such as post-hydropic degeneration, when it may collapse or be pressed sideways until it contacts the wall of the scala vestibuli. This may be due to changes of the osmotic relationships between endolymph and perilymph, but it is not known whether Reissner's membrane is the site of the original lesion leading to the hydropsy, or whether the histological changes in Reissner's membrane are secondary results. The structure, the physical properties, the physiological character and the position of Reissner's membrane in the cochlear duct should certainly all be considered of importance in maintaining the ionic balance of the cochlear endolymph and cochlear perilymph.

REFERENCES

1. BALOGH, K. Jnr. and NOMURA, Y. 1964. *J. Histochem. Cytochem.* **12**, 931
2. BOSHER, S. K. and HALLPIKE, C. S. 1965. *Proc. roy. Soc.* B, **162**, 147
3. CHOU, J. T.-Y. 1961. *Quart. J. micr. Sci.* **102**, 75
4. ——. 1963. *J. Laryng. & Otol.* **77**, 374
5. —— and RODGERS, K. 1962. *J. Laryng. & Otol.* **76**, 341
6. —— and HUGHES, D. 1964. *Biochemie des Hörorgans.* Chap. XIV: Respiormetrie, 446. Stuttgart: Georg Thieme Verlag
7. CITRON, L., EXLEY, D., and HALLPIKE, C. S. 1956. *Brit. Med. Bull.* **12**, 101
8. —— and ——. 1957. *Proc. roy. Soc. Med.* **50**, 697
9. GACEK, R. R., NOMURA, Y., and BALOGH, K. Jnr. 1965. *Acta Oto-laryng.* **59**, 541
10. GREEN, D. 1961. *4th International Congr. Biochem.* Reprint 176 (Moscow)
11. GRÜNEBERG, H., HALLPIKE, C. S., and LEDOUX, A. 1940. *Proc. roy. Soc.* B **129**, 154
12. LEHNINGER, A. L. 1963. *Harvey Lecture* **49**, 176
13. NAFTALIN, L. and SPENCER HARRISON, M. 1958. *J. Laryng. & Otol.* **78**, 118

14. NOMURA, Y., GACEK, R. R., and BALOGH, K. Jnr. 1965. *Acta Oto-laryng.* **57**, 484
15. RODGERS, K. and CHOU, J. T.-Y. 1966. *J. Laryng. & Otol.* **80**, 778
16. SMITH, C. A., LOWRY, O. H., and WU, M-L. 1954. *Laryngoscope* (St. Louis) **64**, 141
17. SPOENDLIN, H. H. and BALOGH, K. 1963. *Laryngoscope* **73**, 1061
18. VOSTEEN, K. H. 1960. *Laryngoscope* **70**, 351
19. WHITTAM, R. 1961. *Nature, Lond.* **191**, 603

Chapter 25

The Mid-Gut Epithelium
of Insects

B. N. SUD

In insects, the mid-gut epithelium is single layered, and generally consists of the principal, or digestive cells, and the regenerative, or replacement cells. The principal cells constitute the functional epithelium. They are constantly subjected to wear and tear due to their double role of secretion and absorption. Thus, they have a short life span, and need to be replaced by new digestive cells from time to time. The latter are formed from regenerative cells which generally occur in groups, called nidi, occurring at intervals between or below the principal cells. In addition to principal and regenerative cells, goblet cells have been described in a few species.

PRINCIPAL CELLS

In most insect species, the principal cells are of fairly uniform structure, except that they may be of different sizes and shapes in different parts of the mid-gut. However, in different orders of insects, in different species of an order, and even at different stages of the life history of a species, the fundamental structure of the principal cells, with regard to the cytoplasmic inclusions, their chemical composition and activities, varies considerably.

Some authors have made distinction between the secretory and absorptive principal cells, but it is likely that the apparently different types are merely different phases of the same cell type, depending on the function actually performed.

Brush-border

The principal cells are lined internally by a continuous brush-border, also termed the striated border (Figs. 25.1–25.9, 25.11, B and C). The nature of the latter remained highly controversial until Beams and Anderson,[1] in *Malacosoma* (larva) and *Melanopus*, and Bertram and Bird,[2] in the female of *Aedes aegypti*, found it, under the electron microscope, to consist of myriad microvilli, representing subcylindrical extensions of the cytoplasm bounded by the cell membrane.

In *Lepisma saccharina*,[35] *Labidura riparia*,[33] the worker *Odontotermes horni*,[34] female of *Culex fatigans*,[40] larva of *Anopheles stephensi*,[39] *Dysdercus cingulatus*,[37] larva of *Callosobruchus analis*,[38] *Poecilocerus pictus*,[41] and *Apis indica*,[36] the brush-border in fixed preparations is seen to consist of numerous closely packed microvilli, lying perpendicular to the surface (Figs. 25.1–25.9). The microvilli are exceptionally tall in *Poecilocerus pictus*, *Apis indica*, and in the larval stomach of *Anopheles stephensi* (Figs. 25.7, 25.9, 25.5, C, 25.11, C). Generally, the detached fresh epithelial cells, under phase-contrast, do not reveal a brush-border, but in *Apis indica* it is retained for a considerable time, and the individual microvilli are clearly made out.

The brush-border, in all the species listed in the preceding paragraph, is intensely stained in PAS preparations, by Rawitz's 'inversion staining' technique, and by acid dyes, but gives only a moderate reaction in Sudan black and acid haematein preparations. In *Lepisma saccharina*,[35] and the worker *Odontotermes horni*,[34] it also reveals metachromasy, suggesting the presence of acid mucopolysaccharides.

In *Poecilocerous pictus*,[41] it gives a positive reaction for alkaline phosphatase. Waterhouse and Stay[45] demonstrated the presence of alkaline phosphatase as well as acid phosphatase in the brush-border of the anterior and posterior mid-gut of the blowfly larva. The brush-border is basiphilic in *Labidura riparia*[33] after Helly fixation, and in *Dysdercus cingulatus*[37] after Flemming fixation. It is blackened in Mann–Kopsch preparations of the worker *Odontotermes horni*[34] and in Hermann–Kopsch preparations of *Lepisma saccharina*.[35]

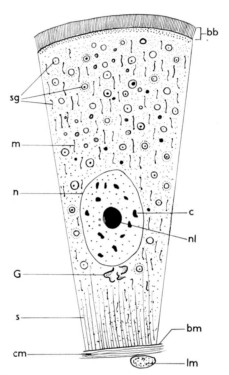

Fig. 25.1   Principal cell of *Lepisma saccharina* (reconstructed from the study by Sud and Marwaha[35]).
        Key to lettering on Figs. 25.1 to 25.9: f = absorbed lipoidal food; z = apical zone rich in phospholipid; bb = brush-border; bm = basement membrane; bs = basiphilic strand; c = chromatin; cm = circular muscle layer; fl = food lipid in lumen; G = Golgi element; gb = goblet; il = intracellular lipid; l = lipochondria; lm = longitudinal muscle bundle; m = mitochondria; mt = malpighian tubule; n = nucleus; nl = nucleolus; pm = peritrophic membrane; s = septa; sb = juxta-nuclear sudanophilic body; sg = secretory globule; sv = secretory vesicle.

Septa

Bartram and Bird,[2] in an electron microscopic study of the stomach principal cells of female *Aedes aegypti*, found the basal cell membrane to invaginate deep into the cell cytoplasm to form an elaborate system of double-walled septa. Waterhouse and Wright,[46] in the mid-gut epithelium of the blowfly larva, under the electron microscope, observed in the 'cuprophilic cells', a labyrinthic complex of septa invaginating deep into the cell from the basal membrane, but found the 'lipophilic cells' to lack septa.

With light microscopy, Newcomer,[22] in the principal cells of *Lepisma* sp., observed in the basal region, a fibrillar or pallisade-like appearance, extending up to the nucleus. In the principal cells of the mid-gut of *Lepisma saccharina*,[35] *Labidura riparia*,[33] the worker *Odontotermes horni*,[34] the larva of *Anopheles stephensi* (only in the stomach),[39] *Poecilocerus pictus*,[41] and *Apis indica*,[36] distinct striations are seen extending from the foot of the cells to as far as the nucleus and lying parallel to the long axis of the cells (Figs. 25.1, 25.2, A, 25.3, A, 25.5, C, 25.9). They invariably give an intense acid haematein reaction (Fig. 25.11, D) and moderate Sudan black colouring and PAS reaction. In *Lepisma saccharina*,[35] they also stain intensely with acid and basic dyes and the dye-lake, iron haematoxylin (Fig. 25.10, A), and show metachromasy. I am of the opinion that the function of these striations, representing the invaginations of the plasma membrane, is to transport the digested food from the principal cell to the haemolymph.

Basiphilia

In the female of *Culex fatigans*,[40] *Lepisma saccharina*,[35] *Dysdercus cingulatus*,[37] *Apis indica*,[36] and the worker *Odontotermes horni*,[34] the cytoplasm of the epithelial cells is evenly basiphilic. In the larva of

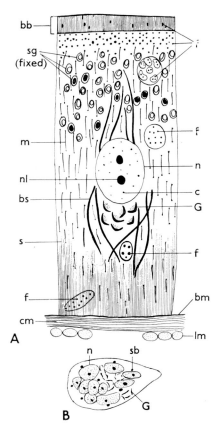

Fig. 25.2 *Labidura riparia*: A, principal cell; B, regenerative cells forming a nidus (reconstructed from the study by Sud and Goyal[33]).

Callosobruchus analis,[38] the basiphilia is mostly restricted to the basal cytoplasm and the cytoplasm on the sides of the nucleus, where numerous thick, intensely basiphilic strands are seen arranged parallel to the long axis of the cells (Fig. 25.8, A). The principal cells of the larva of *Anopheles stephensi*[39] reveal similarly oriented, but uniformly dispersed and less prominent, basiphilic strands (Figs. 25.5, A, B, and C). In *Labidura riparia*,[33] the cytoplasm of the principal cells contains extremely long and thick, intensely basiphilic strands (Fig. 25.2, A). The living principal cells of this species, under phase-contrast, reveal prominent filaments of dark contrast which appear to represent the basiphilic strands (Fig. 25.12, B). The basiphilia of the strands in *Callosobruchus analis*,[38] *Labidura riparia*,[33] and *Anopheles stephensi*,[39] has been proved to be due to RNA.

Bertram and Bird,[2] in the epithelial cells of females of *Aedes aegypti* which had not recently been blood fed, or had been blood starved, recorded, under the electron microscope, whorls of granular endoplasmic reticulum, forming a prominent subovoidal to spherical body, in the vicinity of the nucleus. Sud and Singla,[40] in a light microscopic study of the female of *Culex fatigans*, fixed soon after capture, observed in some of the principal epithelial cells, a prominent spherical or subspherical, homogeneous, highly basiphilic body in the vicinity of the nucleus which appears to represent the body formed by the whorls of the endoplasmic reticulum in the blood-starved females of *Aedes aegypti*.[2]

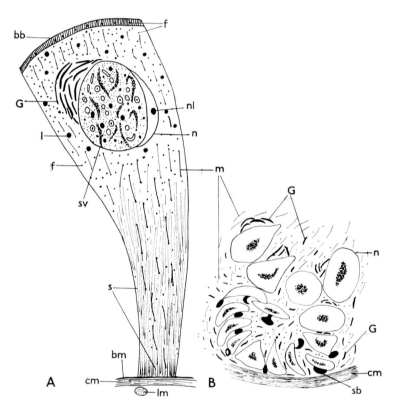

Fig. 25.3   *Odontotermes horni*: A, principal cell; B, regenerative cells forming a nidus (reconstructed from the study by Sud and Mand[34]).

Mitochondria

In the living principal cells, under ordinary positive phase-contrast, mitochondria have been studied in *Lepisma saccharina*,[35] *Labidura riparia*,[33] the worker *Odontotermes horni*,[34] *Culex fatigans*,[40] *Anopheles stephensi* (larva),[39] *Dysdercus cingulatus*,[37] *Poecilocerus pictus*,[41] *Callosobruchus analis*,[38] and *Apis indica*.[36] They are invariably seen as filaments of grey contrast with a granule of dark contrast at each tip, and they are dispersed throughout the cytoplasm and do not reveal any specific orientation. However, in fixed preparations of these species, they are always seen oriented along the long axis of the cells (Figs. 25.1–25.9). In the caterpillar of *Galleria mellonella*, Przełęcka[26] also found mitochondrial cords oriented along the main axis of the cells. In the larval stomach cells of *Anopheles stephensi*,[39] concentric whorls of mitochondria are seen around the nucleus in addition to their orientation along the long axis of the cells. In *Lepisma saccharina*,[35] *Labidura riparia*,[33] and *Poecilocerus pictus*,[41] the mitochondria are more numerous in the apical cytoplasm. In the female of *Aedes aegypti*,[2] the mitochondria tend to concentrate toward the luminal and basal surfaces of the cells, and reveal cristae in high resolution electron micrographs.

The mitochondria stain supra-vitally with Janus green. In fixed preparations, they are best studied by Hermann-Kopsch,[33-35, 37] Helly/Metzner,[34, 35, 37] Regaud/Metzner,[33] Champy/Metzner,[37, 38] Regand iron–haematoxylin,[33, 35] and Helly (postchromed)/iron–haematoxylin[33] techniques. In some cases, they are also seen in Aoyama/silver,[39] Mann–Kopsch,[34, 36] Kolatchev,[34] Rawitz's 'inversion staining',[39] Helly (postchromed)/Sudan black,[41] or formaldehyde–calcium/acid haematein,[33, 39] preparations.

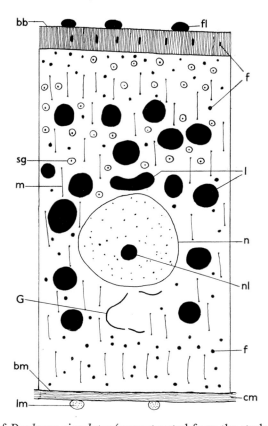

Fig. 25.4 Principal cell of *Dysdercus cingulatus* (reconstructed from the study by Sud and Sarin[37]).

Mitochondria in the mid-gut epithelium of insects have also been studied (from fixed preparations only) in *Bombyx mori*,[28] *Culex annulatus*,[17] *Culex pipiens*,[3, 4, 17] *Drosophila melanogaster*,[30] the grasshopper,[50] *Periplaneta orientalis*,[10] *Periplaneta americana*,[27] and *Passulus cornutus*.[23] Most of this literature has already been reviewed by Gresson,[10] Patterson,[23] and Shay.[27]

Golgi apparatus

The Golgi elements in the principal cells appear to have been described for the first time by Shinoda from Kolatchev preparations. In *Bombyx mori*,[28] he observed them as vesicles, and in *Oedaleus marmoratus*[29] as small crescents 'lying near the nucleus and at the basal portions of the epithelial cells'.

22—C.S.I.

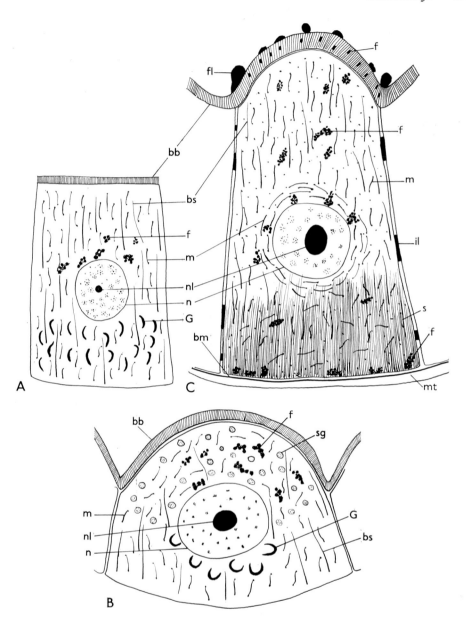

Fig. 25.5 Larva (fourth instar) of *Anopheles stephensi*: principal cell from cardia (A), caeca (B), and stomach (C) (reconstructed from the study by Sud, Satija, and Sud[39]).

Gresson[10] described the Golgi elements of the principal cells of *Periplaneta orientalis*, from Mann–Kopsch preparation, as large granules and rods, some of the curved rods enclosing a lightly impregnated material. He found the Golgi elements in the secretory cells situated in the vicinity of the nucleus, most numerous on the side of the nucleus next to the lumen, and always absent from the basal region. In the absorptive cells he found the Golgi bodies were often smaller, extending further towards the lumen and more numerous in the basal cytoplasm.

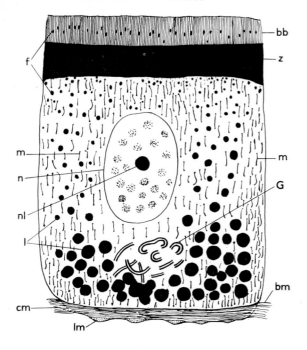

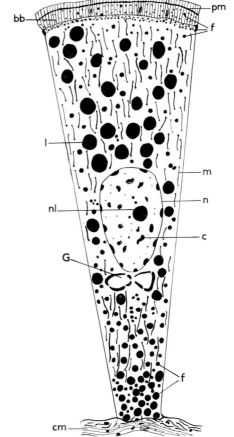

Fig. 25.6 Principal cell of female *Culex fatigans* (reconstructed from the study by Sud and Singla[40]).

Fig. 25.7 Principal cell of *Poecilocerus pictus* (reconstructed from the study by Sud and Walia[41]).

In *Poecilocerus pictus*, Sud and Walia[41] found the Golgi apparatus in the living principal cells, under phase contrast, in a juxta-nuclear position in the basal cytoplasm, in the form of crescents and rods, giving a high phase-change, arranged around a specialized area, giving a slightly lower phase-change than the ground cytoplasm (Figs. 25.7, 25.10, C). The Golgi rods and crescents colour with Sudan black in paraffin sections of Helly post-chromed material, and give a positive acid haematein reaction in the material fixed in chromium trioxide–formaldehyde or Flemming followed by postchroming.

Boissezon[3] located the Golgi apparatus in the apical cytoplasm in the larva of *Culex pipiens*. He found it consisted of one or more vesicles in the secretory cells and numerous vesicles in the absorptive cells.

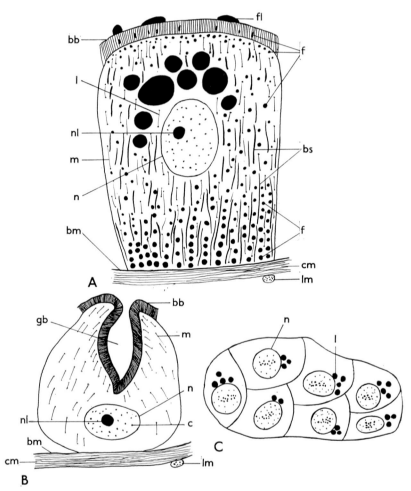

Fig. 25.8   Larva (fourth instar) of *Callosobruchus analis*: A, principal cell; B, goblet cell; C, regenerative cells forming a nidus (reconstructed from the study by Sud, Satija, and Kaur[38]).

In the imago of this species, he found the Golgi vesicles in the apical as well as the basal cytoplasm.[4] Siang-Hsu,[30] in *Drosophila melanogaster*, described the Golgi apparatus in the epithelial cells as osmiophilic spherical and homogeneous bodies, or as osmiophilic rings and crescents enclosing a secretory granule which are finally concentrated towards the lumen end of the cell. In the female *Culex fatigans*,[40] the Golgi apparatus in the principal epithelial cells is in the form of a typical network, located in the basal cytoplasm (Figs. 25.6, 25.10, D). In preparations in which Mann's fixative is used with acetic acid/post-osmication, these strands forming the network reveal canalicular nature with a limiting surface on which osmium is deposited (Fig. 25.6). In the principal cells of the cardia, the Golgi apparatus is not so elaborate as in those of the stomach. Distinct sudanophilia in the Golgi zone could be demonstrated by using unmasking fixatives, such as chromium trioxide–formaldehyde or Aoyama (paraffin sections only). Sud and others[39] were able to locate the Golgi apparatus in the principal cells of the cardia and caeca in the larva of *Anopheles stephensi*, but not in those of its stomach. In the principal cells of the cardia, it is in the form of prominent thick, slightly curved rods, oriented along the long axis of the cells in the basal cytoplasm (Fig. 25.5, A). The rods appear jet black in Kolatchev preparations and are stained deep red in Metzner's preparations. In the principal cells of the caeca, the Golgi apparatus is in the form of horseshoe-shaped bodies, aggregated near the nucleus in the basal cytoplasm (Fig. 25.5, B). In Kolatchev

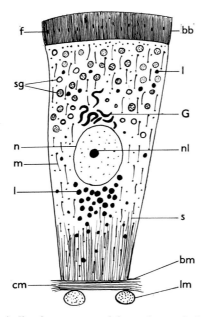

Fig. 25.9 Principal cell of *Apis indica* (reconstructed from the study by Sud and Paul[36]).

and Aoyama preparations, the horseshoe-shaped bodies appear as jet-black objects; in the former, they appear smooth, while in the latter they reveal a granular texture. Bertram and Bird,[2] in the principal cells of the female of *Aedes aegypti*, under the electron microscope, frequently found the Golgi apparatus in the form of dark-walled tubes, associated with vesicles and dense cytoplasm, located around the nucleus. Waterhouse and Wright[46] observed, under the electron microscope, similar but few and less distinct Golgi elements in the blowfly larva.

In *Dysdercus cingulatus*, Sud and Sarin[37] detected intensely sudanophilic filamentous bodies near the nucleus in the basal cytoplasm in material fixed in chromium trioxide–formaldehyde followed by post-chroming (Fig. 25.4). They consider it likely that these filaments may represent the Golgi bodies.

In *Labidura riparia*[33] and *Lepisma saccharina*,[35] the Golgi apparatus occupies a juxtanuclear position in the basal cytoplasm (Figs. 25.1, 25.2, 25.10, A). It is composed of osmiophilic and argentophilic rods and crescents, forming a network. The crescents reveal a distinct chromophobic part. After Sudan black colouring, the rods and crescents appear blue-black but the chromophobic part is only slightly touched. In pyridine or acetone extracted materials, Sudan black fails to colour the Golgi elements, suggesting

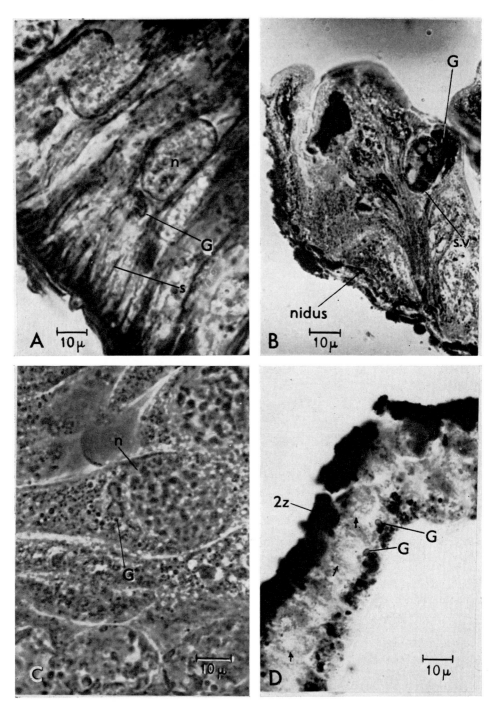

Fig. 25.10 A, *Lepisma saccharina* (transverse section, Regaud postchromed/iron haematoxylin), principal cell showing, in the basal cytoplasm, septa (s) and Golgi elements (G) near the nucleus (n) (after Sud and Marwaha[35]).

B, *Odontotermes horni* (transverse section, Hermann-Kopsch), principal cell showing Golgi elements (G) on the lumen side of the secretory vesicle (sv), the nucleus (not in focus) lies on the peripheral side of the secretory vesicle (after Sud and Mand[34]).

C, *Poecilocerus pictus*, living principal cell under phase contrast showing the Golgi elements (G) near the nucleus (n) in the basal cytoplasm (after Sud and Walia[41]).

D, *Culex fatigans* (transverse section, Mann with acetic acid/postosmication), principal cells showing blackened Golgi elements (G) in the basal cytoplasm and jet-black phospholipid-rich portion (2z) of apical cytoplasm, nuclei indicated by arrows (after Sud and Singla[40]).

their lipoidal nature, and particularly indicating the presence of phospholipid. In *Lepisma*, the Golgi apparatus is stained deep blue in Regaud post-chromed/iron–haematoxylin preparations (Fig. 25.10, A). Acid phosphatase is present in the Golgi elements of *Labidura*.

In *Passulus cornutus*,[23] the Golgi elements in the principal cells are in the form of discrete bodies, occurring as compound rings, variations of rings, and, very frequently, as irregular oval, or almost rectangular structures, mostly near the nucleus but distal to it, i.e. in the apical cytoplasm. Also, in *Apis indica*,[36] the Golgi apparatus in the secretory as well as absorptive cells is seen in the apical cytoplasm, generally occupying a juxtanuclear position (Fig. 25.9). It is in the form of sudanophilic rods, crescents, and wavy filaments which give a positive acid haematein reaction in material fixed in Aoyama's fluid followed by postchroming, suggesting the presence of phospholipid. They are also PAS positive.

In the worker *Odontotermes horni*,[34] the Golgi apparatus is best studied in postosmicated and silvered cells not exhibiting the secretion vesicle. At this stage, it is seen applied to the inner aspect of the nucleus in the form of straight or curved argentophilic or osmiophilic rods associated with argentophobic or osmiophobic material, forming a dense aggregation giving the appearance of a network. The secretory vesicle makes its appearance between the Golgi apparatus and the nucleus, and the latter gradually becomes crescent-shaped due to the pressure of the developing secretory vesicle (Figs. 25.3, A, 25.10, B). The Golgi zone gives an intense homogeneous reaction in acid haematein preparations of material fixed in Aoyama's fluid followed by post-chroming. It is stained homogeneously in Rawitz's 'inversion staining' technique, and in material fixed in Flemming followed by post-osmication. These reactions suggest the presence of phospholipid in the Golgi zone.

Lipochondria

In the larva of *Callosobruchus analis*,[38] the principal cells reveal prominent juxta-nuclear lipochondria in the apical cytoplasm (Figs. 25.8, A, 25.11, B) which, however, tend to spread out when they greatly increase in number. In the anterior portion of the mid-gut, hardly any cells show lipochondria, in the middle region the majority of cells show lipochondria, and in the posterior region, the cells showing lipochondria are comparatively less than in the middle region. In living cells, under positive phase-contrast, the lipochondria appear yellow and are always homogeneous. In fixed preparations, they appear homogeneous in material treated with chrome-osmium or Mann's fixatives, but after treatment with Hermann's fluid, or fixatives containing formalin, they appear homogeneous as well as duplex. In Zenker, Clarke (Carnoy) or Carl fixed material, the lipochondria are represented by their negative images. They are composed of unsaturated lipids, including phospholipids, non-acidic lipids, carotenoids, and some proteins.

Rather similar lipochondria are seen in the principal cells of *Poecilocerus pictus*.[41] They first appear as small spheres, lying mostly near the Golgi zone in the basal cytoplasm. Later, they are seen as large spheres, usually located in the apical cytoplasm (Fig. 25.7). They consist of non-acidic lipids, glycolipids, and carotenoids.

In the principal cells of *Dysdercus cingulatus*,[37] lipochondria occur as voluminous spherical, sub-spherical, oval, crescentic, or kidney-shaped bodies (Fig. 25.4). They are seen in the apical cytoplasm, basal cytoplasm, or both these regions. They are of common occurrence in the principal cells of the first, second, and third ventriculus, and only rarely a few of them may be seen in the principal cells of the fourth ventriculus. In living cells, under positive phase-contrast, the lipochondria always appear spherical or subspherical and consist of a thin cortex of dark contrast, and a homogeneous interior of lighter contrast. They appear homogeneously black in unstained preparations of the chrome–osmium fixed material, but after bleaching and colouring with Sudan black, they reveal a sudanophilic externum, and a sudanophobic internum. However, they colour homogeneously with Sudan black in the material fixed in formaldehyde calcium, Aoyama or chromium trioxide formaldehyde. Further histochemical studies suggested the presence of non-acidic lipids, and phospholipid, restricted to their externum.

In the principal cells of *Apis indica*[36] spherical or oval globules are seen, dispersed throughout the cytoplasm and usually aggregated in the basal cytoplasm near the nucleus (Fig. 25.9). They appear jet black in unstained sections of Flemming fixed material, colour intensely with Sudan black, and give a

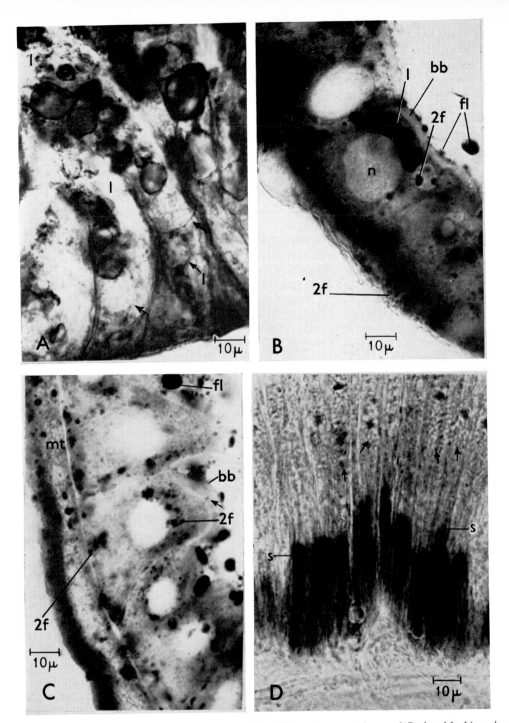

Fig. 25.11 A, *Dysdercus cingulatus* (transverse section, Flemming postchromed/Sudan black), principal cells showing lipochondria (l), nuclei indicated by arrows (after Sud and Sarin[37])

B, Larva (fourth instar) of *Callosobruchus analis* (transverse section, Lewitsky-saline/Sudan black), principal cell showing lipochondria (l) on the apical side of the nucleus (n), absorbed lipid (2f) mostly seen collected at the foot of the cell, food lipid (fl) seen in the lumen and applied to the brush-border (bb) (after Sud, Satija and Kaut[38])

C, Larva (fourth instar) of *Anopheles stephensi* (longitudinal section, chromium trioxide-formaldehyde/Sudan black), principal cells from stomach showing aggregation of absorbed lipid globules (2f) in the apical cytoplasm and at the foot of the cells, food lipid (fl) is seen applied to the brush-border (bb) and passing through (indicated by arrow) the latter, malpighian tubule (mt) surrounds the mid-gut (after Sud, Satija and Sud[39])

D, *Labidura riparia* (transverse section, chromium trioxide-formaldehyde/acid haematein), principal cells showing acid haematein positive septa (s) arising from the foot of the cell and extending toward the nucleus (indicated by arrow) (after Sud and Goyal[33]) positive acid haematein reaction which is negative after pyridine extraction, suggesting the presence of phospholipid. They colour supra-vitally with neutral red, which confirms their phospholipid nature.

Waterhouse and Stay[45] have described in the epithelial cells of the blowfly larva, spheres blackened by osmium in Flemming fixation, and coloured by Sudan black. They contain little, if any, phospholipid, but are rich in triglycerides.

In *Drosophila melanogaster*, the osmiophilic homogeneous spheres, rings, and crescents, described by Siang-Hsu[30] as Golgi elements, may represent the lipochondria.

Secretion

In the majority of insects, the principal epithelial cells perform a dual function of secretion and absorption, but van Gehuchten,[8] in *Ptychoptera contaminata*, and Boissezon,[4] in *Culex pipiens*, claimed to have found distinct secretory and absorptive cells. However, Pradhan,[25] from the study of the mid-gut of *Epilachna indica*, concluded that the two kinds of cells distinguished by van Gehuchten[8] are the same cells at different stages of development. Also, Gresson[10] found that, in *Periplaneta orientalis*, the young cells first act as absorptive cells and later become secretory in function. He never found food material in cells containing secretory granules, and therefore he concluded that secretion and absorption do not take place side by side in the same cell. Also, according to Steudel,[32] the cells showing the formation of secretory globules do not function as absorptive cells, but the secretory phase alternates with the absorptive phase. However, Sud and Goyal,[33] Sud and Mand,[34] Sud and Sarin,[37] and Sud and Paul,[36] in *Labidura riparia*, the worker *Odontotermes horni*, *Dysdercus cingulatus*, and *Apis indica* respectively, invariably found absorbed lipid in the cells showing secretory activity (Figs. 25.2, A, 25.3, A, 25.4, 25.9). Sud and Marwaha,[35] in *Lepisma saccharina*, found that the principal cells invariably contained secretory globules (Fig. 25.1), and concluded that they appear to perform both functions simultaneously.

Shinoda[29] and Woodruff,[50] in Orthoptera, observed small secretory granules stainable with acid fuchsin. Hodge,[16] in *Melanopus*, in addition found clear vacuoles, and dense granules, stainable with haematoxylin and enclosed in vacuoles. In *Periplaneta orientalis*, Gresson[10] described secretion in the form of small granules in Flemming preparations. In addition, he noticed deeply stained spherical bodies in the vicinity of the nucleus, and structures with a darkly stained periphery between the nucleus and the secretory granules, and identified them as developing secretory granules. In old cells, he observed sac-like masses of secretion.

In the worker *Odontotermes horni*[34] a complex secretory vesicles is produced which comes to occupy a major portion of the cells (Figs. 25.3, A, 25.12, A). It contains granules, fibrous material, homogeneous globules, and globules with cortex and medulla with or without a central granule. All the constituents of the secretory vesicle contain phospholipid, are PAS positive, and stain with acid and basic dyes. The homogeneous globules are also Feulgen positive.

In *Lepisma saccharina*,[35] the secretion in living cells, under positive phase-contrast, is in the form of globules with a cortex of dark contrast and a medulla of very light contrast containing a granule of grey contrast (Fig. 25.1). When stained supra-vitally with neutral red, Janus green, Janus black, or methylene blue, their cortex and granule are intensely stained, but the rest of the medulla is only faintly stained. In fixed preparations, stained with basic dyes, acid haematein test or Rawitz's 'inversion staining', some of the globules are homogeneously stained, while only the cortex of other globules is stained. After osmium or silver impregnation, the secretory globules show a black rim and a brown interior. In iron–haematoxylin preparations, only the central granule of the globules is stained. The cortex of the secretory globules gives an intense reaction for alkaline phosphatase.

In *Dysdercus cingulatus*[37] also, the living principal cells, under positive phase-contrast, reveal secretion globules, showing an externum of dark contrast and an internum of light contrast with a granule of dark contrast (Fig. 25.4). In *Apis indica*,[36] the living principal cells, under positive phase contrast, reveal secretion globules as pale spheres which are formed from smaller duplex globules with an externum of dark contrast and internum of light contrast (Fig. 25.9). In *Labidura riparia*,[33] the living secretory globules, under positive phase contrast, always appear duplex, with an externum of dark contrast and internum of very light contrast. In the larva of *Anopheles stephensi*,[39] the living principal cells of caeca

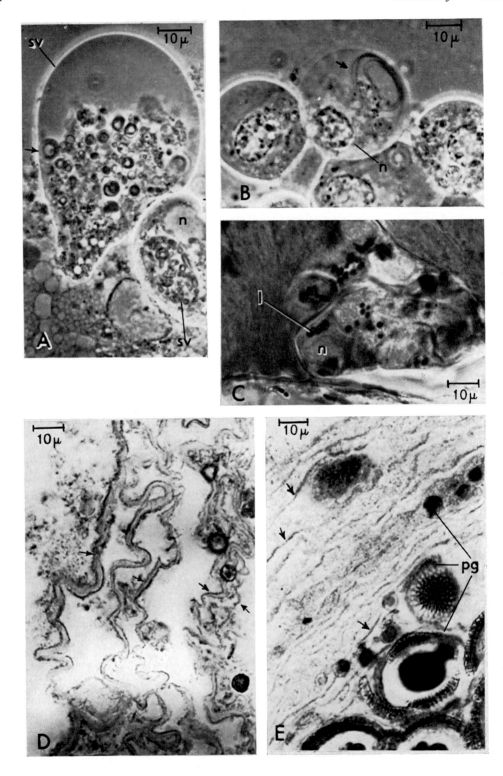

reveal, under positive phase contrast, several refractile secretory globules of light contrast which stain red in Metzner's preparations (Fig. 25.5, B).

It has not been possible to determine the precise role of the cell inclusions in the formation of secretion. Boissezon[3, 4] believed that in the larva and imago of *Culex pipiens*, the secretion of the mid-gut is partly of nuclear and partly of cytoplasmic origin.

Gresson,[10] from a study of the mid-gut of *Periplaneta orientalis*, concluded that the 'material' (precursor of secretion) is formed under the influence of mitochondria, and later on moves to the vicinity of the nucleus where, under the influence of the Golgi elements, it is used in the formation of secretory globules. Siang-Hsu,[30] in *Drosophila melanogaster*, however, found the secretory globules, from the earliest observable beginning, within the Golgi bodies.

In the worker *Odontotermes horni*,[34] the fully differentiated principal cells reveal the Golgi apparatus, in the form of a typical network, closely applied to the lumen side of the nucleus, and the secretory vesicle invariably makes its appearance between these two cell inclusions and remains in close association with them throughout its existence in the cell (Fig. 25.3, A). From this observation, it is obvious that the Golgi apparatus as well as the nucleus, play a major role in the formation of the secretion.

The mode of discharge of the secretion into the lumen is controversial. It has been described as merocrine, holocrine, or a combination of both. It is generally believed that the secretion is merocrine when the cells are not actively secreting, and holocrine in a state of intense secretory activity.

GOBLET CELLS

Goblet cells have been described in species belonging to the orders Ephemerida,[29] Lepidoptera,[29, 42, 13, 14, 49, 6, 43, 31, 26] Coleoptera,[38] and Diptera.[40] They are more or less of the same form as the principal cells, but the striated border of these cells is invaginated to form a deep cavity of variable size, termed the goblet. Due to the presence of the latter, most of the cytoplasm is restricted to the basal region of the cell.

Cytology of the goblet cells has been worked out in the larva of *Callosobruchus analis*[38] and in the adult female of *Culex fatigans*.[40] In *Callosobruchus*, the goblet cells differ from the principal cells in the absence of lipochondria, location of the nucleus nearer the basal cell membrane, and closer packing of the striae of the brush-border (Fig. 25.8, B). In the female of *Culex*, whereas in the principal cells the Golgi apparatus is in the form of an elaborate network, in the goblet cells it could not be detected.

Woke,[49] in the larva of *Prodenia*, considered the principal and goblet cells to arise independently from regenerative cells, but according to Shinoda,[28] in *Bombyx mori* they arise from normal columnar principal cells. Lotmar,[20] however, believes that in *Tineola*, the two types of cells are differentiated before they commence to function.

According to Day,[6] the goblet cells in insects cannot be strictly compared with the cells of the same name in the vertebrate alimentary canal because their contents are not mucoid. Woke[49] has suggested

Fig. 25.12 A, *Odontotermes horni*, living principal cells (phase-contrast). The intact cell shows the secretory vesicle (sv) and nucleus (n); a large secretory vesicle (sv) is seen pushed out of a cell and shows various inclusions. A globule with a cortex of dark contrast, medulla of light contrast, and a central granule of grey contrast is indicated by arrow (after Sud and Mand[34]).

B, *Labidura riparia*, living principal cells (phase-contrast). Long filaments of dark contrast material indicated by arrow (after Sud and Goyal[33]).

C, Larva (fourth instar) of *Callosobruchus analis* (transverse section, Flemming unstained), nidus showing a group of lipochondria (l) blackened by osmium, associated with each regenerative cell nucleus (after Sud, Satija, and Kaur[38]).

D and E, *Apis indica* (transverse section, Zenker/PAS); D shows discarded brush-borders (indicated by arrows) gradually becoming thinner to form the fibrils of the peritrophic membrane, shown in E (indicated by arrows); fragments of the pollen grains (pg) are seen passing through the peritrophic membrane in E (after Sud and Paul[36]).

that the 'cavities' of the goblet cells act as reservoirs of secretion. Waterhouse,[44] in *Tineola* larva, found that the goblet cells are able to accumulate sulphides formed from metals ingested in their food.

REGENERATIVE CELLS

The regenerative or replacement cells are very small and variable in size. When they form nidi, the limits of the individual cells are generally difficult to make out. The nidi mostly lie on the inside of the basement membrane, but in *Hyperaspis vinciguerrae*,[11] some of the nidi are situated outside the circular muscle layer. The latter location of nidi is also commonly met with in the adult specimens of *Callosobruchus analis*.[38]

Mitochondria in the regenerative cells of *Periplaneta orientalis*[10] are in the form of granules, situated in loose clump toward one pole of the nucleus. In *Poecilocerus pictus*,[41] they are also in the form of granules but are dispersed throughout the cytoplasm. Siang-Hsu,[30] in the replacement cells of *Drosophila melanogaster*, described numerous mitochondria in the form of 'chondrioconts' of globular texture throughout the cytoplasm. In the female of *Culex fatigans*,[40] the mitochondria in the regenerative cells are fine granules which assume a thread-like appearance during differentiation of the cells. In the worker *Odontotermes horni*,[34] Helly/Metzner preparations reveal rod-like and granular mitochondria, distributed throughout the regenerative cell cytoplasm (Fig. 25.3, B).

The Golgi elements in the regenerative cells of *Drosophila melanogaster*[30] are spherical or slightly elongated small bodies, not exhibiting any polarization. In *Labidura riparia*,[33] the Golgi elements of the replacement cells are in the form of osmiophilic rods and crescents (Fig. 25.2, B). In the worker *Odontotermes horni*,[34] the regenerative cells reveal prominent osmiophilic straight or curved rods, distributed throughout the cytoplasm. They are thicker than the osmiophilic mitochondria of principal cells and thus they obviously represent the Golgi elements (Fig. 25.3, B). As the regenerative cells differentiate into principal cells, the osmiophilic Golgi rods tend to aggregate at one pole of the nucleus. Similarly, in the regenerative cells of *Apis indica*,[36] the osmiophilic, argentophilic, and sudanophilic rods, representing the Golgi elements, are seen around the nucleus, and in more differentiated cells, they tend to aggregate. Also, in the older replacement cells of *Periplaneta orientalis*, Gresson[10] detected the Golgi elements situated chiefly toward one pole of the nucleus.

Lipochondria have also been found in the regenerative cells of a few species. In the larva of *Callosobruchus analis*,[38] in association with each regenerative cell nucleus in a nidus, is seen a group of prominent juxta-nuclear lipochondria (Figs. 25.8, C, 25.12, C). Like the lipochondria of the principal cells of this species, they contain unsaturated lipids, including phospholipid, non-acidic lipids, and carotenoids.

In *Poecilocerus pictus*,[41] the regenerative cells reveal a juxta-nuclear mass of osmiophilic, argentophilic, and sudanophilic globules. During differentiation, they contain instead a prominent, crescentic, sudanophilic, and PAS positive body, supra-vitally stained with dahlia.

The regenerative cells of the female of *Culex fatigans*[40] reveal juxta-nuclear globules in Mann with acetic acid/post-osmicated preparations. In *Labidura riparia*,[33] a mass of sudanophilic material is invariably noticed associated with each regenerative cell nucleus. The regenerative cells of the worker *Odontotermes horni*[34] reveal a prominent, sudanophilic, juxta-nuclear body of varied shapes (Fig. 25.3, B). In *Lepisma saccharina*, the regenerative cells contain a few sudanophilic globules.

PERITROPHIC MEMBRANE

The mid-gut epithelium of insects, unlike the fore- and hind-gut epithelia, is not internally protected by cuticle, but in most cases it is lined by a delicate sheath, known as the peritrophic membrane. The latter consists of two or three concentric layers in *Lepisma saccharina*,[35] but reveals many layers in the worker *Odontotermes horni*[34] and *Labidura riparia*.[33] In *Poecilocerus pictus*[41] it is composed of a number of concentric lamellae, which develop in succession from the surface of principal cells. In *Apis indica*[36] it is in the form of a thick sheath consisting of concentric wavy fibres embedded in a granular matrix (Fig. 25.12, E). In this species, we have obtained a clear evidence of the origin of the peritrophic membrane from layers of the discarded brush-borders (Figs. 25.12, D and E). In Diptera, e.g. *Psychoda alternata*[12, 18] and *Glossina*,[48] it is believed to be produced by a band of specialized cells, encircling the base of the stomodaeal valve at the anterior end of the cardia. But Bertram and Bird,[2] from electron microscopic study of the mid-gut of the female of *Aedes aegypti*, concluded that it results from the secretory activity of the gut cells, stimulated by the presence of the blood meal. They also suggested the participation of the microvilli in this secretory process since they present well-marked protuberances from their surface, particularly distally.

Under the electron microscope, Mercer and Day[21] found the peritrophic membrane in *Periplaneta americana* to be a complex structure, consisting of a fibrillar network and an associated layer composed of unorganized fibrils embedded in an amorphous ground substance. Giles,[9] from electron microscopic study of the peritrophic membrane in the earwig, *Anisolabis litteria*, concluded that it consists of a fibrillar network with a regular squared arrangement. Under the electron microscope, the peritrophic membrane, in the blowfly larva, has been described by Waterhouse and Wright[46] as structureless, and in the adult female of *Aedes aegypti*, as loosely layered condensed granular material by Bertram and Bird.[2]

In *Lepisma saccharina*,[35] *Labidura riparia*,[33] *Odontotermes horni*,[34] *Poecilocerus pictus*,[41] and *Apis indica*[36] (Fig. 25.12, E), food particles and secretory globules are often seen between the fibres of the peritrophic membranes, suggesting that it is permeable to food and epithelial secretion.

The peritrophic membrane has been studied by a variety of histochemical techniques in *Lepisma saccharina*,[35] *Labidura riparia*,[33] *Odontotermes horni*,[34] *Poecilocerus pictus*,[41] *Apis indica*,[36] and *Callosobruchus analis*.[38] In all these species, it gives a moderate PAS reaction, and in *Lepisma saccharina* it also shows metachromasy. In *Poecilocerus pictus*, it is distinctly stained in mucin carmine preparations. It invariably colours with Sudan black, and gives a positive acid haematein reaction in material fixed in chromium trioxide–formaldehyde followed by post-chroming. In *Apis indica* and *Odontotermes horni*, it gives an intense reaction in Rawitz's 'inversion staining' technique. It is basiphilic in *Lepisma saccharina*, *Labidura riparia*, *Odontotermes horni*, and *Apis indica*. In *Poecilocerus pictus*, the peritrophic membrane is positive to the coupled tetrazonium reaction for tyrosine, tryptophane, and histidine, the Sakaguchi reaction for arginine, and Gomori's alkaline phosphatase technique.

Waterhouse[44] obtained a positive reaction for chitin on the peritrophic membrane of a large number of insects. Hövener[18] also claimed that the peritrophic membrane exhibits two characteristic properties of chitin, double refraction, and resistance to alkalies. Wigglesworth[48] believes that the peritrophic membrane, whenever present, has the same composition as the inner layer of cuticle. In the honey-bee, it is reported to be chitinous by Campbell and von Dehn,[7] and non-chitinous by Pavlovsky and Zarin,[24] Weil,[47] Hering,[15] and Kusmenko.[19] Weil[48] suggested that those who reported the presence of chitin probably got intima of the stomodaeal valve mixed with their test material.

SUMMARY

The literature on the cytology and histochemistry of the mid-gut epithelium of insects is reviewed, with special reference to the work of the author and his collaborators on *Lepisma saccharina* Linn., *Labidura riparia* (Pallas), the worker *Odontotermes horni* Wasmann, *Dysdercus cingulatus* Fabr., *Poecilocerus pictus* Fabr., the female of *Culex fatigans* Wiedemann, the fourth instar larvae of *Anopheles stephensi* Liston and *Callosobruchus analis* Fabr., and *Apis indica* Fab. In each of these species, the cells were studied in living condition by phase-contrast microscopy and supra-vital staining, and in fixed preparations by various cytological and histochemical techniques. Such an integrated approach has not been practised by other workers.

The principal or digestive cells are dealt with in detail and a comparative account of the morphology and histochemistry of the brush-border, basiphilic elements, mitochondria, Golgi apparatus, and lipochondria is presented. The morphology and chemical nature of secretion, and its mode of origin and discharge, are discussed.

The occurrence of goblet cells in some insects, and their cytology as worked out by the author and his collaborators in the larva of *Callosobruchus analis* and the female of *Culex fatigans* are described. The origins of these cells and their analogy with the cells of the same name in vertebrate alimentary canal are discussed.

The cytology of the regenerative cells is described with reference to mitochondria, Golgi elements, and lipochondria. Changes in morphology and distribution of these inclusions during the differentiation of these cells into principal cells are recorded. A comparative account of the morphology, histochemistry, and origin of the peritrophic membrane is presented.

REFERENCES

1. BEAMS, H. W. and ANDERSON, E. 1957. Light and electron microscope studies on the striated border of the intestinal epithelial cells of insects, *J. Morph.* **100**, 601–19
2. BERTRAM, D. S. and BIRD, R. G. 1961. Studies on mosquito-borne viruses in their vectors. 1. The normal fine structure of the mid-gut epithelium of the adult female *Aedes aegypti* (L.) and the functional significance of its modification following a blood meal. *Trans. R. Soc. trop. Med. Hyg.* **55**, 404–23
3. BOISSEZON, P. DE. 1930a. Sur l'histologie et l'histophysiologie de l'intestin de la larve de *Culex pipiens* L. *C. r. Soc. Biol. Paris* **103**, 567–8
4. ——. 1930b. Sur l'histologie et l'histophysiologie de l'intestin de *Culex pipiens* L. (imago) et en particulier sur la digestion du sang. *C. r. Soc. Biol. Paris* **103**, 568–70
5. CAMPBELL, F. L. 1929. The detection and estimation of insect chitin, and the relation of 'chitinization' to hardness and pigmentation of the cuticle of the American cockroach, *Periplaneta americana* L. *Ann. ent. Soc. Am.* **22**, 401–26
6. DAY, M. F. 1949. The occurrence of mucoid substances in insects. *Aust. J. scient. Res.* (B) **2**, 421–7
7. DEHN, M. VON. 1933. Untersuchungen über die Bildung der peritrophischen Membrane bei den Insekten. *Z. Zellforsch. mikrosk. Anat.* **19**, 79–105
8. GEHUCHTEN, A. VAN. 1890. Histology of the gut: peritrophic membrane, *Ptychoptera* larva. *Cellule* **6**, 185–290
9. GILES, E. T. 1965. The fine structure of the peritrophic membrane of the three regions of the mid-gut of earwig *Anisolabis litteria*. *Trans. R. Soc. N.Z. Zool.* **6**, 87–101
10. GRESSON, R. A. R. 1934. The cytology of the mid-gut and hepatic caeca of *Periplaneta orientalis*. *Quart. J. micr. Sci.* **77**, 317–34
11. HAFEZ, M. and EL-ZIADY, S. 1952. On the histology of the alimentary canal of *Hyperaspis vinciguerrae* Capra. *Bull. Soc. Fouad I. Ent.* **36**, 293–310
12. HASEMAN, L. 1910. Histology of the mid-gut of *Psychoda* larva. *Ann. ent. Soc. Am.* **3**, 277–313
13. HENSON, H. 1931. The structure and postembryonic development of *Vanessa urtica* (Lepidoptera). I. The larval alimentary canal. *Quart. J. micr. Sci.* **74**, 321–60
14. ——. 1932. The development of the alimentary canal in *Pieris brassicae* and the endodermal origin of the malpighian tubules of insects. *Quart. J. micr. Sci.* **75**, 283–305
15. HERING, E. M. 1939. The peritrophic membrane in honey-bee. *Zool. J. Anat.* **66**, 129–90
16. HODGE, C. 1936. The anatomy and histology of the alimentary tract of the grass hopper, *Melanopus differentialis* Thomas. *J. Morph.* **59**, 423–34
17. HOSSELET, C. 1931. Contribution à l'étude du chondriome chez les insectes. *Archs. Zool. exp. gén.* **72**, 1–273
18. HÖVENER, M. 1930. Der Darmtraktus von *Psychoda alternata* Say und seine Anhangsdrüsen. *Z. Morph. Ökol. Tiere* **18**, 74–113
19. KUSMENKO, S. 1940. Ueber die postembryonale Entwicklung des Darmes der Honigbiene und die Herkunft der larvalen peritrophischen Hullen. *Zool. J. Anat.* **66**, 463–530
20. LOTMAR, R. 1942. Das Mitteldarmephithel von *Tineola biselliella* (Kleidermotte) während der Metamorphose. *Mitt. schweiz. ent. Ges.* **18**, 445–55
21. MERCER, E. H. and DAY, M. F. 1952. The fine structure of the peritrophic membrane of certain insects. *Biol. Bull. mar. biol. Lab., Woods Hole* **103**, 384–94
22. NEWCOMER, E. J. 1914. Some notes on digestion and the cell structure of the digestive epithelium in insects. *Ann. ent. Soc. Am.* **7**, 311–22
23. PATTERSON, M. F. 1937. The cellular structure of the digestive tract of the beetle, *Passulus cornutus* Fabricus. *Ann. ent. Soc. Am.* **30**, 619–40
24. PAVLOVSKY, E. N. and ZARIN, E. J. 1922. On the structure of the alimentary canal and its ferments in the bee *Apis mellifera*. *Quart. J. micr. Sci.* **66**, 509–56
25. PRADHAN, S. 1936. The alimentary canal of *Epilachna indica*. *J. R. Asiat. Soc. Beng., Sci.* **2**, 127–56
26. PRZEŁĘCKA, A. 1963. Cytochemical investigations on lipid assimilation by the caterpillars *Galleria mellonella* L. *Folia biol. Kraków* **11**, 353–416

27. SHAY, D. E. 1946. Observations on the cellular enclosures of the mid-gut epithelium of *Periplaneta americana*. *Ann. ent. Soc. Am.* **39**, 165–9
28. SHINODA, O. 1926. Contributions to the knowledge of intestinal secretion of insects. I. *Mem. Coll. Sci. Kyoto Univ.* Ser. B, **2**, 93–116
29. ——. 1927. A comparative histology of the mid-intestine in various orders of insects. *Z. Zellforsch. mikrosk. Anat.* **5**, 278–92
30. SIANG-HSU, W. 1947. On the cytoplasmic elements in the mid-gut epithelium of the larva of *Drosophila melanogaster*. *J. Morph.* **80**, 161–84
31. SRIVASTAVA, B. P. 1959. Morphology of alimentary canal of the larva of *Leucinodes orbonalis* (Guen. (Lepidoptera, Pyranstidae). *Proc. natn. Inst. Sci. India Ser. B*, **25**, 188–200
32. STEUDEL, A. 1913. Absorption and secretion in insect gut. *Zool. J. Physiol.* **33**, 165–224
33. SUD, B. N. and GOYAL, A. P. 1966. Cytology and histochemistry of the mid-gut epithelium of the earwig, *Labidura riparia* (Pallas). *Proc. Indian Sci. Congr.* Part III, 331–2 (Abstract)
34. —— and MAND, H. K. Cytology and histochemistry of the mid-gut epithelium of the worker termite, *Odontotermes horni* Wasmann (unpublished)
35. —— and MARWAHA, P. 1965. Studies on the mid-gut of *Lepisma saccharina*. *Proc. Indian Sci. Congr.* Part III, 443–4 (Abstract)
36. —— and PAUL, G. S. 1966. Studies on the mid-gut epithelium of the honey-bee, *Apis indica* Fab. *Proc. Indian Sci. Congr.* Part III, 331 (Abstract)
37. —— and SARIN, A. B. 1966. Cytology and histochemistry of the mid-gut epithelium of *Dysdercus cingulatus* Fabr. *Proc. Indian Sci. Congr.* Part III, 332 (Abstract)
38. ——, SATIJA, R. C., and KAUR, K. 1965. Cytology and histochemistry of the mid-gut of *Callosobruchus analis*. *Proc. Indian Sci. Congr.* Part III, 435–6 (Abstract)
39. ——, ——, and SUD, S. 1966. Cytology and histochemistry of the larval mid-gut epithelium of *Anopheles stephensi* Liston. *Proc. Indian Sci. Congr.* Part III, 332–3 (Abstract)
40. —— and SINGLA, C. L. 1966. Studies on the mid-gut of the female *Culex fatigans*. *Proc. Indian Sci. Congr.* Part III, 330–1 (Abstract)
41. —— and WALIA, R. P. Cytology and histochemistry of the mid-gut epithelium of *Poecilocerous pictus* (unpublished)
42. TCHANG-YUNG-TAÏ. 1929. Sur la localisation de l'absorption intestinale et le comportement des cellules absorbantes chez les chenille d'un lépidoptère (*Galleria mellonella* L.). *C. r. Acad. Sci.* **188**, 93–5
43. WATERHOUSE, D. F. 1952. Studies on the digestion of wool by insects. V. The goblet cells in the mid-gut of larvae of the clothes moth (*Tineola bisselliella* Humm.) and other Lepidoptera. *Aust. J. scient. Res.* B, **5**, 169–77
44. ——. 1957. Digestion in insects. *A. Rev. Ent.* **2**, 1–18
45. —— and STAY, B. 1955. Functional differentiation in the mid-gut epithelium of blowfly larvae as revealed by histochemical tests. *Aust. J. biol. Sci.* **8**, 253–77
46. —— and WRIGHT, M. 1960. The fine structure of the mosaic midgut epithelium of blowfly larvae. *J. Insect Physiol.* **5**, 230–9
47. WEIL, E. 1935. Vergleichendmorphologische Untersuchungen am Darmkanal einiger Apiden und Vespiden. *Z. Morph. Ökol. Tiere* **30**, 438–78
48. WIGGLESWORTH, V. B. 1930. The formation of the peritrophic membrane in insects, with special reference to the larvae of mosquitoes. *Quart. J. micr. Sci.* **73**, 593–616
49. WOKE, P. A. 1941. Structure and development of the alimentary canal of the southern army worm larva. *Tech. Bull. U.S. Dep. Agric.* No. **762**, 1–29
50. WOODRUFF, B. H. 1933. Studies of the epithelium lining the caeca and mid-gut in the grass-hopper. *J. Morph.* **55**, 53–80

Chapter 26

Cellular Differentiation in Ectodermal Explants from Amphibian Gastrulae

F. S. BILLETT

INTRODUCTION

Ectodermal explants derived from amphibian gastrulae provide a relatively simple system in which to study the differentiation of cells. The number of cell types which can arise is limited, and these are fairly easy to recognize. The kind of differentiation obtained from an isolated piece of dorsal ectoderm depends, essentially, on the stage which the donor embryo has reached. Broadly speaking, an explant derived from ectoderm which has not been in contact with archenteron roof, forms a simple epidermal structure, whereas one derived from ectoderm which has been subjected to an inductive stimulus, will form neural structures. Thus in *Xenopus laevis*, for instance, dorsal ectoderm isolated from an early gastrula develops into a relatively simple mass of epidermal-type cells, but when isolated from a late gastrula the ectoderm forms neural structures of the archencephalic type, in addition to epidermal derivatives.

The capacity of the isolated ectoderm to form neural structures, was studied by Chuang,[8] Barth,[2] and Holtfreter.[20,21,22] Since that time, competent ectoderm derived from early gastrulae has been used widely to study the nature of the inductive process (for an account of this work, see, for example, Yamada[42,43]). Such explants, and those derived from later gastrulae, which are capable of self-differentiation, form useful test systems for studying the effect of metabolic inhibitors, either on the inductive process itself, or on the cellular differentiation which results. The small size of the explants facilitates rapid diffusion of substances into and out of the explants when they are alive, and allows for the rapid penetration of fixatives for the preservation of cell structure. This article describes the development of such explants derived from gastrulae of *Xenopus laevis*, and the effects of D-threo chloramphenicol on this test system.

Experiments on explanted chick embryos (Billett, Collini, and Hamilton[4]) have shown that D-threo chloramphenicol produces abnormalities of early development which involve, chiefly, the neural tube, which fails to close, and the blood islands, which are either reduced in number, or completely absent. As the L-threo isomer of chloramphenicol does not produce these effects, it seems reasonable to attribute them to a selective inhibition of protein synthesis in the developing embryo. In view of the difficulty of analysing the effect of chloramphenicol on the rather complicated system of the chick embryo, it was decided to study the effect of the antibiotic on ectodermal explants. The preliminary findings of the experiments presented here, suggest that, at the stages treated, the explants are rather resistant to the action of chloramphenicol, at the concentration used. The main effects of the compound appear to be a partial inhibition of the development of the adhesive gland structure, and an increase in the number of the mitotic figures in the epidermis.

During the course of this work, cytological observations were concentrated on the cellular differentiation which occurs in the epidermal layer of the explants. These observations were made on 1 μ sections

23—C.S.I.

of explants embedded in Araldite (Glauert, Rogers, and Glauert[16]), sectioned on a Huxley ultramicrotome, and stained with toluidine blue. This relatively simple cytological technique gives results for light microscopy which are, in many ways, greatly superior to those obtained by more conventional staining techniques applied to sections cut from material embedded in wax. As Richardson, Jarett, and Finke[36] have emphasized, sectioning at thicknesses of the order of the wavelengths of light is now a routine possibility in ultramicrotomy, and cytologists may readily obtain sections for light microscopy at the thickness particularly recommended by Altmann (Baker[1]). This approach is heartily advocated for embryological studies.

Some of the cytological observations presented in this paper were made in collaboration with Mr. P. Gould, of the Department of Anatomy at the Middlesex Hospital Medical School, and I am extremely grateful to him for allowing me to publish, at this stage, some of our joint work. I am also indebted to Dr. S. M. McGee-Russell for taking the photomicrographs shown in Figs. 26.3, 26.5, and 26.6, and to Miss Hilary A. Belham of the M.R.C. Laboratories at Carshalton for developing and printing most of the photographs.

METHODS

The observations were made using the simplest type of experimental situation. The dorsal ectoderm was removed from mid to late gastrulae of *Xenopus laevis* (Nieukwoop and Faber,[30] stage 11–12), and cultured in a standard amphibian salt medium, on an agar surface. The general procedure for preparing the explants is given in a paper which also described the effect of benzimidazole on the system (Billett and Brahma[3]). Tungsten needles were used to remove the ectoderm from the embryo. The explants were obtained by first making a transverse cut into the remains of the blastocoelic space, then cutting backwards on each side of the embryo, at about the level of the sides of the archenteron roof, and, finally, the ectodermal flap was removed, by making a transverse cut through the ectoderm, above the dorsal lip of the blastopore. In this way approximately one-quarter of the total ectoderm of the embryo was removed.

The explants were cultured under slightly different conditions to those used in the benzimidazole series described in the previous paper. They were cultured at $15°C \pm 2°C$ in Steinberg's medium (Hamburger[17]), containing 200 µg/ml. of sulphadiazine to minimize the risk of infection (Detwiler and Robinson[10]). Under these conditions, provided adequate sterile precautions are taken initially, the explants develop in an archencephalic direction (forebrain and derivatives), and they will survive for seven days or more. As a control tissue, not capable of forming neural structures, ventral ectoderm was taken from gastrulae of the same stage.

The concentration of the D-threo chloramphenicol used ranged from 300–600 µg/ml. This was added to the Steinberg medium, which remained unchanged for the duration of the culture. The maximum period of culture was 15 days. Using a pH meter, the pH of the medium with and without the addition of chloramphenicol was compared, and was found to be identical at 6·9. During the course of the culture, the explants were observed at least daily, and sometimes more frequently. Changes in the external appearance of the explants were noted, and cytological studies were made subsequently of explants 15 hours, 2 days, 4 days, and 6 days in culture.

The majority of the observations reported here were made on material fixed in osmium tetroxide, embedded in Araldite, sectioned at 1 µ, and stained with a 1 per cent solution toluidine blue containing 1 per cent sodium borate. The method is based on that introduced by Trump, Smuckler, and Benditt.[38] After cutting on a Huxley ultramicrotome, the sections were placed on a drop of water on a clean glass slide, and dried down at about 60°C. One or two drops of the toluidine blue solution were applied to the section, and heated at 60°C for 50 sec. Excess stain was removed quickly, with distilled water, and the slide was allowed to dry, in air. Permanent preparations were made by mounting the sections in Araldite, under a coverslip.

A number of comparative observations were made on 6 day explants from materials fixed in Smith's solution, embedded in wax, sectioned at 7–10 µ and stained with Erlich's haematoxylin and eosin.

OBSERVATIONS ON THE EXTERNAL APPEARANCE AND SURVIVAL OF THE EXPLANTS

The development of the untreated explants will be described first, and then an account will be given of the effects of chloramphenicol.

The observations were made at 10–100 times magnification, using a Zeiss stereoscopic microscope (epi-illumination).

Cultured under the experimental conditions, without changing the medium, the isolated dorsal ectoderm survives well for 6 days, but then the explants begin to break up, and few survive beyond the tenth day. Under the same conditions, explants of ventral ectoderm survive for considerably longer periods. Many are intact after 12 days, although few survive beyond the fifteenth day. The survival times of explants made from dorsal and ventral ectoderm are compared in Fig. 26.1. The shorter survival time of the dorsal explants is almost certainly due to the differentiation of neural structures (see below).

With the passage of time, the appearance of the explants changes. These changes are especially note-worthy, as they indicate the nature of the cellular differentiation which is occurring in the explant.

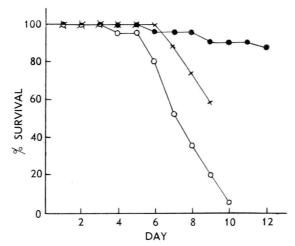

Fig. 26.1 Survival of ectodermal explants cultured at 15°C. —O—O— Dorsal (20 explants); —●—●— Ventral (28 explants); —X—X— Dorsal treated with 300–600 μg/ml. D-threo chloram-phenicol (27 explants).

Very soon after the ectodermal pieces have been removed from the gastrulae, they roll up, so that the cut surfaces are apposed, and the internal surface is shut off from the exterior. After 6–12 hr, the external surface of the explants is uniformly brown and smooth, and shows little or no sign of the originally cut, and subsequently apposed surfaces. For the next 24 hr or so, little further change occurs, although during this time, loose cells may be shed from the surface of the explants.

By the end of the second day, the surface of the explants becomes wrinkled, and, in some of them, the movement of small particles in the medium near the explant surface, suggests that ciliated cells have appeared. The most obvious feature of the explants, at this stage, is the developing adhesive (cement) gland. This is frequently cone shaped, and may occupy as much as one-fifth of the total surface area of the explant. Soon after the gland appears, it begins to secrete a sticky material which it continues to form throughout the life of the explant.

By the end of the third day, some of the explants are sufficiently strongly ciliated to propel themselves across the surface of the agar. The number of explants which move in this way increases on subsequent days, reaching a peak on the sixth day, when usually about 60 per cent of the explants will be moving around.

During the fourth day, many of the explants develop a single vesicle which is seen as a small transparent blister beneath the stretched surface of the epidermis. This internal space grows larger during the fifth and sixth days. Vesiculation is probably a good indication of the formation of neural structures inside the explant (Johnen[25]). When the vesicles attain their maximum size, their walls are frequently transparent, and it is then often possible to see a pigmented eye structure, and small, dense, whitish masses of neural tissue inside the explant.

In a few cases, prominent melanocytes appear on the wall of the vesicles between the fifth and seventh days.

By the sixth day several of the explants usually show degenerative changes. The commonest form of breakdown is the expulsion of a mass of cells from the interior. This seems to correspond to the 'explosion phenomenon' described by Boterenbrod,[7] and results in the loss of the neural material. Explants may 'survive' this loss, in the sense that the epidermal part remains intact, and moves around until about the tenth day. However, following the loss of the inner cell mass, degenerative changes take place in the epidermis of the surviving fragment, and in the epidermis of those explants which never developed neural structures. During the seventh and eighth days, areas of cells on the surface of the explants assume a pale glistening appearance, and, eventually, individual cells become glassily transparent, resembling minute oil droplets. It is assumed that these cells are dead. Viable groups of cells, recognized by their ciliary activity, break away in small clumps from time to time, but these disintegrate within about 24 hr.

During the first 2 days of culture, control explants, prepared from ventral ectoderm, undergo the same initial changes as those described for the dorsal explants, but no adhesive gland appears. Out of 52 ventral explants prepared, only one developed an adhesive gland. The development of cilia, and consequently, of movement of the explants over the agar surface, is similar to that seen in the dorsal explants. The increased survival time of the ventral explants, compared with the dorsal, can be related to the fact that neural tissue does not develop in the former. Ventral explants are not, therefore, subject to the risk of 'explosive' disintegration.

At the concentrations used, D-threo chloramphenicol appeared to have no dramatic effect on the general appearance of explants prepared from dorsal ectoderm. Several minor differences were noted between treated and untreated explants. Sometimes the development of the adhesive gland in the treated explants was poor, and the area covered by the gland was relatively small. The secretion of the gland was often less copious, and less sticky, than that of the controls. Vesiculation was often not so marked in the treated explants in the initial stages, although it often became as marked in the later stages. As indicated in Fig. 26.1, the chances of survival of the dorsal explants beyond the sixth day appear to be increased by treatment with chloramphenicol. This may account for the fact that most of the explants which developed melanocytes in the wall of the vesicle were in the chloramphenicol group; melanocytes only appeared in the untreated dorsal explants on the sixth–seventh day.

The above observations suggest that treatment with chloramphenicol prevents the fullest development of the adhesive gland, but does not prevent the formation of neural structures, although this may be retarded.

CYTOLOGICAL OBSERVATIONS

The following remarks apply to dorsal ectoderm, on which most of the observations were made.

Most of the cells contained in the dorsal explants may be classified into one of three groups: epidermal, neural, and ectomesenchymal. In most cases if cell types other than these occur, such as notochord, it may be assumed that cells from the archenteron roof have been included in the original explant.

In this contribution emphasis will be placed on the cell types which develop in the epidermal covering of the explant. These cell types are illustrated in Fig. 26.2, and indicate the degree of cellular differentiation in a 2 day explant. Typically, the epidermis is a two-layered structure, consisting of an outer layer of cells which tend to be columnar, and an inner layer of cells which are flatter and broader than the covering layer (Fig. 26.5, B). The exact geometry of the cells depends on the degree of stretch in the outer layer. Three cell types are found in the outer layer: (1) The commonest type is similar to that seen in the larval epidermis (Pflugfelder and Schubert[32]). It is mucus secreting, columnar in shape and characterized by an outer basiphilic border. (2) Seen in section, about 1 in 10 of the cells in the outer layer is ciliated. Apart from the presence of cilia, these cells are distinguished from their neighbours by their broadness and basiphilia, and by the fact that they lack pigment granules (Fig. 26.4, B). The upper part of these cells contains a large number of small spaces giving a honeycomb-like structure. (3) The third cell type makes up the adhesive gland. These cells have elongated necks which are orientated towards the external surface. The necks of these cells are strongly basiphilic and are loaded with pigment granules. In their general structure they resemble the cells seen in the dorsal lip of the blastopore. The packing of pigment granules in the narrow necks of the adhesive gland cells gives rise to the pigmented

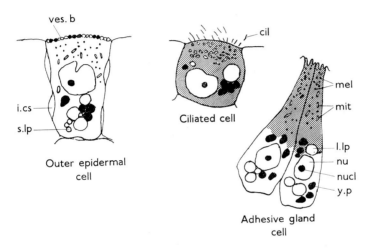

Fig. 26.2 Cell types of the epidermal layer. *Abbreviations.* cil. cilia; i.cs. intercellular space; l.lp. large lipid body; s.lp. small lipid body; mel. melanin granule; mit. mitochondrion; nu. nucleus; nucl. nucleolus; ves.b. vesicular border of epidermal cell; y.p. yolk platelet.

area seen on the surface of the explant (the pigmented lip of blastopore is produced in the same way). An ultrastructural study of these cells has recently been published by Perry and Waddington.[32] Eakin[11] has also made a study of these cells in the tree frog.

A number of cell structures are clearly preserved and revealed by the osmium tetroxide fixation and the toluidine blue staining. These are illustrated in Fig. 26.2. The overall background staining is pale blue and this contrasts sharply with the intense blue staining of the yolk platelets (yp), and the yellow bodies (slp, llp) which are presumably lipid in nature. It is interesting to note that, in the unstained preparation, these bodies are pale brown in colour. Apart from the yolk platelets, which possess the strongest affinity for the toluidine blue, many other cell structures are stained with varying degrees of intensity. These include the cell membranes, the nuclear membrane, the nucleoli, the chromosomes of dividing cells, and the mitochondria.

The neural tissue, and in the older explants, the eye structures, are easy to recognize, by the ordered arrangements and small size of the cells. These cells possess fewer yolk platelets and lipid bodies than do the epidermal cells, and have a larger nuclear-cytoplasmic ratio.

Apart from the neural tissue, groups of mesenchymal cells are a common internal feature of the dorsal explants. For the most part these cells form randomly arranged groups. which are in close contact with the cells of the inner layer of epidermis, which they resemble. In some areas the mesenchyme cells merge

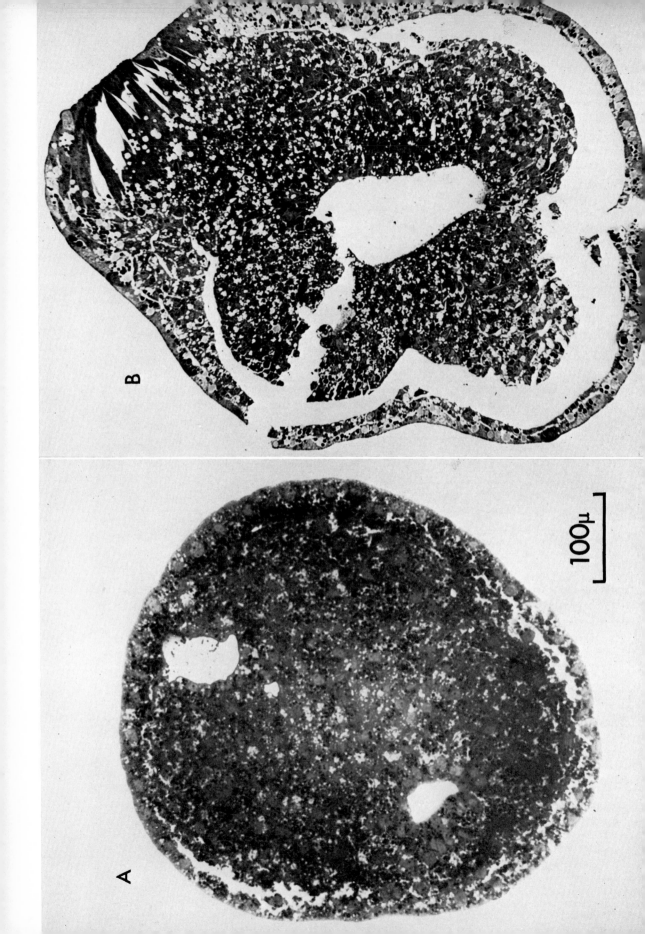

with the inner layer of the epidermis so that it is impossible to distinguish between the two tissues. Placodes are easy to recognize, as they form beautifully rounded clumps of cells having few yolk platelets, and prominent nucleoli. Melanocytes, seen only rarely, are usually sandwiched between the epidermal cell layers and the neural tissue.

Explanation of photographic figures

Figures 26.3, A and B, 26.4, A and B, 26.5, B, 26.6, A and B are araldite sections stained with toluidine blue. Figures 26.5, A, and 26.6, C and D, are paraffin sections stained with haemotoxylin and eosin.

Abbreviations: adh. adhesive gland; b.g. border granule; b.v. border vacuole; cil. cilia; ics. intercellular space; l.lp. large lipid body; s.lp. small lipid body; mit. mitochondrion; neur. neural structure; nu. nucleus; nucl. nucleolus; ves.b. vesiculated border of epidermal cell; ep.cl. epidermal cell layer.

15 hr explants

The general appearance of these explants is shown in Fig. 26.3, A. At this stage the explants are virtually a solid mass of cells. The outer layer of cells is more lightly stained than the inner mass. Cell boundaries are not very distinct, and the nuclei of most cells are large and irregular. Nucleoli are infrequent. The cells contain many small lipid bodies and yolk platelets. Larger lipid bodies, which are a common feature of the cells in the later explants, are not seen at this stage. Apart from the slight difference in the staining reaction of the outer cell layer and the inner cell mass, no sign of cellular differentiation can be detected. The cells of the outer layer do not possess the external boundary structure seen in the later explants, and the cytoplasmic inclusions of these cells (yolk platelets, lipid bodies and pigment granules) appear to be randomly arranged. No difference was observed between the treated and untreated explants.

2 day explants

After 48 hr culture, sections of the explants reveal that cellular differentiation is well under way, especially in the outer cell layers. A typical 2 day explant is shown in Fig. 26.3, B.

The outer layer now forms a distinct structure, clearly consisting of a double cell layer in many places. The outer cells of the epidermal layer are usually distinctly columnar, and their contents show some degree of polarization, inasmuch as the yolk platelets and lipid bodies tend to accumulate between the base of the cell and the nucleus, which lies in the upper half of the cell. The pigment granules of these cells are usually more numerous in the upper half of the cell than in the lower half. The outer-most boundary of the cells is now distinguished by a vesiculated border (Fig. 26.4, A). In many cases the vesicles contain granules possessing a strong affinity for toluidine blue. The nuclei of the cells of the epidermal layer are large, tend to be sharply indented, and sometimes possess prominent nucleoli. In the outer cell layer the numerous small lipid bodies have been replaced by larger bodies of the same kind. At this stage there are usually three or four large lipid bodies per cell. In contrast to the earlier explants, there are prominent spaces between the cells of the outer layer. An occasional mitosis is seen in the outer layer.

In some explants, at this stage, the adhesive gland cells are readily distinguished. These cells are much taller than they are broad. Apart from the melanin granules, which are confined to the relatively narrow neck of the cells, the bulk of the cytoplasmic contents, including the nucleus, tends to lie in the lower half of the cell. Each of these cells contains several yolk platelets, and, typically, three or four large lipid bodies.

Ciliated cells are also well developed at this stage. Although these cells tend to be broader than they are long, their shape is variable. Typically, most of the upper border of the cell is exposed to the external medium, but sometimes, only a small part of this outer edge is exposed. The nucleoli are prominent and

Fig. 26.3 A, Explant after 15 hr culture; B, Explant after 2 days culture.

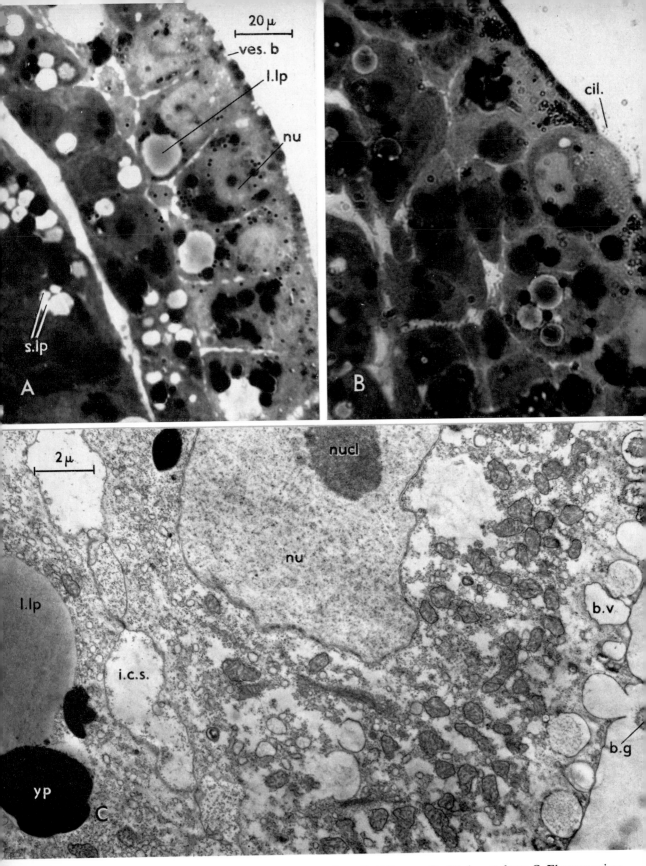

Fig. 26.4 A, Epidermal cell layers of 4 day explant; B, Ciliated cell of 6 day explant; C, Electron micrograph of part of outer epidermal cell in a 4 day explant.

the cytoplasmic inclusions lie in the lower two-thirds of the cell. The number of yolk platelets and lipid bodies varies greatly. In some cells there are few inclusions of this kind, whereas others may contain as many as ten lipid bodies and a single yolk platelet, or many yolk platelets and a few lipid bodies. The ciliated cells are invariably isolated from each other, and appear to be regularly spaced over the epidermal surface.

The appearance of the inner cell mass is variable. Two types of cell arrangement are often present. One consists of a single mass of closely packed small cells, and the other forms a loosely packed network of larger cells. The cells of the closely packed mass are about half the size of the epidermal cells, contain nuclei with prominent nucleoli, and a few small yolk platelets. The loose network consists of larger cells which contain yolk platelets, and sometimes large numbers of the smaller lipid granules. Mitoses are fairly common in the closely packed mass.

Explants treated with chloramphenicol present the same general appearance as untreated ones. However, they appear to differ in two significant ways. In treated explants the adhesive gland cells are either missing, or relatively few in number, and the outer epidermal cell layers are not so clearly defined. In the latter case the distribution of the cytoplasmic contents may be random, as at 15 hr, and no clearly defined vesicular boundary structure can be discerned. Because of the variable appearance of the inner cell mass in untreated explants, it is difficult to assess any change which might have been brought about by the chloramphenicol treatment. In general, however, the yolk platelets seem to be more numerous in treated explants.

4 day explants

In their general appearance these resemble the 2 day explants. A 4 day explant is shown in Fig. 26.4, A. The epidermis is now more clearly defined as a two-layered structure, and the adhesive gland shows further development. Both the epidermal cells, and those of the adhesive gland, show a further reduction in number and an increase in size of the lipid bodies. In many cases, the cells of the outer epidermal layer now contain only a single large lipid body, placed basally, and often partly surrounded, by yolk platelets. Mitochondria can be detected easily in many cells as they may be in the 2 day explants. In the outer epidermal layer they are most numerous between the nucleus and the outer border of the cell. This is shown in the electron micrograph in Fig. 26.4, C.

The closely packed inner cell mass is now obviously the precursor of neural structures, as it now presents a more clearly orderly arrangement of the pallisade type. The loose network of the cells appears to give rise to the mesenchymal elements of the explant.

Apart from the less well-developed adhesive gland, the treated explants do not differ markedly from the untreated ones.

6 day explants

At 6 days many dorsal explants reach their maximum development, and possess quite a complicated internal structure. The adhesive gland is prominent at this stage, and, like the epidermal cells, the cells of the gland almost invariably possess a single, basally placed, large lipid sphere. In a transverse section these spheres form a crescent-shaped array at the base of the gland.

Internally, neural structures occur which are most frequently of the archencephalic type. These are usually associated with well-developed eye structures. Such differentiation has been described many times before (Fig. 26.6, C). Well-defined mesenchymal cells are also present. The neural cells possess few yolk platelets and the lipid bodies in these cells are small and tend to occur in groups.

An obvious difference between the treated and control explants at this time is the development of the adhesive gland. This is poorly developed in many cases in the treated explants, and is sometimes partially enclosed by the outer epidermal cell layer (ep.cl in Fig. 26.6, B).

At 6 days, control explants prepared from ventral ectoderm show none of the differentiation exhibited by the dorsal explants, apart from the development of ciliated cells. Sections of ventral explants are fairly uniform in appearance. The explants consist of cells of epidermal type, but the outer cells do not form the two-layered structure so typical of the dorsal explants. The surface of the ventral explants is

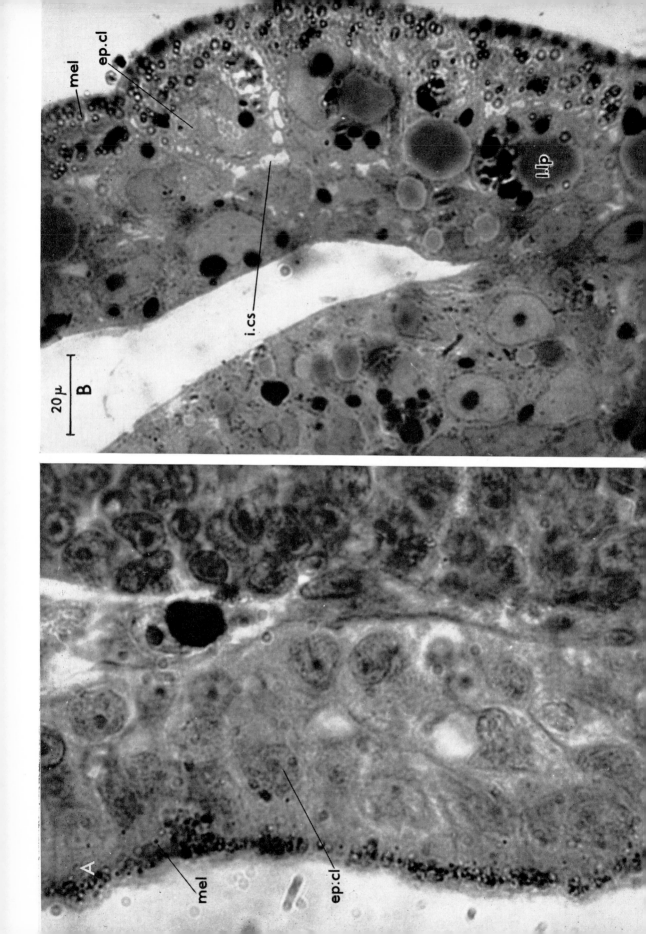

much indented, compared with the surface of the dorsal explants. A section through a ventral explant is shown in Fig. 26.6, D.

At 6 days, mitoses are still common in both dorsal and ventral explants. They are particularly obvious in the outer epidermal layer in 10 μ paraffin sections stained with haematoxylin and eosin. An initial impression that the mitotic figures are more frequent in explants treated with chloramphenicol was confirmed by counting the number of these figures, and the total number of cells in the outer epidermal layer, in alternate sections of explants. The results, given in Table 26.1, show that there are more mitotic figures in treated explants.

Table 26.1 Comparison of mitoses in outer epidermal layer of treated and untreated explants

		No. of explants examined	No. of cells counted	No. of mitotic figures	Mitotic figures /1000 cells	Type of mitotic figures (percentage)			
						P	M	A	T
Explants from ventral	U	6	2996	58	19	47	47	3	3
ectoderm	T	6	2695	78	29	38	49	10	3
Explants from dorsal	U	9	5960	33	8	43	47	10	0
ectoderm	T	7	5585	82	15	27	58	10	4

U = Untreated explants. T = Explants treated with 400 μg/ml. Chloramphenicol. P = Prophase. M = Metaphase. A = Anaphase. T = Telophase.

DISCUSSION

The lack of shrinkage produced by osmium tetroxide fixation and the embedding procedures of electron microscopy, plus the thinness of the Araldite sections, together give one a remarkably good picture of cytoplasmic detail by light microscopy. It is obviously superior to that obtained after the procedures hitherto usual for fixing and embedding amphibian embryo material. A comparison of Fig. 26.5, A and B shows that the nuclei of neural tissues preserved by acetic acid-dichromate (Smith's fixative) are about two-thirds the size of those preserved by the osmium fixative. The degree of shrinkage is also apparent when whole sections prepared by the two methods are compared.

The mucus containing border structure of the epidermal cells is clearly preserved in the 1 μ Araldite sections. Detail of this kind cannot be seen in paraffin sections. The electron micrograph in Fig. 26.4, C shows this structure in more detail. The origin and function of the vesicular border is the present subject of a separate study (Billett and Gould[5]).

Techniques for studying intracellular lipid, and the importance of lipid inclusions, have always been emphasized by Baker.[1] A striking feature of amphibian embryonic cells in the osmium-fixed tissue is the abundance of lipid bodies. In paraffin sections these bodies, when small, cannot be detected at all, and the presence of the larger lipid bodies, also lost in the conventional procedure, is only revealed by spaces in the cells. These are clearly visible as holes in the basal region of cells of the outer layer (Fig. 26.6, D).

The lipid inclusions of amphibian embryo cells have been studied by Holtfreter[23, 24] and Karasaki.[27] Holtfreter called the larger bodies liposomes, and the smaller ones lipochondria. He observed that treatment of freshly separated cells with alkali and acid caused the lipochondria to coalesce into the larger

Fig. 26.5 A comparison of paraffin (A) and Araldite (B) embedded specimens.

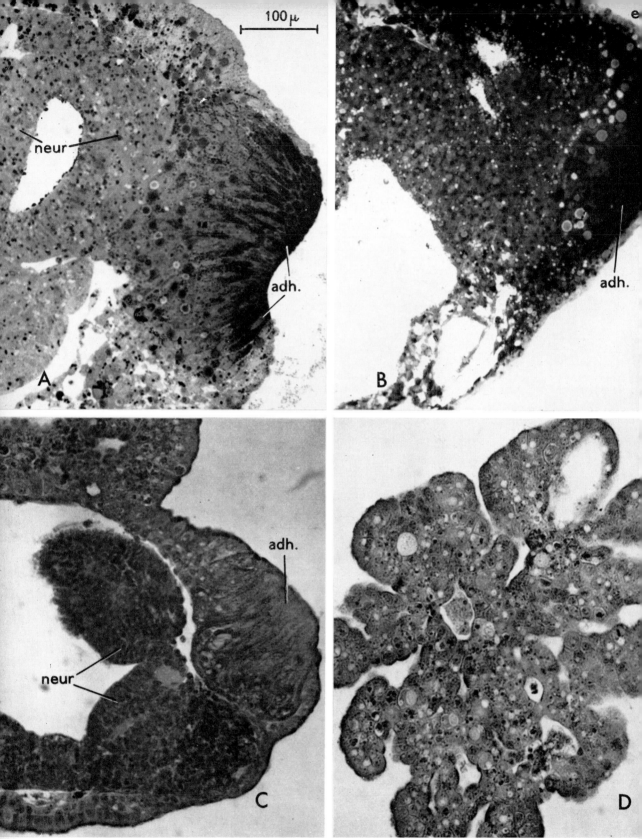

Fig. 26.6 A, Adhesive gland structure in a 6 day untreated explant; B. Adhesive gland structure in a 6 day explant treated with chloramphenicol; C, An explant of dorsal ectoderm after 6 days' culture; D, An explant of ventral ectoderm after 6 days' culture.

liposomes, and that such a transformation also occurred in the tissues of young amphibian larvae. In the latter case he suggested that the phenomenon is associated with yolk utilization. He also noted that, under certain conditions, the liposomes underwent shrinkage, and vacuolization, to form structures which resembled the Golgi system. Karasaki in his electron microscope studies refers to all the lipid bodies as lipochondria, and distinguishes two types. Type A is the most frequent, and it is characteristically more dense than type B.

In the present study the lipid inclusions have been simply called lipid bodies. They vary in size from about 0·5 to 5 μ in diameter and appear to form a continuous size range. Some of Holtfreter's lipochondria were smaller than those seen in the present observations. In certain cells, it is clear that, as differentiation proceeds, the smaller lipid bodies coalesce to form larger ones, and, in most of the outer epidermal cells, eventually form a single sphere in each cell. This process corresponds to that described by Holtfreter. At the level of the light microscope it does not appear possible to distinguish the two types of lipochondria described by Karasaki.

In the explants, cellular differentiation seems to be accompanied by the loss of some yolk platelets. Although no quantitative measurements have been made, yolk platelet breakdown seems to be slightly less in the chloramphenicol treated explants, at least in the early stages. Such a retardation might be expected, in view of the finding of Miciarreli and Pinamonti[29] that chloramphenicol retards yolk utilization in disembryonated chick blastoderms.

The relationship between the lipid bodies and the yolk platelets is obscure. In some cells the disappearance of the yolk platelets is accompanied by an increase in the number of lipid bodies and it is tempting to believe that at least some of these bodies arise as the result of yolk breakdown. Suzuki (unpublished observations, quoted by Ohno and others[31]) noted that the outer layer of yolk platelets possesses an affinity for toluidine blue which he suggested was due to the presence of RNA, the loss of the affinity representing loss of RNA. However, the experiments of Ohno, Karasaki, and Takata[31] cast some doubt on the specificity of the toluidine blue staining for RNA. Yolk platelet utilization in amphibian embryos has been the subject of a recent study by Selman and Pawsey.[37] The osmium/Araldite/toluidine blue method does not reveal the stages in yolk breakdown depicted by these authors.

Apart from the disappearance of yolk platelets, and the coalescence of lipid bodies, another sign of differentiation, seen in the outer epidermal layer, is a polarization of cell contents. After two days this polarization is less marked in the explants treated with chloramphenicol.

The absence of melanin granules in ciliated cells is interesting, as the neighbouring epidermal cells contain many granules of this kind. If the ciliated cells arise from those already on the surface of the explant, they must lose their melanin granules fairly quickly at an early stage of differentiation. The alternative is to imagine that the precursors of the ciliated cells exist, initially, below the surface of the outer epidermal layer. It may be significant, in this connection, that in some cases the ciliated cells are only exposed to the external medium over a very small area.

The main effect of the chloramphenicol appears to be inhibition of the development of the adhesive gland. The differentiation of the neural tissue is not affected by the compound. The other effects, retardation of development at two days, better survival, and increase in the number of mitotic figures, are comparatively small. They nevertheless confirm that the compound penetrates the explant.

The increase in the number of mitotic figures could be the result of either the arrest of mitosis at a particular stage, or of its stimulation. It should be possible to decide between the two alternatives, by comparing the frequency of the different stages of mitosis. This comparison is made in Table 26.1, and, although the figures are not very conclusive, the higher percentage of metaphase figures in the treated dorsal explants indicates a retardation or a complete arrest of mitosis at this stage.

The reason why chloramphenicol affects some developing embryonic tissues and not others merits further discussion. Chloramphenicol is a well-known broad-spectrum antibiotic, which acts by an inhibition of bacterial protein synthesis (Gale and Folkes[15]). The compound also inhibits protein synthesis in cell-free systems derived from bacteria. Some metazoan cells and cell systems appear to be susceptible to chloramphenicol, others not. For instance, the compound does not inhibit protein synthesis in fractions derived from mammalian cells (Rendi;[34] Ehrenstein and Lipmann[12]). On the other hand, clinical use of the drug is sometimes associated with undesirable side effects, indicating a measure of cytotoxicity towards human cells (Woodward and Wisseman[41]). The compound also affects the uptake of labelled amino-acids in planarians (Flickinger[14]), and produces abnormalities in chick embryos (Blackwood;[6] Colombo and Micciarelli;[9] Billett, Collini, and Hamilton[4]).

A molecular basis for this difference in susceptibility towards chloramphenicol is provided by the experiments of Vasquez,[39] which suggest that the resistance of certain cells towards chloramphenicol is due to the fact that their ribosomes possess little or no affinity for the antibiotic.

At the concentrations used in the experiments on explanted chick embryos, not all the tissues are affected by the chloramphenicol. Despite the fact that the compound is applied before any of the major organ systems appear, the heart, the somites, and the notochord, develop fairly normally. Clearly, the abnormalities which are produced cannot be due to a general inhibition of protein synthesis in all the differentiating cells. Two possible explanations for the teratogenic action of chloramphenicol may be advanced. Firstly, the failure of the neural tube to close, and the extent of this damage in relation to the time of application of the compound, may be the result of a weakening of inductive interaction between the mesoderm and the overlying epiblast. Secondly, it may be the case that only some of the cell types in the embryo are susceptible to the action of chloramphenicol.

No matter whether the inductive stimulus, the competence of reacting cells, or the differentiation of sensitive cells is involved, the molecular mechanism of action of chloramphenicol could be that proposed by Julian and Jardetsky,[26] and supported by the experiments of Vasquez. In embryonic tissue, it may be imagined that ribosomes capable of binding chloramphenicol are only found in certain cells, at certain times, during development. Cells containing ribosomes capable of binding chloramphenicol would fail to form protein, and, for this reason, might be rendered incapable of taking part in an inductive interaction. If the combination of messenger ribonucleic acid with ribosomes makes them less likely to interact with chloramphenicol, then the selective sensitivity of embryonic cells may depend on the extent to which this RNA is bound to the ribosomes. Vasquez[40] has emphasized that chloramphenicol is bound by 70S but not by 80S ribosomes, and this may mean that the selective sensitivity of embryonic cells is dependent on the presence of the smaller ribosomes.

Only the D-threo isomer of chloramphenicol produces a significant inhibition of protein synthesis (Rendi and Ochoa[35]), and only this isomer possesses any significant antibacterial activity (Maxwell and Nickel[28]). Where both the D and the L forms are active, some other basis for the activity of the compound should be sought. For instance, both threo isomers inhibit the uptake of certain ions by plant cells (Ellis[13]), an effect which may be due to the uncoupling of oxidative phosphorylation (Hanson and Hodges;[18] Hanson, Stoner, and Hodges[19]). Thus although the treatment of explants with D-threo chloramphenicol does affect them, it is not possible, at this time, to say that the changes which occur are definitely due to an inhibition of protein synthesis, until the action of the L-threo isomer upon explants is also known. Analysis of the control systems in differentiating explants through the use of pharmacological agents, coupled with detailed cytological studies of the molecular morphology, at all microscopical levels, should prove a field of some interest in embryology.

The use of metabolic inhibitors with a known molecular site of action is now an accepted approach for the analysis of the control mechanisms in differentiating cells. Such studies are of value to developmental biologists in general and of particular interest to those concerned with the genesis of embryonic abnormalities. It is worth while to emphasize from time to time that such investigations gain much in value when they are coupled with detailed cytological studies at all levels of microscopy.

SUMMARY

A description is given of the cellular differentiation which occurs in ectodermal explants derived from the gastrulae of *Xenopus laevis*. Particular emphasis is laid on the differentiation of the epidermal cell types in explants derived from dorsal ectoderm. Most observations were made on 1 μ sections of Araldite-embedded material stained with toluidine blue. Compared with more conventional procedures used for the light microscopical examination of amphibian embryonic material, the method appears to offer considerable advantages. The lipid structures of the cells are especially well preserved. Treatment of the explants with D-threo chloramphenicol partially inhibits the development of the adhesive (cement) gland and increases the number of mitotic figures in the epidermal cell layer.

REFERENCES

1. BAKER, J. R. 1950. *Cytological Technique*. Methuen Monographs on Biological Subjects. London: Methuen
2. BARTH, L. G. 1941. Neural induction without organiser. *J. Exp. Zool.* **87**, 371–84
3. BILLETT, F. S. and BRAHMA, S. K. 1960. The effect of benzimidazole on the differentiation of ectodermal explants from the gastrulae of *Xonopus laevis*. *J. Embryol. exp. Morph.* **8**, 396–404
4. ——, COLLINI, R., and HAMILTON, L. 1965. The effects of D- and L-threo chloramphenicol on the early development of the chick embryo. *J. Embryol. exp. Morph.* **13**, 341–56
5. —— and GOULD, P. 1967. Differentiation of the epidermis during the early development of *Xenopus laevis*. *J. Anat.* **101**, 833–834
6. BLACKWOOD, U. B. 1962. The changing inhibition of early development on the general development of the chick embryo by 2-ethyl 5-methyl benzimidazole and chloramphenicol. *J. Embryol. exp. Morph.* **10**, 315–36
7. BOTERENBROD, E. C. 1962. On pattern formation in the prosencephalon. An investigation of the disaggregated and reaggregated presumptive prosencephalic material of the neurali of Triton embryos. Ph.D. Thesis, University of Utrecht
8. CHUANG, H. H. 1939. Inductionsleistungen von frischen und gekochten Organteilen (Niere, Leber) und ihre Verpflanzung in Explante und verschiedene Wirtsregionen von Tritonkeimen. *Roux Archiv Entw. mech. Org.* **140**, 25–48
9. COLOMBO, G. and MICCIARELLI, A. 1964. Richerche sugli effeti del chloramphenicolo di polo disebrionati espianti *in vitro*. *Acta embryol. exp. Morph.* **7**, 89
10. DETWILER, S. R. and ROBINSON, C. O. 1945. On the use of sodium sulfadiazine in surgery on amphibian embryos. *Prof. Soc. exp. Biol. & Med.* **59**, 202–6
11. EAKIN, R. M. 1963. Ultrastructural differentiation in the oral sucker of the tree frog, *Hyla regilla*. *Devel. Biol.* **7**, 169–79
12. EHRENSTEIN, G. VON and LIPMANN, F. 1961. Experiments on haemoglobin synthesis. *Proc. Nat. Acad. Sci. Wash.* **47**, 41–50
13. ELLIS, R. J. 1963. Chloramphenicol and the uptake of salt by plants. *Nature, Lond.* **200**, 596–7
14. FLICKINGER, R. A. 1959. A gradient of protein synthesis in planaria and reversal of axial polarity in regenerates. *Growth* **23**, 251–71
15. GALE, E. F. and FOLKES, J. P. 1953. Assimilation of amino acids by bacteria 15. Action of antibiotics on nucleic acid and protein synthesis in *Staphylococcus aureus*. *Biochem. J.* **53**, 493–8
16. GLAUERT, A. M., ROGERS, G. E., and GLAUERT, R. H. 1956. A new embedding medium for electron microscopy. *Nature, Lond.* **178**, 803
17. HAMBURGER, V. 1960. *A Manual of Experimental Embryology*. University of Chicago Press
18. HANSON, J. B. and HODGES, T. K. 1963. Uncoupling action of chloramphenicol as a basis for the inhibition of ion accumulation. *Nature, Lond.* **200**, 1009
19. ——, STONER, C. D., and HODGES, T. K. 1964. Chloramphenicol as an inhibitor of energy-linked processes in maize mitochondria. *Nature, Lond.* **203**, 258–61
20. HOLTFRETER, J. 1938a. Differenzierungspotenzen isolierter Teile der Urodelengastrula. *Roux. Arch. Entw. Mech.* **138**, 522–656
21. ——. 1938b. Differenzierungspotenzen in isolierten Teilen der Anurengastrula. *Roux Arch. Entw. Mech.* **138**, 657–738
22. ——. 1945. Neuralisation and epidermalisation of gastrula ectoderm. *J. exp. Zool.* **98**, 161–209
23. ——. 1946a. Experiments on the formed inclusions of the amphibian egg. I. The effect of pH and electrolytes on yolk and lipochondria. *J. exp. Zool.* **101**, 355–405
24. ——. 1946b. Experiments on the formed inclusions of the amphibian egg. II. Formative effects of hydration and dehydration on lipid bodies. *J. exp. Zool.* **102**, 51–108
25. JOHNEN, A. G. 1956. Experimental studies about the temporal relationships in the inductive process. *Kononkl. Ned. Acad. Wetenschap. Proc.* Ser. c. **59**, 554–61
26. JULIAN, G. R. and JARDETSKY, O. 1964. Studies on the mechanism of action of chloramphenicol. *Proc. 6th Int. Congr. Biochem.* N.Y. Abs. **1**, 92
27. KARASAKI, S. 1959. Electron microscope studies on cytoplasmic structures of ectodermal cells of the *Triturus* embryo during the early phase of differentiation. *Embryologia* **4**, 247–72
28. MAXWELL, R. E. and NICKEL, V. S. 1954. The antibacterial activity of chloramphenicol. *Antib. and Chemo.* **4**, 289–95

29. MICCIARELLI, A. and PINAMONTI, S. 1964. Richerche sugli effeti del chloramfenicoli sull entroblasto di blastodischi di pollo disembrionati, od espianti *in vitro*. *Bell. di Zool.* **31**

30. NIEUKWOOP, P. D. and FABER, J. 1956. *Normal Table of Xenopus laevis*. Amsterdam: North Holland Publishing Co.

31. OHNO, G., KARASAKI, S., and TAKATA, K. 1964. Histo- and cytochemical studies on the superficial layer of yolk platelets in the *Triturus* embryo. *Exp. cell Res.* **33**, 310–18

32. PERRY, M. M. and WADDINGTON, C. H. 1966. The ultrastructure of the cement gland in *Xenopus laevis*. *J. Cell. Sci.* **1**, 193–200

33. PFLUGFELDER, O. and SCHUBERT, G. 1965. Electronmikroskopische Untersuchungen an der Haut von Larven und Metamorphosenstadien von *Xenopus laevis* nach Kaliumperchloratbehandlung. *Z. fur Zellforsch.* **67**, 96–123

34. RENDI, R. 1959. The effect of chloramphenicol on the incorporation of labelled amino acids into proteins by isolated subcellular fractions from rat liver. *Exp. cell Res.* **18**, 187–9

35. —— and OCHOA, S. 1962. Effect of chloramphenicol on protein synthesis in cell-free preparations derived from *Escherichia coli*. *J. Biol. Chem.* **237**, 3711–3

36. RICHARDSON, K. C., JARETT, L., and FINK, E. H. 1960. Embedding in epoxy resins for ultrathin sectioning in light microscopy. *Stain Tech.* **35**, 313–23

37. SELMAN and PAWSEY. 1966. The utilization of yolk platelets by tissues of *Xenopus* embryos studied by a safranin staining method. *J. Embryol. exp. Morph.* **14**, 191–212

38. TRUMP, B. F., SMUCKLER, E. A., and BENDITT, E. P. 1961. A method for staining epoxy sections for light microscopy. *J. Ultrastruct. Res.* **5**, 343–5

39. VASQUEZ, D. 1964. Uptake and binding of chloramphenicol by sensitive and resistant organisms. *Nature, Lond.* **203**, 257–8

40. ——. 1966. Mode of action of chloramphenicol and related antibiotics. In *Biochemical Studies of Antimicrobial Drugs* (edited by Newton and Reynolds), pp. 169–91. 16th Symposium of the Society for General Microbiology. London: Cambridge University Press

41. WOODWARD, T. E. and WISSEMAN, C. L. 1958. *Chloromycetin (chloramphenicol)*. Antibiotics monograph No. 8. New York: Medical Encyclopaedia Inc.

42. YAMADA, T. 1960. A chemical approach to the problem of the organizer. *Adv. in Morphogenesis* **1**, 1–50

43. ——. 1962. The inductive phenomenon as a tool for the understanding of the basic mechanism of differentiation. *J. cell. comp. Physiol.* **60**, 49–64

Chapter 27

A Problem posed by the Structure of Lampbrush Chromosomes

H. G. CALLAN.

There are many cases of related organisms which have, or may be suspected to have, grossly different C-values of DNA (i.e. quantities of DNA per haploid set of chromatids), and the Amphibia provide a striking example. Anura have C-values of less than 10 μμg, whereas Urodela have C-values ranging from about 30 (*Triturus* species) to over 100 μμg (*Necturus*).[8] It appears most unlikely that ten times the *diversity* of genetic information required to specify a frog is needed to produce a mud-puppy (or nearly thirty times that needed for man), and two alternative suggestions have been put forward to account for this anomaly.[21]

One of these proposals is based on the notion that chromosomes are multistranded, and it accounts for diverse DNA values by diversity of strandedness. This explains satisfactorily differences in DNA content per nucleus in somatic tissues whose chromosomes are known, from other evidence, to be polytene; but it is not acceptable where gametic quantities of DNA are being contrasted. First, mutation and recombination are, indisputably, phenomena shown by chromatids, not sub-units of chromatids. Second, the replication of chromosomal DNA is just as semi-conservative[19] as bacterial or 'chromonemal'[22] DNA,[3,14] and the rule which governs the segregation of chromatids is as applicable to organisms with large chromosomes as to those with small.[16]

The third line of evidence against multistrandedness comes from kinetic analyses of DNase action. Gall[7] found that when DNase breaks the lateral loops of lampbrush chromosomes, and the fibril connecting adjacent chromomeres, the rates of increase of breaks with time are such that substantially two sub-units are involved in the former, and four in the latter. Thomas[20] found that DNase cleaves the two polynucleotide chains of DNA independently and at random in an *in vitro* system. It therefore seems reasonable to equate each of the DNase-sensitive sub-units demonstrated by Gall with a single polynucleotide chain.

The alternative explanation for diverse C-values is based on the idea that the chromosome complements of related organisms have similar quantities of genetically potent DNA, but may have diverse quantities of that more or less impotent DNA known by the infelicitous term 'heterochromatin'.[1] There are situations where this explanation certainly holds, e.g. within grasshopper species where individuals may have varying numbers of supernumerary chromosomes, or none, and similarly in maize plants where B-chromosomes may be present or absent. But the explanation can hardly have wide applicability, and it evidently breaks down when lampbrush chromosomes are considered.

RNA is synthesized on the lateral loops of lampbrush chromosomes.[9] The full significance of this synthesis is not yet appreciated, but it is likely to be the synthesis of specific and diverse molecules, for the morphologies of some of the lateral loops, largely expressed in protein, are themselves specific and they are inherited.[4,6] The loops sprout from compact DNA-containing chromomeres where RNA synthesis cannot be detected, and thus one might be tempted to envisage chromomeric DNA as 'heterochromatin' diffusely spread over the entire chromosome complement. But this idea does not stand up to close scrutiny. There is a general inverse correlation between the lengths of the lateral loops and the size of the chromomeres in oocytes at different developmental stages, which implies that loop axis DNA extends at the expense of chromomeric DNA, and vice versa. This is not to say that variations in the size of a single chromomere at the base of a pair of loops have been detected, and related directly to loop length; for any such relationship is confused by amalgamation between neighbouring chromomeres when the lateral loops are small or non-existent. That there is a generally applicable rule of inverse correlation, however, is manifested in particularly striking fashion when the lampbrush chromosomes of *Triturus cristatus* and *Ambystoma mexicanum* are compared. In *T. cristatus*, when the general run of lateral loops are at their longest, the chromomeres are still resolvable with a phase microscope. But in *A. mexicanum*[5] there are many more, and longer lateral loops—the total length of lateral loops of the complement could well be an order of magnitude greater than the corresponding length in *T. cristatus*—and the chromomeres for the most part cannot be resolved. Yet the number of foci from which loops arise appears to be much the same in *Ambystoma* as in *Triturus*, and the C-values of these two urodeles are not so very dissimilar (about 30 μμg for *T. cristatus*, about 40 μμg for *A. mexicanum*); evidently the lengths of the lateral loops are not directly related to C-values.

Isawa, Allfrey, and Mirsky[11] have claimed that the lampbrush chromosome complement of *Triturus viridescens* contains four times more DNA than the 4 C quantity to be expected of meiotic chromosomes prior to the reduction divisions. If this assertion were to be substantiated—the evidence is not acceptable yet, and further assays are in progress—the lack of correlation between total loop length and C-value might be ascribed to variations between species in the amount of DNA synthesis occurring in meiotic prophase, and be otherwise unremarkable. But synthesis of DNA (other than nucleolar) on a substantial scale *during* meiosis has not been detected in any organism, including some with large lampbrush chromosomes. In short, it is probable that urodele chromosomes during the lampbrush phase of meiosis, like the meiotic chromosomes of other organisms, will prove to contain the 4 C quantity of DNA, and that chromomeric DNA and loop axis DNA are different states, inactive and active respectively in regard to RNA synthesis, of one and the same material.

Comparison of the lampbrush chromosomes of sub-species of *Triturus cristatus* makes the idea that loop-bearing chromomeres consist of 'heterochromatin' still less plausible. One of the sub-species, *T.c. karelinii*, possesses centromeres that are flanked on either side by dense axial bars containing DNA but regularly bare of lateral loops. The other sub-species lack these axial bars. In the lampbrush chromosomes of F_1 hybrids the *karelinii* centromeres retain their neighbouring axial bars and, by comparing chromomere patterns around the centromeres of the bivalents, the axial bars are seen to be *extra* features of the *karelinii* chromosomes. Inheritance of these accessory DNA-containing structures has been followed into backcross generations, and although they must make only a marginal contribution to total DNA per chromosome complement, they evidently represent 'heterochromatin'. An analogous feature distinguishes the polytene chromosomes of two sub-species of *Chironomus thummi* from one another.[12]

If neither degree of strandedness nor variation in amount of 'heterochromatin' are acceptable explanations for grossly dissimilar C-values amongst urodeles, then it may be advantageous to examine the hypothesis that potentially specific coding is repeated serially at each gene locus, to different degrees in different species. This suggestion was originally put forward[6] to account for the asymmetrical accumulation of RNP on lateral loops. The assumption was made that loop axis extends from one side of a parent chromomere, engages in the synthesis of RNP while extended, and is later retracted into the other side of the chromomere; the degree to which RNP has accumulated then reflects the length of time a given portion of loop axis has engaged in synthesis. The anticipated movement of RNP from the thin towards the thicker part of one lateral loop, the giant granular loop of chromosome XII of *Triturus cristatus* was thereafter demonstrated,[9] a demonstration made possible by the fact that this loop is exceptional in synthesizing RNA in a restricted zone at its thin end, not over its entire length as is the case with other lateral loops. These other lateral loops show an asymmetrical distribution of RNP similar to that of the giant granular loops, so movement of RNP is likely to be occurring in these also, though it has not been demonstrated. But a more consequential unanswered question is whether there is movement of RNP relative to a stationary loop axis, or a moving loop axis which transports attached RNP.

The former possibility does not necessarily demand a serial repetition of genetic information, for the entire length of loop axis might be differentially coded. But this is tantamount to admitting that diversity of genetic information is related to total lateral loop length at maximum loop extension, e.g. that *Ambystoma* has perhaps an order of magnitude more diversity of genetic information per chromosome complement than *Triturus*. Furthermore this explanation is challenged by the fact that RNP of characteristic morphology may be equally present on a short example of a given loop, or on a large one.

The latter possibility, that loop axes move, does require serial repetition of coded information, for at different times different parts of the loop axis derived from a certain chromomere are engaged in RNP synthesis, yet the characteristic morphology of that RNP remains unchanged. Now if the coded information is serially repeated, mutations could presumably occur independently in different functional units of a 'single' gene, leading to a multiplicity of related proteins assembled according to the coding of these different units. Ingram[11] has proposed just such an origin of the codes which determine the related amino-acid sequences of myoglobin and the α, β, γ, and δ chains of human haemoglobin, and Smithies, Connell, and Dixon[18] have argued along similar lines regarding the genetic determinants of the human serum haptoglobin system.

A further line of evidence supports the contention that there may be serial repetition of coding units on chromosomes, and this comes from studies of nucleoli. Beermann,[2] in a beautiful series of experiments involving *Chironomus* species hybrids, demonstrated first that the absence of a nucleolar organizer from the *Chironomus* genome is lethal, and second, that either of two fragments of nucleolar organizer (produced in the first instance by irradiation), when introduced into an otherwise nucleolus-less genome, is sufficient to sustain normal development. If either part of a divided nucleolar organizer can substitute for an intact nucleolar organizer, *both* parts must carry the information essential for the synthesis and vital functioning of nucleoli.

Ritossa and Spiegelman[17] hybridized *Drosophila* ribosomal RNA with the DNA of strains of *Drosophila* having from 1 to 4 nucleolar organizers. The amount of DNA capable of hybridizing with ribosomal RNA was found to be proportional to the number of nucleolar organizers present in the various strains, and in the case of normal *Drosophila* 0·27 per cent of the total DNA has this capacity. Working from estimates of the C-value of *Drosophila* and the molecular weight of ribosomal RNA, Ritossa and Spiegelman computed that the cistrons for ribosomal RNA must be serially repeated 100 or more times along the chromosome.

In the lampbrush chromosome complement of *Ambystoma mexicanum* the nucleolar organizer locus is deceptively variable in appearance. It may be occupied by a relatively large chromomere without lateral attachments, it may have one or several laterally attached ring-shaped nucleoli (i.e. loops), like the hundreds of free nucleoli in these oocytes, or it may exist as a double bridge spanned by opened-out loops of nucleolar material.[5] Now in this latter condition the DNA axes through the nucleoli[13, 15] bear exactly the same structural relationship to the nucleolar chomomere as do the axes of any other pair of lateral loops to their chromomere of origin. If nucleolar organizer DNA consists of a serially repeated sequence of coding units, perhaps many other sites along the chromosomes do too.

REFERENCES

1. BAKER, J. R. and CALLAN, H. G. 1950. Heterochromatin. *Nature, Lond.* **166**, 227
2. BEERMANN, W. 1960. Der Nukleolus als lebenswichtiger Bestandteil des Zellkernes. *Chromosoma* **11**, 263–96
3. CAIRNS, J. 1963. The bacterial chromosome and its manner of replication as seen by autoradiography. *J. molec. Biol.* **6**, 208–13
4. CALLAN, H. G. 1963. The nature of lampbrush chromosomes. *Int. Rev. Cytol.* **15**, 1–34
5. ——. 1966. Chromosomes and nucleoli of the axolotl (*Ambystoma mexicanum*). *J. Cell Sci.* **1**, 85–108
6. —— and LLOYD, L. 1960. Lampbrush chromosomes of crested newts *Triturus cristatus* (Laurenti). *Phil. Trans. R. Soc.* Ser. B **243**, 135–219
7. GALL, J. G. 1963. Kinetics of deoxyribonuclease action on chromosomes. *Nature, Lond.* **198**, 36–8
8. —— (personal communication)

9. —— and CALLAN, H. G. 1962. H³ uridine incorporation in lampbrush chromosomes. *Proc. nat. Acad. Sci. U.S.A.* **48**, 562–70
10. INGRAM, V. M. 1961. Gene evolution and the haemoglobins. *Nature, Lond.* **189**, 704–8
11. ISAWA, M., ALLFREY, V. C., and MIRSKY, A. E. 1963. Composition of the nucleus and chromosomes in the lampbrush stage of the newt oocyte. *Proc. nat. Acad. Sci. U.S.A.* **50**, 811–17
12. KEYL, H.-G. and STRENZKE, K. 1956. Taxonomie und Cytologie von zwei Subspecies der Art *Chironomus thummi. Z. Naturf.* **11**b, 727–35
13. KEZER, J. (personal communication)
14. MESELSON, M. and STAHL, F. W. 1958. The replication of DNA in *Escherichia coli. Proc. nat. Acad. Sci. U.S.A.* **44**, 671–82
15. MILLER, O. C. 1964. Extrachromosomal nucleolar DNA in amphibian oocytes. *J. Cell Biol.* **23**, 60A
16. PRESCOTT, D. M. and BENDER, M. A. 1963. Autoradiographic study of chromatid distribution of labeled DNA in two types of mammalian cells *in vitro. Expl. Cell Res.* **29**, 430–42
17. RITOSSA, F. M. and SPIEGELMAN, S. 1965. Localization of DNA complementary to ribosomal RNA in the nucleolus organizer region of *Drosophila melanogaster. Proc. Nat. Acad. Sci. U.S.A.* **53**, 737–45
18. SMITHIES, O., CONNELL, G. E., and DIXON, G. H. 1962. Chromosomal rearrangements and the evolution of haptoglobin genes. *Nature, Lond.* **196**, 232–6
19. TAYLOR, J. H., WOODS, P. S., and HUGHES, W. L. 1957. The organization and duplication of chromosomes as revealed by autoradiographic studies using tritium-labeled thymidine. *Proc. nat. Acad. Sci. U.S.A.* **43**, 122–8
20. THOMAS, C. A. 1956. The enzymatic degradation of desoxyribose nucleic acid. *J. Am. chem. Soc.* **78**, 1861–8
21. WHITE, M. J. D. 1954. *Animal Cytology and Evolution*, 2nd edition, p. 368. Cambridge University Press
22. WHITEHOUSE, H. L. K. 1965. *Towards an Understanding of the Mechanism of Heredity*, p. 167. London: Arnold

NOTE ADDED IN PROOF

The argument presented in this essay has been extended in a paper by the author: 'The organization of genetic units in chromosomes' published in the *Journal of Cell Science*, **2**, v–7 (1967)

Chapter 28

Mitochondria in Differentiation and Disease

S. A. SHAFIQ and M. GORYCKI

During the past twenty years striking advances in knowledge of the structure and physiology of mitochondria have been made. The pioneering work of Palade[31] and Sjöstrand[39] established the distinctive structure of the mitochondria as organelles bounded by two membranes: the outer limiting the organelle, and the inner forming regular infoldings, the cristae mitochondriales. Extensive research on different cell types of plant and animal origin has revealed variations in the conformation of the cristae,[29,22] but the basic fine structure appears to be essentially the same. Functionally, also, the mitochondria appear to be the main site of oxidative phosphorylation in cells in general, indicating that they may be truly homologous organelles. Recently, 80–100 Å 'elementary particles' on the cristal membranes, postulated to contain the complete respiratory chain, have been described.[10,6] Suspicion that these particles may be artifacts has been raised,[39] but they have now been seen in a variety of mitochondria, after a number of preparative techniques,[32,41,11,3,26] so that their occurrence as a distinctive feature of mitochondrial cristae, also seems to be well established. Recent work, however, indicates that these particles may not contain the respiratory chain,[8,44] and, perhaps, only part of the phosphorylating system.[34,20] Currently, correlated biochemical and morphological studies are in progress, in several laboratories, and a wealth of new information is becoming available, e.g. on the distribution of respiratory enzymes and other proteins in the mitochondrial membranes, on active ion translocations, and on energy-linked mechanochemical changes in the mitochondria. Comparatively little is known, however, about changes in mitochondria during differentiation, and in disease. In this paper, some morphological problems in these fields will be discussed, with special reference to observations on muscle mitochondria.

MITOCHONDRIA IN DIFFERENTIATION*

Changes in the mitochondria during differentiation of the fibrillar flight muscles of *Drosophila melanogaster* are described below. The results are essentially in agreement with a recently reported study on the mitochondria of the honey bee.[17]

Figure 28.1, A, shows a myoblast from an early pupal stage when the myoblasts are still separate cells, with only a few isolated myofilaments in their cytoplasm. The mitochondria are sparsely distributed in the cytoplasm; they are spherical in shape, and measure about $0.3\ \mu$ in diameter. Figures 28.1, B, and C are pictures of young muscle fibres formed by the fusion of the myoblasts.[36] Many thin fibrils, and an enormous number of mitochondria, are seen in them. The mitochondria now generally have the shape of elongate rods, and are usually aligned parallel to the myofibrils; in diameter they are comparable to those of the myoblasts. Figure 28.1, D is a picture of the flight muscle of the adult fly, to illustrate the great increase in the diameters of the mitochondria, and also of the myofibrils, over those in the early pupa. The increase in the size of the mitochondria is probably due to growth, as well as fusion of

---

* [*Editorial note:* The reader is referred to the three separate aspects of differentiation, defined on pp. 278–279.]

individual mitochondria. The latter process is suggested by the decrease in the ratio of mitochondria to fibrils, during the late pupal stages (cf. Figs. 28.1, C and 28.1, D), and also by the presence in these stages of many mitochondrial profiles (Figs. 28.2, A, B, C) which seem to represent fused mitochondria. During the differentiation of the flight muscle mitochondria, an increase in the number of cristae also seems to occur. This is readily seen by comparing Figs. 28.2, D and 28.2, D, the former being of mitochondria from a late pupa, and the latter, from an adult fly. Figure 28.2, F, shows the fine structure of a cristal membrane of a mitochondrion from a stage corresponding to that of Fig. 28.2, E. The size, structure, and distribution of the elementary particles seen in the figure are highly comparable to those in the muscle mitochondria of the adult flies. Pupae are being studied early in development to see whether any stages in the differentiation of the fine structure of the cristae can be detected.

The main features in the differentiation of the flight muscle mitochondria, then, are great multiplication in the early stages, followed by growth and also possibly some fusion, and a large increase in the number of cristae in late stages of development. Multiplication of mitochondria is of widespread occurrence, in embryonic development, especially at the onset of gastrulation,[14, 37] in certain experimental conditions,[45, 14] and, indeed, in dividing cells in general. Multiplication by growth and division of pre-existing mitochondria appears to be well established. This process has been observed by direct microscopy of living cells (e.g. reference 12), and the studies of Luck[23] on the mitochondria of *Neurospora*, and, more recently, of Vogel and Kemper[52] on cultured mushroom mitochondria, are striking evidence for this phenomenon. The presence of DNA in mitochondria of both animal and plant cells (e.g. references 24, 4, 28), and the considerable biosynthetic abilities of mitochondria,[18, 21] would also support this view. Also, in the early pupae of *Drosophila*, dumbbell-shaped mitochondria, which possibly represent division, are commonly seen. However, reports claiming the origin of new mitochondria from diverse membrane systems of cells continually appear in the literature,[7, 35, 4, 53] and the problem of the origin of mitochondria in various cell types has, as yet, not been satisfactorily resolved. The true homology of mitochondria can be established only when a common mode for their developmental origin has been demonstrated.

Increase in the number of the cristae in mitochondria is also a common feature, for example, in embryonic development,[5] and in the life cycles of certain protozoans.[51] Formation of new cristae in growing mitochondria probably involves the development of new membrane material, rather than simply a folding of the pre-existing membrane. In such growing mitochondria, the fine structure of the cristae, especially with reference to the elementary particles, should be of interest, and is being studied. Also, in spermatogenesis, where mitochondrial cristae undergo profound structural changes,[1] such studies should be especially rewarding.[48]

MITOCHONDRIA IN DISEASE

A large number of muscle biopsy specimens of patients with different diseases have been examined in our laboratory. Some mitochondrial alterations observed in these studies are described below.

For comparison Fig. 28.3, A illustrates the structure of mitochondria in a normal red fibre of the human vastus lateralis muscle. The mitochondria are about $0.3\ \mu$ in diameter and are generally in register

Fig. 28.1 A. Myoblast from an early pupa of *Drosophila* showing sparsely distributed mitochondria. Arrows indicate myofilaments. n, nucleus.

B. A young fibre of developing flight muscle showing numerous filamentous mitochondria; the thin myofibrils are indicated by arrows.

C. Transverse section of a fibre comparable to that in Fig. 28.1, B. The myofibrils are indicated by arrows. Mitochondria/myofibril ratio is high.

D. Transverse section of flight muscle of the adult fly. The mitochondria are larger and mitochondria/myofibril ratio is lower than in Fig. 28.1, C.

The scales on this page represent $1.0\ \mu$.

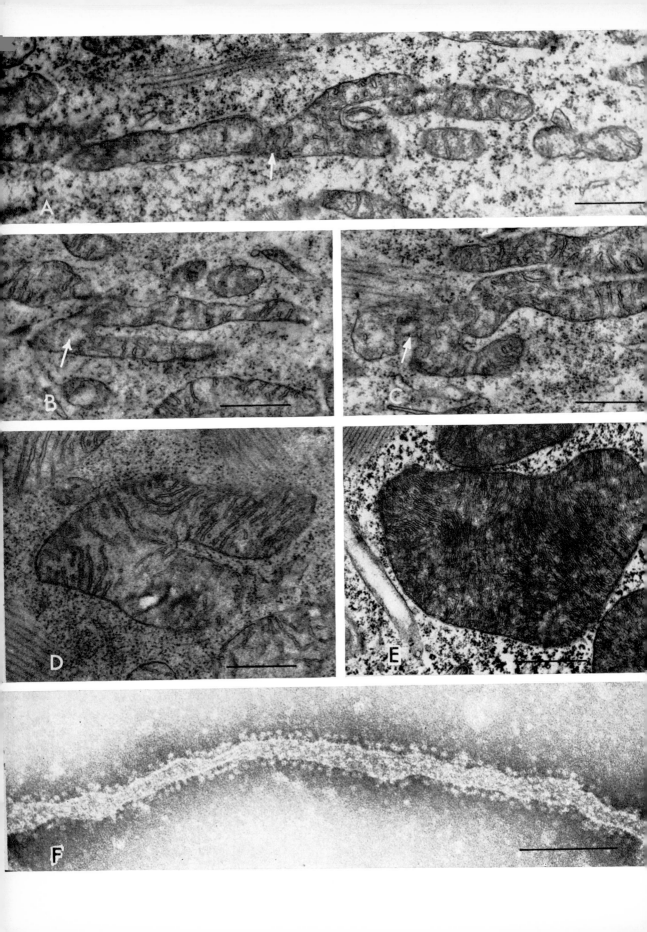

with fibre striations; some others (not seen in the picture) occur in subsarcolemmal aggregates. Figures 28.3, B–F, 28.4, A–D, illustrate the different types of mitochondrial alterations. Figure 28.3, B, shows swollen mitochondria which appears to be an unspecific feature found in many muscle diseases as also occasionally in normal muscle. It should be noted that the swelling seen here is limited to the mitochondrial matrix; this is unlike the swelling in thyrotoxic conditions[13] or in aflatoxin poisoning,[49] in which spaces inside the cristae also become much expanded. Moderate swelling of mitochondria accompanied by a great reduction in the number of the cristae is seen in Fig. 28.3, C; the muscle specimen was from a patient with neurogenic muscular atrophy. The swelling in the last two pictures should be distinguished from a true hypertrophy of the mitochondrion as seen in Fig. 28.3, D. The mitochondrion (of a patient whose other mitochondria are illustrated in Figs. 28.4, A, and 28.4, B) shows a real increase in size; its matrix is of normal density and there are a greater than normal number of cristae in it. Figure 28.3, E, illustrates another mitochondrial alteration in muscle of a patient with polymyositis. There is an increase in the density of the matrix similar to that in liver mitochondria after vitamin-E deficiency,[46] and the cristae have become indistinct. Figure 28.3, F, shows degeneration of mitochondria in an autophagic vacuole inside a muscle fibre (of a patient with childhood polymyositis); the picture is comparable to that seen in pathological kidney and liver cells.[30,2]

Two types of mitochondrial modifications, illustrated in Figs. 28.4, A–D, deserve special mention. The one seen in Figs. 28.4, A, and 28.4, B, involves deposition of paracrystalline material inside certain cristae, which in sections then commonly appear in rectangular outline. Such mitochondria usually occurred in aggregates under the sarcolemma, though some were also found in between the myofibrils; the other fibre constituents were generally normal. This mitochondrial alteration has now been found in four patients with myopathy. It seems to differ from accumulations of diverse materials in liver mitochondria[19,47,43,27] in that, in the latter, the deposits are in the matrix. Intracristal deposits are especially prominent in the mitochondria of amphibian oocytes, during yolk synthesis,[54] and the altered mitochondria described above are closely comparable to them. Figures 28.4, C, and 28.4, D, illustrate the other striking modification of mitochondria, in a patient with a myopathy of undetermined origin. The altered mitochondria appear as long threads (about 4 μ long), and are seen in transverse and oblique sections in Figs. 28.4, C, and 28.4, D. The cristae are arranged in concentric layers and at higher magnification many of them appear to be joined by dense periodicities similar to those in the mitochondria of cat retina.[33] The abnormal mitochondria occurred only under the sarcolemma, whereas those in between the myofibrils were normal; the other constituents of the fibres were also normal.

Figs. 28.2 A, B, C. Sections of young fibres of *Drosophila*. Arrows indicate regions of possible fusion of mitochondria.

D. Mitochondria from flight muscle of a late pupa. The mitochondria are larger than those in early pupae (e.g., in Figs. 28.1, B–C) but the number of cristae in the mitochondria is still rather low.

E. Mitochondrion from the flight muscle of the adult fly. The cristae are much more numerous than in pupal muscles.

F. Cristal membrane from osmotically disrupted mitochondria (of fibre corresponding to that of Fig. 28.2, D) prepared by negative staining technique (Potassium phosphotungstate, pH 6·8).

The scales on this page represent 0·5 μ except that of Fig. 28.2, F which equals 0·1 μ.

Fig. 28.3 A. Longitudinal section of a red fibre of human vastus lateralis muscle to show structure and distribution of the normal mitochondria.

B. Mitochondria with swollen matrix also from muscle specimen of a normal subject.

C. Swollen spherical mitochondria with greatly reduced number of cristae from a patient with neurogenic muscular atrophy.

D. Hypertrophic mitochondrion with increased number of cristae from the same patient as of Figs. 28.4, A–B (unidentified myopathy).

E. Mitochondria with dense matrix from a patient with polymyositis. The cristae are indistinct.

F. Autophagic vacuole in a muscle fibre from a patient with childhood dermatomyositis. The degenerate mitochondria inside the vacuole are indicated by arrows.

The scales on this page represent 1 μ. (*see over*)

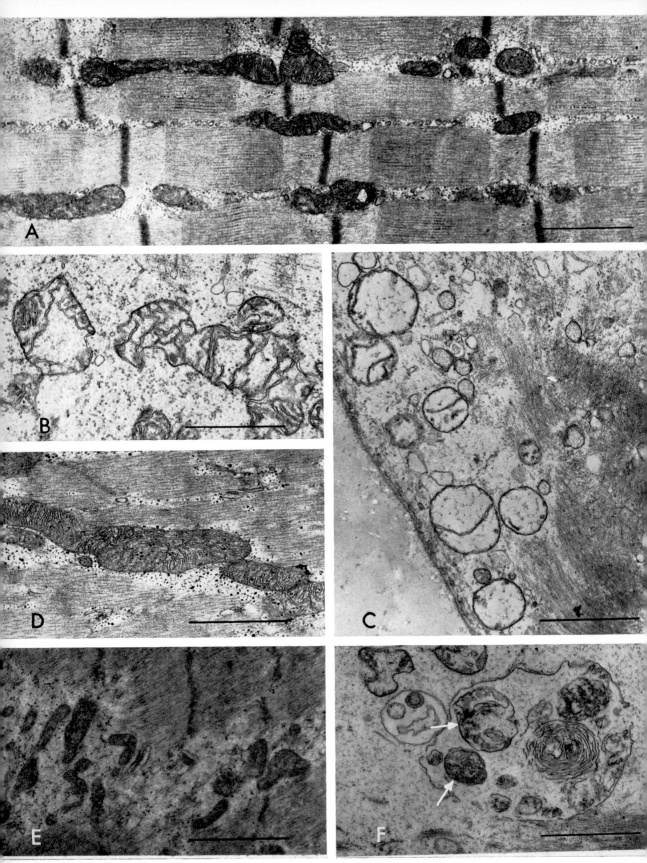

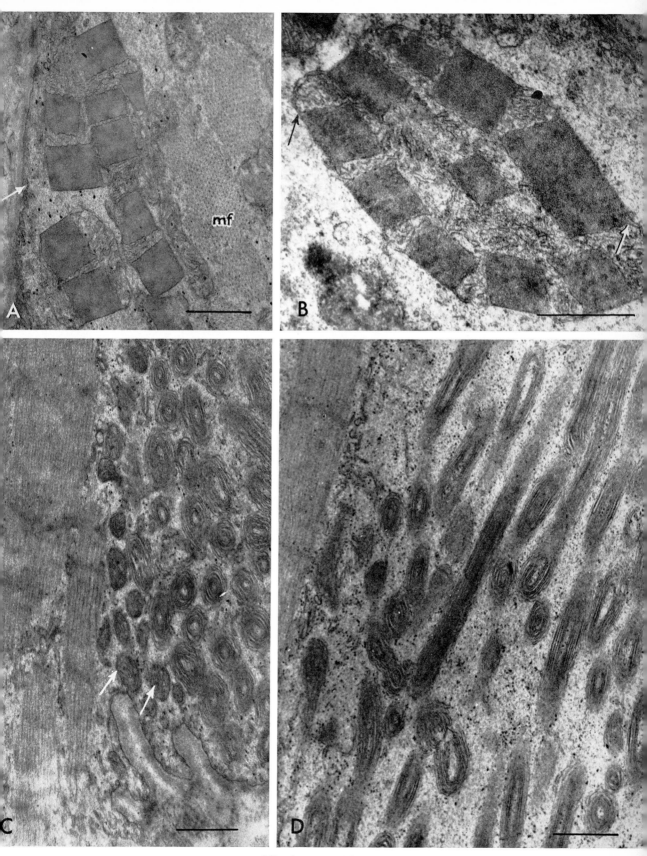

[Caption on p. 368]

The various changes in the muscle mitochondria described above are difficult to interpret. The complexities of mitochondrial reactions have been well demonstrated by recent studies on transformations *in vitro* in mitochondria,[16] and on effects of thyroxine on rat skeletal muscle *in vivo*.[15] Thyroxine in toxic doses causes swelling of mitochondria (as also described by Greenawalt and colleagues[13]), although in the physiological range, in both hypo- and hyperthyroid states, it causes an actual increase in the number of mitochondria and in their cristae.

Trump and Ericsson[50] have recently published a comprehensive review of the literature on mitochondrial changes in various pathological states. Most such changes are generally considered as non-specific degenerative reactions. In the case of skeletal muscle, the swelling of mitochondria, increase in the density of their matrix, decrease in the number of cristae, and mitochondrial degeneration in autophagic vacuoles, are probably such reactions. There is no apparent specificity with respect to the type of muscle disease. Such mitochondria changes usually occurred in fibres in which the other fibre constituents also showed degenerative changes. The two types of mitochondrial alteration illustrated in Figs. 28.4, A–D, however, do not appear to belong to this category, and may represent specific mitochondrial defects, as postulated for the cases described by Luft and colleagues,[25] and by Shy and Gonatas,[38] to which our cases show some similarities. The full morphological details, together with the clinical data on these patients, will be published elsewhere.

Finally, it should be mentioned that the pathogenesis of mitochondrial alterations is little understood. Smuckler and colleagues[42] consider the mitochondrial alterations in carbon tetrachloride poisoning to be related to altered protein synthesis. Ernster[9] is of the opinion that the mitochondrial defects in their case of hypermetabolism[25] are probably due to alterations in the organization of the enzyme patterns. At the present time (1965), however, opinions are rather speculative, and much work remains to be done.

REFERENCES

1. ANDRÉ, J. 1962. Contribution á la connaissance du chondriome. *J. Ultrastruct. Res.* Suppl. **3**, 1–185
2. ASHFORD, T. P. and PORTER, K. R. 1962. Cytoplasmic components in hepatic cell lysosomes. *J. Cell Biol.* **12**, 198–202
3. ASHHURST, D. E. 1965. Mitochondrial particles seen in sections. *J. Cell Biol.* **24**, 497–9
4. BELL, P. R. and MÜHLETHALER, K. 1964. Evidence for the presence of deoxyribonucleic acid in the organelles of the egg cells of *Pteridium aquilinum*. *J. Mol. Biol.* **8**, 853–62
5. BERG, W. E. and LONG, N. D. 1964. Regional differences of mitochondrial size in the sea urchin embryo. *Exp. Cell Res.* **33**, 422–37
6. BLAIR, P. V., ODA, T., GREEN, D. E., and FERNÁNDEZ-MORÁN, H. 1963. Studies on the electron transfer system. LIV. Isolation of the unit of electrical transfer. *Biochemistry (Wash.)* **2**, 756–64
7. BRANDT, P. W. and PAPPAS, G. D. 1959. II. The nuclear-mitochondrial relationship in *Pelomyxa carolinensis* Wilson (*Chaos chaos L.*). *J. Biophys. Biochem. Cytol.* **6**, 91–6
8. CHANCE, B., PARSONS, D. F., and WILLIAMS, G. R. 1964. Cytochrome content of mitochondria stripped of inner membrane structure. *Science* **143**, 136–9
9. ERNSTER, L. 1965. Control of cell metabolism at the mitochondrial level. *Fed. Proc.* **24**, 1222–36
10. FERNÁNDEZ-MORÁN, H. 1962. Cell-membrane ultrastructure. Low-temperature electron microscopy and x-ray diffraction studies of lipoprotein components in lamellar systems. *Circulation* **26**, 1039–65

Fig. 28.4 A. Mitochondria with deposits of paracrystalline materials in the cristae. The fibre plasma membrane (indicated by the arrow) and myofibrils (mf) have normal structure. The patient was biopsied for unidentified myopathy.

B. Higher magnification of a mitochondrion with dense deposits from the same case as of Fig. 28.4, A At points indicated by arrows the continuity of the cristae with the inner mitochondrial membrane is seen.

C. Transverse section of thread-like mitochondria with concentric arrangement of cristae from another case of unidentified myopathy. In the mitochondria indicated by arrows the cristae are comparatively normal.

D. Longitudinal and oblique sections of the thread-like mitochondria from the same patient as of Fig. 28.4, C.

The scales on this page equal 0·5 μ. (*see previous page*)

11. FERNÁNDEZ-MORAN, H., ODA, T., BLAIR, P. V., and GREEN, D. E. 1964. A macromolecular repeating unit of mitochondrial structure and function. *J. Cell Biol.* **22**, 63–100
12. FREDERIC, J. 1958. Recherches cytologiques sur le chondriome normal ou soumis à l'experimentation dans des cellules vivantes cultivées in vitro. *Arch. Biol. (Liège)* **69**, 167–350
13. GREENAWALT, J. W., FOSTER, G. V., and LEHNINGER, A. L. 1962. The observation of unusual membranous structures associated with liver mitochondria in thyrotoxic rats. In *5th International Congress for Electron Microscopy* (edited by S. S. Breese), Vol. 2, pp. 1–5. New York: Academic Press
14. GUSTAFSON, T. and LENICQUE, P. 1952. Studies on mitochondria in the developing sea urchin egg. *Exp. Cell Res.* **3**, 251–74
15. ——, TATA, J. R., LINDBERG, O., and ERNSTER, L. 1965. The relationship between the structure and activity of rat skeletal muscle mitochondria after thyroidectomy and thyroid hormone treatment. *J. Cell Biol.* **26**, 555–78
16. HACKENBROCK, C. R. and BRANDT, P. W. 1965. Reversible ultrastructural changes in mitochondria with changes in functional state. *J. Cell Biol.* **27**, 40A–41A
17. HEROLD, R. C. 1965. Development and ultrastructural changes of sarcosomes during honey bee flight muscle development. *Develop. Biol.* **12**, 269–86
18. HÜLSMANN, W. C. 1962. Fatty acid synthesis in heart sarcosomes. *Biophys. Biochem. Acta* **58**, 417–29
19. JEZEQUEL, A. M. 1959. Dégénérescence myelinique des mitochondries de foie humain dans un epithelioma du choledoque et un ictère viral. *J. Ultrastruct. Res.* **3**, 210–15
20. KAGAWA, Y. and RACKER, E. 1965. A factor conferring oligomycin-sensitivity to mitochondrial ATPase. *Fed. Proc.* **24**, 363
21. KROON, A. M. 1965. Protein synthesis in mitochondria. III. On the effects of inhibitors on the incorporation of amino acids into protein by intact mitochondria and digitonin fractions. *Biophys. Biochem. Acta* **108**, 275–84
22. LEHNINGER, A. L. 1964. *The Mitochondrion.* New York: W. A. Benjamin
23. LUCK, D. J. L. 1965. Formation of mitochondria in *Neurospora crassa. J. Cell Biol.* **24**, 461–70
24. —— and REICH, E. 1964. DNA in mitochondria of *Neurospora crassa. Proc. Natl. Acad. Sci.* **52**, 931–8
25. LUFT, R., IKKOS, D., PALMIERI, G., ERNSTER, L., and AFZELIUS, B. 1962. A case of severe hypermetabolism of nonthyroid origin with a defect in the maintenance of mitochondrial respiratory control: a correlated clinical, biochemical, and morphological study. *J. Clin. Invest.* **41**, 1776–1804
26. MOOR, H., RUSKA, C., and RUSKA, H. 1964. Elektronenmikroskopische Darstellung tierischer Zellen mit der Gefrierätztechnik. *Z. Zellforsch.* **62**, 581–601
27. MUGNAINI, E. 1964. Filamentous inclusions in the matrix of mitochondria from human livers. *J. Ultrastruct. Res.* **11**, 525–44
28. NASS, M. M. K., NASS, S., and AFZELIUS, B. A. 1965. The general occurrence of mitochondrial DNA. *Exp. Cell Res.* **37**, 516–39
29. NOVIKOFF, A. B. 1961. Mitochondria (chondriosomes). In *The Cell* (edited by J. Brachet and A. E. Mirsky), pp. 299–422. New York: Academic Press
30. —— and ESSNER, E. 1962. Cytolysomes and mitochondrial degeneration. *J. Cell Biol.* **15**, 140–6
31. PALADE, G. E. 1953. An electron microscope study of the mitochondrial structure. *J. Histochem. Cytochem.* **1**, 188–211
32. PARSONS, D. F. 1963. Mitochondrial structure: two types of subunits on negatively stained mitochondrial membranes. *Science* **140**, 985–7
33. PEASE, D. C. 1962. Demonstration of a highly ordered pattern upon a mitochondrial surface. *J. Cell Biol.* **15**, 385–9
34. RACKER, E., CHANCE, B., and PARSONS, D. F. 1964. Correlation of structure and function of submitochondrial units in oxidative phosphorylation. *Fed. Proc.* **23**, 431
35. ROBERTSON, J. D. 1959. The ultrastructure of cell membranes and their derivatives. In *The Structure and Function of Subcellular Particles*, 16th Biochemical Society Symposium (edited by E. M. Crook), pp. 3–43. Cambridge University Press
36. SHAFIQ, S. A. 1963. Electron microscopic studies on the indirect flight muscles of *Drosophila melanogaster*. II. Differentiation of myofibrils. *J. Cell Biol.* **17**, 363–73
37. SHAVER, J. R. 1956. Mitochondrial populations during development of the sea urchin. *Exp. Cell Res.* **11**, 548–59
38. SHY, G. M. and GONATAS, N. K. 1964. Human myopathy with giant abnormal mitochondria. *Science* **145**, 493–6
39. SJÖSTRAND, F. S. 1953. Electron microscopy of mitochondria and cytoplasmic double membranes. *Nature, Lond.* **171**, 30–2
40. ——, CEDERGREN, E. A., and KARLSSON, U. 1964. Myelin-like figures formed from mitochondrial material. *Nature, Lond.* **202**, 1075–8
41. SMITH, D. S. 1963. The structure of flight muscle sarcosomes in the blowfly *Calliphora erythrocephala* (Diptera), *J. Cell Biol.* **19**, 115–38

42. SMUCKLER, E. A., ISERI, O. A., and BENDITT, E. P. 1964. Studies on carbon tetrachloride intoxication. II. Depressed amino acid incorporation into mitochondrial protein and cytochrome c. *Lab. Invest.* **13**, 531–8

43. SORENSON, G. D., HEEFNER, W. A., and KIRKPATRICK, J. B. 1964. Experimental amyloidosis. II. Light and electron microscopic observations of liver. *Amer. J. Path.* **44**, 629–44

44. STASNY, J. T. and CRANE, F. L. 1964. The effect of sonic oscillation on the structure and function of beef heart mitochondria. *J. Cell Biol.* **22**, 49–62

45. STEGWEE, D., KIMMEL, E. C., DE BOER, J. A., and HENSTRA, S. 1963. Hormonal control of reversible degeneration of flight muscle in the Colorado potato beetle, *Leptinotarsa decemlineata* Say (Coleoptera). *J. Cell Biol.* **19**, 519–27

46. SVOBODA, D. J. and HIGGINSON, J. 1963. Ultrastructural hepatic changes in rats on a necrogenic diet. *Amer. J. Path.* **43**, 477–95

47. —— and MANNING, R. T. 1964. Chronic alcoholism with fatty metamorphosis of the liver. *Amer. J. Path.* **44**, 645–62

48. TAHMISIAN, T. N. 1965. The locus and structure of mitochondrial elementary particles as seen in *Helix aspersa* spermatocytes. *J. Cell Biol.* **27**, 103A

49. THERON, J. J. 1965. Acute liver injury in ducklings as a result of aflatoxin poisoning. *Lab. Invest.* **14**, 1586–1603

50. TRUMP, B. F. and ERICSSON, J. L. E. 1965. Some ultrastructural and biochemical consequences of cell injury. In *The Inflammatory Process* (edited by B. W. Zerfach, L. Grant, and R. T. McCluskey), pp. 35–120. New York: Academic Press

51. VICKERMAN, K. 1965. Polymorphism and mitochondrial activity in sleeping sickness trypanosomes. *Nature, Lond.* **208**, 762–6

52. VOGEL, F. S. and KEMPER, L. 1965. Structural and functional characteristics of deoxyribonucleic acid-rich mitochondria of the common meadow mushroom. *Agaricus campestris*. II. Extracellular cultures. *Lab. Invest.* **14**, 1868–93

53. WALLACE, P. G. and LINNANE, A. W. 1964. Oxygen-induced synthesis of yeast mitochondria. *Nature, Lond.* **201**, 1191–4

54. WARD, R. T. 1962. The origin of protein and fatty yolk in *Rana pipiens*. II. Electron microscopical and cytochemical observations of young and mature oocytes. *J. Cell Biol.* **14**, 309–41

Chapter 29

The Electron Microscopy of Experimental Degeneration in the Octopus Brain

E. G. GRAY and J. Z. YOUNG.

INTRODUCTION

There have been several experimental studies with the electron microscope on degeneration in the vertebrate central nervous system, but so far, apparently, only one[11] on an invertebrate nervous system. In vertebrates at least three types of change have been described. Agglutination and loss of synaptic vesicles, mitochondria, and surface membrane of end-bulbs in the ventral acoustic nucleus of the guinea-pig, has been reported within 24 hr of unilateral destruction of the cochlea.[5] A second type was described in the avian optic tectum[8] after unilateral removal of the eye. Here the vesicles and mitochondria soon become depleted and lost, but the surface membrane of the end-bulb remains intact for several days, and the presynaptic cytoplasm becomes filled with neurofilaments. This is correlated with the appearance of argyrophilic boutons in light microscopy. Similar electron and correlated light microscopic observations have been described for degenerating endings in the lateral geniculate nucleus of mammals.[4, 10] A third type of degeneration was described in the mammalian cerebral cortex, after undercutting.[3, 2] Here, the end-bulbs begin to degenerate within 12 hr, and form compact dense masses with no proliferation of neurofilaments. These masses are phagocytosed by astrocytes in the subsequent 2–5 days. Some observations showed that part of the post-synaptic dendrite (spine tip) adhered to the phagocytosed bulb, and was therefore interpreted as a transneuronal degenerative change (but see the Discussion for some reservations).

We have studied degeneration of the superior frontal fibres of the octopus, partly out of interest in the cytological changes that result, and partly to gain more information about the synaptic connections made by the superior frontal fibres, in the vertical lobe. The above account of central nervous degeneration is very brief, because of space restrictions. A more detailed review has been made by Gray and Guillery.[7]

METHODS

Octopuses were anaesthetized with urethane, and the superior frontal lobe either removed, or sectioned through its hilum. In this way the axons running into the vertical lobe are severed away from their terminal regions, so that vascular disturbances, and direct mechanical damage to the synapses, are avoided. Animals were allowed to survive for 13, 33, 36 hr, and 3, 5, 7, and 10 days, before fixing the vertical lobe for electron microscopy. The temperature varied from 18 to 26°C at different times of the year. Temperature control was impracticable, since most of the animals were operated upon in Naples, and transported by air to London, where they were maintained in an aquarium for the appropriate time intervals.

The vertical lobes of normal and operated animals were fixed for 3 hr, in 1 per cent osmium tetroxide buffered at pH 7·3 with veronal acetate. On immersing a lobe in fixative it was immediately cut into thin transverse sections with a razor blade. After dehydrating with ethanol, the pieces were stained with 1 per cent phosphotungstic acid and embedded in Araldite for sectioning.

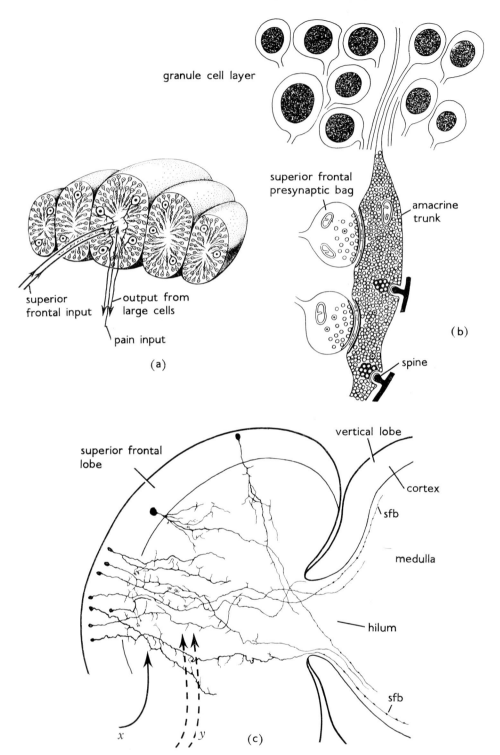

Fig. 29.1 (a) Diagrammatic representation of the five segments of the vertical lobe. The anterior part has been removed by a transverse cut. (b) Diagram of small (amacrine) cells of vertical lobe. The trunk of one cell is shown with its synapses. (c) Diagram of Golgi preparation showing the axons of the superior frontal fibres running into the outer zone of the medulla.

For light microscopy, normal and operated material was examined, using the Nissl, Golgi, and Cajal reduced silver methods. Phase-contrast light microscopy was used on Araldite sections as an indispensable aid for orientating the blocks, so that the relevant region of the vertical lobe could be sectioned for electron microscopy.

OBSERVATIONS

Normal material

The light microscopy of the vertical lobe and its connections from the superior frontal lobe, have been described in detail by Young,[16, 17] and the electron microscopy by Gray and Young.[9] The cortex of the vertical lobe (Fig. 29.1) is composed of numerous perikarya of small neurons (amacrines), which send their processes into the medulla or synaptic zone of the lobe. They do not proceed beyond the lobe. The output is through fewer large cells, with perikarya in the innermost layer of the cortex. They send processes also into the medulla for synaptic interaction, and then continue out through the deep surface of the vertical lobe to the subvertical and lateral superior frontal lobes. The cell bodies of the median superior frontal lobe send their axons as long, parallel bundles of fibres, into the outermost zones of the medulla of the vertical lobe (Fig. 29.1), where they meet, at right angles, the processes of the small cells. These superior frontal fibres show no ring-shaped or reticulated bulbs in Cajal preparations (see Discussion), but in Golgi preparations they can be seen to have numerous varicosities along their length. These varicosities are interpreted as presynaptic dilatations making *en passant* contacts with the descending processes of the small (amacrine) neurons.

Since this marginal zone of the medulla is easily located for electron microscopy, and the superior frontal fibres can be sectioned experimentally without much difficulty, this zone of the vertical lobe is thus especially suitable for studies of synaptic organization and degeneration.

Examination of this outer zone of the medulla has shown a characteristic organization,[9] interpreted as a serial synaptic arrangement. Thus, in the normal animal (Figs. 29.2, A and B), the amacrine trunks (amt) have a dense appearance, because of the close packing of vesicles (sv) contained in them. These trunks have a special cluster of dense-walled vesicles (dv), grouped against membrane 'thickenings' (t) made with apposed clear processes. These clear processes (s) are probably the spines or collaterals of the trunks of the large cells mentioned above. The amacrine trunks are themselves contacted by bags (sfb), which are much clearer in appearance since they contain vesicles and mitochondria (m) that are much less tightly packed. Membrane 'thickenings' (t) also occur between them and the apposed regions of the amacrine trunks. For reasons given in detail elsewhere,[9] these pale bags are thought to be the varicose bulbs of the fibres from the superior frontal lobe, making synaptic contact (either terminal or *en passant*) with the amacrine trunks, which in turn synapse with the clear (i.e. non-vesicular) spines. If, in fact, the pale bags are the 'terminals' of the superior frontal fibres then these would be expected to show degeneration after the operation.

Operated animals

Within 13 hr of the operation the pale bags show marked degenerative changes (Fig. 29.2, B). The synaptic vesicles (sv) are drastically reduced in number and form isolated clusters. The mitochondria lose their organization and break up into loose membranous configurations (lm). The surface membranes and thickenings (t) show discontinuities and become ill-defined. The amacrine trunks (amt) differ little from normal at this stage.

At 33 hr and 36 hr after operation, many of the pale (presumed superior frontal) bags have apparently completely vanished, and conspicuous clear zones (cz, Figs. 29.3, A and B) can be seen surrounding the dark amacrine trunks, in the places originally occupied by the pale bags. These clear zones are interpreted as extracellular compartments.

At this stage (33–36 hr) the dark processes (presumed amacrine trunks) are themselves showing degenerative changes (amt, Figs. 29.3, A and B). The vesicles (sv) and mitochondria (m) become ill-defined (Fig. 29.3, A) and apparently fuse into dense spherical granular masses (gm) and membranous whorls (mw) (Fig. 29.3, B). This is presumed to be transneuronal degeneration.

25—C.S.

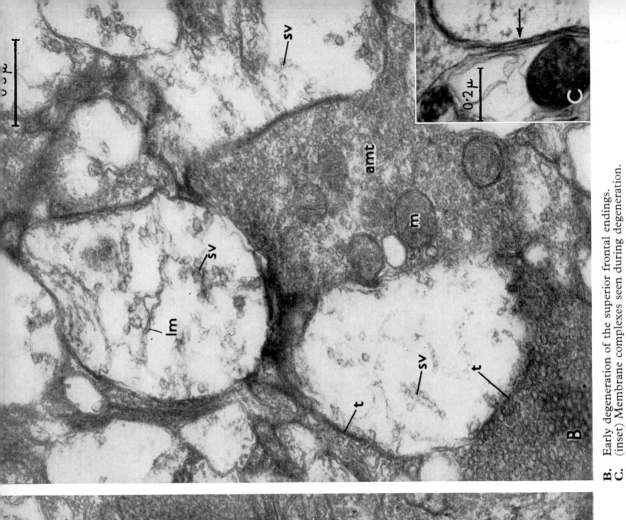

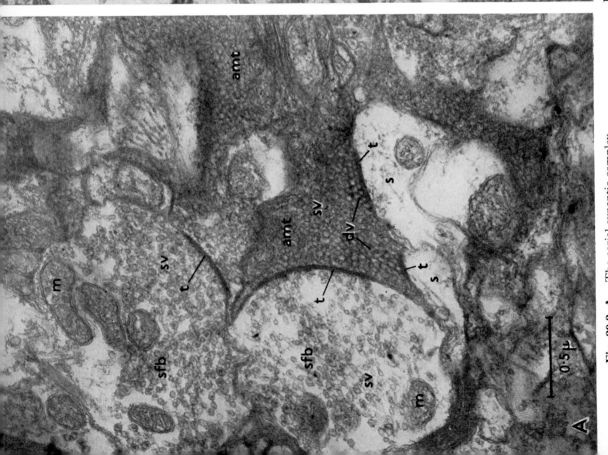

Fig. 29.2, A. The serial synaptic complex.

B. Early degeneration of the superior frontal endings.
C. (inset) Membrane complexes seen during degeneration.

At this stage, and the 3–5 day stage, numerous phagocytic cells appear in this marginal zone of the medulla, and the degenerating globular amacrine fragments can be seen lying inside the phagocytes (Fig. 29.3, C). The two ingested masses in the left in Fig. 29.3, C have been degraded into a complex membranous phase, whilst the two masses on the right resemble those described in Figs. 29.3, A and B.

At 3 days the spaces left by the pale bags are no longer conspicuous, since they have been invaded and occupied by the phagocytes (Fig. 29.4, B, ph). In this figure, two bundles of apparently intact amacrine trunks (amt) can be seen flanked (where pale bags would normally be found) with phagocytes containing ingested material. The nucleus (nuc) of the phagocyte above appears in the plane of section.

Examination of the perikaryal cortical zone at the 3–5 day stages by phase-contrast microscopy (Fig. 29.4, C) and electron microscopy (Fig. 29.4, A) showed that a proportion of the cells had developed numerous granule-containing vacuoles (vac). These are thought to be non-neural reactive cells. At 7–10 days the granules could only be rarely detected in cells of the cortex. Very occasionally, a cell body could be seen in an advanced state of degeneration. One is illustrated at very low magnification in Fig. 29.5, D. The nucleus (nuc) shows fragmentation, the surface membrane shows discontinuities, and a few dense patches represent the remains of the cytoplasmic organelles.

At the third- and fifth-day stages, the outer zone of the medulla contains, in addition to the synaptic bags in various stages of degeneration, and the phagocytes containing ingested material, granule-containing cells identical to those described in the cortex. These granulated cells seem to be a distinct cell type from the phagocytes, but occasionally a granulated cell showed ingested debris.

At 3, 5, and 7 days dense, apparently extracellular, material could be seen lying between cell processes, and extending within invaginated channels in the granule-containing cells (Fig. 29.5, B, em). Some sections (Fig. 29.5, C) showed extracellular dense material to consist of a regularly packed array of small particles similar to those described as haemocyanin molecules.[1] Whether this is the same material as that lying in the channels invaginated into the granular cells remains to be determined.

At 7 and 10 days the material ingested by the phagocytes showed almost complete degradation into masses of fine membranous whorls (Fig. 29.5, A).

DISCUSSION

The rapid degeneration (within 12–24 hr) of a proportion of the 'pale' bags in the margin of the medulla of the vertical lobe confirms that these are in fact the presynaptic bulbs of the superior frontal fibres, as postulated by Gray and Young.[9] The clumping and loss of their vesicles, and disruption of the surface membrane (compare reference 6), soon results in a complete disappearance of the bulbs, so that only structureless extracellular spaces remain to mark their position. There is no evidence that any of the material of these bags is phagocytosed by glia. This must mean that the bulbs are destroyed either by autolytic enzymes, or enzymes supplied from the extracellular spaces, and presumably secreted by neighbouring glial cells or from the vascular system. This change conforms to the first type mentioned in the Introduction. There is no persistence of the surface membrane to maintain an intact lining round the bulb (although 'isolated' membrane portions sometimes appear between two profiles, Fig. 29.2, C, arrow) and no internal proliferation of neurofilaments. This accounts for the absence of light microscopic argyrophilic boutons in the Cajal silver preparations in this zone under the same experimental conditions.[15] Sections of the mantle connective fibres in *Octopus* does result in the appearance of argyrophilia and/or increase of argyrophilia of boutons in the stellate ganglion,[13] so that in other regions of the octopus nervous system, degeneration conforms to the second type mentioned in the Introduction. One would expect to detect proliferation of neurofilaments at these degenerating endings in the stellate ganglion, with electron microscopy, and Lund[12] has in fact shown that this is so. The persistence for some days of these endings, and the proliferation of neurofilaments within them, is not well understood, but Guillery[10] regards it as the reflection of an abortive attempt to regenerate.

Subsequent to the degeneration of the superior frontal bags, a proportion of the dense vesicle-packed processes, identified as amacrine trunks,[9] show degenerative changes. These are observed within 33–36 hr after the operation. The trunks form spherical masses of dense granular and whorled membranous

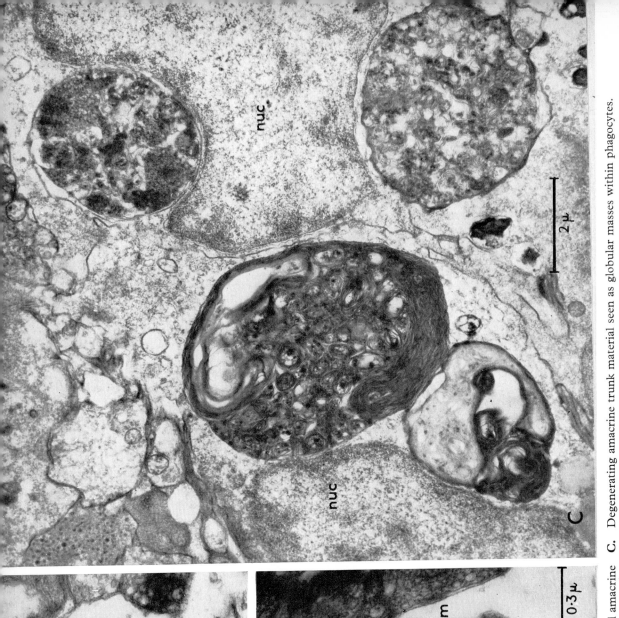

Fig. 29.3, A. (above) Early degeneration of a presumed amacrine trunk. **B.** (below) Later stage in degeneration of an amacrine trunk. **C.** Degenerating amacrine trunk material seen as globular masses within phagocytes.

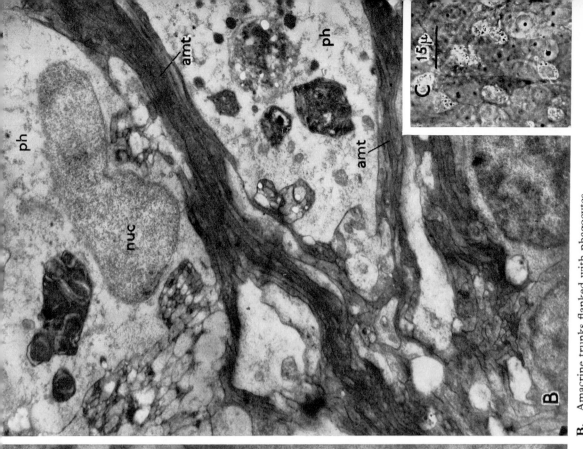

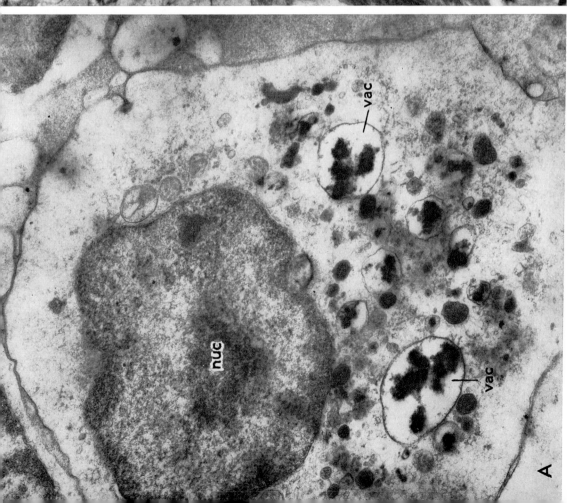

Fig. 29.4, A. Non-neural cells with granule-containing vacuoles.

B. Amacrine trunks flanked with phagocytes.
C. (inset) Phase-contrast light micrograph of plastic section showing reactive cells with the granules.

material derived from the breakdown of the synaptic vesicles and mitochondria. Within 3 days numerous glial cells are found in the degenerating zone.[13] These non-neural cells move into and occupy the extracellular spaces left by the superior frontal fibres, and they phagocytose the degenerating spherical masses, formed from the amacrine trunks. The masses are further degraded, and at the 7 and 10 day stage they can be seen as balls of fine membranes, presumably of lipo-protein, within the glia.

This form of degeneration and phagocytosis shows a close resemblance to that described in the mammalian cerebral cortex by Colonnier[2]—the third type mentioned in the Introduction. However, in Colonnier's material, the ingested presynaptic bulbs were presumably those with their axons severed by undercutting. In the octopus, however, the amacrine trunks are post-synaptic to the axon-sectioned superior frontal presynaptic bulbs, and the amacrine cells are intrinsic, having their perikarya in the cortex of the vertical lobe. It is therefore postulated that amacrine-trunk degeneration is transneuronal.

This could be explained in two different ways. The amacrine trunks degenerate perhaps because they no longer receive synaptic bombardment from the superior frontal fibres, i.e. disuse. Alternatively, the enzymes that destroyed the superior frontal fibres could be so powerful that they act indiscriminately, and also destroy the amacrine trunks. Not all the amacrine trunks degenerate, however, but it could be that the enzymes only attack those trunks that have been weakened, at the points where they have become denuded of the superior frontal endings.

The only other electron microscopic evidence for transneuronal degeneration in the central nervous system is that of Colonnier and Gray[3] and Colonnier,[2] where portions of (post-synaptic) dendritic spines were apparently adherent to the degenerating presynaptic bulbs, and were carried in with them into phagocytosing astrocytes. However, there is now some doubt as to whether this should be regarded as true transneuronal degeneration (involving only the small spinous portions of the post-synaptic neuron) since the axons of these cortical cells were also severed by undercutting, and it could hence be regarded as a retrograde effect of these cells operating concurrently with the degeneration of the contacts on the spines. The doubt is strengthened since Westrum[14] has shown that when the lateral olfactory tract in the rat is sectioned, the endings in the related cortex degenerate and are phagocytosed, just as described by Colonnier and Gray, but the spines they contact remain intact and are not nipped off and phagocytosed, the point being, in Westrum's experiments, that the region of section avoided destruction of the axons of the post-synaptic cells.*

The problem of transneuronal degeneration is interesting and complex.[7] As we have seen, it may not simply be the result of disuse. It was suggested by Guillery (quoted in reference 6; see also references 4 and 10), that transneuronal atrophy or degeneration in the mammalian lateral geniculate nucleus may result from phagocytic attack (presumably either by direct ingestion, or by enzymes, or both), stimulated initially by the degeneration of the presynaptic bulbs resulting from section of the contralateral optic nerve. The loss of dendritic arborizations of the post-synaptic cell, detectable by light microscopy, could then be accounted for in this way, and not necessarily as a 'disuse' mechanism. It was partly for this reason that Guillery studied degeneration in the lateral geniculate nucleus with the electron microscope, but, apparently, he has so far been unable to show degeneration of the post-synaptic dendrites, although presynaptic degeneration could be clearly seen.

The appearance of granule-containing 'vacuoles' in certain cells of the cortex of the octopus vertical lobe at 3 days needs further investigation. For various reasons these cells are best interpreted as non-neural cells. They bear a close resemblance to blood amoebocytes. The granules may represent a stage in the formation and liberation of destructive enzymes and the appearance round them of dense material in the extracellular clefts, and in channels running 'from' these cells, may indicate enzyme concentrations *en route* to the degenerating zones just deep to the cell bodies (compare reference 11).

Finally, we must emphasize that the observed degenerative changes are complex, and difficult to understand, and our interpretation, given above, may need modifying in the light of future work. Certainly, serial synapses can still be seen in the outer medullary zone, up to 10 days after the operation,

* However, more recent work by Colonnier (personal communication) suggests that spine tips might still be nipped off in the heocortex after a lateral geniculate lesion.

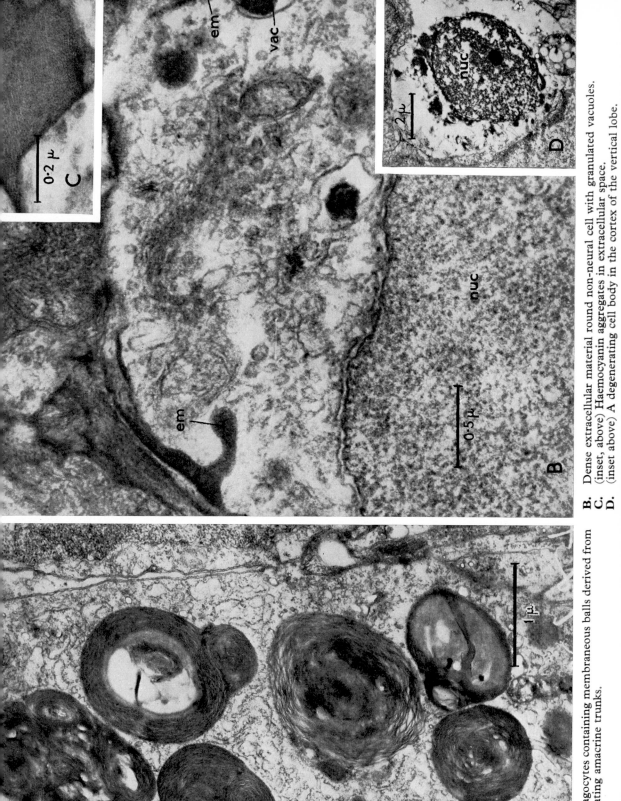

Fig. 29.5, A. Phagocytes containing membraneous balls derived from ingested degenerating amacrine trunks.

B. Dense extracellular material round non-neural cell with granulated vacuoles.

C. (inset, above) Haemocyanin aggregates in extracellular space.

D. (inset above) A degenerating cell body in the cortex of the vertical lobe.

apparently completely intact. These are probably the connections made by 'pain' fibres, ascending from the subvertical lobe, with certain of the amacrine trunks.[16,17]

Many problems remain for investigation. For example: we do not know what happens to the pale spines that are post-synaptic to the (degenerated) amacrine trunks; nor what happens to the superior frontal axons (probably they just disappear by autolysis like the bulbs they give rise to); nor have we studied what changes take place in the deeper, more central regions of the medulla of the vertical lobe—a comparatively small region when seen with the light microscope, but under the electron microscope a vast and highly complex synaptic field.

SUMMARY

The degenerative changes induced after section of the superior frontal fibres have been studied by electron microscopy. Within 12 hr the pale presynaptic bulbs degenerate presumably by enzyme destruction. These are thought to be the 'terminals' of the superior frontal fibres. Within 3 days some of the amacrine cell trunks also degenerate and are rapidly phagocytosed by reactive cells. This form of degeneration is postulated to be transneuronal. Retrograde changes in the cell bodies of these amacrine cells have not been detected. Certain presumed non-neural cells develop granulated vacuoles at 3 days.

ACKNOWLEDGEMENTS

We are indebted to Miss P. Stephens and Miss T. Charlton for technical assistance. Mr. S. Waterman made the photographic prints and Mrs. J. Astafiev drew the text figure. The work has received financial support from the Nuffield Foundation, and the United States Office of Aerospace Research, Grant No. EOAR 65-72.

REFERENCES

1. BARBER, V. C. and GRAZIADEI, P. 1965. The fine structure of cephalopod blood vessels. 1. Some smaller peripheral vessels. *Z. Zellforsch.* **66**, 765–81
2. COLONNIER, M. 1964. Experimental degeneration in the cerebral cortex. *J. Anat.* **98**, 47–53
3. —— and GRAY, E. G. 1962. Degeneration in the cerebral cortex. In *Electron Microscopy* (edited by S. S. Breese), Vol. 2, p. U.3. New York: Academic Press
4. —— and GUILLERY, R. W. 1964. Synaptic organization in the lateral geniculate nucleus of the monkey. *Z. Zellforsch.* **62**, 333–55
5. DE ROBERTIS, E. 1956. Submicroscopic changes of the synapse after nerve section in the acoustic ganglion of the guinea pig. An electron microscope study. *J. Biophys. Biochem. Cytol.* **2**, 503–12
6. GRAY, E. G. 1964. The fine structure of normal and degenerating synapses of the central nervous system. *Arch. Biol.* **75**, 285–99
7. —— and GUILLERY, R. W. 1965. Synaptic morphology in the normal and degenerating nervous system. *Int. Rev. Cytol.* **19**, 111–82
8. —— and HAMLYN, L. H. 1962. Electron microscopy of experimental degeneration in the avian optic tectum. *J. Anat.* **96**, 309–16
9. —— and YOUNG, J. Z. 1964. Electron microscopy of synaptic structure of *Octopus* brain. *J. Cell Biol.* **21**, 87–103
10. GUILLERY, R. W. 1965. Some electron microscopical observations of degenerative changes in central nervous synapses. In *Progress in Brain Research*, Vol. 14, pp. 57–76. Amsterdam: Elsevier
11. HESS, A. 1960. The fine structure of degenerating nerve fibres, their sheaths, and their terminations in the central nerve cord of the cockroach (*Periplaneta americana*). *J. Biophys. Biochem. Cytol.* **7**, 333–44
12. LUND, R. W. 1966. (in preparation)
13. SERENI, E. and YOUNG, J. Z. 1932. Nervous degeneration and regeneration in cephalopods. *Pubbl. Staz. Zool. Napoli.* **12**, 173–208
14. WESTRUM, L. E. 1966. Electron microscopy of degeneration in the plepyriform cortex. Proceedings of Anatomical Society. *J. Anat.* **100**, 683–85
15. YOUNG, J. Z. 1960. (unpublished observations)
16. ——. 1963. Some essentials of neural memory systems. Paired centres that regulate and address the signals of the results of action. *Nature, Lond.* **198**, 626–30
17. ——. 1965. The organization of a memory system. *Proc. Roy. Soc.* B. **163**, 285–320

Chapter 30

Initial Post-mortem Changes in Mitochondria and Endoplasmic Reticulum*

S. K. MALHOTRA

INTRODUCTION

In Part 1 of this book, the possibility has been considered that the mitochondria, *in vivo*, have a membranous system which shows a quintuple-layered pattern in electron micrographs, and resembles in structure the tight-junctions between two adjacent plasma membranes, or a lamella of myelin sheath. These observations are based on a study of various tissues by freeze-substitution,[10,11] and freeze drying.[25,32] The theoretical superiority of these methods over direct chemical fixation, for preservation of cells and their spatial relationships, is, to some extent exhibited in the appearance of the radial repeat period of the myelin sheath, and the entirety of the unit membrane structure of the 'disks' in photo-receptors, which are more life-like in frozen tissue than in the preparations chemically fixed by OsO_4 (see pp. 11–20 for details).

It was noticed, during an investigation of central nervous tissue in which freezing (in freeze-substitution) had been delayed purposely for several minutes (eight) after decapitation of the animal, that the membranous system of many mitochondrial cristae appeared in micrographs as if the tissue had been fixed, directly, by OsO_4. This appearance is in contrast to the five-layered pattern consistently seen in tissues frozen within $\frac{1}{2}$ min after decapitation. Similar observations were made later on pancreatic exocrine cells. These *post-mortem* alterations, demonstrable by electron microscopy within a few minutes after sacrifice of the animal, indicate possible physico-chemical changes that a tissue may undergo after death. For example, electrical impedance measurements on cerebral cortex indicate that the resistance of the tissue greatly increases within 8 min after decapitation.[31,33] This may indicate that the electrolytes, that is, Na^+ and Cl^-, undergo redistribution, and become inaccessible to the measuring current. The movement of electrolytes will be accompanied by movement of water, to maintain osmotic equilibrium. Such movements are likely to induce swelling at some locations, and shrinkage at others. If deprivation of oxygen, effected by decapitation, can result, within a few minutes, in gross alterations in the tissue, it is likely that application of chemical fixatives may produce similar alterations, by setting up diffusion currents, which may affect the *in vivo* distribution, and the concentrations of chemical substances in cells.

The *post-mortem* changes produced in the pancreatic exocrine cells of the mouse, during the first 2 hr after decapitation, have been studied. Endoplasmic reticulum, ribosomes, and mitochondria, showed marked changes from their appearance in comparable preparations of normal cells. The initial observations on mitochondria were made on central nervous tissue of the mouse, and are briefly mentioned in the text. The investigation is based on a study of tissues prepared by freeze-substitution technique, which is elaborated in Part 1 of this book. In order to study possible *post-mortem* alterations, pieces of the pancreas were frozen, at varying periods, up to 2 hr after the animal was decapitated, instead of being frozen immediately after decapitation. Tissue of the latter type is, henceforth, referred to as 'fresh frozen tissue', and is considered to have undergone the least changes from its normal *in vivo* structure. Central

* Supported by grants from the National Science Foundation (GB2055).

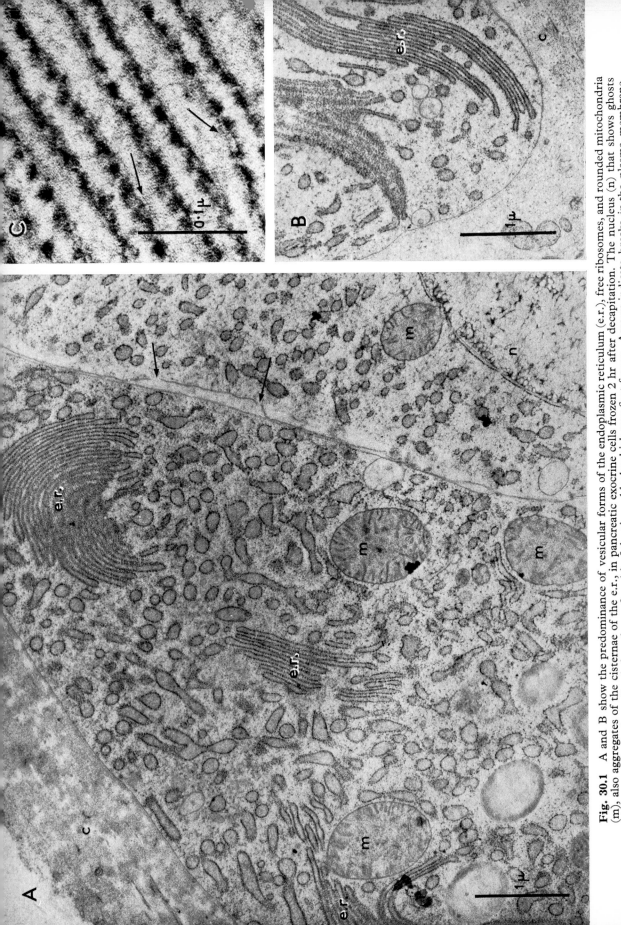

Fig. 30.1 A and B show the predominance of vesicular forms of the endoplasmic reticulum (e.r.), free ribosomes, and rounded mitochondria (m), also aggregates of the cisternae of the e.r., in pancreatic exocrine cells frozen 2 hr after decapitation. The nucleus (n) that shows ghosts of ice-crystals is about 7 μ below the surface (c) of the tissue block which was first frozen. Arrows indicate breaks in the plasma membrane, probably due to the preparative technique. c, connective tissue capsule.

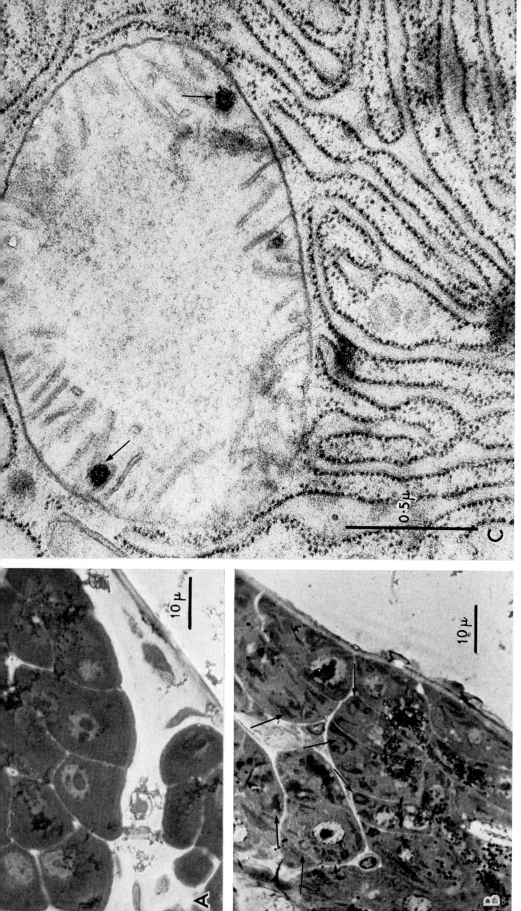

Fig. 30.2 A and B. A comparison of the organization of exocrine cells of the pancreas frozen immediately after decapitation (A), and frozen 2 hr after decapitation (B). The most conspicuous change at the light microscope level is seen in the e.r., which is relatively uniformly dispersed in the cytoplasm of normal freshly frozen cells, but breaks up into aggregates of varying shapes and sizes (arrows), as a result of *post-mortem* alterations.

C, shows a swollen mitochondrion with short cristae and low density matrix, from an exocrine cell of the pancreas frozen ½ hr after decapitation. Arrows indicate dense granular material, which are probably calcium deposits. Note the rather normal organization of the cisternae of the e.r. [Reproduced with permission of the Editors of the *Journal of Ultrastructure Research*.]

nervous tissue was frozen while still intact in the skull, either immediately after decapitation, or 8 min after decapitation. Details of the procedure are given elsewhere.[33] The tissue, after freeze substitution, was embedded in Maraglas, or Araldite, and ultra-thin sections were stained with the usual lead or uranium salts.

A part of the results included in this paper has already been reported elsewhere.[10]

RESULTS

Changes in endoplasmic reticulum

It has been abundantly emphasized (see Part 1, pp. 11–20) that the electron microscopic structure of various intracellular organelles, seen in fresh frozen tissues, is similar to that produced by fixation in OsO_4, apart from the structure of the mitochondrial membrane system. When the structure of pancreatic exocrine tissue, frozen 2 hr after decapitation, is compared with the structure of the normal tissue, it becomes obvious that the cells have undergone changes. The characteristic appearance of parallel arrays of cisternae of the endoplasmic reticulum, in the basal region of the cell, is generally absent. Instead, vesicles (mostly $\sim 0.1\ \mu$ in diameter), and tubules, with ribosomes scattered on their surfaces, are predominant throughout the cytoplasm (Fig. 30.1, A). Aggregates of cisternae of the endoplasmic reticulum, of various shapes and sizes, are also seen dispersed in the cytoplasm (Fig. 30.1, A, B). Some of these aggregates are large enough to be resolved, in approximately $1\ \mu$ thick sections, stained by methylene blue, and examined by light microscopy (Fig. 30.2, B). In contrast, similar preparations of normal pancreatic tissue show homogeneously stained ground cytoplasm without revealing the presence of uniformly dispersed underlying arrays of characteristically disposed cisternae of the endoplasmic reticulum (Fig. 30.2, A). It therefore seems that the uniformly dispersed cisternae of the endoplasmic reticulum tend to clump into new aggregations, and eventually break up into vesicles, presumably by a process of 'pinching off'. These vesicles are reminiscent of the microsomal fraction which represents the endoplasmic reticulum in situ.[16,17] The membranes of the vesicles, tubules, and cisternae of the endoplasmic reticulum show typical unit membrane structure (as shown in Fig. 30.2, C). The ground cytoplasm also shows an abundance of ribosomes, which are freely dispersed, and not associated with the membranous elements of the endoplasmic reticulum (Fig. 30.1, A). This is in contrast to the rather sparse distribution of free ribosomes in the normal cells. The increase in free ribosomes is probably achieved by detachment of ribosomes from the cisternae of the endoplasmic reticulum, as the latter breaks up into vesicles.

The normal pattern of the endoplasmic reticulum is retained in most cells, in the pancreas frozen within $\frac{1}{2}$ hr after decapitation (Fig. 30.2, C), and as such is largely indistinguishable from the fresh frozen tissue. Only slight displacement in the parallel distribution of the cisternae is noticeable (Fig. 30.2, C).

Changes in mitochondrial membranes

In fresh frozen tissue, the mitochondrial membrane system consistently shows fusion of apposed unit membranes, resulting in the formation of a (~ 130 Å) five-layered pattern (Fig. 30.3, B). The five-layered pattern represents two fused unit membranes; this is indicated by the separation of the two membranes at the middle (dense) line, seen along a part of the length of a crista (Fig. 30.3, C, D) in a very small number of mitochondria, in tissues frozen immediately after decapitation. The five-layered pattern is shown both in the cristae, and in the limiting membranes of the mitochondria. The possibility that the appearance of the mitochondrial membrane system seen in OsO_4 fixed tissue may represent a swollen state of the in vivo situation, was first suggested by the results of an investigation of the cerebellar cortices of mice, which had been decapitated 8 min before freezing. In micrographs of these cortices, many more mitochondria showed separation of the five-layered pattern into two distinct unit membranes, with an intervening electron-transparent gap (henceforth referred to as the gap) in the cristae than in the fresh frozen tissue (Fig. 30.3, D).

A comparison of the mitochondria in fresh frozen pancreatic exocrine tissue, with mitochondria in the same tissue, frozen 2 hr after sacrifice of the animal, shows that they undergo alterations. The mitochondria round up and become more spherical (Fig. 30.1, A) than their normal cylindrical form. This

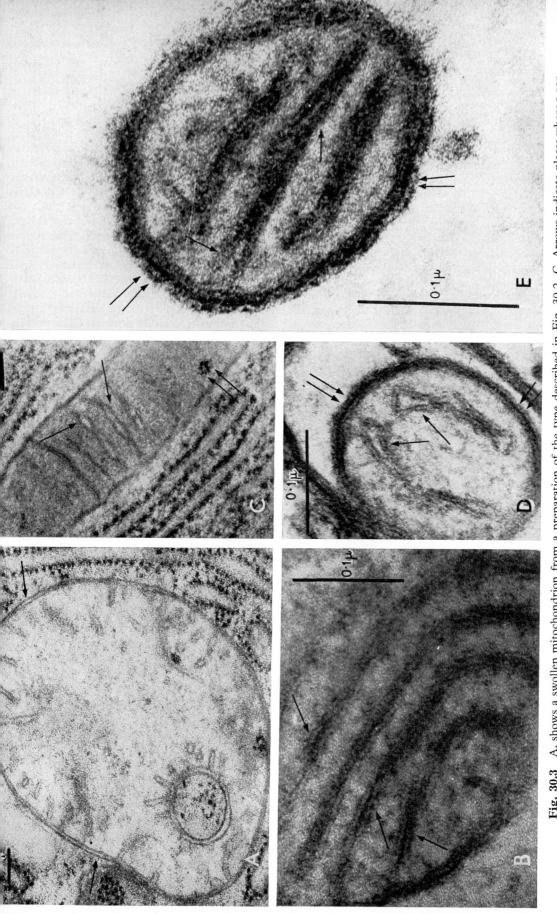

Fig. 30.3 A, shows a swollen mitochondrion from a preparation of the type described in Fig. 30.2, C. Arrows indicate places where a gap is shown between the two limiting membranes of the mitochondrion. The lower half of the mitochondrion shows the unusual appearance of what seems to be a cross section of ground cytoplasm which has pushed in the limiting membranes of the mitochondrion in a finger-like projection. B, shows the five-layered pattern (arrows) in the cristae and at the surface of a mitochondrion commonly seen in fresh frozen tissues. C, D, and E. Single arrows indicate where a gap is present along a part of the length of the mitochondrial cristae. Double arrows indicate the five-layered pattern at the surface of the mitochondria. C, from pancreas frozen $\frac{1}{2}$ hr after decapitation. D, from cerebellar cortex frozen 8 min after decapitation. E, from medulla oblongata frozen 8 min after decapitation.

transformation of shape has been confirmed by examination of a few of the mitochondria in serial sections. Such a change of shape indicates that the mitochondria have undergone swelling.[28] The cristae appear to be shorter in length in this material, than in mitochondria in fresh frozen pancreas. Mitochondria in pancreatic exocrine tissue frozen 2 hr after decapitation show cristae similar in structure to those seen in tissue fixed, without delay, in OsO_4. This is the commonly known appearance of cristae, and it is predominant in most of the mitochondria in such preparations of pancreas. The cristae of the mitochondria consist of two symmetrical unit membranes, with an electron-transparent gap between them (Fig. 30.4, A). The width of the gap often varies, from approximately 30 to 100 Å in most mitochondria. Each of the two unit membranes measures approximately 60 Å in width, which makes the thickness of a crista vary from about 150 to 200 Å, or more. In some mitochondria, the cristae have swollen so much that they have been transformed into tubular or vesicular structures, and appear circular in cross sections. The five-layered pattern of the membranes is very rarely seen in the cristae of mitochondria in pancreatic tissue frozen 2 hr after decapitation (Fig. 30.4, C), as compared with its consistent appearance in fresh frozen material. The limiting membranes of the mitochondria in fresh frozen pancreas tissue, on the other hand, do not, generally, show the pattern seen in most of the cristae, that is, two unit membranes with an intervening gap, but the five-layered pattern *is* discernible in the limiting membranes of many mitochondria in exocrine cells of the pancreas frozen 2 hr after decapitation (Fig. 30.4, B). In nervous and non-nervous cells of the cerebellar cortex and medulla oblongata, frozen 8 min after decapitation, the five-layered pattern is predominant in the limiting membranes of the mitochondria, though their cristae may show the common appearance seen after routine fixation in OsO_4 (Fig. 30.3, D, E).

In pancreatic tissue frozen 10 min to $\frac{1}{2}$ hr after decapitation, many mitochondria do not notably differ from mitochondria in fresh frozen material. However, many mitochondria are seen in the former tissue, which are similar to those commonly encountered in tissue frozen 2 hr subsequent to decapitation (Fig. 30.2, C; 30.3, A). Cristae, showing features of OsO_4-fixation, and of freeze-substitution, are seen in mitochondria in tissues in which freezing has been delayed after the death of the animal, as well as in the fresh frozen material. But cristae with such dual features are few (less than 5 per cent counted in micrographs) in fresh frozen material. The freeze-substitution pattern of the cristae, in the pancreatic tissue, is more or less completely transformed to the OsO_4-fixation pattern as a result of *post-mortem* changes, within 2 hr after sacrifice of the animal (Fig. 30.4, A).

The matrix of many mitochondria in the pancreatic exocrine cells frozen $\frac{1}{2}$ hr to 2 hr after decapitation appears more electron-transparent than in the mitochondria in fresh frozen material. These observations, correlated with change of shape of the mitochondria and shortening of their cristae, indicate that the mitochondria have undergone *post-mortem* swelling. Some of the mitochondria, particularly in tissue frozen 2 hr after decapitation, show patches of electron-dense material, which is aggregated into small granular formations, approximately 500 to 1000 Å maximum diameter (Fig. 30.2, C; 30.3, C). They appear to be formed of minute granules. Judging by their density, and general distribution, these granules are similar to the electron-dense deposits which have been identified as calcium deposits in rat liver mitochondria.[4, 21] * Mitochondria within a cell show great variations in degree of swelling, and it is not rare to find mitochondria in tissue frozen up to $\frac{1}{2}$ hr after decapitation, which are indistinguishable from those in fresh frozen material.

The Golgi apparatus is the only other cellular system that seems to undergo *post-mortem* changes in the pancreatic tissue investigated in this study. The Golgi apparatus shows a swollen appearance in tissue frozen 2 hr after decapitation.

DISCUSSION

The fresh frozen tissues and the tissues in which freezing was delayed after decapitation, were both subjected to the same preparative procedure, of freeze-substitution and subsequent embedding. The

* Trump and others[30] have described similar dense aggregations in liver mitochondria during the development of necrosis *in vitro* which they consider to be not connected with calcium accumulations. The authors suggest that the dense aggregations may be characteristic of mitochondria in dead or dying cells.

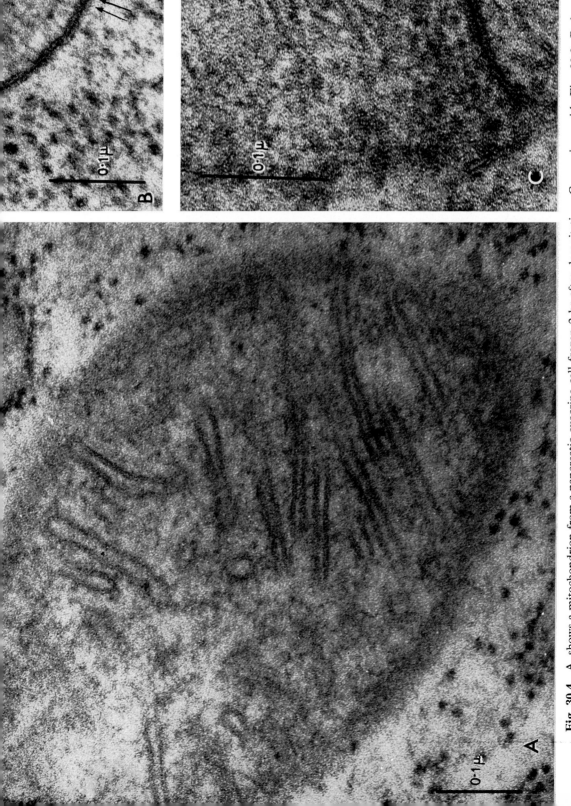

Fig. 30.4 A, shows a mitochondrion from a pancreatic exocrine cell frozen 2 hr after decapitation. Comparison with Fig. 30.3, B shows that the cristae have a distinct electron transparent gap (a characteristic of OsO$_4$ fixation), that is not demonstrated in fresh frozen tissues (Fig. 30.3, B). B, shows the persistence of the five-layered pattern (double arrows) at the surface of a mitochondrion from a preparation similar to the one described in A. Note the presence of a narrow gap in the only crista in this micrograph and also its symmetrical unit membrane structure (arrow) [Reproduced with permission of the Editors of the *Journal of Ultrastructure Research*]. C, shows the persistence of the five-layered pattern (double arrows) in a crista from a mitochondrion from a preparation similar to the one described in A.

differences in the organization of the endoplasmic reticulum and mitochondria shown in the electron micrographs can therefore be assumed to indicate differences in the state of the tissue at the time it was frozen. This is the most direct explanation of the differences in the micrographs. The cisternae of the endoplasmic reticulum are known to be very susceptible to breakdown into vesicles, when the cell membranes are damaged during excision of the tissue, or disrupted by shearing during tissue homogenization.[16] This is a possible reversal of the suggested process of formation of the cisternae of the endoplasmic reticulum, by coalescence of vesicles, described in the cnidoblasts of *Hydra*.[26] The retention of the cisternae in a more or less normal organization for at least $\frac{1}{2}$ hr after the death of the animal, is an indication of the relative stability of the organization of the cisternae of the endoplasmic reticulum, as compared to the mitochondrial membranous system.

The fusion (close-apposition) of the two unit-membranes of mitochondria has been observed as a common feature of fresh frozen tissues, namely nervous and non-nervous cells of the cerebral and cerebellar cortex[11] and medulla oblongata of the mouse,[13] dorsal roots[12] and retina (unpublished observations) of the rabbit, and hepatic cells and exocrine cells of the pancreas of the mouse.[11] The same technique of freezing, subsequent substitution, and embedding shows an electron-transparent gap between apposed membranes in many mitochondria, in tissues in which freezing has been delayed subsequent to sacrifice of the animal. This appearance of the mitochondria, is shown by direct fixation in OsO_4, and has been considered generally as the *in vivo* structure of mitochondria. It is unlikely that the mitochondrial membrane system retains a life-like organization in the pancreases of mice for 2 hr before freezing, while tissue frozen immediately (within $\frac{1}{2}$ min)[*] by the same technique, after decapitation, shows an artificial appearance. The evidence presented in Part 1 (p. 11) of this book, on the preservation of the radial repeat period of the myelin sheath, and the completeness of the disk membranes in the outer segments of the retinal photoreceptors, indicates that the technique of freeze-substitution applied in this investigation, produces a more representative, and life-like distribution of water, than routine chemical fixation. Indirect evidence in favour of this conclusion is also available from a study of extracellular space in central nervous tissue.[33] Electron micrographs of fresh frozen tissue show large areas of extracellular space (about 24 per cent of the tissue volume) in the molecular layer, particularly dispersed amongst groups of axons.[†] If, however, the tissue is frozen 8 min after decapitation, these large areas of extracellular space disappear and the plasma membranes of adjacent axons form tight-junctions. The possibility should, however, be borne in mind, that the electron microscopic appearance of the mitochondrial membrane system in frozen tissues is somewhat shrunken from the original state at the time of freezing. In this context, it may be recalled that the fusion of apposed membranes, seen in mitochondria in fresh frozen tissues, is not a feature due only to the preparative procedure of freeze-substitution. A similar five-layered pattern of the mitochondrial membrane system is also seen by freeze-drying (references 25, 32, and Part 1), and by fixation in permanganate,[1, 22, 24, 25] or in formaldehyde (with no post-fixation in OsO_4).[10] It is known that if fixation in formaldehyde or glutaraldehyde is followed by fixation in OsO_4, the mitochondrial membrane system appears as if the tissue had been fixed directly by OsO_4.

The mitochondrial membrane system seems to be highly susceptible to alterations, particularly in central nervous tissue. Webster and Ames[35] found that isolated retinae of rabbit deprived of oxygen and glucose, for only 3 min, showed swelling of mitochondria, as evidenced by decrease in electron density of matrix, and increase in the width of the gap between the two surface membranes, and in the cristae. (These changes are reversible, see also reference 30.) A similar swelling of mitochondria, as a result of

---

[*] There is evidence from impedance measurements on the cerebellar cortex that no considerable change in distribution of electrolytes and water takes place during $\frac{1}{2}$ min of oxygen deprivation caused by circulatory arrest (see reference 33). One would therefore expect the organization of the tissue at the time of freezing to be that of the living material.

[†] These observations support the physiological data that there is a relatively large extracellular space in the C.N.S. (see reference 31), whereas routine chemical fixation shows a small (about 5 per cent of the tissue volume) extracellular space in electron micrographs. The electrophysiological evidence is, however, still debated, as the glia may function as 'extracellular space'.

anoxia, has been recorded in liver cells.[14] Much more pronounced swelling of mitochondria, indicated by their enlargement, disappearance of the cristae, and disruption of their membranes, has been shown in cardiac muscle, liver, and the superior cervical ganglion of young rats subjected to an environment of decreased oxygen.[27] Since the presence of a 50 to 100 Å wide gap between the two membranes has been considered an *in vivo* feature of mitochondria, its appearance in micrographs in the initial stages of degeneration of cells has not often been considered a change from the normal structure of mitochondria. Mitochondria have therefore been described as resistant to *post-mortem* alterations, for relatively prolonged periods.[6,30] Swelling of mitochondria is also caused as an indirect effect of poisoning with carbon tetrachloride.[20] It has been suggested[20] that administration of carbon tetrachloride results in increase in permeability of plasma membranes to calcium, which enters the cell and is bound to the mitochondria. The initial accumulation of calcium in mitochondria does not impair mitochondrial respiration, but its continued accumulation results in failure of their respiration, and ultimate disintegration.[21] Trump and colleagues[30] studied the possible correlation of function with alterations of mitochondrial structure during necrosis in liver, *in vitro*. They investigated succinic dehydrogenase, succinoxidase, and glutamic dehydrogenase activity, and found little change in these enzyme systems during the first 2 hr, even though the mitochondria exhibited swelling.

The present investigation indicates that the appearance and enlargement of the gap between the two unit-membranes of the cristae is the first of the *post-mortem* alterations that mitochondria undergo, and is detectable by electron microscopy. Subsequent alterations lead to the swelling of mitochondria, which is indistinguishable from swelling *in vitro* (see e.g. references 7, 9). This swelling results in rounding up of the mitochondria, which is accompanied by reduction in length and, probably, in number of the cristae, and a loss of electron density of the matrix. These alterations make their appearance gradually after death. There seems to be great variation in the degree of swelling shown by the mitochondria of a cell. The gap between the two limiting membranes may not be observed although it is shown in the cristae, despite the known continuity between the gap in the limiting membranes and the gap in the cristae. The gap is often not seen in the limiting membranes of the mitochondria of pancreatic tissue, even 2 hr after death. There are no known structural differences between the membranes of the cristae, and the limiting membranes, that are routinely demonstrable in sections by electron microscopy. But an explanation for the difference in behaviour may perhaps be found to be related to differential concentrations of enzymes in the two locations. The differential physiological properties of the two membranes are suggested by the results on retinae of rabbit, isolated in a physiological solution, and supplied with oxygen.[2] Such retinae respond to light,[2] and electron micrographs prepared from them by freeze-substitution technique show the existence of a gap in most of the cristae, whereas a five-layered pattern is demonstrated in the limiting membranes of the mitochondria (unpublished observations).

Mitochondria of liver and kidney are known to swell reversibly to more than twice their original volume *in vitro*,[7] and to more than four times normal, before they become leaky to solutes.[28] There is no clear evidence, provided by the present investigation, that any discontinuities in either the inner or the outer membrane are caused during the swelling of mitochondria, at least during the first 2 hr of post-mortem alterations in the pancreas. Watson and Siekevitz[34] did not observe any breaks in the membranes of isolated liver mitochondria swollen to five times their original diameter. But Trump and colleagues[30] noticed interruptions in the continuity of the outer membrane of mitochondria 1 hr after necrosis began in liver. However, for swelling to be reversible, the organizational integrity of the mitochondrial membranous system must be retained. It therefore seems that a mechanism for increase in membrane during swelling must reside in the mitochondrion itself. If the cristae unfold during swelling, they presumably make up for the increase in the inner limiting membrane. It is, so far, not clear how the outer limiting membrane stretches so much (45 per cent increase in diameter for doubling the volume—Lehninger[7]), particularly *in vitro*.* However, evidence available from measurements of electric resistance, and electric

---

* It is conceivable that a direct association between mitochondria and membranes of the endoplasmic reticulum in intact cells (for which there is some evidence, insufficient to be generally accepted, see references 5, 22) could provide membranes for incorporation into mitochondria, without the need to synthesize new ones. There is, however, no evidence available to support such an interpretation.

26—c.s.

capacity[19] and charge density,[29] rules out any considerable stretching of membranes, during swelling of mitochondria. If the cristae unfold during swelling, perhaps a part of the membrane material is also transferred to the outer limiting membrane. This transference of the membrane material could be effected easily, especially if the outer limiting membrane is in contact with the inner membrane, as is consistently seen in frozen tissues. Some degree of increase in a membrane made up of a bimolecular leaflet of lipid (covered with thin layer of protein or non-lipid material) could be obtained by the stretching of hydrocarbon chains of phospholipid molecules (if they are folded). Such a stretching of hydrocarbon chains could change the membrane thickness by about 20 per cent, which is not easily detectable by electron microscopy. It has been suggested[24] that quite a considerable increase in surface area of a membrane could be achieved by dispersal and re-orientation of lipid molecules, to form a bimolecular leaflet from globular micelles, without any marked change in the thickness. Such a transformation, from globular micellar structure to a bimolecular leaflet, could account for the manifold swelling of mitochondria.[24] The presence of globular subunits has been demonstrated, by electron microscopy of sections of membranes of mitochondria.[3,10,24,25] Similar globular subunits have also been observed in micrographs of the membranes of the endoplasmic reticulum and Golgi apparatus,[24,25] in 'disks' in the outer segments of the retina,[15] and in the chloroplasts.[18,36] The existence of discrete globular subunits laterally aligned in the lamellar leaflets has been demonstrated by negative staining of preparations of lecithin dispersed in water.[8] Observations regarding the existence of sub-units within the unit membrane should be cautiously interpreted, as Robertson[23] has shown that the artifacts of overlapping images can produce the appearance of transverse sub-divisions in the unit membranes of the 'disks' in the outer segments of the retinal rods. Whether a mechanism of transformation from globular micelles to bimolecular leaflet is operative during swelling at the outer limiting membrane of mitochondria, is difficult to decide, in view of the existing uncertainty of the behaviour of this membrane during swelling,[9] and our limited knowledge of the structure and properties of biological membranes.

SUMMARY

Mitochondrial membranes show tight-junction-like five-layered structure after tissues are subjected to a freeze-substitution technique. If freezing of tissues is delayed for some time (8 minutes) after the animal is sacrificed, some of the mitochondrial cristae now show their seven-layered appearance, which is characteristic of OsO_4 fixation, i.e., two unit membranes separated by a 50 to 100 Å wide electron transparent gap, while most of the cristae and membranes at the surface still show their five-layered pattern. As the interval between sacrifice of the animal and freezing of the tissue is prolonged, the five-layered pattern is gradually replaced by seven-layered pattern. After about two hours *post-mortem*, most of the mitochondria show a seven-layered pattern in their membranes. Hardly any cristae that show a five-layered pattern are seen in electron micrographs. By this time many mitochondria show features that indicate swelling, i.e., rounding up of the mitochondria, decrease in electron density of the matrix, and retraction of the cristae to the periphery. Besides these alterations in mitochondria, early *post-mortem* alterations involve the endoplasmic reticulum, which begins to break up into vesicles. Qualitative observations suggest an increase in the freely dispersed ribosomes that are not associated with the membranes of the endoplasmic reticulum.

REFERENCES

1. AFZELIUS, B. J. 1962. Chemical fixatives for electron microscopy. *Symp. Internat. Soc. Cell Biol.* **1**, 1
2. AMES, A. and GURIAN, A. S. 1963. Electrical recordings from isolated mammalian (rabbit) retina mounted as a membrane. *Arch. Ophthal.* **70**, 837
3. FERNÁNDEZ-MORÁN, H., ODA, T., BLAIR, P. V., and GREEN, D. E. 1964. A macromolecular repeating unit of mitochondrial structure and function. *J. Cell Biol.* **22**, 63
4. GREENWALT, J. W., ROSSI, C. S., and LEHNINGER, A. L. 1964. Effect of active accumulation of calcium and phosphate ions on the structure of rat liver mitochondria. *J. Cell Biol.* **23**, 21
5. HAY, E. D. 1958. The fine structure of blastema cells and differentiating cartilage cells in regenerating limbs of Amblystoma larvae. *J. Biophys. Biochem. Cytol.* **4**, 583

6. ITO, S. 1962. Light and electron microscopic study of membranous cytoplasmic organelles. *Symp. Internat. Soc. Cell Biol.* **1**, 129

7. LEHNINGER, A. L. 1964. *The Mitochondrion.* New York: W. A. Benjamin

8. LUCY, J. A. and GLAUERT, A. M. 1964. Structure and assembly of macromolecular lipid complexes composed of globular micelles. *J. Mol. Biol.* **8**, 727

9. MALAMAD, S. 1965. Structural changes during swelling of isolated rat mitochondria. *Z. Zellforsch.* **65**, 10

10. MALHOTRA, S. K. 1966. A study of the structure of mitochondrial membrane systems. *J. Ultrastruct. Res.* **15**, 14

11. —— and VAN HARREVELD, A. 1965. Some structural features of mitochondria in tissues prepared by freeze-substitution. *J. Ultrastruct. Res.* **12**, 473

12. —— and ——. 1965. Dorsal roots of the rabbit investigated by freeze-substitution. *Anat. Rec.* **152**, 283

13. —— and ——. 1966. Distribution of extracellular material in central white matter. *J. Anat.* London **100**, 99

14. MOLBERT, E. and GUERRITORE, D. 1957. Elektronenmikroskopische untersuchungen am Leberparenchym bei akutur Hypoxie. *Beit. Pathol. Anat. alleg. Pathol.* **117**, 32

15. NILSSON, E. E. G. 1964. Receptor cell outer segment development and ultrastructure of the disk membranes in the retina of the tadpole (*Rana pipiens*). *J. Ultrastruct. Res.* **11**, 581

16. PALADE, G. E. 1961. The secretory process of the pancreatic exocrine cell. *Symp. Anat. Soc. Great Britain*, p. 176. London: Edward Arnold

17. —— and SIEKEVITZ, P. 1956. Pancreatic microsomes. An integrated morphological and biochemical analysis. *J. Biophys. Biochem. Cytol.* **2**, 671

18. PARK, R. B. 1965. Substructure of chloroplast lamellae. *J. Cell Biol.* **27**, 151

19. PAULEY, H. and PACKER, L. 1960. The relationship of internal conductance and membrane capacity to mitochondrial volume. *J. Biophys. Biochem. Cytol.* **7**, 603

20. REYNOLDS, E. S. 1963. Initial alterations of the cell following poisoning with carbon tetrachloride. *J. Cell Biol.* **19**, 139

21. ——. 1965. The nature of calcium-associated electron opaque masses in rat liver mitochondria following poisoning with carbon tetrachloride. *J. Cell Biol.* **25**, 53

22. ROBERTSON, J. D. 1964. *Cellular Membranes in Development* (edited by M. Locke), p. 1. New York: Academic Press

23. ——. 1965. Problems of unit membrane substructure. *J. Cell Biol.* **27**, 86A

24. SJÖSTRAND, F. S. 1964. *Cytology and Cell Physiology* (edited by G. H. Bourne), p. 311. New York: Academic Press

25. —— and ELFVIN, L. G. 1964. The granular structure of mitochondrial membranes and of cytomembranes as demonstrated in frozen-dried tissue. *J. Ultrastruct. Res.* **10**, 263

26. SLAUTTERBACK, D. B. and FAWCETT, D. W. 1959. The development of the cnidoblasts of Hydra. *J. Biophys. Biochem. Cytol.* **5**, 441

27. SULKIN, N. M. and SULKIN, D. F. 1965. An electron microscopic study of the effects of chronic hypoxia on cardiac muscle, hepatic, and autonomic ganglion cells. *Lab. Invest.* **14**, 1523

28. TEDESCHI, H. 1959. The structure of the mitochondrial membrane. Inferences from permeability properties. *J. Biophys. Biochem. Cytol.* **6**, 241

29. THOMPSON, T. E. and MCLEES, B. D. 1961. An electrophoretic study of suspensions of intact mitochondria and fragments of mitochondrial membranes. *Biochim. Biophys. Acta* **50**, 213

30. TRUMP, B. F., GOLDBLATT, P. J., and STOWELL, R. E. 1965. Studies on necrosis of mouse liver *in vitro*. *Lab. Invest.* **14**, 343

31. VAN HARREVELD, A. 1962. Water and electrolyte distribution in central nervous tissue. *Fed. Proc.* **21**, 659

32. —— and MALHOTRA, S. K. 1966. Demonstration of extracellular space by freeze-drying in the cerebellar molecular layer. *J. Cell Science* **1**, 223

33. ——, CROWELL, J., and MALHOTRA, S. K. 1965. A study of extracellular space in central nervous tissue by freeze-substitution. *J. Cell Biol.* **25**, 117

34. WATSON, M. L. and SIEKEVITZ, P. 1956. The separation and identification of a membrane fraction from isolated mitochondria. *J. Biophys. Biochem. Cytol.* **2**, 639

35. WEBSTER, H. DE F. and AMES, A. 1965. Reversible and irreversible changes in the fine structure of nervous tissue during oxygen and glucose deprivation. *J. Cell Biol.* **26**, 885

36. WEIER, T. E., ENGELBRECHT, A. H. P., HARRISON, A., and RISLEY, E. B. 1965. Subunits in the membranes of chloroplasts of *Phaseolus vulgaris*, *Pisum sativum*, and *Aspidistra* sp. *J. Ultrastruct. Res.* **13**, 92

Part 4

The Future for the Interpretation of Cell Structure

To the Editors, the future of cytology lies in the areas of qualitative morphology, and quantitative measurements of cell activity and structure. With present cytological methods it is possible to study the dynamic patterns of change in cell structures involved in physiological maintenance, developmental growth, health and sickness, genetics and evolution, down to the molecular level, where the biochemist becomes a morphologist. As methods are improved, dynamic cytology must grow, and, as data collection and analysis become ever more sophisticated, will become an easier field in which to work. The final article in this book, by Professor Hirsch, reminds us, quite rightly, of the importance of time, and of what is possible with present methods for those with the necessary will to endeavour in dynamic cytology.

Chapter 31

The Golgi Field and Secretion: An Example of the Dynamic Interpretation of Cell Structure

G. C. HIRSCH

The untiring investigations of John R. Baker from 1944 onwards,[2-11] have helped greatly in our understanding of the Golgi field. It was his wise decision to study only a limited number of objects that could at that time be thoroughly investigated from all points of view by the methods then available. It was chiefly the spermatocytes and spermatids of the snail (*Helix aspersa*), the intestinal epithelial cells of the common newt (*Triturus vulgaris*), and the neurons of the snail and rabbit that J. R. Baker and his students observed. They studied living cells, and tissues treated by 'impregnation' techniques, and also worked with a wide variety of histochemical methods, including those devised by Baker himself. They avoided the production of artifacts, so far as possible, by concentrating on the use of living cells, un-stained, or coloured by vital dyes.

Much of our present knowledge of the Golgi field was brilliantly conjectured by J. R. Baker in his 1944 paper,[2] written under the limitations of the light microscope. His schematic figure (Fig. 31.1) of the Golgi field, derived from the study of only three kinds of cells, comes closer to reality than I myself could come in my 1939 monograph[24] on the 'Golgi-Körper', which was based on nearly 200 different kinds. He showed[9] that some of the objects considered by myself as Golgi fields were, in fact, only random deposits of osmium on various objects in cells—for not everything blackened by osmium techniques is a Golgi body. Baker realized that to identify homologous cell-constituents it was necessary to study cells during life and to use histochemical and ontogenetic methods; and he preferred to make detailed studies of only a limited number of cells.

A study of Fig. 31.1 shows that the component parts of the Golgi field do not constitute a network, and therefore Golgi's expression *apparato reticolare interno*, as well as Peroncito's *Dictyosom* (Peroncito, 1909), was an incorrect description. The appearance of a network is actually the effect of over-impregna-tion, as was shown by Baker and also by Hirsch:[24] it is thus an artifact. Baker recognized the separate bodies that contain lipid (Golgi-externum), and his figure indicates that these produce the vacuoles (Golgi-internum). The vacuoles, which contain cell-products, then separate from the bodies that con-tain lipid. The sequence of events is shown in Fig. 31.1, beginning at the bottom and ending at the top.

Baker also pointed out the peculiarities of certain other cell-constituents. Fig. 31.2, E, shows the *lepidosomes* of Parat (1926) and Fig. 31.2, C, the bodies that Baker called *lipochondria*. (The latter must not be confused with the lipochondria of E. Ries (1933–5), which are actually pigment-particles, as was shown by O. Järvi[37, 38] and J. W. Sluiter.[55]) Baker[6] showed in 1953 that the lepidosomes could be dyed vitally like mitochondria, but not with neutral red. Lipochondria (in his usage of the word), on the con-trary, could be coloured in life by the latter dye. They can easily be revealed as separate objects (Baker[6]) by use of his Sudan black technique,[3] but the standard Golgi methods tend to clump them together or fill the spaces between them with a precipitate, thus forming an artificial network. *Osmiophil platelets* (Fig. 31.2, D) are really spherical when alive; they may perhaps represent a special kind of lipochondria.

In his paper of 1959, Baker[10] reviewed the status of the Golgi bodies in the light of what had been revealed by the electron microscope. He preferred at that time to use the term 'Dalton-complex' for the group of cell-constituents numbered 8, 9, and 12 in Fig. 31.6 of the present paper, because he considered the 'Golgi apparatus' of other authors was not sufficiently clearly defined. Various cell-constituents, including the nuclear membrane, are often blackened by the techniques supposed to be more or less specific for the 'Golgi apparatus', and as a result unrelated objects often received a name that grouped them as homologous with one another. Various kinds of lipid droplets had been confused with the Dalton-complex, which was genuinely homologous from cell to cell.

In 1963 Baker[11] once more considered the results of recent electron microscopical and other studies bearing on the Golgi problem.[43, 45–49, 53, 54] This time he concentrated on the neurons of vertebrates and

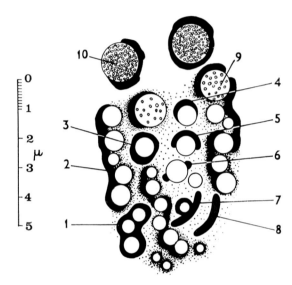

Fig. 31.1   Diagram illustrating J. R. Baker's opinion[2] (in 1944) on the structural plan of the fully developed Golgi element, as seen in a thin section. (1) A complete investment of two or more vacuoles. (The blackened substance corresponds with the saccules shown in Fig. 31.6, **8**, of the present article.) (2) A partial investment of two or more vacuoles. The blackened substance here appears as irregular strands running from vacuole to vacuole. (3) A complete investment of a separate vacuole. (4) A partial investment or 'cap'. (5) A crescent-shaped rod on the surface of a vacuole. (6) A ring passing right round a vacuole. (7, 8) Curved rods. (9, 10) Vacuoles containing secretion-product. From Baker.[2]

of the insects *Locusta* and *Schistocerca*. He now accepted the use of the term 'Golgi apparatus' for the group of cell-constituents previously referred to by him as 'Dalton-complex'. He was influenced by the work of Novikoff and Goldfischer[50] on the identification of the 'apparatus' by the use of a histochemical test for thiamine diphosphatase.

In many kinds of cells, this method revealed the same cell-constituents as those blackened by the classical Golgi techniques. When the test was applied to the Purkinje cells of the owl, which had been Golgi's object in his first paper on the apparatus, an appearance was given that very closely resembled Golgi's own figure of the apparatus as revealed by his silver technique. The use of electron microscopy, in conjunction with Novikoff and Goldfischer's histochemical technique, made it possible for the first time to identify the Golgi apparatus objectively. It is unfortunate that we still have no real understanding of the significance of the presence of thiamine diphosphatase in this particular cell-constituent.

In his 1963 paper Baker[11] gave a precise definition of his 'lipochondria', and distinguished them clearly from the Golgi apparatus.

The importance of observing *stages of activity* in the cell was stressed by Baker[6] in his paper of 1953. He clearly realized that *time* is a factor of fundamental importance in the study of constantly-changing structures.

The idea of studying a cell at work during various stages was first put forward by the Göttingen philosopher and physiologist, H. Lotze, in 1896.[44] Lotze compared the activity in a living cell or other vital system to a whirlpool in flowing water. Quantities of water are drawn in and appear to form stable

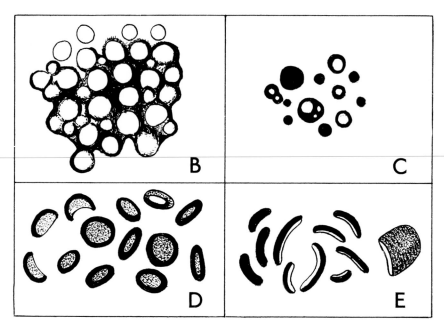

Fig. 31.2   Diagram of various types of bodies in the Golgi field. B, the field as seen in Fig. 31.1. C, lipochondria. D, osmiophil platelets. E, lepidosomes. From Baker.[6]

configurations, although the water that has entered is continually being cast out. I shall give several examples of this dynamic interpretation of what appear to be static structures.

A step-by-step investigation helped to explain the process of pancreatic secretion after excitation by pilocarpine. The stages are explained in the legend to Fig. 31.3.

The pancreas cell demonstrates *polyphase secretion*,[14–36, 39, 40] in which each cell can carry out similar secretory phases repeatedly. A cell that can produce its secretion only once provides an example of *monophase secretion*. The mid-gut gland of *Astacus*[19] and the sebaceous glands of mammals are of this type. The differences can only be established by using step-by-step methods. One must work either with single living animals under normal conditions for a number of hours, or else with a number of animals caught at different stages of secretion after given periods of starvation.

Perpetual secretion is distinguished by the continuity of the process, even though variations in the amount secreted may differ at different times. Examples are provided by the gall-bladder, by glycogen-production, by the secretion of saliva in the sheep's parotid gland, by the stomach of the 'black-leaf' louse, and by secretion in the transplanted Brunner's gland of the duodenum.[25]

Periodic secretion is also continuous, but there is a certain degree of periodicity, which is *independent* of feeding stimuli. Examples are provided by the isolated intestine of dogs, and also by the secretion of the pancreas during starvation.

Excitation of an organ by a prolonged, gradually decreasing stimulus leads to *rhythmic secretion*, which appears when cells exhibit a fixed refractory (non-secretory) period alternating with an active phase with all the cells secreting in unison. The whole gland must act rhythmically when the cells work synchronously and each cell exhibits a refractory period. Examples are provided by the mid-gut glands and fore-gut glands of *Pleurobranchaea* (Opisthobranchiata), *Murex*, and *Helix*,[42] the mantle-glands of *Haliotis*, the intestinal gland of the larva of *Deilephila euphorbiae* after ingestion of food, the stomach-cells of *Aphis fabae* (Hemiptera), and many other glands of vertebrates and invertebrates that secrete fats or proteins, but mostly enzymes. The stomachs of many insects and crabs have been particularly studied, but only in a few cases statistically analysed (for instance, in *Astacus*[17]). In the case of *Astacus* the rhythmic secretion of enzymes (amylase, casein-proteinase, guiacol-peroxidase) was demonstrated, and so was the rhythmic fluctuation of the cells of the mid-gut tubules. From the distal and proximal parts of each tubule the embryonic cells wander to the intermediate part and are here changed into 'fibrillar' cells and then into

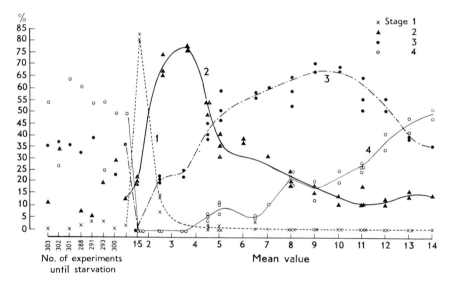

Fig. 31.3   Graph showing the sequence of four restitution-stages in a *living* pancreas of the mouse. The *ordinate* represents the percentage of cells that have reached a particular stage in the secretory cycle (stage 1, 2, 3, or 4). The *abscissa* represents the number of hours that have elapsed since the injection of pilo-carpine into a starved animal.
    Stage 1. The first visible signs of excitation in the pancreatic cell.
    Stage 2. Two rows of new zymogen granules appear at the apex of the cell.
    Stage 3. The zymogen granules reach from the apex of the cell to the nucleus.
    Stage 4. The granules now extend beyond the nucleus, towards the base of the cell.
    The figures are based on averages obtained from experiments on four or five animals. From Hirsch.[23]

'bladder-cells' which store up the enzyme. This is an example of a simultaneous enzymic and morpho-logical study with analysis by time-phases. The whole rhythmical secretion of the glandular tubules depends on the rhythmical transformation and wandering of the embryonic cells. The glandular tubules respond to a continuing stimulus, starting at time '0', by a *rhythmic production of mitoses*, from which then the rhythmic secretion of the bladder cells results. The formation of mitotic figures is significantly higher and starts sooner in spring than in summer or autumn. The periodic regeneration of cells has been observed in many glands of insects, but statistical analysis has not been applied.

Finally, *non-rhythmic secretion* has been observed. From asynchronous activity of the cells or absence of a refractory period, there results a non-rhythmic secretory process that rises only once and then falls away. The early studies of W. Grosser in 1905 and B. P. Babkin in 1912[1] showed this in three different mammals, with very different feeding stimuli. The curve showing the amount of secretion produced by the stomach rises steeply in the first hour or two, and then subsides gradually during a period of 6 to 13 hr.

The non-rhythmic secretion of the exocrine pancreatic cell has been studied particularly fully. B. P. Babkin[1] was the first to assay the amount of pancreatic juice in the dog after a single stimulus with

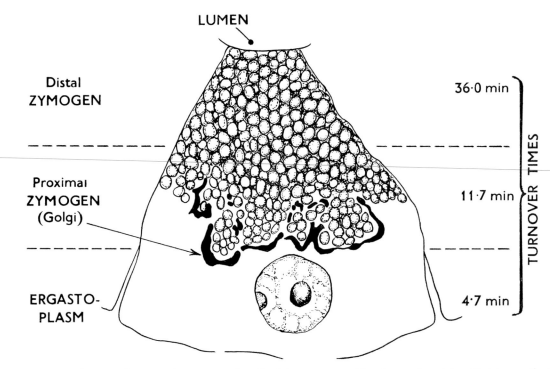

Fig. 31.4 Diagram of an exocrine pancreatic cell of the rat, treated by the osmium tetroxide/zinc method for the Golgi apparatus. The ergastoplasm, with the nucleus situated in it, is at the base of the pyramidal cell. The osmiophil elements of the Golgi field are located at the junction between the ergastoplasm and the zymogen granules; some elements lie between the granules. The zymogen granules fill the rest of the cell, extending right up to the lumen of the acinus. The division of the cell into ergastoplasm and proximal and distal zymogen regions is indicated at the left of the figure. The estimated 'turnover times' of the proteins in each region are shown at the right side of the figure. From Warshawsky, Leblond and Droz.[57]

meat, bread, or milk. He found a sharp rise up to 1 or 2 hr (with milk up to 3 hr); subsequently there was a plateau in the curve, and finally a slow fall-off to the initial level. The various stages in the pancreas were then analysed microscopically in the living mouse, up to 10 hr after several stimulations with pilocarpine. Four stages are briefly characterized in the legend to Fig. 31.3. It will be noticed in this figure that stage 1 reaches its maximum extent (80 per cent of all cells) in 1·5 hr; stage 2 has its maximum at 3·5 hr, stage 3 at 9 hr, and stage 4 at 14 hr. It must be remembered that several doses of pilocarpine were given.

Further, the time required for the restitution of an enzyme-granule in a pancreatic cell can be observed. This can happen by observation in a living cell in situ,[23] or by autoradiography carried out on several animals at various stages.[57] The first technique was used to determine the time that the proteins need to be in the ergastoplasm.[20] We know today that proteins originate on the ribosomes, are sluiced into the canals of the endoplasmic reticulum by active transport, and are passed through these canals to the Golgi field. In this apically directed movement the protein may appear in two forms. Normally the proteins are in suspension in the canals and as a result are transported rapidly; but when restitution is violent—for instance, after a strong dose of pilocarpine,[23] or as a result of congestion in the canals caused by sudden feeding of an animal after 2 days' starvation, or after ligature of the pancreatic duct,[56]

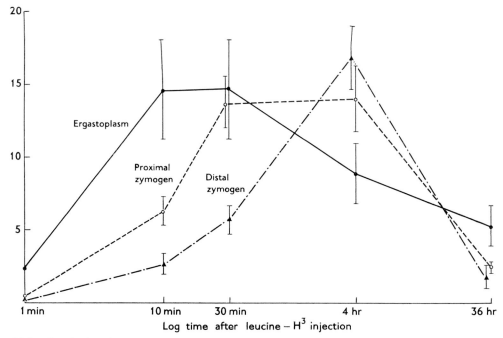

Fig. 31.5   Graph showing the specific activity of the proteins in the pancreatic exocrine cell of the rat. The circles represent activity in the proximal region, and the triangles represent activity in the distal zymogen region.

The ordinate gives the measurements of radioactivity concentration as the number of silver grains per unit area, yielding the mean specific activity of proteins in various regions of the acinar cells.

The abscissa gives the time that has elapsed after the injection of leucine-H[3]. The times are plotted on a logarithmic scale.

Each point on the graph shows the mean specific activity at a particular time, and the standard error of the mean is in each case represented by a vertical line.

(The 1-minute points were determined graphically by extra-polation on a non-logarithmic scale from 0 to 10 minutes. Reliance was placed on the fact that when x is very small, log x may be considered nearly equal to x.)

The peak of specific activity is reached first in the ergastoplasm, then in the proximal, and finally in the distal zymogen regions. From Warshawsky, Leblond, and Droz.[57]

granules are formed. Consequently I have called these bodies *condensation-granules*. They are so large that they are visible by optical microscopy. This has the advantage that they can be observed in the living animal after pilocarpine stimulation.[20] It appears from these observations that a single granule moves by transport in the canals of the reticulum to the Golgi field with a speed of 1 μ in 7 to 13 min.

The measurement of this time has been still more accurately studied by C. P. Leblond in 1963 and a mathematical analysis provided.[57] (Figs. 31.4 and 31.5). The conclusion is as follows. The origin of the secretory proteins in the ergastoplasm (ribosomes and endoplasmic reticulum) takes, on the average, 4·7 min, with a peak 10 to 30 min after the administration of leucin-H³, glycocoll-H³, or methionin-H³. The number of silver granules was determined by autoradiography in a uniform area, and the relative

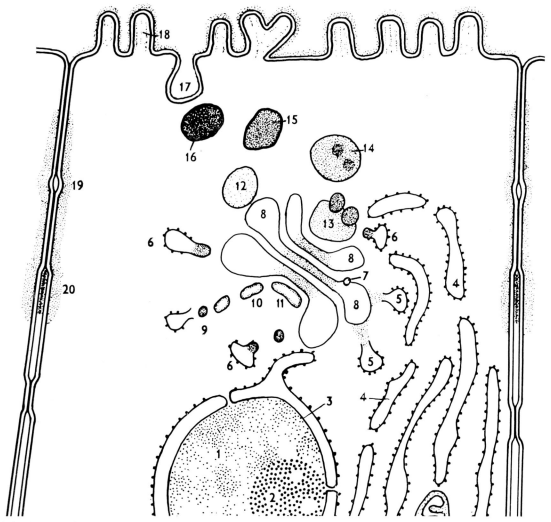

Fig. 31.6 Diagram of the apical part of an exocrine pancreatic cell. (An interpretation based on an electron-micrograph of Johannes Rhodin, New York 1963)

(1) Nucleus. (2) Nucleolus. (3) Nuclear membrane, continuous with the endoplasmic reticulum. (4) A tubule of the endoplasmic reticulum here opens by a 'mouth' towards the Golgi field. (Possibly proteins may escape through this mouth and enter through 'pores' (7) into a Golgi saccule (8).) (6) The end of a tubule of the endoplasmic reticulum in the neighbourhood of the Golgi field. There is a store of concentrated protein in a 'bulge' or protrusion that is directed towards the centre of the Golgi field. (7) A pore in a Golgi saccule (8). (9) An X-body has broken off from the endoplasmic reticulum. It will enlarge to become a vesicle (10). (11) A vesicle, formed in this way, stretches itself longitudinally and places itself parallel with and close to a saccule in a stack of saccules. (12) A Golgi vacuole has separated from the stack. (13) An intermediate stage, in which two X-bodies, laden with protein, enter a Golgi vacuole. (14) Young zymogen granule. (15) Later stage in the development of a zymogen granule. (16) Ripe zymogen granule. (17) Remnant of the membrane after the extrusion of a zymogen granule. (18) Microvilli projecting into the lumen of the acinus. (19) Upper and (20) lower desmosomes (*Kittleiste*) connecting two cells.

protein concentration established by microspectrography. The first radiographically recognizable zymo-gen granules appeared in the Golgi field after 7 or 8 to 15 min (mean 11·7 min). The average sojourn of the stacked granules amounted to 36 min. It follows that the average life-span of the newly-formed secretory proteins was 52 min, and that of the zymogen granules 48 min. Condensation granules move rather more slowly than suspended proteins in the canals of the reticulum, owing to their size. When one remembers this fact, the two sets of data for the time occupied, correspond fairly well.

Finally, light has been thrown on the fate of the proteins in the Golgi field (Fig. 31.6), and their activity there. G. E. Palade[51,52] supposed that the condensation-granules in the guinea-pig are trans-formed directly into zymogen granules, but this is an oversimplification. On the contrary, the conden-sation granules are all dissolved, as soon as they come into the vicinity of the Golgi field. This field appears to be constructed in a complicated way, which differs in the course of the phases of secretion, and differs also in the cells of animals and plants. Externally it is not bounded by a membrane, but lies free in the cytoplasm (Fig. 31.6). The canals of the endoplasmic reticulum possess a limiting layer, and they incline at their blind ends towards the middle of the Golgi field (Fig. 31.6, Nos. 5, 6). Often the ends of the canals show a storage place for proteins at a slight terminal 'bulge' or protrusion (Fig. 31.6, 6). These protrusions separate themselves by constriction and a small 'X-body' (Fig. 31.6, 6, 9) is thus produced, which is at first filled with protein. The X-body enlarges and places itself parallel to the Golgi saccules (Fig. 31.6, 10, 11), and an internal light spot develops in it. Then the X-body grows out to become a long saccule, surrounded by a membrane (Fig. 31.6, 8). In the middle, this saccule is filled with a somewhat denser substance, but at its ends it widens into bladders (Fig. 31.6, 8); these then nip them-selves off from the saccules (Fig. 31.6, 12). From these bladders or 'vacuoles' the intermediate bodies originate (Fig. 31.6, 13, 14). I have often observed that small X-bodies, stuffed full of protein, are absorbed by intermediate bodies (Fig. 31.6, 13, 14). Then the zymogen granules develop from the intermediate bodies (Fig. 31.6, 16). The zymogen granules are stored, and are extruded on stimulation.

Perhaps there is another possible relationship between the endoplasmic canals and the saccules. Many of the canals have a 'mouth', which stands open towards the middle of the Golgi field (Fig. 31.6, 5). Here proteins might emerge from the canals and be taken up into the sacs through pores (Fig. 31.6, 7).

I shall here only add that the theoretical methodology of this *Stufenuntersuchung*, and an introduction to the study of the time-factor in the changes undergone by tissues and cells, have been presented in 1929 by G. C. Hirsch,[16] in 1943 by H. W. Chalkley,[13] in 1962 by R. J. Britten and B. J. McCarthy,[12] and in 1966 by H. U. Koecke.[41]

I shall conclude with a word of thanks to my friend John R. Baker, who permitted me to work in his laboratory in 1948. I have learnt much from him. The author is grateful to the Forschungs-gemeinschaft Godesberg for financial assistance.

REFERENCES

1. BABKIN, B. P. 1912. *Die äussere Sekretion der Verdaungsagrudsen*, 2nd edition. Berlin
2. BAKER, J. R. 1944. The structure and chemical composition of the Golgi element. *Quart. J. micr. Sci.* **85**, 1
3. ——. 1949. Further remarks on the Golgi Element. *Quart. J. micr. Sci.* **90**, 293
4. ——. 1950. Studies near the limit of vision with the light microscope, with special reference to the so-called Golgi-bodies. *Proc. Linn. Soc.* **162**, 67
5. ——. 1951. The Golgi substance. *Nature, Lond.* **168**, 1089
6. ——. 1953. Nouveau coup d'œil sur la controverse du Golgi. *Bull. de Microscopie appliquée* (2) **3**, 1–2
7. ——. 1954. A study of the osmium techniques for the Golgi apparatus. *Quart. J. micr. Sci.* **95**, 383
8. ——. 1955. What is the Golgi controversy? *J. Roy. micr. Soc.* **124**, 217
9. ——. 1957. The Golgi controversy. *Symp. Soc. exp. Biol.* **10**, 1
10. ——. 1959. Towards a solution of the Golgi problem: Recent developments in cytochemistry and electron microscopy. *J. Roy. micr. Soc.* **77**, 116
11. ——. 1963. New developments in the Golgi controversy. *J. Roy. micr. Soc.* **82**, 1
12. BRITTEN, R. J. and MCCARTHY, B. J. 1962. The synthesis of ribosomes in Escherichia coli II. *Biophys. J.* **2**, 83

13. CHALKLEY, H. W. 1943. Method for the quantitative morphologic analysis of tissues. *J. nat. Cancer Inst.* **4**, 47–53
14. HIRSCH, G. C. 1915. Ernährungsbiologie fleischfressender Gastropoden. I. Bau, Nahrung, Nahrungs-aufnahme, Verdauung, Sekretion. *Zool. Jahrbuch, Abt. Phys.* **35**, 357
15. —— and JACOBS, W. 1928. Arbeitsrhythmus der Mitteldarmdruse von Astacus. I. Methodik und Technik, Beweis der Periodizität. *Z. f. Vergl. Physiol.* **8**, 102–44
16. ——. 1929. Dynamik organischer Strukturen. *Roux Arch. Ent.-mech.* **117**, 511–61
17. —— and JACOBS, W. 1930. Arbeitsrhytmus Astacus II. Wachstum als primärer Faktor des Rhytmus. *Z. f. Vergl. Physiol.* **12**, 524–58
18. —— and BUCHMANN, W. 1940. Oxydations—Reduktions-System und Sekretion. *Z. f. Vergl. Physiol.* **12**, 559–78
19. ——. 1931. Theory of fields of restitution with special reference to secretion. *Biol. Reviews, Cambridge* **6**, 88–131
20. ——. 1931. Lebendbeobachtung des Pankreas. I. Restitution. *Z. f. Zellforsch.* **15**, 37–68
21. ——. 1931. II. Analyse des Sekretmaterials mittels Röntgenstrahlen. *Roux Arch. Ent.-mech.* **123**, 792–821
22. ——. 1931. III. Wechselnde Permeabilität der Pankreaszelle. *Z. f. Zellforsch.* **14**, 517–43
23. ——. 1932. IV. Restitution der Drüse als Ganzes. *Z. f. Zellforsch.* **15**, 290–310
24. ——. 1939. *Form- und Stoffweichsel der Golgi-Körper.* Berlin: Bornsträger
25. ——. 1946. Daten zum Arbeitsrhytmus der Drüsenzellen und Drüsen. *Tabulae Biologicae* **23**, 82–136
26. ——. 1948. Dynamik der Sekretions-Systeme. *Verhandl. d. Dtsch. Zool. Ges.* **5**, 226–32
27. ——, JUNQUERIA, L. C. U., ROTHSCHILD, H. A., and DOHI, S. R. 1957. Pankreasekretion der Ratte. I. Die kontinuierliche, irreguläre Hungersekretion und ihre Ursachen. *Pflügers Arch. d. ges. Physiol.* **264**, 78–87
28. ——. 1958. Der Arbeitszyklus im Pankreas und die Entstehung der Eiweisse. Versuch eines synthet-ithschen Bildes nach der Elektronen-Mikroskopie und Biochemie. *Naturwissenschaften* **45**, 349–58
29. ——. 1959. Die Entstehung der Verdauungsenzyme im exokrinen Pankreas der Säugetiere. *Materia Medica Nordmark* **11**, 149–77
30. ——. 1959. Dynamik der tierischen Zelle, I. Teil. *Handb. der Biologie*, Bd. **1**, S. 219–86. Konstanz: Athenaion-Verlag
31. ——. 1960. Die Fliessbandarbeit der exokrinen Pankreaszelle bei der Produktion von Enzymen. Mit einem Exkurs über Ergastoplasma und Golgi-Körper. *Naturwissenschaften* **47**, 25–35
32. ——. 1961. Lamellen-Vakuolen-Systeme. *Handb. der Biologie*, Bd. **1**, S. 409–52. Konstanz: Athenaion-Verlag
33. ——. 1961. Suggestions for a new look at the lamellar-vacuolar field or 'Golgi-complex' and their pattern of lower parts. *Acta Med. Okayama* **15**, 289
34. ——. 1962. Das Lamellen-Vakuolen-Feld. *Handb. der Biologie*, **1**, S. 409–52. Konstanz: Hachfeld
35. ——. 1964. Konstruktion und adaptive Umkonstruktion in den Zellen des exokrinen Pankreas. *Materia Medica Nordmark* **49**, 1–39
36. ——. 1964. 'The Golgi apparatus' or the lamellar-vacuolar field in the electron microscope. In *Symposium April 1963 Biwako Hostel, Ohtsu* (edited by E. V. Cowdry, S. Seno, S. Katsunuma). Tokyo: Maruzen Co.
37. JÄRVI, O. 1939. Restitution des Sekretstoffes in der Unterzungendrüse der Katze. *Z. f. Zellforsch.* **30**, 98–195
38. ——. 1940. 'Lipochondrien' von Ries und ihre Beziehungen zu den Pigmenten. *Z. f. Zellforsch.* **31**, 1–42
39. JUNQUEIRA, L. C., HIRSCH, G. C., and ROTHSCHILD, H. A. 1955. Glycine-C^{14} uptake by the proteins of the rat pancreatic juice. *Biochem. J.* **61**, 275
40. —— and HIRSCH, G. C. 1956. Cell secretion. A study on pancreas and salivary glands. *Intern. Review of Cytology.* London. **5**, 323–64
41. KOECKE, H. U. 1966. Die Anwendung des Elektronenmikroskopes bei der Untersuchung entwicklungs-physiologischer Probleme. *Information Zeiss.*
42. KRIJGSMAN, B. J. 1924. Arbeitsrhythmus der Verdauungsdrüsen bei Helix I. *Z. f. Vergl. Physiol.* **3**, 264–96
43. LACY, D. and CHALLICE, C. E. 1956. Studies on the Golgi apparatus by electron microscopy with particular reference to Aoyama's technique. *J. Biophys. Biochem. Cytol.* **2**, 395
44. LOTZE, H. *Gründzüge der Naturphilosophie.* Leipzig (1889). *Mitrokosmos*, 5. Aufl. Leipzig (1896)
45. MALHOTRA, S. K. and MEEK, G. A. 1960. An electron microscope study of some cytoplasmic inclusions of the neurones of the prawn, *Leander serratus. Quart. J. micr. Sci.* **80**, 1
46. NAGATANI, YOSHIMI. 1963. Blind point in the Golgi controversy. 1. Reticular silvered image and granular silvered image. *Science Reports of Yamaguchi University* **14**, 17

47. —— and SUSUMU TOMONAGA. 1963. Cytological natures of the pre-existing structures in the maize root-tip cells and their relation to the silvered Golgi apparatus. *Science Reports of Yamaguchi University* **14**, 1–16

48. —— and IKUO YAMAOKA. 1963. Problem in the Golgi apparatus revealed by electron microscopy. *The Bulletin of the Marine Biological Station of Asamushi, Tohoku University* **11**, 229

49. —— and AKIRA TANAKA. 1965. Relation between the so-called Golgi apparatus and the Dalton-Felix complex. *Science Reports of Yamaguchi University* **15**, 63

50. NOVIKOFF, A. B. and GOLDFISHER, S. 1961. Nucleosidediphosphatase activity in the Golgi apparatus and its usefulness for cytological studies. *Proc. Nat. Acad. Sci.* **47**, 802

51. PALADE, J. E. 1956. Intracisternal granules in the exocrine pancreas. *J. Biophys. Biochem. Cytol.* **2**, 417

52. ——, SIEKEVITZ, P., and CARO, L. J. 1962. Structure, chemistry and function of the pancreatic exocrine cell. *Ciba Foundation Symposia on the Exocrine Pancreas.* London: J. and A. Churchill

53. PALAY, S. L. and PALADE, M. D. 1955. The fine structure of neurones. *J. Biophys. Biochem. Cytol.* **1**, 69

54. SHAFIQ, S. A. and CASSELMAN, W. G. B. 1954. Histochemical investigations, with special reference to the lipochondria. *Quart. J. micr. Sci.* **95**, 315

55. SLUITER, J. W. 1944. Restitutionsproblem in der Pankreaszelle. I. Die Bedeutung des Golgi-Apparates. *Z. f. Zellforsch.* **33**, 187–224

56. SUZUKI, I. 1959. *Jap. J. of Gastroentology* **56**, 155 (Japanese)

57. WARSHAWSKY, H., LEBLOND, C. P., and DROZ, B. 1963. Synthesis and migration of proteins in the cells of the exocrine pancreas as revealed by specific activity determination from radiomicrography. *J. Cell Biol.* **16**, 1

Appendices

Appendix I

JOHN BAKER, CYTOLOGIST
(An Appreciation)

W. G. B. CASSELMAN

'Science has its cathedrals, built by the efforts of a few architects and many workers . . . these great structures are . . . the result of giving ordinary human effort a direction and a purpose.'* Modern cytology is such a structure. John Baker is one of its great contemporary architects. Through his personal example and his rare ability to inspire others, he has given direction and purpose to the efforts of many. For this reason, he has a special place in the life of each of us who has contributed to this volume.

John Baker is exceptional as a person. Certain of his qualities have impressed and influenced all of us, especially his integrity, his motivation, and his self-discipline. His insistence upon the truth leaves no place for false concepts. His self-discipline and his intense motivation give his work and his life a unique singleness of purpose. Preciseness is 'an obvious necessity' in his every thought and act. Qualities such as these, together with his power of perception and his incisiveness of mind, have resulted in important contributions to the advancement of science, recorded in more than 190 publications over the past 44 years.

John Baker is outstanding among scientists. His life is devoted to the quest of finding out what nature is really like. He has pursued that quest with constant inquiry, constant seeking of objective data and their meaning, and constant demanding of verification. He has brought to the quest not only rational thought and a deep appreciation of the logic of scientific inquiry, but also a wide knowledge and fine technical skill. These have given him exceptional command over the problems he has studied.

As a man of science, John Baker has always been ahead of his time. Seldom has the full value of his concepts and his contributions been recognized at once, as, for example, during the early days of modern histochemistry, or during the Golgi controversy, or in discussion of the significance of technical artifacts, especially in present-day electron microscopy.

John Baker also contributed importantly to the development of cytology as an editor of the *Quarterly Journal of Microscopical Science* from 1946 to 1964. It seems more than mere coincidence that as a cytologist he has spanned the transition from classical cytology to modern cell biology, and that as an editor, he spanned the transition from the classical ' *Q.J.*' to the new *Journal of Cell Science*. To his work as an editor he brought not only broad interests, great technical ability, and wide experience but also unfaltering judgement and perception. In dealing with authors, as in dealing with others, he could be confronting where necessary, but he was always helpful and understanding. Toward young contributors and students he showed a great humanness and acceptance.

* From *Thermodynamics*, by G. N. Lewis and M. Randall, p. vii. McGraw-Hill Book Co., 1923.

Upper Photograph   John Baker delivering the Presidential Address to the Royal Microscopical Society on the 5th January, 1966

Lower Photograph   John Baker lecturing to the Royal Microscopical Society on the 6th January, 1965

John Baker stands out as a teacher. There are as many measures of a teacher as there are objectives of teaching. An important one is the influence that a great teacher can have upon the whole life of a student, even upon succeeding generations of students.

The significance of great teachers is often overlooked, yet it is they who form the essential nucleus of any great centre of learning. It is they alone who lead a student into a new world, who challenge his imagination, who sometimes seem to demand the impossible, yet unfold for him the path by which it can be attained. Such a teacher is John Baker.

The ultimate reward of a true teacher comes when one or two, or perhaps half a dozen, students who have come under his guidance, catch the fire of inspiration, become impatient with the meagre fragments of contemporary knowledge, and start out on their own to push back the barriers of the unknown. Those of us who have contributed to this book have felt that impatience and that urge. We hope that in these essays John Baker will find assurance that his life and his teaching have been fruitful, and that some of his students will carry the torch to succeeding generations.

Appendix II

Bibliography of the Published Writings (1924-1968) of

JOHN RANDAL BAKER, D.Sc. (Oxon.), F.R.S.

Compiled by R. M. Park, A.L.A.

PREFACE

In compiling this bibliography of 196 books, journal articles and letters, and broadcast talks, published over 44 years, on a wide range of subjects, I have been fortunate in obtaining their author's assistance. Nevertheless, the existence of *six* scientists with the same surname *and* initials has necessitated much checking to ensure accuracy. A further helpful factor has been Dr. Baker's meticulous listing of references.

His methodical presentation, whether of research findings or of philosophical or sociological argument, is a delight to follow (even for the layman), and an excellent model for all who aim to transmit intelligibly the results of their research.

For omitting to list his reviews of books, or to do more than mention his editorship of the *Quarterly Journal of Microscopical Science* (from 1946 to 1964), I must crave the author's indulgence, and that of all readers of this *Festschrift*.

I am particularly grateful to the library staffs of the Science Museum, the London School of Economics and the British Museum for their assistance. Above all, thanks are due to Miss Judith Astin, formerly the librarian in the Department of Zoology, Oxford, who compiled a list of the references that formed the core of this bibliography from the private records of Dr. J. R. Baker, who kindly allowed her access to them, at the request of the Editors.

Anyone interested in only one, or a few, of Dr. Baker's several spheres of activity should find the general index useful. The chronological arrangement by publication date will, it is hoped, help trace the author's development of various lines of research. This arrangement has not, however, been pursued *within* each year, because of the varying time-lags between the submission of papers and their publication. (Where appreciable and ascertainable, this time-lag has been indicated.)

SELECT LIST OF JOURNALS

AND SERIES TO WHICH THE AUTHOR HAS MOST FREQUENTLY CONTRIBUTED

Title	*Abbreviation*
British Journal of Experimental Biology	*Brit. J. exp. Biol.*
Eugenics Review	*Eugen. Rev.*
Geographical Journal	*Geogrl J.*
†*Journal of Contraception*	*J. Contracept.*
Journal of Hygiene [*Cambridge*]	*J. Hyg.*
Journal of the Linnaean Society (Zoology)	*J. Linn. Soc. (Zool.)*
Journal of the Quekett Microscopical Club	*J. Quekett microsc. Club*
Journal of the Royal Microscopical Society	*Jl R. microsc. Soc.*

† Continued, from 1940, as *Human Fertility*.

Nature [London]	*Nature*
Proceedings of the Royal Society	*Proc. R. Soc.*
Proceedings of the Zoological Society of London	*Proc. zool. Soc. Lond.*

[N.B.—Volume-numbers in references prior to 1938 signify the *physical* volume within that year, since no *running* volume-numbers are quoted for those years.]

†*Quarterly Journal of Microscopical Science*	*Q. Jl microsc. Sci.*
School Science Review	*Sch. Sci. Rev.*
Society for Freedom in Science. *Occasional Pamphlets.*	

ABBREVIATIONS

diag.	= diagram(s)	phot.	= photograph(s)
gr.	= graph(s)	pl.	= plate(s)
il.	= illustration(s)	R	= references
incl.	= including	repr.	= reprinted
OUP	= Oxford University Press	tab.	= table(s)

NOTES

* marks works for which the author hopes to be remembered.

Heavy type is used for years and titles and editions of books; also for volume-numbers of journals.

Titles of journals are abbreviated in accordance with the *World list of scientific periodicals*, 4th ed. 1965 [omitting distinguishing *Camb.* and *Lond.* from *Journal of Hygiene* and *Nature* respectively].

John Wiley is American publisher of all books in English herein listed, unless otherwise stated.

1924

An hermaphrodite dogfish (*Scyliorhinus canicula*). (*with* P. D. F. Murray) *J. Anat.* **58** (4), 335–9 (5 il. 4R)

1925

A coral reef in the New Hebrides. *Proc. zool. Soc. Lond.*, (vol. 2), 1007–19. [Paper 40] (map, diags., phots. 3R)

*On sex-intergrade pigs: their anatomy, genetics, and developmental physiology. *Brit. J. exp. Biol.* **2** (2), 247–63 + 2 pl. (10 diags., tab. 11R)

On the descended testes of sex-intergrade pigs. *Q. Jl microsc. Sci.* **69** (276), 689–701. (8 diags. 7R)

1926

Asymmetry in hermaphrodite pigs. *J. Anat.* **60** (4), 374–81. (4 diags., 2 tabs. 15R)

Sex in man and animals, *etc.* Routledge. xvi, 175 p. + 4 pl. (22 il. 19R) [Preface by Julian S. Huxley]

1927

Temperature and enzyme activity. *J. mar. biol. Ass. U.K.* **14**, 723–7. (4 gr. 3R)

1928

Depopulation in Espiritu Santo, New Hebrides. *J. R. anthrop. Inst.* **58**, 279–303 + 1 pl. (map, tabs., gr. 7R)

The influence of age at castration on the size of various organs. *Brit. J. exp. Biol.* **5** (3), 187–95. (gr., tabs. 6R)

The influence of high temperatures on the testis. [Dog, at 39·5–40°C] *J. Hyg.* **27**, 183–5. (2R) [received, 21.10.27]

Natural pyramids [of pebbles] on a beach in the New Hebrides. [letter] *Nature* **122** (3083), 843–4. (2 diags.)

A new type of mammalian intersexuality. *Brit. J. exp. Biol.* **6** (1), 56–64 + 1 pl. (5 diags. 6R)

The non-marine vertebrate fauna of the New Hebrides. *Ann. Mag. nat. Hist.* series 10, **2** (9), 294–302. (+ list. 6R)

Notes on New Hebridean customs, with special reference to the intersex pig. *Man* **28** (81), 113–18. (il. 2R) [Visits, incl. Banks Islands, 1922–3 and 1927, under Percy Sladen Trust]

† Continued, from March 1966, as the *Journal of Cell Science.*

1929

Man and animals in the New Hebrides. Routledge. xiv, 200 p. + 17 pl. (43 il. 23R) [Percy Sladen Trust expeditions]

The northern New Hebrides. *Geogrl J.* **73**, 307–25 + map.

On the zonation of some coral reef Holothuria. *J. Ecol.* **17**, 141–3. (tab. 2R)

1930

The breeding-season in British wild mice. *Proc. zool. Soc. Lond.* (vol. 1), 113–26 + 1 pl. (9 gr., 10 tabs. 2R) [received, 18.2.29]

A fluid for mammalian sperm-suspensions. *Q. Jl exp. Physiol.* **20**, 67–70

The spermicidal powers of chemical contraceptives:

    1. Introduction, and experiments on guinea-pig sperms. *J. Hyg.* **29**, 323–9 + 1 pl. (tabs.) [received, 10.8.29]

1931

An improved fluid for mammalian sperm-suspension. *Q. Jl exp. Physiol.* **21**, 139–40. (4R)

The spermicidal powers of chemical contraceptives:

    2. Pure substances. *J. Hyg.* **31**, 189–214. (1 phot., tabs. 5R) [received, 2.9.30]

    3. Pessaries. *J. Hyg.* **31**, 309–20. (2R)

The stimulation of spermatozoa by drugs. [letter] *Nature* **127**, 708. [Under auspices of BIRTH CONTROL INVESTIGATION COMMITTEE]

1932

The breeding-season of the blackbird (*Turdus merula* Linn.). (*with* Ina BAKER) *Proc. zool. Soc. Lond.* (Vol. 2), 661–7. [Paper 30] (map, 2 gr., tab. 1R)

Factors affecting the breeding of the field-mouse (*Microtus agrestis*). (*with* R. M. RANSON):

    *1. Light. *Proc. R. Soc.* B, **110** (767), 313–22. (gr., tabs. 5R)

    2. Temperature and food. *Proc. R. Soc.* B, **112** 774, 39–46

Fixation of mitochondria. [letter] *Nature* **130**, 741. (3R)

A new method for mitochondria. [letter] *Nature* **130**, 134. (1R)

The spermicidal powers of chemical contraceptives:

    4. More pure substances. *J. Hyg.* **32**, 171–83. (tabs., diags. 5R) [received, 21.7.31]

    5. A comparison of human sperms with those of the guinea-pig. *J. Hyg.* **32**, 550–6. (gr., tabs. 2R)

1933

Biology in everyday life. (*with* John Burdon Sanderson HALDANE) Allen & Unwin. 123 p. [6 talks broadcast in Spring, 1933.—BAKER: *A biologist's view of everyday life*; *Social life in animals*; *The determination of sex*; *The quality and quantity of mankind*; *War, disease and death*. (pp. 13–100) HALDANE: *Biology and statesmanship*.] [repr. 1945, vi, 83 p. (17R); + Forces Issue, W.O. 68.]

*****Cytological technique.** Methuen. xi, 131 p. (3 il. 120 R) (*Methuen's monographs on biological subjects*)— [2nd ed., 1945; 3rd ed., 1950; Chinese ed. (*pirated*), 1953; 4th ed., 1960; 5th ed., 1966]

Factors affecting the breeding of the field-mouse (*Microtus agrestis*) (*with* R. M. RANSON):

    3. Locality. *Proc. R. Soc.* B, **113**, 486–95

What is man ? (*with* Julian S. HUXLEY) *in* ADAMS, Mary, *ed.* **Science in the changing world.** [Collection of broadcast talks by Sir Thomas Henry HOLLAND and others] Allen & Unwin. 286 p. [BAKER: *Our place in nature*; *Missing links*; *The evolution of mind*; *The control of development*. (pp. 131–85)

1934

Measurement of ultra-violet light [in New Hebrides]. [letter] *Nature* **134**, 139–40. (1 il.)

The spermicidal powers of chemical contraceptives:

    6. An improved test for suppositories. (*with* R. M. RANSON) *J. Hyg.* **34** (4), 474–85. (tabs. 4R)

1935

Apparatus for measuring the drying power of the atmosphere. *J. scient. Instrum.* **12** (7), 214–16. (3 diags. 1R)

The chemical control of conception, with a chapter by H. M. CARLETON [and a postscript by J. R. BAKER *and* R. M. RANSON]. Chapman & Hall. x, 173 p. + 4 pl. (il., tabs. 28R)

*Espiritu Santo, New Hebrides. *Geogrl J.* **85**, 209–33 + maps. [1933 survey, with T. F. BIRD, Tom HARRISSON, and Mrs. BAKER]

Laboratory research in chemical contraception. [lecture, 16.4.35, with demonstration by R. M. Ranson]
  Eugen. Rev. **27** (2), 127–31
A note on stereoscopic photography. *Nature* **136**, 193. (diag.)
Stereoscopic photography. [letter] *Nature* **136**, 551

1936

Methods of stereophotomicrography. *Photogr. J.* **76**, 275–9. (1 diag. + 4 stereo pairs. 14R) [received, 16.5.35]
Nomograms for saturation deficiency. *J. anim. Ecol.* **5** (1), 94–6 + 1 pl. (3R)
★The seasons in a tropical rain-forest (New Hebrides) [communicated by Prof. E. S. Goodrich]:
    1. Meteorology. (*with* Tom H. Harrisson) *J. Linn. Soc. (Zool.)* **39**, 443–63. (map, 6 gr., tab. 16R)
    2. Botany. (*with* Ina Baker) *J. Linn. Soc. (Zool.)* **39**, 507–19 + 1 pl. (diag. 20R)
    3. Fruit-bats (*Pteropidae*). (*with* Zita Baker) *J. Linn. Soc. (Zool.)* **40**, 123–41. (gr., tabs. 25R)
    4. Insectivorous bats (*Vespertilionidae* and *Rhinolophidae*). (*with* T. E. Bird) *J. Linn. Soc. (Zool.)* **40**, 143–61 + 1 pl. (gr., tabs. 36R)

1937

Further research on chemical contraception. [incl. nomenclature] *Eugen. Rev.* **29** (2), 109–12. (2R)
Light and breeding seasons. *Nature* **139**, 414
The nomenclature of chemical contraceptives. *J. Contracept.* **2**, 225 (1R) [Birth Control Clinic Research Bureau]
Nomenclature of the seasons. [letter] *Nature* **140**, 890–1
The Sinharaja rain-forest, Ceylon. *Geogrl J.* **89**, 539–51. (3 phots., 2 maps, diag. 9R)
The spermicidal powers of chemical contraceptives:
    ★7. Approved tests. (*with* R. M. Ranson *and* J. Tynen) *J. Hyg.* **37**, 474–88. (tabs. 3R)

1938

★The breeding seasons of southern hemisphere birds in the northern hemisphere. (*with* R. M. Ranson) *Proc. zool. Soc. Lond.* **108**, series *A*, 101–41. [Paper 8] (tabs.: pp. 112–41. 192 R) [received, 20.5.37]
The evolution of breeding seasons (pp. 161–77. 2 gr. 27R) *in* De Beer, *Sir* Gavin Rylands, *ed.* **Evolution: essays on aspects of evolutionary biology;** presented to Prof. E. S. Goodrich on his 70th birthday. OUP
Latitude and egg-seasons in old-world birds. *Tabul. biol.* **15**, 333–70. (197R)
The nematocysts of hydra. *Sch. Sci. Rev.* **20** (78), 261–7. (diag. 8R)
A new chemical contraceptive. (*with* R. M. Ranson *and* J. Tynen) *Lancet* **235** (2), 882–5. (2 tabs. 6R); + 1381 [letter]
The production of a vaginal epithelium suitable for the testing of chemical contraceptives in the laboratory for harmful effects. *J. Contracept.* **3**, 105 (1R)
Rain-forest in Ceylon. [Sinharaja] *Kew Bull.*, No. 1, 9–16 + 2 pl. (tab. 13R) [Royal Botanic Gardens, Kew. *Bulletin of miscellaneous information*, 1938, No. 1. HMSO, 1939.]
★The relation between latitude and breeding seasons in birds. *Proc. zool. Soc. Lond.*, **108**, series *A*, 557–82. [Paper 34] (16 gr., tabs. 27R)

1939

The breeding seasons of birds, with special reference to the need for further data from Australia. *Emu* **39**, 33–8. (gr. 4R)
★The chemical composition of the Volpar contraceptive products. (*with* R. M. Ranson *and* J. Tynen):
    1. Phenyl mercuric acetate as a spermicide. *Eugen. Rev.* **30** (4), 261–8. (gr., tabs. 8R)
    2. Vehicles for phenyl mercuric acetate. *Eugen. Rev.* **31** (1), 23–31. (tabs. 1R)
Collection and examination of semen specimens. *J. Contracept.* **4** (6), 127–9. (12R) [*see also* preceding paper: Tynen, John. Observations on specimens of human semen. pp. 125–7. (tab. 1R), first published as leaflet, March 1939, by English Association of Clinical Pathologists]
Increasing winter egg-production in Spain more than a hundred years ago. [letter] *Nature* **143**, 477. (1R)
★Counterblast to Bernalism. *New Stateman and Nation*, **18** (440) 174–5

1940

Science in the USSR. [Letter opposing J. D. Bernal] *New Statesman and Nation*, **19** (471), 276
★The seasons in a tropical rain-forest (New Hebrides):
    5. Birds (*Pachycephala*). (*with* A. J. Marshall *and* Tom H. Harrisson) *J. Linn. Soc. (Zool.)* **41**, 50–70 + 1 pl. (gr., tabs. 72R)

The sex-ratios in the wild animal populations of the New Hebrides (*with* A. J. MARSHALL). *Proc. Linn. Soc. New South Wales* **65** (5–6), 565–7. (7R)

1941

Chlorazol black *E* as a vital dye. [letter] *Nature* **147**, 744. (2R)

A fluid for softening tissues embedded in paraffin wax. *J. R. microsc. Soc.* **61**, 75–8. [*Trans.*, 5] (4R)

Gel-suppositories and spermicidal action. [letter] *Hum. Fert.* **6** (1), 29–30

1942

Chemical composition of mitochondria. *Nature* **149**, 611–12. (6R)

The free border of the intestinal epithelial cells of vertebrates. *Q. Jl microsc. Sci.* **84**, 73–103 + 3 pl. (diags. 37R) [*incl.* a review of the literature, pp. 74–9]

The future of research. [letter replying to leading article] *Endeavour* **1** (3), 90–1

The scientific life. Allen & Unwin. 154 p. (152R) [+ repr. 1944] [*see also* **Freiheit und Wissenschaft, 1950**]

Some aspects of cytological technique. (pp. 1–27 + 1 pl.; pp. 261–2. 78R) *in* BOURNE, G., *ed.* **Cytology and cell physiology.** OUP. [+ corrected reprint, 1945; 2nd ed., **1951**]

Tube-length in photomicrography. *J. R. microsc. Soc., 3rd series* **62**, 112–15. [*Trans.*, 8] (1R)

1943

Colour of red blood corpuscles. [letter] *Nature* **152** (3855), 331

Curdled milk for supporting tissues in celloidin embedding. *Stain Tech.* **18** (3), 113–15. (4R)

†The discovery of the uses of colouring agents in biological microtechnique. *J. Quekett microsc. Club, series 4*, **1** (6), 256–75. (83R) [repr. **1945**]

1944

★The structure and chemical composition of the Golgi element. *Q. Jl microsc. Sci.* **85**, 1–71 + 2 pl. (diags. 83R)

1945

★**Cytological technique. 2nd ed.** Methuen. vii, 211 p. (147R)

The discovery of the uses of colouring agents in biological microtechnique. Williams & Norgate. 22p. (QUEKETT MICROSCOPICAL CLUB. Monograph No. 1) [repr. from *Journal*, **1943**]

Freedom in science. [letter] *Time & Tide* **26** (50), 1052–3

Hot baths and human fertility. [letter] *Lancet* **248** (1), 771–2. (4R)

★**Science and the planned state.** Allen & Unwin. 120 p. (110R) [*see also* **La Science et l'état planifié. 1946;** *and* **Freiheit und Wissenschaft. 1950**]

1946

The course of the controversy on freedom in science. [letters] (*with* Arthur G. TANSLEY) *Nature* **158**, 574–6; 796–7. (1R) [*see also* editorial, 565–7]

A critique of materialism. *Hibbert J.* **45** (1), 31–7. (11R)

Establishment of cytochemical techniques. [letter] (*with* F. K. SANDERS) *Nature* **158**, 129. (1R)

The growth of biological ideas. (pp. 144–72, chap. 7. 3R) *in* MEES, Charles Edward Kenneth. **The path of science.** Chapman & Hall. xii. 250 p.

★Histochemical recognition of lipine. *Q. Jl microsc. Sci.* **87**, 441–70 + 1 pl. (14R)

La Science et l'état planifié. Paris, Librairie de Médicis. 132 p. (*Collection 'Civilisation'*) [Translation of **Science and the planned state. 1945**, with preface by Jean THIBAUD]

1947

★Further remarks on the histochemical recognition of lipine. *Q. Jl microsc. Sci.* **88**, 463–5. (il. 3R) [*see also* CAIN, A. J. Baker's acid hematein test for phospholipines. *ibid.*, 467–78. (8R)]

★The histochemical recognition of certain guanidine derivatives. *Q. Jl microsc. Sci.* **88**, 115–21. (il. 12R)

★The seasons in a tropical rain-forest [final papers, read 7.6.45]:

      6. Lizards (*Emoia*). *J. Linn. Soc.* (*Zool.*) **41**, 243–7. (gr., tabs. 5R)

      7. Summary and general conclusions. *J. Linn. Soc.* (*Zool.*) **41**, 248–58. (gr. 11R)

   † also item 567 in MORTON, Leslie T. *Garrison and Morton's medical bibliography*, 2nd ed., 1965 (p. 50, *under* Histology).

1948

*The cell-theory: a restatement, history, and critique:
    1. *Q. Jl microsc. Sci.* **89** (1), 103–25. (91R)

A new method for oblique microscopical illumination. *Q. Jl microsc. Sci.* **89**, 233–8. (diags. 1R)

Proposed central publication of scientific papers. (*with Sir* George P. THOMSON) *Nature* **161**, 771–2. [*Delivered at* Conference on Science Information Services, Royal Society; *see also* pp. 469–70]

Scientific basis of kindness to animals. UNIVERSITIES' FEDERATION FOR ANIMAL WELFARE. [UFAW] 10p. (3 il.)

A simple method for phase-contrast microscopy. (*with* D. A. KEMPSON *and* O. L. THOMAS) *Q. Jl microsc. Sci.* **89** (7), 351–8. (diags. 3R)

†The Soviet genetic controversy. *Time & Tide* **29** (51), 1297–8

*The status of the protozoa. *Nature* **161**, 548–51 (12R); 587–9. (20R)

1949

*The cell-theory: a restatement, history, and critique:
    2. *Q. Jl microsc. Sci.* **90** (1), 87–108. (3 il. 84R); + Addendum, *ibid.*, **90** (3), 331. (3R)

Further remarks on the Golgi element. *Q. Jl microsc. Sci.* **90** (3), 293–307. (diag. 25R)

A simple method for phase-contrast microscopy: improvements in technique. (*with* D. A. KEMPSON *and* P. C. J. BRUNET) *Q. Jl microsc. Sci.* **90** (3), 323–9.† (tab. 2R)

The sticking of crabs. *UFAW Courier* **3**, 17–18

1950

*Cytological technique: the principles and practice of methods used to determine the structure of the Metazoan cell. 3rd ed.** [2nd ed. repr.] Methuen. vii, 211 p. (4 il. 147R)

Freiheit und Wissenschaft. Bern, A. Francke. 163 p. (177R) [Selections from **Science and the planned state. 1945** and **The scientific life. 1942**; translated, with additions, by Heinrich SCHWARZ]

'Heterochromatin'. [letter re undesirability of this word] (*with* H. G. CALLAN) *Nature* **166**, 227–8 (9R)

Notes on phase-contrast. [letter] *Microscope* **8**, 103–4

Studies near the limit of vision with the light microscope, with special reference to the so-called Golgi-bodies. *Proc. Linn. Soc. Lond.* **162**, 67–72 (30R) + discussion, 82–3. [*Part of* A discussion on morphology and fine structure (*with* C. F. A. PANTIN *and* L. E. R. PICKEN), *ibid.*, 65–83]

1951

*The absorption of lipoid by the intestinal epithelium of the mouse. *Q. Jl microsc. Sci.* **92**, 79–86 + 1 pl. (20R)

Cytological techniques:—A. Preparation of tissues for microscopical examination and histochemistry. (pp. 1–19. 60R) *in* BOURNE, G., *ed.* **Cytology and cell physiology. 2nd ed.** OUP. 540 p.

*Experiments on the illumination of microscopic objects. (*with* A. Stewart BELL) *J. Quekett microsc. Club*, series 4, **3** (4), 261–75. (tabs., diags. 12R)

The 'Golgi substance'. *Nature* **168**, 1089–90. (3R)

Note on certain recent papers on the so-called Golgi apparatus. *J. R. microsc. Soc.* **71**, 94–6. [*Trans.*, 3] (diag. 13R)

Remarks on the discovery of cell-division. *Isis* **42** (4), 285–7. (diag. 17R)

Remarks on the pronunciation of scientific terms. *Sch. Sci. Rev.* **32** (118), 284–6

*OXFORD. UNIVERSITY. *University of Oxford Exploration Club.* **New Hebrides papers: scientific results of the Oxford University expedition to the New Hebrides, 1933–34;** articles extracted from periodicals and issued in a collective volume; edited by Charles Sutherland ELTON with the assistance of J. R. BAKER. Oxford, 25s. 259 p. [for original papers, *see* **1935, 1936** (1–4), **1938, 1940** (5), **1947** (6–7)]

1952

*Abraham Trembley of Geneva, scientist and philosopher, 1710–1784.** Edward Arnold, 35s. xix, 259 p. (52 il. + port. 193R)

*The cell-theory: a restatement and critique:
    3. The cell as a morphological unit. *Q. Jl microsc. Sci.* **93** (2), 157–90. (7 diags. 172R)

Remarks on the effect of the aperture of the condenser on resolution by the microscope. *Q. Jl microsc. Sci.* **93** (4), 375–7. (tab. 2R)

Surveying in the New Hebrides. [letter] *Geogrl J.* **118** (1), 114–15

---

† Reprinted 1949 in SOCIETY FOR FREEDOM IN SCIENCE, *Occasional Pamphlet* No. 9. (pp. 10–14)

1953

*The cell-theory: a restatement, history, and critique:
    4. The multiplication of cells. *Q. Jl microsc. Sci.* **94**, 407–40. (21 il. 116R)
Coup d'œil sur la controverse du 'Golgi':
    2. La question des homologies. (translated by Prof. A. POLICARD) *Bull. Microsc. appl.* **3**, 96–101.
    (23R)
Cytological technique [in Chinese; pirated edition, based on 2nd ed., **1945**] (2), v, 167 p. (4 figs., tab.)
The expressions 'Golgi apparatus', 'Golgi body', and 'Golgi substance'. *Nature* **172**, 617–18. (10R)
Freedom and authority in scientific publications. Oxford. 13, (1)p. (SOCIETY FOR FREEDOM IN SCIENCE.
    Occasional Pamphlet No. 15) [Originally given at conference on *Wissenschaft und Freiheit*, Hamburg,
    Jan. 1952; published by permission of Congrès internationale pour la Liberté de la Culture]
Golgi bodies in the male germ-cells of *Vaginula maculata*. *Nature* **172**, 690. (10R)
Miscellaneous contributions to micro-technique. (*with* Barbara M. JORDAN) *Q. Jl microsc. Sci.* **94** (3),
    237–42. (14R)
Nouveau coup d'œil sur la controverse du 'Golgi':
    1. Les techniques du 'Golgi' et les objets qu'elles révèlent. (translated by Prof. A. POLICARD) *Bull.
    Microsc. appl.* **3** (1), 1–8. (diags. 23R)

1954

The mirror-cover: an aid to axial illumination. *J. Quekett microsc. Club, series 4* **4** (2), 110–12. (phot. 3R)
A study of the osmium techniques for the 'Golgi apparatus'. *Q. Jl microsc. Sci.* **95** (3), 383–8 + 1 pl. (8R)
What is the 'Golgi' controversy? *J. R. microsc. Soc.* **74**, 217–21. [*Trans.*, 18] (15R)

1955

*The cell-theory: a restatement, history and critique:
    5. The multiplication of nuclei. *Q. Jl microsc. Sci.* **96**, 449–81. (14 diags. 116R)
The cytoplasmic inclusions of a mammalian sympathetic neurone: a histochemical study. (*with* W. G.
    Bruce CASSELMAN) *Q. Jl microsc. Sci.* **96**, 49–56. (diag. 27R)
*English style in scientific papers. *Nature* **176**, 851–2. (5R)
Experiments on the humane killing of crabs. *J. mar. biol. Ass. U.K.* **34** (1), 15–24 + 2 pl. (8R)
A new microscope-lamp. *J. R. microsc. Soc.* **75**, 43–7 + 1 pl. (6R)
Remarks on the pointolite as a light-source in microscopy. *J. R. microsc. Soc.* **75**, 96–9. [*Trans.*, 9] (3R)
A simple pyronine/methyl green technique. (*with* Barbara M. JORDAN) *Q. Jl microsc. Sci.* **96**, 177–9. (6R)
Sir Arthur Tansley, 1871–1955. [A tribute] Oxford. (3) p. (SOCIETY FOR FREEDOM IN SCIENCE. *Occasional
    Pamphlet* No. 16)

1956

The histochemical recognition of phenols, especially tyrosine. *Q. Jl microsc. Sci.* **97**, 161–4. (16R)
The house-cricket (*Acheta domestica*) as a laboratory animal. (*with* Barbara M. JORDAN) *Entomologist* **89**
    (1115), 126–8. (7R)
Improvements in the Sudan black technique. *Q. Jl microsc. Sci.* **97**, 621–3. (4R)
Mitochondrial *Nebenkern* of the water-boatman. *Nature* **177**, 1039–40. (2 phots. 4R)
Nomenclature in microscopy. [letter] *Nature* **177**, 194.
The pronunciation of scientific terms. *Sch. Sci. Rev.* **37** (132), 201–5. (2R)

1957

The effect of acetic acid on cytoplasmic inclusions. *Q. Jl microsc. Sci.* **98**, 425–9 + 1 pl. (21R)
Freedom in research. [Paper originally delivered at Uganda Scientists' Club, 6th Aug., at Makerere
    College, Kampala] Oxford. (5) p. (SOCIETY FOR FREEDOM IN SCIENCE. *Occasional Pamphlet* No. 17)
The Golgi controversy. *Symp. Soc. exp. Biol.* **10**, 1–10. (pl., diags. 24R) (*Mitochondria and other cyto-
    plasmic inclusions*)
Lipid globules in cells. *Nature* **180**, 947–9. (3 figs. 18R)

1958

Fixation in cytochemistry and electron microscopy. *J. Histochem. Cytochem.* **6** (5), 303–8. (diags. 34R)
    (*Symposium on problems of fixation in histochemistry*)
Note on the use of bromophenol blue for the histochemical recognition of protein. *Q. Jl microsc. Sci.* **99**,
    459–60. (6R)

*Principles of biological microtechnique, a study of fixation and dyeing.** Methuen, 45s. 357 p.
  (26 il., *incl.* 9 pl. (2, col.) 562R) [Appendices include: *Use of the word 'chromatin'; Notes on spelling*]
Science and the sputniks. Oxford. 6 p. (SOCIETY FOR FREEDOM IN SCIENCE. *Occasional Pamphlet* No. 18)

1959

Å or Mμ in electron microscopy? [letter] *Nature* **183**, 416
Towards a solution of the Golgi problem: recent developments in cytochemistry and electron microscopy.
  J. R. microsc. Soc. **77**, 116–29 + 1 pl. [*Trans.*, 6; received 10.6.58] (diag. 82R)
The use of the Abbe substage in phase-contrast microscopy. *J. R. microsc. Soc.* **77**, 144–5. [*Trans.*, 11;
  received 18.10.57] (3R)
The use of the Philips 'mirror-condenser lamp' in microscopy. (*with* Wilfrid LLOWARCH) *Q. Jl microsc.
  Sci.* **100**, 321–4. (diag. 5R)

1960

*Cytological technique: the principles underlying routine methods. 4th ed.** [rewritten] Methuen,
  15s. xv, 150 p. + 1 pl. (6 diags. 190R)
*Experiments on the action of mordants:
  1. 'Single bath' mordant dyeing. *Q. Jl microsc. Sci.* **101**, 255–72. (gr., diags., tab. 15R)

1961

Baker and Ruyonga. [Outlines Samuel Baker's adventures with Uganda chieftains, 1864–1872, and author's
  meeting with Ruyonga's descendants, 1957] *Uganda J.* **25** (2), 214–16 + 1 pl. (9R)
The controversy on freedom in science in the nineteenth century. (pp. 89–96. 19R) *in* **The logic of personal knowledge: essays presented to Michael Polanyi on his seventieth birthday.** Routledge.
  [published separately, **1962**]
Freedom in science in South Africa. Oxford. (2) p. (SOCIETY FOR FREEDOM IN SCIENCE. *Occasional Pamphlet*
  No. 21]

1962

The controversy on freedom in science in the nineteenth century. Oxford. (13) p. (SOCIETY FOR FREEDOM
  IN SCIENCE. *Occasional Pamphlet* No. 22] [first published, **1961**]
*Experiments on the action of mordants (effect of mordanting):
  2. Aluminium-hematein. *Q. Jl microsc. Sci.* **103**, 493–517 + 1 pl. (gr. 23R)

1963

The fine structure produced in cells by primary fixatives:
  1. Mercuric chloride. (*with* Barbara M. LUKE, *née* JORDAN) *Q. Jl microsc. Sci.* **104**, 101–6 + 1 pl. (3R)
A further note on the use of the Philips 'mirror-condenser lamp' in microscopy. *Q. Jl microsc. Sci.* **104**,
  279–80. (1R)
*New developments in the Golgi controversy. *J. R. microsc. Soc.* **82** (2), 145–57. (8 diags. 49R)
The use of the word 'dense' in micro-technique. *Q. Jl microsc. Sci.* **104**, 107–8. (2R)

1964

A substitute for Ehrlich's haematoxylin and similar dyes used in biological microtechnique. *Sch. Sci. Rev.*
  45 (156), 400–1. (5R)

1965

*The fine structure produced in cells by fixatives. *J. R. microsc. Soc.* **84**, 115–31. (8 phots., diag. 42R)
The fine structure produced in cells by primary fixatives:
  2. Potassium dichromate. *Q. Jl microsc. Sci.* **106**, 15–21 + 2 pl. (25R)
'Rionga's Island': a note on Samuel Baker's first expedition to Central Africa. *Geogrl J.* **131** (4), 526–8.
  (map. 4R)
Samuel Baker's route to the Albert Nyanza. [1864–5] *Geogrl J.* **131** (1), 13–20. (2 maps. 2R)
*The use of methyl green as a histochemical reagent. (*with* Elizabeth G. M. WILLIAMS) *Q. Jl microsc. Sci.*
  106, 3–13. (36R)

1966

Charter centenary of the Royal Microscopical Society. *Nature* **210**, 564–5. (10R)
*Cytological technique. 5th ed.** [partly rewritten; incl. more electron microscopy, and chapter on
  mordants revised] Methuen, 25s. xi, 149 p. (6 diags. 209R) [also published in paperback]

*The design of the biological student's microscope. *J. R. microsc. Soc.* **86** (1), 59–67. (diag., 2 phots., 3R)

*Experiments on the function of the eye in light microscopy. [*Based on* Presidential address, 5.1.66] *J. R. microsc. Soc.* **85** (3), 231–54. (phot., diags., gr., tabs. 26R)

The fine structure resulting from fixation by formaldehyde: the effects of concentration, duration, and temperature. (*with* J. M. McCRAE) *J. R. microsc. Soc.* **85** (4), 391–9. (2 phots., 10R)

*The ultrastructure of the 'Podura' scale (*Lepidocyrtus curvicollis*, Insecta, Collembola), as revealed by whole mounts. (*with* J. M. McCRAE) *J. ultrastruct. Res.* **15** (5–6), 516–21. (5 phots. 19R) [received, 8.10.65]

1967

A further study of the *Lepidocyrtus* ('Podura' scale (Insecta, Collembola). (*with* J. M. McCRAE) *J. ultrastruct. Res.* **19**, 611–5. (4 phots. 9R)

The student's microscope. (Ergonomic study, concerned with convenience of student and demonstrator) *Nature* **215** (5098), 237–8. (phot. 4R)

1968

*Cro-Magnon man, 1868–1968. *Endeavour* **27** (101), 87–90. (3 figs., 48R)

Observations on the cranium of Broken Hill man, *Homo rhodesiensis* Woodward. *Zeitschr. Morph. u. Anthrop.* **60** (2) *in the press*

*The student's microscope: a new prototype. *Proc. R. microsc. Soc.* **3** (1), 4–11. (3 phots., 2 figs. 3R)

Appendix III

To Advanced Students about to undertake Research under my Supervision

JOHN R. BAKER

It will, I believe, be helpful to my research-pupils and to myself if, before we start working together, I make a few comments and suggestions.

It appears to me that my main formal duties are the following:

(1)   To propose subjects of research and to advise on the suitability of subjects suggested by pupils.
(2)   To teach techniques connected with the research.
(3)   To put pupils in touch with the literature of the subject and (if possible) with other people carrying out research on allied subjects.
(4)   To watch the progress of the research and advise changes of plan in accordance with the way in which it develops.
(5)   To check the validity of discoveries claimed by pupils, and to maintain high standards of accuracy.
(6)   To advise when the time has come to write up discoveries for publication; to suggest the lines on which papers should be composed; to suggest improvements in the first drafts of papers submitted to me in typescript; and to advise on suitable journals for publication.
(7)   To indicate when the time to hand in the 'thesis' has arrived.

I am anxious to devote a considerable amount of time to each pupil. I would ask for consideration in the matter of choice of time for discussions. As a general rule, research-pupils have more time at their disposal than they have ever had since they were small children, and more than they will ever have again until they retire from active life. In this they differ very markedly from their supervisor, who has lectures and demonstrations to prepare and give, and examinations to set and correct, as well as other departmental duties (staff-meetings to attend, departmental apparatus to obtain and maintain in good order, official letters to write, testimonials to compose, etc.). In my own case, it is also necessary to devote a lot of time to editorship of the *Quarterly Journal of Microscopical Science* and to acting as Secretary of the Society for Freedom in Science. I am also engaged on a history and critique of the cell-theory. I must also do a good deal of reading to keep myself up to date in the subjects that I teach. Beyond all this, I have the responsibilities of a large family. The various activities leave me very much less time than I would wish for what is to me the most precious part of my scientific work—my own research.

It is not possible to carry out research or to write books and papers if one is being continually interrupted. Unbroken periods of two or three hours or even more are often necessary for consecutive thought. I greatly prefer to have discussions with pupils at pre-arranged times, and in general I think that a few long talks are preferable to many short ones, both from my pupils' and from my own point of view. I like sometimes to invite pupils to my house for prolonged discussions of their work. I often visit the cytological laboratory for chats with pupils, but I think it only fair that I should choose the times for these chats. I would ask pupils not to 'catch' me when I am entering or leaving my laboratory. In particular, I never want to have a discussion before I have read my letters on arrival at the laboratory in the morning, nor shortly before giving a lecture, nor between 1 and 2 p.m. (when I eat my sandwiches in the laboratory, and relax). I would ask pupils who want to see me, but have not arranged a special time, to come between 3 and 4 p.m.

At long intervals it happens that a really exciting discovery is made, which demands the immediate attendance of the supervisor. Under these circumstances I am willing to come at once, whatever the time may be.

I hope that research-pupils will accept these suggestions in the spirit in which they are offered. I believe that if their own time for research were as limited as mine, they would guard it just as jealously and try just as hard as I do to protect themselves from unnecessary interruptions.

John R. Baker.

Index

Figures in italic type refer to topics discussed in J. R. Baker's published writings (Appendix II), and the figures in brackets after these entries are the number of publications on this topic that appear on the page referred to. Figures in bold type refer to illustrations.

absolute quantitative measurements, inaccuracy of, 278
absorption, 252
accuracy, absolute, of phase change measurement, 275
  of individual nucleolar dry mass values, 279–287
  of linear measurement under a microscope, 275
acetone, 116, 117, 119, 120, 122, 132
acid haematein method, 150
  on cerebrosides, 151
  effect of methanol-chloroform on, 151–152
  mechanism of, 151
  and pyridine extraction test, 152
  results of tests on pure substances, 150
  significance of double bonds in, 151
  use of bromination as control, 152
acid haematein test, applied and nuclear phospholipids, 158, 161–164
acid phosphatase, 138, 170, 172–180, **173**, **175**, **177**, 232
  in acridine orange particles, 91
  in neutral red granules, 91
acridine orange, 176, **177**
  particles, in HeLa cells, 88, 91, 92, 93
Actinosphaerium, 87
active transport, 315
'activity', 68, 73
  biological, 185
  enzyme, 'latent', 265, 267
    'manifest', 265, 267
  metabolic, 185
  structural, of phospholipids, 118
adaptation, 7
adenosine triphosphatase (ATPase), 170, **173**, 174, **175**, 179
  triphosphate, 318
adhesive gland, in chick embryo explants, 343, 344, **345**, **352**
adsorbed dyes, 61
Aedes aegypti (female), peritrophic membrane in mid-gut, 337
  principal (digestive) cells in mid-gut, 321, 322, 324, 325, 329
'affinity', definition of, 67
  of dyes, 67 *et seq.*
  standard, 67
aggregation of adsorbed dyes, 61
  of dye molecules, 68, 75
  of dyes in aqueous solution, 60
  number, 68
  alcian blue, 70, 71
alcohol, 116, 119, 122, 132
  removal of dye by, 73, 74
aldehyde, fixation, 124, 126, 191
  fuchsin, 70
  reaction, 131
alizarin red S, 195, 197, 198, 205
alligator, 204
amacrine cells, 372–380
Amblystoma, 238, 240
  mexicanum, 358, 359
amino acid content, 184
amino acids, 126

'amitotic' division, 278
ammonium molybdate, 290
Amoeba, 178
amoebocyte, 254
Amphibia, 238
amylase, 191
anabiosis, 4
  anabiotic state, 9
analogy, 6
  functional, 10
ancestry, common, 9
aniline blue orange G method, 150
animals, humane treatment of, *414–415 (3)*
  laboratory, *410 (2)*, *414–415 (4)*
Anisolabis litteria, peritrophic membrane in mid-gut, 337
Annelid, 237, 241
annulate lamellae, 204, 205
Anopheles stephensi (larva), principal (digestive) cells in mid-gut, 321–325, **326**, 329, **332**, 333, 335
anoptral-contrast, 191, 194
  objectives, 188, 189
antemedium, 117, 122
Antheraea pernyi (Assam silk moth), 186, 189, 191, 192, 193, 195, 197, 205
Anthozoa, 8
anthropology, *410–411 (3)*, *417 (2)*
'anucleolate' mutant *Xenopus laevis*, 281
aorta, 238, 247
Aoyama Golgi technique, 79
apatite, 256, 258
apical disk, 252, **254**
Apis indica, peritrophic membrane in mid-gut, 336, 337
  principal (digestive) cells in mid-gut, 321–325, **329**, 331, 333
  regenerative (replacement) cells in mid-gut, 336
Apis mellifica, **165**
Aplysia, 174
apocrine secretion, 227, 240, 247
apparato reticulare interno (Golgi), 395
Arachnid, 237
aragonite, 251, 258
Araldite, 184, 187
Araldite-embedded sections, advantages of for light microscopy, 342, 354–355
areolar tissue, 212, 213
argentophilia, 80, 81
arginine, estimation by microdensitometry, 266
  histochemistry, 40
Aristotle's lantern, **254**, 256
arteriosclerosis, 155
artifact(s), 3, 7, 8, 19, 97, 115, 183
  concept, 115
  sectioning, 186
arthritis, 225
Astacus, 397
Asteroidea, 251
astrocytes, 371
atom(s), 7
  carbon, 10
  hydrogen, 7
automatic photometric scanning, 290

28—C.S.I.

autophagic vacuoles, 178
'autophagic vesicles', 91
autoradiography, 226
Azure A, 72, 73, 76
    B, 290

background density, 295
    gradient, 295
bacteria, 5
ball of reagent, 188
Baker, John R., an appreciation, 407–408
    Samuel, *416* (*3*)
Balbiani rings, 270
basement lamella, 240
basic fuchsin, 186, 198
basiphil strands, principal (digestive) cells of insect mid-
    gut, 322, **323**, 324, **326**, **327**
basiphilia, 79–81
    principal (digestive) cells of insect mid-gut, 323,
        324
bathochromasy, 290
Beer's law, 71
benzimidazole, 342
benzopurpurine, 60, 61
benzpyrene, 149
Bernal's ideas for the organization of science, opposition
    to, *412* (*2*), *see also* Society for Freedom in
    Science, *413*
bichromate-silver technique, 106, 108, **110**, **113**
biliary canaliculi, 112
biochemical definitions of cellular structures, fallacies
    relating to, 178
biochemistry and morphology, a unity, 394
    'non-disruptive', 272
biography, Abraham Trembley, *414*
    Sir Arthur Tansley, *415*
biology, *410–411* (4), *413*
birth control, *410–413* (*22*)
bismarck brown, 70, 87
black-leaf louse, 397
blebbing, 200
blood, coagulation, 9
    vessels, 185
blowfly (larva), peritrophic membrane in mid-gut, 337
    principal (digestive) cells in mid-gut, 322, 329
Bombyx mori, goblet cells in mid-gut, 335
    principal (digestive) cells in mid-gut, 325
Bodian's technique, 98
bone, 5, 237, 238, **255**, 256, 258
    dinosaur, 184
Bouin's fluid, 'weak', 152
bound lipids, 152–155
brachiopod, 258
brain, 128, 247
    human, 3
breeding seasons, *411–414* (*20*)
brilliant scarlet, 3R, 127
Brunner's gland, 397
brush-border, insect mid-gut epithelium, histochemical
    reactions, 321, 322
    morphology, 321, **322**, **323**, 324, **325**, **326**, **327**, **328**,
        329, **332**, 335
buffer, cacodylate, 116, 117, 119, 126, 128
    phosphate, 126
    storage, 116, 117, 120, 128
buffered osmium tetroxide, as fixative, 51, 54
Bürker count (with haemocytometer), 308
    factor (for haemocytometer), 308
    haemocytometer, 307
byssus threads, 237

Cajal preparations, 373
calcamphora(e), 200, 203, 205
calcampulla(e), 200, 203, 205
calcification, 195, 197, 200, 203, 252
calcite, 251–260, **253**, **255**, **257**, **259**
calcium, 195, 197, 198, 203, 205
    accumulation, sequestration, 200, 203
        sites of, 197, 198
    deposits, in mitochondria, *post mortem*, **383**, 386
    histochemical methods for, 195
    carbonate, 251–260
    chloride, 117, 128
    ferricyanide, 128
calcogenesis, 198, 205
calcospherite(s), 198, 200, 203, 205
    secretion of, 198
Callosobruchus analis (larva), goblet cells in mid-gut, **328**,
    335
    peritrophic membrane in mid-gut, 337
    principal (digestive) cells in mid-gut, 321, 323–325,
        328, 331, **332**
    regenerative (replacement) cells in mid-gut, **334**, 335,
        336
capillaries, localizations of TPPase, **108**
capsular layer, 192
Carausius morosus, 165
carbohydrates, in connective tissue ground substance,
    221
carbon tetrachloride poisoning, 389
carrageenin granuloma, 238, 240
Cartesian diver technique, 316
cartilage, 3, 238, 240, 241
    fibro-, 215
    hyaline, 215
    spongy, 216
    staining of, 72, 76
casein, 161
cell area, from contiguous densitometer traces, 295, 300,
    301, 302
    breakdown, damage, 193, 195
    membrane, 240, 241, 243, 247
    structure, classified solely by enzymological activity,
        178
    surface, roughness of, 305–312
cell-theory, *413–415* (*9*)
cells, chemical constituents, 9
    fusing, 278, 284, 285, 291, **293**, 296, 298, **299**, 300, 301
    ganglion, 9
    living, 3, 4, 8, 9
    membrane(s), 8
    nerve, 7
    organelles, 4, 9
    surfaces, 4
    uni-nucleate, 291, **293**, **299**, 300, 301
cellular differentiation, 341–355
centrifugation, differential, 7
Cepaea hemotalis, **255**, 256
cephalin, 161
cephalines, 119
cerebral cortex, 378
cerebral ganglia, *Lumbricus terrestris*, 108, **109**
cerebroside, 150
cetyl pyridinium chloride, 141
Ceylon, rain-forest, *412* (*2*)
charges, negative, 118
    positive, 118
chemical activity, 126
chick, 238, 240
Chironomus sp., 389
chitin, 258

chloramphenicol, 341, **343**, **352**, 353, 354, 355
    D and L-threo isomers, 354
cholesterol, 119, 120, 132, 150
    oleate, 150
cholinesterase, 138–139
chondroblasts, 225, 238
chondrocytes, 240
chondroitin sulphate, 237, 240, 241
chromatin, masses, 193
    nuclear, 131
    staining of, 72, 73, **74**, 75
chromatography, thin layer, 120
chromium, reaction with phospholipids, 151
chromosomes, 204
    polytene or 'giant', 278
    strandedness, 357
Cidaris, 256
ciliated cells, 345, 353
cinemicrography, 284–285
Clausius-Clapeyron equations, 33
cleavage, crystalline, 254
    plane, 252
    steps, **255**, 256
clonal culture technique of Konigsberg, 296
Cnidaria, 6
coating, metallic, 256
cochlea, 315, 318, 371
cochlear endolymph, 318
    ganglion, 318
    perilymph, 318
    wall, 315
coefficient, activity, 68
    distribution, 68
Coelenterate, 237
coelomocyte, 251
Coleoptera, goblet cells in mid-gut, 335
collagen, 3, 4, 8, 225–234, 237–247, **253**, 258, 260
    fibres, 'sewing' echinoderm skeleton together, **252**, 260
    solubility, 238
colloidal metals, 141, **142**
colour saturation, 189
colouring procedures, selective, differential, 184
combined observations, 184, 186, 187, 191, 197, 205
comparative quantitative measurements, biological information from, 277, 280–302
comparison eyepiece, 72
conchiolin, 258
Congo red, 60, 61
connective tissue, 237–247
    cells, 192, 195
    changing concepts of, 209–222
    collagenous, 219
    description, Bichat, 210
        Kolliker, 212
        Maximow and Bloom, 219
        modern, 220
        Rollet, 218
        Schwann, 211
        Virchow, 214
    elastic, 216, 219, 220
    ground substance, 220
    of insects, 192
contraception, *410–413 (22)*
contractile machinery, 5
    structures, 8
    tissue, 5
contrast, 115, 116, 124, 126, 127, 128, 131
    differential, 115
    electron, 127, 131, 200, 203

contrast—*contd.*
    enhancement of, 118, 127, 128, 185
    equal, 116, 122, 131
    inherent, 117
    measurable increase in, 144
    negative, 116, 120, 122, 124, 131
    positive, 116, 122, 124, 128, 131
    positive membrane, 117, 124
    relative, 189
    selective, 128
control procedures, 191, 192
constant, equilibrium, 67
    gas, 67
    temperature chamber, around microscope, 28
copper ions, in silver techniques, 98, 99
coral, 258
cornea, 237
Corti's organ, 318
'coupled tetrazonium test', 140
crabs, humane killing of, *415*
Crinoidea, 251
cross-linking, 126, 131
cryptobiosis, 4
crystal, 251–260
    strength, 258
    violet, 61, 76
crystals, 7
crystallites, 252, 256
crystallographic axis, 252
cryostat, 23–49
    anti-roll plate, 23, **24**, **26**, 29–30, 31
    definition, 23
    description, 23–25, **24**
    design of, 25
    environmental control in, 25–27
    fixed tissues, 23, 44, **45**
    freezing-drying, 23, 30–31
    heterothermic cutting conditions, 25–27, 30, 31, 49
    homoeothermic cutting conditions, 25, 26, 30, 31
    microtome, 27–28
        Cambridge rocker, 28
        knife, 28–29
        lubrication, 27
        Minot, 27
    sections, 197, 198
    ultra-thin sectioning, 31
    unfixed tissues, embedding, 32
        fixation and mounting, 39–40
        freezing, 32–35
        handling of, 31–44
        post-fixation, 40, **41**
        protective media, 38–39
        thawing, 35–38
cuirass, 251
Culex annulatus, principal (digestive) cells in mid-gut, 325
Culex fatigans (female), goblet cells in mid-gut, 335
    principal (digestive) cells in mid-gut, 321, 323, **327**, 328, 329, **330**
    regenerative (replacement) cells in mid-gut, 336
Culex pipiens, principal (digestive) cells in mid-gut, 325, 328, 333, 335
cultured cells, 176, 178
curing, 118
cuticle, 237
Cuverian tubules, 237
Cynips folii, 164, **165**
cytoblastema, 211
cytochemistry, *413*, *415 (2)*
cytology, *411*, *413–414 (4)*, *416 (2)*

'cytolysomes', 91
cytoplasm, in cells of middle ear, 316
    ultra-violet absorbing, 299
cytoplasmic inclusions, in cells of middle ear, 316
cytosomes, 178

dahlia violet, 316
Dalton complex, 396
Darwin, 5
degeneration, of neurons, 371–380
    transneuronal, 371, 375–380
dehydrating agents, 117, 118, 119, 120, 132
    fluids, 118
dehydration, 116, 117, 118, 122, 124, 131, 132, 192
    acetone, 124
    GMA, 122, 124
dehydrogenases, 270
    quantitative histochemistry of, 270–271
Deilephila euphorbiae, 398
dendritic spines, 371, 378
dense fibrous masses, in 'anucleolate' cells, 281
densitometer, 275, 294
    double beam automatic recording, 294
    traces, **292**, 294, 295
densitometric analysis, 290
    measurements, 127, 275, 276, 277, 292–302
densitometry, for basic proteins, 266–267
    in histochemistry, 265–267
    review of some of the literature, 266
    with interference microscopy, 275, 276
    for nucleic acids, 266
    scanning, 266
    standard densities for calibration, 294
    ultraviolet, 266, 277
density, of background, 295
    differences, 294
    distribution charts, **293**, 295, **296**
'density maps' of stained cells, **293**, **296**, 300
depth of focus of 4 mm objectives, 294
desmosomes, **401**
detergent, uptake by lysosomes, 92
diastase, 191
diazo coupling techniques, 138
diazopthalocyanins, 140
dictyosomes, 169–172, **171**, **173**, **175**, **177**, **179**, 395
dielectric constant(s), 127
differentiation, cellular, 6
    mitochondrial changes associated with, 361, **362**, 363, **364**, 365
    definition of, 278–279
    of tissues, role of RNA in, 277–278
diffraction plate, 188
digestive gland of snail, 197
dimensionless quantities, 305
dinitrofluorobenzene, 139
Dionaea muscipula, 5
Diptera, goblet cells in mid-gut, 335
    peritrophic membrane in mid-gut, 337
double bonds, 126
DNA, 4, 5, 7
    abnormality in, 286
    C-values, 357–359
        definition of, 357
        electron histochemistry of, 139
        hybridization, 359
    interphase, 166, 278
    molecule, structure and function, 277
    protein-complexes, 266
    quantitative estimations of, 266
    staining of, 76

DNA—contd.
    stoichiometry of, 266
    synthesis, 358
DNase, 357
drag coefficient, 305, 306, 308, 309
Drosophila sp., 359
    melanogaster, **165**
        principal (digestive) cells in mid-gut, 325, 328, 333, 335
        regenerative (replacement) cells in mid-gut, 336
    muscle mitochondria of, 361, **362**, 363
dry mass, of cells, by interference microscopy, 275
    of nucleolar material, 279–281, **280**, **286**, **287**, 288–289
    of nucleoli, by interference microscopy, 279–282, 286–288
    total values only provided by interference microscopy, 279
Durcupan A, 116, 119, 120, 122, 124, 132
dye-lake, 197
    method, 198
dyes, 413 (3), 415–416 (3)
    acid and basic, 127
    aggregation of, 59, 60–61
    as electrolytes, 59–60
    metachromasy, 61
dyestuffs, light microscopical, 185
dynamic cytology, 195–197, 277–303, 394, 395–402
dystrophic mice, Bar Harbor strain, 287
dystrophia myotonica, 286
dystrophy, muscular, characteristics of, 286
Dysdercus cingulatus, principal (digestive) cells in mid-gut, 321–324, **325**, 329, 331, **332**, 333

Earle's fluid, 307
earthworm, 237
echinoderm, 204
    skeleton, 251–260
echinoderms, 237
Echinoidea, 251, **253**, **254**, **255**, **257**, **259**
Echinus esculentus, **253**, **254**, **255**, 256, **257**, **259**
    miliaris, 254
ecology, 410–412 (4), 413
ectodermal layer of cells of middle ear, 315
elastin, 225
electron(s), 7
    autoradiography, 238
    beam, 124
    contrasting reagents, 185
    histochemistry, 184
        acid phosphatase, 136, **137**
        alkaline phosphatase, 136
        approaches, 136
        carbohydrates, 140–141
        cholinesterases, 138
        dehydrogenases, 138
        enzymes, 136–139
        future prospects, 144–145
        Gomori-type methods, 136–138, 144
        noradrenalin, 141
        nucleic acids, 139
        nucleoside diphosphatases, 138
        problems and limitations, 136, 141–144
        protein groups, 139–140
        reagent penetration, 143
        sulphydryl groups, 139
    micrographs, 8
    microscope, contrast enhancement by low voltages, 144
        differentiation of chemical elements, 135
        limitations, 135
    microscope, serial section reconstruction, 135

electron(s)—*contd.*
  microscopy, criticism of, 221
    octopus brain, 371–380
  scattering power, capacity, 115, 116, 120, 122, 127, 128
'electron stain', 76
embedding, 116
  media for cryostat work, 32, 44
  medium, 117, 118, 124
embryo, chick, 280, 282, 283, 341, 354
empiricism, 95–101
endolymph, 315
endolymphatic space, 318
endoplasmic reticulum, 124, 226, 238, 240, 247, **401**, 402
  cisternae, 238, 241, 247
  in freeze-dried tissues, 15
  post-mortem changes in, 381–391
  smooth, 200, 204
  TPPase activity, 174, **177**
  vesicles, 238, 240, 241, 243, 247
end-product, 122
energy, 3, 4, 6, 9
  free, 67
enhancement effect, 188, 189
enzymatic extraction, 191, 192
enzyme activity, 184
  in cells of middle ear, 316
  latent, 265, 267
  manifest, 265, 267
  membrane control of, 265
enzyme histochemistry, 25, **36**, **37**, 40–44, **45**, 184, 266, 270–271
  importance of mitochondrial membranes in, 271
  retention of unfixed enzymes in, 171
  statistical methods in, 271
  stoichiometry in, 270
enzymes, 6
Ephemerida, goblet cells in mid-gut, 335
epidermal layer, from chick embryo explants, cells in, 344, **345**, **346**, 347, **348**, 349, 350
epidermis, 341
Epilachna indica, principal (digestive) cells in mid-gut, 333
epithelium, 260
Epon, 184–189, 191, 192, 197, 198
  C, 117, 124, 126
epoxy resin, 118
equilibrium constant, 67
ergastoplasm, 399
erosion, skeletal, 260
ester, osmium dioxide, 126
ethanol, 118, 119
evolution, *412*
exocrine pancreas, 237, 238
  mouse, 79
explant, 280, 282, 283, 288
'explosive' disintegration of tissue, 344
extracellular matrix, 240, 241, 243
  zones, octopus nervous system, 375
extraction, 118, 122, 124, 126

'fashions' in scientific research, 149
fast freezing or slow? 34–35
fat, staining of, 68, 75
'fats', 149, 155
femur, **255**
Feulgen reaction, 276
fibrils, 237, 240, 241, 243, 247
  collagen, intracellular and extracellular, 230–234
  intercellular, 240, 243, 247
  intracellular, 240, 241, 243, 247

fibrin films, staining of, 75
fibroblast, 225, 232
fibroblasts, 237–247, 296, 300
  invertebrate, 237, 238, 241–247
  vertebrate, 237–241, 247
fibrocyte, 230, 232
fibrous phase, 258
filament, mesogloeal, 8
finder grids, 186
fine structure, 251–260, *414*, *416–417 (4)*
five-layered mitrochondrial membrane, 13, 19
fixation, 3, 5, 7, 9, 10, 96–99, 116, 117, 118, 120, 122, 124, 126, 131, 132
  Bouins fluid, 163
  chemistry of, 126, 131
  formaldehyde, 7, 8, 96
  glutaraldehyde, 232
  of lysosomes, 268
  by osmium tetroxide, 8, 12, 51–56, 136
fixatives, *411*, *415–417 (6)*
flaws, development of, in echinoderm skeleton, 258, 260
fluorescent antibodies, 25
focus, optimum, 116
formalin, *see* formaldehyde
formaldehyde, 93, 138, 270
formazans, 93, 138, 270
fossil calcite, **253**
fossilization, 252, 254
fracture, surface, 256
freedom in science, *412–416 (18)*
freeze drying, 144, 388
  criticism of, 222
  in electron microscopy, 11
  technique, 12
freeze substitution, 11, 12, 388
  in electron microscopy, 11
  technique, 12
frontal lobe, superior, Octopus, 371
frost damage, 33–35, **37**, 38–39, 42–43
  intra-cellular ice, 33–35, 38–39
  salt-concentration, 33–39
  sulphydryl oxidation (Levitt's damage), 33–34, 42
'functional structure' in cells, 272
fusing myoblasts, 278, 284, 285, 291, 293, 296, **297**, 298, **299**, 300, 301
future prospects of cytological studies, 394

Galleria mellonella, 159–166, **160**, **162**, 241, 243
  egg vesicle, phospholipids in nuclei, 159, **160**, **162**
  principal (digestive) cells of mid-gut, 324
gamma-cytomembranes, 111
ganglion(ia), 186, 189, 191, 192, 193, 197, 205
'gap', in mitochondrial membrane structure, 18, 19, 386, **387**, 388
gas constant, 67
gastric glands, parietal cells, 112, **113**
gastropod, **255**, 256
gegenion(en) (counter-ions), 60, 70
gelatine embedding, 32
'genetic code', 277, 357–359
  abnormalities in, 288
genetics, *410*, *414 (2)*
geography, *410–412 (5)*, *414*, *416 (3)*
'giant' chromosomes, 278
Glees's method for tracing terminal degenerations, 100, 101
glial cells, 170, **173**, 174, **177**
globules, compound, 198
  homogeneous, 198
  lipid, 198

Glossina sp., peritrophic membrane in mid-gut, 337
glycerophosphate, 150
glycine p., 266
glycogen, 140, 194, 195, 197
  'rosettes', 140, 195, **196**, 197
glycolipoprotein granules, 176
glycol methacrylate (GMA), 116, 117, 118, 119, 120, 122, 124, 126, 128, 131, 132
glutaraldehyde, 116, 117, 122, 126, 128, 131
goblet cells, insect mid-gut, 321, **328**, 335, 336
Goethe, quotation from, 115
Gomori technique, 268
Gomori-type reactions, 144
Golgi apparatus, 106, 110, 226, 232, 396, 397, **399**
  bodies, 396
  complex, 240, 243
  controversy, *413–416 (14)*
  elements, principal (digestive) cells of insect mid-gut, **322, 323, 324, 325, 326, 327,** 328, **329, 330,** 331
  regenerative (replacement) cells of insect mid-gut, **323, 324,** 336
  externum, 395
  field, 395–404
  , apparatus, etc., 169–174, **171, 173, 175,** 177–179, **177,** 195, 198, 200
  internum, 395
  Körper, 395
  saccules, **401**
  system, 353
  techniques, 79
  vacuole, **401**
Golgi–Holmgren canals, 105, 106
grasshopper, principal (digestive) cells in mid-gut, 325
graticule, transparent, 295
grids, finder, 186
Grimley's trichrome method, 186
ground substance, 225, 252
  of connective tissue, carbohydrates in, 220–221
  importance of, 220–221
growth, cellular, 6
  spicule, 252
guinea-pig, 238, 240
gum tragacanth, 305, 311

haemalum, 188
haemochromatosis, relationship to neutral red granule formation, 84
haemocytometer, 307
haemoglobin α, β, γ and δ chains, 359
Hale reaction, 140
Haliotis, 398
heart, 5, 240
HeLa cells, staining of, 76
Helix aspersa, 9, 92, 197, 225, **228, 229,** 230–234, **233,** 243, 247, 395
  pomatia, 197
  sp., 172, 174, **175**
Heterocentrotus mammillatus, **253,** 256
'heterochromatin', 357–358
hereditary muscular disease, 288
histochemical method(s), 184, 185, 205
  reaction, technique, 184, 188
  validity, 191
histochemistry, *413–416 (7)*
  definition of, 185
  extractive, 185
  substantive, 185, 192, 195, 197, 198
  vivicative, 185
histone, staining of, 76

Hirudo medicinalis, 227, 247
Holmes's technique, 98
Holmgren 'juice canals', 105, 106
holocrine secretion, 234
Holothuria, 238, 240
Holothuroidea, 251
homology(ies), 5, 6, 7, 8, 10
honey-bee, peritrophic membrane in mid-gut, 36, 37
hot plate, miniature, 186
hyaluronic acid, 237
Hydra, 6
hydrogen peroxide, 191
hydrolase histochemistry, 43–44, **45**
hydropsy, 318
hydroxy proline, 226, 237
Hymenoptera, 237
Hyperaspis vinciguerrae, regenerative (replacement) cells in mid-gut, 336

ice crystals, avoidance of, 11, 12, 13, 33–35, 38–39, 42–43
image(s), comparison of, 183, 184
  double, 185
  electron microscope, 115, 116, 117, 118, 122, 131, 132
  final, 118, 120, 124, 131, 132, 184
  intensity, 276
  optical, 186
  positive phase contrast, 189
  quality of, 185, 189
image-shearing eyepiece, 307
impedence, electrical, *post mortem* increase in, 381, 388
impregnation, 95
improbability, 9
indifferent salts, added in fixatives, 56
indigo-carmine, 69
inert dehydration, 144
infiltration, 118, 131
infra-red analysis, 126
inner ear, 315–318
insect, glands, regeneration of, 398
  tissues, lipids in nuclei, **165**
  prothoracic glands, 170
insects, 237, 241, 247
integrated phase change, 275
interaction, tissue and fixative, 118, 124, 132
interambulacral plate, **254, 255, 257**
intercellular canals, **105**
interference microscopy, 267, 275–277, 279–290
  Barer's 'double immersion' method, 290, 294
  limitations of, 275–276, 279–281
intersexuality, *410 (7), 413*
intracellular canals, 103, **104, 105,** 114
  and extracellular collagen fibrils, 230–234
invertebrates, 237, 238, 241–247
ionic concentration in cells of middle ear, 318
ions, calcium, 128, 131
  ferricyanide, 131
  positive and negative, 127, 128
  uranyl, 122, 127
  transport, in cells of middle ear, 318
irradiation, 124
irrational methods, v, 95
isotherm, Freundlich, 68, 69, 75, 76
  Langmuir, 69, 75, 76

'junket' embedding medium (Baker), 413
Janus black, 316
  green, 76, 316

KB cells, 176, **177**
Kupffer cells, 81

Labidura riparia, peritrophic membrane in mid-gut, 336, 337
  principal (digestive) cells in mid-gut, 321, 322, **323**, 324, 325, 329, 331, **332**, 333, **334**, 335
  regenerative (replacement) cells in mid-gut, 336
'lace holes', in echinoderm skeleton, 260
laked blood injection method, 105
lamina osceca spiralis, 315
lampbrush chromosomes, 357–359
Lampsilis ventricosa, 247
Landschütz ascites tumour cells, 305, 307, 312
larva(e), Chironomid, 4
  Cypris, 5
  echinoderm, 251, 252
latency, structure-linked, 179
lead, 124, 128, 185
  citrate, 192
Leander, 174, 176
Leberwurst embedding, 44
lecithin(s), 119, 127, 128, 150, 161
  hydrogenated, 150
leech, 237, 247
length, 9
Lepidoptera, 195, 197, 241
  goblet cells in mid-gut, 335
lepidosomes (Parat), 395
Lepisma saccharina, peritrophic membrane in mid-gut, 336, 337
  principal (digestive) cell in mid-gut, 321, 322, **322**, 323–325, 329, **330**, 331, 333, 336
  sp., principal (digestive) cells in mid-gut, 322
  regenerative (replacement) cells in mid-gut, 336
lesion, 286
levels, 183, 191
  of analysis, 9, 10, 115
  cellular, 3, 9, 10
  of contrast, 122
  of discussion, 116
  gross, 10
  of integration, 5, 6, 7, 178
  molecular, 6, 8, 10
  of organization, 5, 6, 8, 9
  of structure, 9
Leucophaea, 174
ligands, 118, 122, 127
  negative or positive, 127
  specific, 116, 127
light, *411*, *414–416* (7)
  monochromatic, 71
  scatter, 185, 186, 188
limits, of detection, 189
  of resolution, 192, 275
Limnea stagnalis, 92
Limulus, 174
linear shrinkage, 284, 285
  measurements, under a microscope, limitations in accuracy of, 275
lipid(s), 3, 4, 118, 126, 149, 198
  bodies, 345
  bound, 152–155
  demonstration by benzpyrene, 149
  droplets, 195
  extraction of, 154
  histochemical proof of, 153
  histochemistry, 39–40
  masked, 152
  in nucleohistone, 153
  and phospholipid content of nuclei, 158
  vacuole, 120
lipids/lipines/lipoids, *413–415* (*4*)

lipid-protein complexes, 153–155
  significance of, 155
  suggested structure for, 154–155
  unsaturated bonds in, 154
lipochondria, 169–180, **171**, **173**, **175**, **177**, 351, 395
  Baker's definition of, 397
  principal (digestive) cells of insect mid-gut, **324**, **325**, **327**, **328**, **329**, 331, **332**, 333
  regenerative (replacement) cells of insect mid-gut, **323**, **324**, **328**, 336
lipoprotein, 194
liposomes, 351
liver, 185
  cells, plasma canaliculi, 103, **104**
  mouse, 117, 128, 131
locust, 241, 243
Locusta, 170, 195, 396
London–van der Waals forces, 127
Lumbricus, 174, 176
Lymnaea, 174
lyophillization, 31, 38
lysochromes, 75
lysolecithin, 150
lysophosphatidyl ethanolamine, 150
lysosomes, **137**, 169–180, **173**, **175**, **177**, 232, 267–269
  activation of, 268
  cultured cells, 176, **177**
  effect of fixation on, 268
   of pathological conditions on, 268
  formation, 84
  fragility, 267
  in gastropod molluscs, 172, 174, **175**, **177**
  histochemistry of, 268–269
  latency, 179
  latent activity in, 269
  leucocyte, 178
  Lumbricus, 176
  macrophage, 178
  in orthopteran insects, 169, 170, **171**, **173**
  phagocytic activity, 92, 93
  significance of membrane, 267
  substrate specificity, 179
  as 'suicide capsules', 267
  vertebrate, 176, 178, 179
  vital staining of, 91–93

magnesium carbonate, 251, 252
Malacosoma sp. (larva), brush-border of mid-gut epithelium, 321
mammary gland, lactating mouse, 116, 120, 122, 124
Mann–Kopsch Golgi technique, 79
masked lipids, 158, 159
mass, 9
  action, law of, 68, 69
mast-cell granules, staining of, 72
material, living, 115
  osmium reacting, 117
matrix, mitochondrial, 116, 120, 124
  printing, scanning microdensitometer, 290
  skeletal, 258
matter, 3, 4, 6, 9
Maxtaform grids type, H6, 186, **187**
medium, embedding, 184
Melanoplus, 170, **171**, **173**, 174
  sp., principal (digestive) cells in mid-gut, 321, 333
membrane(s), mitochondrial, 116, 117, 118, 120, 122, 124, 131
  configurations, 200
  contrast, 120, 128
  cytoplasmic, 193, 200

membrane(s)—*contd.*
  polar part of, 116
  smooth, 198, 200, 203
mercuric bromophenol blue, 139
mercuric-formol, 291
merocrine secretion, 227, 240, 247
mesenchyme, 345
mesothelial layer, 315
'metabolic initiative', 298
metabolism, 6
methacrylate, 118, 124, 184, 187, 195, 203
methanol, 119
metachromasy, 61–64
  current hypotheses of, 63
  induced by chromotropic substances, 64
metallic cations, uptake by lysosomes, 92, 93
method for handling thin sections, 188
methyl green, 70
  mercuric chloride, 139
methylation, 140
methylene blue, 60, 64, 70, 75, 76, 178, 184, 316
  red forms of, 64
mitochondria, *411 (2)*, *413*, *415*
mitochondria, 116 117, 118, 120, 122, 128, 132, 195, 197,
      198, 204, 205
  beef heart, 118
  in cells of middle ear, 316
  cristae, 361, 363, 368
  degenerating, 373–380
  in developing *Drosophila* muscle, 361–363
  in diseased muscles, 365–368
  elementary particles, 265, 361, 363
  in normal human muscle, 363
  paracrystalline materials in, 365, **367**, 368
  post mortem, calcium deposits in, **383**, 386
    swelling of, 388
  (principal digestive) cells of insect mid-gut, histo-
      chemical reactions, 325
  (principal digestive) cells of insect mid-gut, morpho-
      logy, **322**, **323**, **324**, **325**, **326**, **327**, **328**, **329**
  regenerative (replacement) in cells of insect mid-gut,
      324, 336
  structure in freeze-dried tissues, 15
    freeze-substituted tissues, 13
  thread-like, 365, **367**, 368
mitochondrial, membrane, 5 layered, 13, 19
  membranes, five-layered, 384, **385**, 386, **387**, 390
    seven-layered, 384, **385**, 386, **387** 390
  system *post-mortem* changes, 381–391
  permeability, 271–272
    changes in cell damage, 272
    as indication of their structural state, 271
mitosis, arrest of, 353
  frequency of, 351
  rhythmic production of, 398
Micraster cortestudinarium, **253**
microdensitometer, *see* densitometer
microdensitometry, 71, 73, *see also* densitometry
micrographs, light, electron, 194
microprobe analyser, 135
microscope design, *414*, *417 (3)*
microscope, electron, 183 185–187, 204
  light, 183–188, 204, 205
microscopy, *412–417 (20)*
microscopy, 184, 194, 197, 198
  anoptral-contrast, 188, 205
  optical, direct, 188, 189, 191, 205
  phase-contrast, *414 (3)*, *416*
  phase contrast, 184, 188, 205
microtechnique, *411*, *413–417 (21)*

microtechnique, 184
microtubules, 240 ,243
microvilli, 305, 306, **401**
middle ear, 315–318
molecular biology, 5
  boundary, 7
  configuration(s), 5
  machinery, 4
  structure, 5
'molecular cohesion' in protoplasm, 271
molecule(s), 3, 7, 10
  biological, 185
  nucleic acid, 4, 7
mollusc, 237
  skeleton, 254, **255**, 256, 258
monophase secretion, 397
mordants, *416 (3)*
morphogenesis, morphogenetic processes, 6, 7
morphology and biochemistry, a unity, 394
moth, 241, 243, 247
mouse, 238, 240, 286–288
mucin, 161
  staining of, 70, 71, 72
mucopolysaccharides, 225, 237–243
  acid, 237, 238, 240, 241, 243
  neutral, 237, 243
mucus, 345
multi-nucleate syncytia, 278
Murex, 398
muscle, 4, 6
  developing, as material for studying differentiation,
      278
  diseases, mitochondrial alteration associated with, 363,
      364, 365, **366**, **367**, 368
  fibres, 8
  filaments, 8
  striated, 6
'muscle strap', 278, 290, **296**, 298
muscular dystrophy, 288
mushrooms, mitochondria of, 363
myelin figures, 193
  sheath, in freeze-substituted tissues, 17
myoblasts, actively fusing, 278, 284, 285, 291, 293, 296,
      297, 298, **299**, 300, 301
  binucleate, **280**, 281, **282**, **283**, 298, 302
  and fibroblasts, methods and merits of distinguishing,
      296, 304
myogenesis, review of literature on, 300
myopathy (unidentified), 365, **366**, **367**, 368
myosin, 232
  synthesis, 304

naphthol yellow S, 75
natural calcite, **255**, 256
  selection, 5, 6, 7, 10
Necturus, 204, 355
needles, calcite, 258, **259**
nematocysts, 6
nematode, 237
negative stain, 230
nerve, 4, 7
  axon, 95–101
  cell(s), 189, 193, 195
    bodies, 192
    cytoplasm, 194
    Golgi apparatus, **107**, **108**
    processes, 192
  cord, 241
  net, 8

nervous system, 241, 247
  degeneration, 371–380
  regeneration, 375–380
neural lamella, 192, 193, 194, 237, 241, 243
neural structures, differentiation, 343
neurofibril concept, 97, 99
neurofibrils, 97–101
neurofilaments, 99–101, 371, 375
neurogenic muscular atrophy, 365, **366**
neuron(s), 193, 194, 198
  cat, 176
  gastropod mollusc, 172, 174, **175**
  insect, 169, 170, **171**, **173**
  opisthobranch mollusc, 174
  pigeon, 176
  rabbit, 176
  rat spinal ganglia, 170, 179
Neurospora, mitochondria of, 363
neutral red, 79, 176, **177**, 178, 316, 395
  affinity for basiphilic cells, 79
  effect on protein synthesis 79
  electron microscopy, 81
  granules, 79, 178–180
    cycle, 79
    in exocrine pancreas, 81
    formation, cellular capacity, 79
    in liver, 81
    in mouse lymph node 88
    in mouse pancreas, 87, 88, **89**, **90**, 91
    lysosomes, 81
New Hebrides, *410–414 (16)*
Nile blue, 252, 316
  A, 184, 186
nomenclature, scientific, *412 (2)*, *414–415 (2)*, *416 (2)*
'non-disruptive biochemistry', 272
non-rhythmic secretion, 399
nuclear envelope, 128
  hypertrophy, 204
  membrane, 193, 203
  phospholipids, acid haematein test's specificity, 161–164
nuclease, 192
nuclei, lipid and phospholipid content of, 158
  staining of, 72, 73, 74, 75, 76
nucleic acid(s), 4, 10
  histochemistry, 40, **41**
  indium as a stain for, 139
  iron as a stain for, 139
nucleohistone, high lipid content of, 153
  staining of, 76
nucleolar dry mass, 279–281, 288
  abnormally low values in dystrophic myoblasts, **286**, **287**, 288–289
  accuracy of, 279–281
nucleolar organizer, 359
nucleoli, 193
  abnormal, 288
  diameter of, changes in, 284–287, **284**, **285**
  dry mass of, 279–281, **280**, **286**, **287**, 288–289
  percentage of solid material in, 281–282, **282**, 288–289, **289**
  volumes of, 282–284, **283**
nucleoprotein, 5, 161
nucleoside phosphatases, 106, 110
nucleus, 120, 128, 131, 188, 192, 193, 203
  chemical composition, 157, 166
  lipid autoradiography, 166
    content, 157, 158, 166
    localization, 158, 159
  variability of lipid pattern, 159, **160**, 164
numerical aperture, and depth of focus, 294

object, living, 115
octopus, brain, 371–380
Odontotermes horni (worker), peritrophic membrane in mid-gut, 336, 337
  principal (digestive) cells in mid-gut, 321–323, **324**, 325, **330**, 331, 333, **334**, 335
  regenerative (replacement) cells in mid-gut, 336
Oedaleus marmoratus, principal (digestive) cells in mid-gut, 325
oil red O, 149
oleic acid, 150
oocytes, 162–165, 197, 198, 203–205
Opalina, 87
open steady state, 3, 4, 5, 9
Ophiuroidea, 251
optic tectum, avian, 371
  tentacles, gastropod, 174
optical axis, 252
  densities, 276
  density maps of cells, **293**, **296**, 300
  microscope, 260
  techniques, sophisticated, 189
organic envelope, 252
  matrix, 252
organism, living, 7
organization, living, 9
Orthoptera, principal (digestive) cells in mid-gut, 333
osmication, 127
osmium black, 145
osmium/ferricyanide solution, 117, 128, 131
osmium fixation, 124, 128
  reacting component, 122
  reaction, 120, 128, 131
    product, 127
  post-stabilization, 122
  tetroxide, 116, 117, 122, 124, 126, 127, 128, 131
    dissolved in distilled water, 51–56
      in sea water, 51–56
    first introduced to microtechnique, 97
    solution, 186, 195
    vapour, 118
osmiophil platelets, 395–396
osmiophilia, 170
osmoregulation, 56
ossicles, 251–260
osteoblasts, 225, 238
ovary of sea urchin, staining of, 75
ovotestis, 197
oxidative phosphorylation in cells of middle ear, 318
oxides, of osmium, 128
oxidoreductase histochemistry, 25, **36**, **37**, 40–43
oxygen, effect of deprivation of, on nerve cells, 381, 388

pain fibres, octopus brain, 378
palmitic acid, 150
pancreas, 398–402
  cells, intracellular canals 104, **105**
    secretory capillaries, **105**, **109**, **111**
  exocrine cell diagrams of, **399**, **401**
parabasal bodies, 3
parasites, 5
Paravorticella terebellae, 51–56
paraldehyde poisoning, yielding 'perfect' Golgi preparations, 98
PAS method, reaction(s), 185, 187, 191, 192, 193, 195, 205, 221, 230
  positive, 188, 189, 192–195
Passulus cornutus, principal (digestive) cells in mid-gut 325, 331
pathological damage, 200, 203

patterns, ordered, repetitive, 9
Pelagothuria, 251
pepsin digestion, 164
perception, 9
periodic acid, 191
periodic secretion, 397–398
periodicity, of banding, in collagen fibres, 227
  in developing muscle cells, 232
perilymph, 315
Periplaneta americana, peritrophic membrane in mid-gut, 337
  principal (digestive) cells in mid-gut, 325
Periplaneta orientalis, principal (digestive) cells in mid-gut, 325, 326, 333, 335
  regenerative (replacement) cells in mid-gut, 336
peristome, 252
peritrophic membrane, insect mid-gut, chemical nature, 337
  insect mid-gut, morphology, 322, **327**, **334**, 335–337
  origin, 336, 337
permeability, of mitochondria, as indicator of their structural state, 271
  of tissues, 70
perpetual secretion, 397
petroleum ether, 120
pH, of fixatives, 51, 54
  of toluidine blue solution, 291
Phages, 5
phagocytosis, 371, 373–380
'phagosomes', 91
phase annulus, 188, 189
phase change measurements, integrated, 275
  with interference microscopy by densitometry, 275
    by phase modulation, 275
    by visual means, 275
phase-contrast microscopy, *414 (3)*, *416*
phase-contrast, microscopy, 198, 252
  negative, 188
  objectives, 188, 189
  positive, 204
phase modulated light methods of measuring phase changes, 275
philosophy, *411 (2)*, *413*, *416*
phosphatases, 200
phosphatidic acid, 150
phosphatidyl ethanolamine, 150
  serine, 150
phosphine 3R, 150
phospholipid(s), 9, 118, 119, 120, 122, 124, 126, 127, 131, 149–155, 169, 170, 174, 176, 178, 180
  content, 117, 119, 120
  layers, 9
  membranes, 169, 170, 174, **175**, **177**
  and nuclear structure, 157–166
  phosphate groups, 128
  reaction with dyes, 92, 93
phospholipoproteins, 157
phosphoprotein, 203
phosphorus content, 127
  determination, 120
phosphotungstic acid (PTA), 116, 230
  staining, 371
photography and silver methods, 95, 96
photomicrography, *412 (2)*, *413*
physiological cytology, 183
pial vessels, 247
pinocytosis, 227
planarians, 237
Planorbis, 174, **175**, **177**, 179
plants, insectivorous, 5

plasma, of cock, 290
  clot, 290
  membrane, 103, 104, 110, 114, 243
plastic, embedded tissues, 184, 192
  embedding media, 191
  sections, 185, 189, 191, 192, 195, 197, 198, 204, 205
  of tissue, colouring of, 184, 188
plastics, water soluble, 184
plateau of dye uptake, 69
plates, echinoid, 252–260, **253**, **257**, **259**
Pleurobranchia, 398
ploidy, 266
plot, reciprocal, 70
  Scatchard, 70, 73, **74**, 76
pluteus larva, 251, 254
Poecilocerus pictus, peritrophic membrane in mid-gut, **327**, 336, 337
  principal (digestive) cells in mid-gut, 321, 325, **327**, 328, **330**, 331
  regenerative (replacement) cells in mid-gut, 336
polycrystal, 254
polymerization, 124, 131
  damage, 124
  with UV, 124
polymyositis, 365, **366**
polypeptides, 126
polyphase secretion, 397
polytene chromosomes, 278
polytetrafluoroethylene (PTFE), as lubricant in cryostats, 30
polythene micro-cups, 186
polyvinyl alcohol (PVA), in histochemistry, 271
positive colouration(s), 191, 192
  phase-contrast microscopy, 315
post-fixation, 40, 122
post hydropic degeneration, 318
post-mortem, changes in cells, 381–390
post-osmication, 232
potassium ferricyanide, 117, 128
  permanganate, 128
precipitation, of calcium, 200
  seed foci of, 200, 203
'pre-synaptic bag', **372**, **374**, 378
primary amines, histochemistry, 39
  calcite, 252, **253**
primordium, 252
primuline, 76
principal (digestive) cells, insect mid-gut, 321, **322**, **323**, **324**, **325**, **326**, **327**, **328**, **329**, 330, 331, **332**, **333**, **334**, 335
principles, classificatory, 6
  engineering, 5
procedures, light microscopical, 127
Prodenia sp. (larva), goblet cells in mid-gut, 335
pronase, 192
pronunciation, scientific, *414–415 (2)*
propane, 1,2,epoxy, 117, 119, 120, 122, 124, 132
Protargol, 96, 98, 99
proline, 226, 238 240
'protective agent', 271
  media', 32, 38–39, 271
proteins, 3, 126, 131
protein synthesis, 237, 238, 277–278, 298, 304, 341, 353–354
protoplasm, 184
Psammechinus miliaris, 254
Psychoda alternata, peritrophic membrane in mid-gut, 337
Ptychoptera contaminata, principal (digestive) cells in mid-gut, 333

'puffs', 278
pupa(e), 186, 192, 193, 195
pure substances, acid haematein reaction on, 161
purine/pyrimidine linkages, 277
pyknometer, 306, 307
pyridine, 98
    extraction test, Baker's, 152
pyronine, 290

quantitation in histochemistry, 265–272, 275–303
quantitative investigations on cells, difficulties discussed,
        275–276
    cell surface 'roughness', 305–312
quenching, 32–35

rabbit, 240
radula sac, 241
    cartilage, 230
rain-forest (Sinharaja), Ceylon, *412 (2)*
Raoult's law, 33
rat, 240
    protein starved, 298
reaction product(s), 115, 116, 122, 124, 126, 127
    stable, 122
reactivity, of tissue, 127
reagents, contrasting, 117
    light microscopical colouring, 185
refractile granules, in cells of middle ear, 316
refractive index, 189, 279
regenerating limb, 238, 240
regenerative (replacement) cells, insect mid-gut, 321,
        323, 324, 328, 330, 334, 335, 336
rehydration, 122, 124
Reissner's membrane, 315–318
    high metabolic rate of, 315, 317, 318
relationship, between object optical density and density
        of its photographic image, 275
    between optical density and phase-change, 275
    evolutionary, 9
Reynolds number, 305, 306, 308, 309
replica, 254, 256
replication, 10
reproduction, 7
'residual bodies', 91, 178
resorption, skeletal, 252, 254, 260
respiration rate, in cells of middle ear, 316
reticulin, 225
rheumatism, 225
rhodamine B, 76, 150
ribosomes, 131, 192, 227, 232, 238, 241, 243, 278, 354
    post-mortem changes in, 381, **382**, 390
ring and cup technique, 186
RNA, 4, 80
    cytoplasmic, 278, 289, 290, 298, **299**, 300–302
        abnormally low values in dystrophic myoblasts, 289
    deficiency of, 289
    importance of, in metabolism and growth, 298, 299,
        300
    'messenger', 204, 277, 354
    nucleolar, 289, 298, 302
    in nucleoli, unreliable values for, 296, 300, 301, 302
    quantification of, 290–302
    quantitative cytochemistry of (Swift), 291
    ribosomal, 281, 354
    'soluble', 277, 278
    staining of, **72, 74**, 75
    synthesis, 298
        abnormalities in, 289
    'transfer', 277, 278
    with yoke platelets, 353

RNAase (ribonuclease), 192
    extraction with, 290, 291, 294
RNP, 285, 359
roughness, of cell surfaces, 305–312
    index, 306, 309, 310, 311
        related to age of tumour cells, 311
Rotifera, 4
round window in middle ear, 315
rhythmic production of mitoses, 398
    secretion, 398

Sacculina, 5
safranin, 70
salamander, 204
saliva, 191
salt, inhibition of staining by, 71
scala media, 315
    vestibuli, 315
scanning electron micrograph, **255, 257, 259**
    microscope, 254, 260
Scatchard plot, 70, 73, **74**, 76
Schiff's reagent, 188, 191
Schistocerca gregaria, 241
    sp., 396
Schwann cells, 234, 247
science, 'applied', 288
    'pure', 288
scientific papers, unreliability of, in indicating currently
        accepted ideas, 209
scrubbing, of sections, 186
seagull, 204
sea-urchin, 252, 256
sea water, as solvent for fixative, 51–56
secondary calcite, 252, **253**
secretion, calcite, 251, 252
    non-rhythmic, 399
    periodic, 397
    perpetual, 397
    polyphasic, 397
    principal (digestive) cells of insect mid-gut, **322, 323,
        324, 325, 326, 329, 330**, 333, **334**, 335
    staining of, 72
secretory capillaries, exocrine glands, 106, **109, 110**
section, cut surface of, 185
    of mouse heads, 185
    plastic, 116
    semi-thin, 186
    ultra-thin, 185–188
    wrinkle-free, importance of, 186
segresomes, 178
selective colourations, 188, 191
selectivity, 124, 184, 185, 192
septa, principal (digestive) cells of insect mid-gut, histo-
        chemical reactions, 323
    morphology, **322, 323, 324, 326, 329, 330, 332**
shaker-1, genetic strain of mouse, 318
shape, of cells, *re* roughness index, 305
Sharpey-Schafer's demonstration of intra-cellular canals,
        103, 104, 105
sheath cells, of connective tissue, 195
shell, 197
significance, functional, 8
silica gel, 120
silk, 237
    staining of, 69
silver impregnation techniques, 375
'silver proteinates' (Protargol), 96
single-phase skeleton, 251, 256, 258
sites, dye-binding, 69

skeletal phase, 251
  repair, 260
skeleton, 251–260
  external, 197
  internal, 197
  maintenance of, 256–260
skin, 238, 240
  wounds, 238, 240
smooth muscle cells, 247
snail, 204, 205, 241, 243, 247
sodium alizarin sulphonate solution, 197
solution, ideal, 68
  solid, 68
Solway ultra blue B, 61
spatangoid, echinoid, **253**
species, 5, 6, 7
specific reactions, 127
  refraction increment, 279
specificity, 128, 184, 191
spermatoza, staining of, 76
sperm-suspension, *411 (2)*
spheres, standard, for roughness index measurements, 305, 307
spherical vacuoles, 316
spherulites, 198
sphingomyelin, 150, 161
spicules, 251, 252
spinach leaves, 118
spines, dendritic, 371, 378
  echinoid, 252, **253**, **254**, 256, 260
sponge, 237, 241, 247
stability, 124
stabilization, 124, 127
staining, 67–76
  activation energy of, 67
  affinity of, 67
  differential, 191
  electron contrast, 116
  entropy of, 67
  heat of, 67
  with heavy metals, 184
  methods, 184
  thermodynamics of, 67
  visual assessment of intensity, 71
state, anabiotic, 9
  cryptobiotic, 4, 5
  living, 8
  standard, 67, 68
stereophotomicrography, *412 (3)*
Stereoscan electron microscope, 256
stoichiometry, 266, 270
  discussion of, 276
strain, 258
stress, 258
stria vascularis, 315
Strongylocentrotus droebachiensis, 252
structure, cellular, 5
  gross, 5
  molecular, 5
style, scientific, *415*
substantive histochemistry, 185, 192, 197, 204, 205
sucrose solution, 126
Sudan black, 149
  B, 75, 161, 184, 186
  technique, 395
Sudan colourants, 149
  effect of partition coefficient, 149
Sudan III, 75
Sudanophilia, 169, 170
sulphatide, 150

surface flaws, **257**, 258, **259**, 260
  replica, 254
synapse, 371–380
synaptic endings, 100–101
  structure, 100
  vesicles, 371–380
syncytium, 251, 278
synomous terms? lipochondria, neutral red granules and lysosomes, 169–180
system(s), classificatory, 6, 10
  dynamic, 9
  haemal, 7
  living, 1, 3, 4, 9
  nervous, 7, 8
  vascular, 5, 7
systematic errors, in interference microscopy, 275
$S^{35}O_4$, 240

taenidia, 193
Tansley, Sir Arthur, tribute to, *415*
tectorial membrane, 318
temperature, *410–411 (3)*, *413*, *417*
'templates', DNA, 277
tendon, 237, 238
Terebella lapidaria, 51–56
tetrazolium salts, 93, 138, 270
Tethya lyncurium, 247
textbooks, as reliable indicators of currently accepted ideas, 209
thiamine pyrophosphatase, 106, 110, 170, 172, **173**, 174, **177**, 179
thickness, of living myoblasts, 294
  of sections, optimum recommended by Altmann, 342
thorium dioxide, 141
thymus, staining of, 76
thyroid colloid, staining of, 72
thyroid cells, intra-epithelial vessels, 104
  intrinsic capillaries, 104
Thyone briareus, 204
tight-junctions, as artifacts, 19
time, 9
time-lapse cine-micrography, 284
Tineola sp. goblet cells in mid-gut, 335, 336
'tissu cellulaire' (Voltaire), 209
tissue, contractile, 5
  culture of myoblasts, 278, 288, 290
  fixed, 120
  fresh, 120
  glutaraldehyde fixed, 117
  insect, 185
  living, 6
  mollusc, 185, 197
  plastic embedded, 184, 185, 192
  residual, 119
  vertebrate, 185
tissues as compatible embedding media, 32
toluidine blue, 70, 75, 76, 184, 185, 186, 194, 290, 291, **293**, 295, **297**, 342, 353
tooth, molar, 7
tracheolar cell cytoplasm, 192
tracheoles, 193
transneuronal degeneration, 371, 375–380
traumatized muscle, growth of myoblasts from, 288
Trembley, Abraham, biography, *414*
Trichonympha, 3
tri-complex, 127, 128, 131
triglycerides, 119, 161
triolein, 150
tristearin, 150

tritiated leucine, 400
tritium, 226
Triturus cristatus, 358
   karelinii, 358
   viridescens, 358
Triturus sp., 357, 358, 359, 395
trophocytes, 161, 163
tropocollagen, 226, 237, 240, 241
true syncytium (*re* Baker's definition of), 278
trypsin digestion, 163
turbulence of flow, 305
turgor, loss of, 6
turnover, 3
tyrosine histochemistry, 40

ultrastructure, *417 (2)*
ultra-violet absorption, 266
   measurements, 290
ultra-violet microspectroscopy, 277
unit membrane, 200
   structure, 128, 131
'unmasking' of lipid, 163
uranyl, acetate, 116, 122, 124, 128, 185, 186, 189, 192
uranyl salts, 76
   staining, 120
utricular endolymph, 318

vacuoles, empty, 198, 200
variation, 5, 184
vascular system, reaction to degeneration, 375

veronal buffer, 186
vesicles, membrane-bounded spherical, 200
vestibular lip, 315
Vestopal, 184, 187
virus(es), 4, 9, 185
viscometer, 305, 307, 308
vital dye uptake, 169, 170, 172, 176–180, **177**
vital staining, 87–93
   first use of, 87
   types of, 87
vitellogenesis, 159
Voltaire, quotation from, 209

water, 4, 9, 184, 185
'water soluble methacrylates', 144
'wet mass', 281
whorls, membraneous, 200

X-bodies, **401**, 402
Xenopus laevis, 281, 342
X-ray, 254
X133/2097 (*see also* Durcupan A), 116, 119, 120

yolk, platelets, 353
   utilization, 353

zymogen, 'condensation granules', 400
   granules, **399, 401**, 402
   secretion, 399, **400**
zone of fusion, 26, 28

044/55
co/66
S. A Shafiq

Robert B. McKay.

J. P. Young

W. J. Morgan.

Bill Goldsmith

Bhupinder (N. Ind)

R. M. Park.

George Dan-di

Johnson. Chen

S. Bradbury.

H. G. Callان

J. Ahrelmann

D. J. Goldste

Jennifer M. Gregory